# MINERAL TOLERANCE OF DOMESTIC ANIMALS

SUBCOMMITTEE ON MINERAL TOXICITY IN ANIMALS

Committee on Animal Nutrition
Board on Agriculture and Renewable Resources
Commission on Natural Resources
National Research Council

NATIONAL ACADEMY OF SCIENCES

Washington, D.C. 1980

NOTICE: The project that is the subject of this report was approved by the Governing Board of the National Research Council, whose members are drawn from the councils of the National Academy of Sciences, the National Academy of Engineering, and the Institute of Medicine. The members of the committee responsible for the report were chosen for their special competencies and with regard for appropriate balance.

This report has been reviewed by a group other than the authors according to procedures approved by a Report Review Committee consisting of members of the National Academy of Sciences, the National Academy of Engineering, and the Institute of Medicine.

This study was supported by the Bureau of Veterinary Medicine, Food and Drug Administration of the U.S. Department of Health and Human Services; by Agricultural Research, Science and Education Administration of the U.S. Department of Agriculture; and by the National Feed Ingredients Association.

**Library of Congress Cataloging in Publication Data**

National Research Council. Subcommittee on Mineral Toxicity in Animals.
Mineral tolerance of domestic animals.

Bibliography: p.
1. Veterinary toxicology. 2. Minerals in animal nutrition. I. Title.
SF757.5.N27 1980 636.089'59 80-15466
ISBN 0-309-03022-6

*Available from*

National Academy Press
National Academy of Sciences
2101 Constitution Avenue, N.W.
Washington, D.C. 20418

Printed in the United States of America

# Preface

The Subcommittee on Mineral Toxicity was requested by the Committee on Animal Nutrition to review and evaluate the literature relating to mineral tolerance of animals and to set maximum tolerable levels of dietary minerals for domestic animals. Information for 35 minerals, including the essential elements and those traditionally considered as "toxic," was reviewed. The report emphasizes the potential adverse effects of minerals on animals, with only limited discussion of other aspects, such as essentiality and metabolism. The information should be useful to those persons involved with the formulation of diets for domestic animals.

The subcommittee was fortunate in having the advice and guidance of many scientists in the review of this report. The following individuals contributed significantly to the work of the subcommittee by critically reviewing some or all chapters of the report and suggesting ways in which the report could be more useful to the scientific community: Richard A. Anderson, David H. Baker, V. R. Bohman, Ian Bremner, C. C. Calvert, Rufus L. Chaney, David C. Church, George K. Davis, Richard M. Forbes, Douglas V. Frost, Howard E. Ganther, Robert A. Goyer, Samuel L. Hansard II, Vernon R. Heaton, Roger W. Hemken, Charles H. Hill, William A. House, Norman L. Jacobson, Leo S. Jensen, P. Larvor, Roland M. Leach, Jr., Orville A. Levander, Kathryn R. Mahaffey, Walter Mertz, Elwyn R. Miller, James G. Morris, Talmadge S. Nelson, Forrest H. Nielsen, Boyd L. O'Dell, Loyd Poitevint, Bobby L. Reid, G. Stanley Smith, Sedgwick E. Smith, Roy

F. Spalding, James F. Standish, John W. Suttie, Roger A. Teekell, and H. J. Weeth.

The subcommittee is also indebted to Philip Ross, Executive Secretary, and Selma P. Baron, Staff Officer, of the Board on Agriculture and Renewable Resources for their assistance in the production of this report, to the members of the Committee on Animal Nutrition and reviewers from the Board on Agriculture and Renewable Resources and the Commission on Natural Resources for their comments and suggestions, and to Pamela R. Henry for technical assistance in the preparation of the manuscript.

SUBCOMMITTEE ON MINERAL TOXICITY

CLARENCE B. AMMERMAN, *Chairman;* University of Florida
JOSEPH P. FONTENOT, VPI and State University
MATTIE RAE SPIVEY FOX, Food and Drug Administration, DHHS
HAROLD D. HUTCHINSON, Moorman Manufacturing Company
PAUL LEPORE, Food and Drug Administration, DHHS
HOWARD D. STOWE, Michigan State University
DAVID J. THOMPSON, International Minerals and Chemical Corporation
DUANE E. ULLREY, Michigan State University

## COMMITTEE ON ANIMAL NUTRITION

JOSEPH P. FONTENOT, *Chairman*; VPI and State University
CARL E. COPPOCK, Texas A&M University
RICHARD D. GOODRICH, University of Minnesota
PAUL W. MOE, USDA Nutrition Institute
QUINTON R. ROGERS, University of California, Davis
GARY L. RUMSEY, Tunison Laboratory of Fish Nutrition
VAUGHN C. SPEER, Iowa State University
JOHN D. SUMMERS, University of Guelph
DUANE E. ULLREY, Michigan State University

## BOARD ON AGRICULTURE AND RENEWABLE RESOURCES

CHESTER O. McCORKLE, JR., *Chairman;* University of California, Davis
GEORGE K. DAVIS, *Vice Chairman;* University of Florida
JOHN D. AXTELL, Purdue University
THADIS W. BOX, Utah State University
ROBERT E. BUCKMAN, USDA Forest Service
NEVILLE P. CLARKE, Texas A&M University
ELLIS B. COWLING, North Carolina State University
SALLY K. FAIRFAX, University of California, Berkeley
JOHN E. HALVER, University of Washington
HELEN M. INGRAM, University of Arizona
RALPH J. McCRACKEN, USDA Science and Education Administration
BERNARD S. SCHWEIGERT, University of California, Davis
GEORGE R. STAEBLER, Weyerhaeuser Company

# Contents

# Introduction

All mineral elements, whether essential or nonessential, can adversely affect an animal if included in the diet at excessively high levels. Undoubtedly, tolerance levels vary from animal to animal and even from day to day in a single animal. Many factors, such as age and physiological status of the animal (growth, lactation, etc.), nutritional status, levels of various dietary components, duration and route of exposure, and biological availability of the compound, influence the level at which a mineral element causes an adverse effect.

There are several reasons for concern about the possibility of excessive mineral intake by animals. The levels of some minerals in plant tissues vary greatly due to soil factors and the quantity and availability of the mineral present in the soil. Manganese, selenium, and molybdenum, for example, may occur in forages at such elevated levels that, when consumed, result in adverse effects on the animal. The leaves and stems of forages can also become contaminated with minerals such as cadmium, lead, or fluorine from industrial-processing plants, resulting in amounts of the contaminating element sufficient to cause harm to animals consuming the forage. Natural water supplies may contain excessive levels of sulfur, fluorine, sodium, magnesium, or iron. In addition, some minerals may be contributed to the water by industrial wastes and other sources of pollution. Mineral supplements are provided frequently to correct the deficiencies in diets formulated from natural feedstuffs. These supplements contain variable amounts of elements other than those of primary interest. The amounts of "other" elements depend on the native material from which the mineral supplement is obtained and the type of processing that it undergoes.

Increased recycling of animal wastes results in higher mineral intakes by animals. As animals digest and utilize the nutrients in their diet, there is a proportionately greater disappearance of organic nutrients with a consequent increase in the concentration of mineral elements in the excrement. In addition, some by-product feedstuffs, such as the residue remaining from the fermentation of molasses, may have high levels of minerals.

The accidental administration of excessively high levels of minerals is another reason for concern about the effect of excess minerals in domestic animals. Such administration can result in acute signs of toxicosis, which may be quite different from the chronic effects displayed after minerals have been fed at higher than normal levels over an extended period of time.

The words "mineral" or "minerals" as used in this publication refer to the elements rather than to an ore of geologic origin.

# Maximum Tolerable Levels

The mineral tolerances of animals have been investigated in a wide variety of studies in which graded levels of the element were offered and specific effects examined. Ideally, long-term feeding studies of 1 year or more should be conducted with domestic animals, with additional studies in laboratory animals involving two or more generations. Extensive studies of this type generally have not been made and, depending on the element, may not be necessary. Studies of much shorter duration have been conducted in which the criteria included feed intake, growth rate, biochemical or morphological lesions, mortality, and deposition of the element in meat, milk, or eggs. Sensitive indices of adverse health effects were not used in most studies. Highly soluble, purified forms of the test element typically were used to identify adverse effects and to assess their severity. The results obtained may be different when the element is supplied as feedgrade salts or incorporated during growth into plant and animal products used in feed.

In this publication, *maximum tolerable level* is defined as that dietary level that, when fed for a limited period, will not impair animal performance and should not produce unsafe residues in human food derived from the animal. The form of element, length of study, criteria for response, and species of test animal have all been considered in developing the suggested maximum tolerable levels of dietary minerals for domestic animals presented in Table 1. Data obtained by routes of administration other than oral were used when information with the oral route was lacking. Particular consideration was given to food residues from cadmium, lead, and mercury.

Although tolerance values for minerals will vary with age and physiological condition of the animal, only one maximum tolerable level has been listed for each species, except in a few cases where information was considered adequate. The discussions in the chapters for individual elements may be helpful in developing tentative levels for other species. The available information for most of the elements is less than desired, and it is likely that these tolerance levels will need to be modified. Problems that may arise when two or more elements are present at high levels represent a virtually unexplored area.

It is in keeping with good nutritional practice to maintain the mineral intake at required levels, which generally are well below the maximum tolerable levels. Greater sensitivity to high mineral levels can be expected in animals that are young, pregnant, lactating, malnourished, or diseased. The amounts of certain elements, such as cadmium, lead, and mercury, should always be maintained as far below the maximum tolerable level as feasible to minimize the contributions to the human diet.

The maximum tolerable levels shown in Table 1 are expressed in terms of either parts per million (ppm) or percent (%) in the total diet. In discussions and tables for the individual elements, intake of the element is usually expressed either as a concentration in the total diet (ppm or %) or as a quantity per unit of body weight. Information on the expected dietary intakes of domestic animals as influenced by body weight and other factors is presented in reports of the Nutrient Requirements Series for the various species of domestic animals published by the National Academy of Sciences and may be helpful in making use of the values presented in the present publication.

TABLE 1 Maximum Tolerable Levels of Dietary Minerals for Domestic Animals[a]

| Element | Species | | | | | |
|---|---|---|---|---|---|---|
| | Cattle | Sheep | Swine | Poultry | Horse | Rabbit |
| Aluminum,[b] ppm | 1,000 | 1,000 | (200) | 200 | (200) | (200) |
| Antimony, ppm | — | — | — | — | — | 70–150 |
| Arsenic, ppm | | | | | | |
| Inorganic | 50 | 50 | 50 | 50 | (50) | 50 |
| Organic | 100 | 100 | 100 | 100 | (100) | (100) |
| Barium,[b] ppm | (20) | (20) | (20) | (20) | (20) | (20) |
| Bismuth, ppm | (400) | (400) | (400) | (400) | (400) | 2,000 |
| Boron, ppm | 150 | (150) | (150) | (150) | (150) | (150) |
| Bromine, ppm | 200 | (200) | 200 | 2,500 | (200) | (200) |
| Cadmium,[c] ppm | 0.5 | 0.5 | 0.5 | 0.5 | (0.5) | (0.5) |
| Calcium,[d] % | 2 | 2 | 1 | Laying hen 4.0<br>Other 1.2 | 2 | 2 |
| Chromium, ppm | | | | | | |
| Chloride | (1,000) | (1,000) | (1,000) | 1,000 | (1,000) | (1,000) |
| Oxide | (3,000) | (3,000) | (3,000) | 3,000 | (3,000) | (3,000) |
| Cobalt, ppm | 10 | 10 | 10 | 10 | (10) | (10) |
| Copper, ppm | 100 | 25 | 250 | 300 | 800 | 200 |

TABLE 1 *Continued*

| Element | Species: Cattle | Sheep | Swine | Poultry | Horse | Rabbit |
|---|---|---|---|---|---|---|
| Fluorine,[e] ppm | Young 40; Mature dairy 40; Mature beef 50; Finishing 100 | Breeding 60; Finishing 150 | 150 | Turkey 150; Chicken 200 | (40) | (40) |
| Iodine, ppm | 50[f] | 50 | 400 | 300 | 5 | — |
| Iron, ppm | 1,000 | 500 | 3,000 | 1,000 | (500) | (500) |
| Lead,[c] ppm | 30 | 30 | 30 | 30 | 30 | (30) |
| Magnesium, % | 0.5 | 0.5 | (0.3) | (0.3) | (0.3) | (0.3) |
| Manganese, ppm | 1,000 | 1,000 | 400 | 2,000 | (400) | (400) |
| Mercury,[c] ppm | 2 | 2 | 2 | 2 | (2) | (2) |
| Molybdenum, ppm | 10 | 10 | 20 | 100 | (5) | 500 |
| Nickel, ppm | 50 | (50) | (100) | (300) | (50) | (50) |
| Phosphorus,[d] % | 1 | 0.6 | 1.5 | Laying hen 0.8; Other 1.0 | 1 | 1 |
| Potassium, % | 3 | 3 | (2) | (2) | (3) | (3) |
| Selenium, ppm | (2) | (2) | 2 | 2 | (2) | (2) |
| Silicon,[b] % | (0.2) | 0.2 | — | — | — | — |

| | | | | | | |
|---|---|---|---|---|---|---|
| Silver, ppm | — | — | (100) | 100 | — | — |
| Sodium Chloride, % | Lactating<br>4<br>Nonlactating<br>9 | 9 | 8 | 2 | (3) | (3) |
| | | — | — | — | — | — |
| Strontium, ppm | 2,000 | (2,000) | 3,000 | Laying hen<br>30,000<br>Other<br>3,000 | (2,000) | (2,000) |
| Sulfur, % | (0.4) | (0.4) | — | — | — | — |
| Tin, ppm | — | — | — | — | — | — |
| Titanium,[g] ppm | — | — | — | — | — | — |
| Tungsten, ppm | (20) | (20) | (20) | 20 | (20) | (20) |
| Uranium, ppm | — | — | — | — | — | — |
| Vanadium, ppm | 50 | 50 | (10) | 10 | (10) | (10) |
| Zinc, ppm | 500 | 300 | 1,000 | 1,000 | (500) | (500) |

[a] The accompanying text should be consulted prior to applying the maximum tolerable levels to practical situations. Continuous long-term feeding of minerals at the maximum tolerable levels may cause adverse effects. The listed levels were derived from toxicity data on the designated species. The levels in parentheses were derived by interspecific extrapolation. Dashes indicate that data were insufficient to set a maximum tolerable level.

[b] As soluble salts of high bioavailability. Higher levels of less-soluble forms found in natural substances can be tolerated.

[c] Levels based on human food residue considerations.

[d] Ratio of calcium to phosphorus is important.

[e] As sodium fluoride or fluorides of similar toxicity. Fluoride in certain phosphate sources may be less toxic. Morphological lesions in cattle teeth may be seen when dietary fluoride for the young exceeds 20 ppm, but a relationship between the lesions caused by fluoride levels below the maximum tolerable levels and animal performance has not been established.

[f] May result in undesirably high iodine levels in milk.

[g] No evidence of oral toxicity has been found.

# Aluminum

Aluminum (Al) is the third most abundant element in the earth's crust, exceeded only by oxygen and silicon. However, it is found only in trace amounts in biological organisms. Its concentration in soils varies geographically, and this variation is reflected in the levels in plants and animals. Most water supplies contain little aluminum. In soils, much of it occurs in clays and feldspars, complex minerals composed primarily of aluminum, silicon, and oxygen. Certain of these clays, such as kaolin and bentonite, are added to animal diets for the following reasons: (1) as pellet binders, (2) as external dusts to prevent clumping of pellets, or (3) for suggested beneficial effects in the gastrointestinal tract. Any effects of the clays on the animals are likely due to the properties of the clays per se and not to their aluminum content.

Aluminum has been the subject of excellent reviews by Tracor-Jitco, Inc. (1973), Sorenson *et al.* (1974), the Life Sciences Research Office (LSRO, 1975) of the Federation of American Societies for Experimental Biology, and Valdivia-Rodriguez (1977).

## ESSENTIALITY

Aluminum has not been proven essential to animals (Schroeder and Nason, 1971), but indirect evidence suggests it may be (Sorenson *et al.*, 1974). Its concentration in tissues changes in a circadian rhythm and with other changes in biological activity (Sorenson *et al.*, 1974). Aluminum accumulates in regenerating bone (Belous, 1961), stimulates certain enzyme systems involved with succinate metabolism (Hutchinson, 1943), and is reported essential for fertility in female rats (Lauro and Giornelli, 1963). Aluminum increased the growth rate of poults, but this may have resulted from prevention of the absorption of high levels of fluorine (Struwe and Sullivan, 1975). It may be important in the immune response, since certain of its compounds are effective adjuvants and since it (along with iron) increases in concentration in the spleen and bone marrow during immunization (Kotter *et al.*, 1966). If a biological requirement for aluminum exists, it has not been quantified.

## METABOLISM

Rats and mice presented with a moderately high dietary aluminum content (160 to 335 ppm) simply excreted most of it in the feces. A larger dose fed to rats overwhelmed their resistance to absorption and resulted in increased urinary, as well as fecal, excretion (Ondreicka *et al.*, 1966). The importance of urinary excretion of aluminum is also demonstrated by the effects of partial nephrectomy in rats, including enhancement of aluminum accumulation in bone (Thurston *et al.*, 1972) and of susceptibility to aluminum toxicity (Berlyne *et al.*, 1972). However, in spite of regulation of the aluminum content of the body by the intestines and kidneys, the amount of aluminum retained is positively related to the amount consumed (Tracor-Jitco, Inc., 1973; LSRO, 1975). Retained aluminum in rats, mice, and dogs is deposited in the liver, skeleton, brain, and probably other tissues (those mentioned include spleen, skeletal muscle, heart, kidney, thyroid, testis, and blood), but there is no good agreement on which of these other tissues accumulate significant quantities (Ondreicka *et al.*, 1966; Berlyne *et al.*, 1972; Tracor-Jitco, Inc., 1973). Elevated levels of parathyroid hormone increase aluminum retention and alter its distribution among tissues (Mayor *et al.*, 1977). Liver aluminum levels are decreased by excess dietary fluorine (Kortus, 1966). Furthermore, aluminum is bound *in vitro* by transferrin or ovotransferrin (Tomimatsu and Donovan, 1976) and seems to be transported and deposited in an inverse relationship with iron (Butt *et al.*, 1956).

## SOURCES

Aluminum is generally not intentionally added to animal diets except as clays such as kaolin and bentonite, as mentioned earlier. Grazing animals ingest considerable amounts of soil, sometimes over 10 percent of their total dry matter intake (Field and Purves, 1964). Intake of this much of some soils very high in aluminum could result in aluminum consumption as high as 1.5 percent of the diet dry matter (Valdivia-Rodriguez, 1977). Aluminum is a significant contaminant in sewage grown algae precipitated with alum (up to 8 percent) (Grau and Klein, 1957), in soft phosphate (7 percent) (Harmon *et al.,* 1970), and in certain other mineral supplements at lower levels (Ammerman *et al.,* 1977). Storer and Nelson (1968) were unable to detect aluminum in a semipurified diet for chicks, but Valdivia-Rodriguez (1977) found from 147 to 210 ppm in basal ruminant diets made largely from conventional ingredients.

## TOXICOSIS

Aluminum toxicosis is expressed largely as a secondary phosphorus deficiency, presumably because it binds phosphorus in an unabsorbable complex in the intestine. High levels of dietary aluminum have resulted in phosphorus deficiency signs in sheep (Valdivia-Rodriguez, 1977), chicks (Deobald and Elvehjem, 1935; Storer and Nelson, 1968), rats (Street, 1942; Thurston *et al.,* 1972), and mice (Ondreicka *et al.,* 1966).

Table 2 summarizes pertinent reports of aluminum toxicosis in animals. Studies involving the dietary inclusion of clays and soils are not included, because the effects may not be due to the aluminum content, even though there is some indication that a high aluminum content in soil may cause problems (Valdivia-Rodriguez, 1977).

### LOW LEVELS

Chronic aluminum toxicity (see Table 2 for specific levels) produces signs of a secondary phosphorus deficiency such as reduced growth rate and feed efficiency in chicks (Grau and Klein, 1957; Storer and Nelson, 1968) and rats (Street, 1942; Thurston *et al.,* 1972), reduced phosphorus absorption in sheep (Valdivia-Rodriguez, 1977), reduced phosphorus retention in mice (Ondreicka *et al.,* 1966), lower bone ash content in chicks (Deobald and Elvehjem, 1935; Storer and Nelson, 1968), and reduced serum phosphorus in chicks (Deobald and Elvehjem, 1935) and in rats (Street, 1942). Low to moderate levels of aluminum (up to 240 mg/kg per day of body weight) as the chloride or sulfate caused various disturbances in carbohydrate and phosphorus metabolism, including a reduction of the ATP:ADP ratio (Ondreicka *et al.*, 1966; Tracor-Jitco, Inc., 1973). Dietary aluminum at a low level reduced the working capacity of rabbits (Nekipelov, 1966), perhaps related to a reduction of hemoglobin levels (Tracor-Jitco, Inc., 1973).

Lesions have been reported in chick muscles (Pragay, 1962) and in the gastrointestinal tract and ovaries of mice (Schaeffer *et al.*, 1928). The ovarian lesions were associated with reproductive failure. Microconcretions were found in the renal tubules of female but not male rats and only rarely in dogs of both sexes (Industrial Bio-Test Laboratories, Inc., 1972a,b,c,d,e). An increased incidence of tumors was seen in male but not female rats consuming a very low level of aluminum throughout their lifetime (Schroeder and Mitchener, 1975). This needs verification. An increased incidence of perosis in the early work of Insko *et al.* (1938) resulted from the addition to the diet of a combination of aluminum and zinc salts and thus may not have been due to the aluminum. The effect was only seen when the manganese level was low.

### HIGH LEVELS

The primary clinical sign of phosphorus deficiency resulting from acute aluminum toxicosis is rickets in chicks (Deobald and Elvehjem, 1935; Pragay, 1962; Storer and Nelson, 1968) and rats (Thurston *et al.*, 1972; Tracor-Jitco, Inc., 1973). Other signs include weakness in chicks (Williams and Rodbard, 1957), tetany in rats (Tracor-Jitco, Inc., 1973), periorbital bleeding in rats (Berlyne *et al.*, 1972), and death in chicks (Williams and Rodbard, 1957; Storer and Nelson, 1968), rabbits (Tracor-Jitco, Inc., 1973), and mice (Ondreicka *et al.*, 1966). Histological changes in the cornea and reduced oxygen uptake in liver homogenates have been reported in rats (Berlyne *et al.*, 1972). A high aluminum concentration in the brain can damage that organ (Crapper *et al.*, 1973; Crapper and Tomko, 1975; Alfrey *et al.*, 1976).

Depending on the dosage, aluminum chloride given intraperitoneally to pregnant rats resulted in maternal death, maternal liver damage, growth retardation of the dam and offspring, skeletal defects in the offspring, and increased incidence of fetal deaths and resorption (Benett *et al.*, 1975).

Studies have failed to find a carcinogenic (Sorenson *et al.*, 1974) or teratogenic (LSRO, 1975) effect of high dosages of aluminum, except for possibly the long-term study by Schroeder and Mitchener mentioned above and for a report of sarcomas in a single rat following subcutaneous implantation of aluminum foil (O'Gara and Brown, 1967). Mutagenicity tests resulted in a marginal response to acidic sodium aluminum phosphate in some but not all tests (Litton Bionetics, Inc., 1975). Neither sodium aluminum sulfate nor aluminum ammonium sulfate was mutagenic.

## FACTORS INFLUENCING TOXICITY

The primary factors affecting the oral toxicity of aluminum are the phosphorus level in the diet and the solubility of the aluminum source. As stated above, aluminum toxicosis is largely expressed as a secondary phosphorus deficiency. In fact, supplementation with phosphorus to correct this deficiency can eliminate the toxic effects of aluminum in chicks (Deobald and Elvehjem, 1935) and rats (Street, 1942; Thurston *et al.*, 1972). Intramuscular injection of a phosphate solution twice daily improved the survival rate of chicks already showing muscular weakness due to aluminum toxicity, and phosphorus has likewise been shown to have therapeutic action against aluminum toxicosis in humans (Tracor-Jitco, Inc., 1973). Phosphorus administration would seem the therapy of choice, since several signs of aluminum toxicosis appear to result from a secondary phosphorus deficiency.

Thompson *et al.* (1959) suggested that ruminants are less vulnerable to the phosphorus deficiency resulting from excess aluminum intake. They reasoned that organic anions in the rumen may form complexes with the aluminum, preventing it from precipitating with a phosphate radical. Partially nephrectomized rats, a model for chronic renal failure in humans, were much more susceptible to aluminum toxicity than were controls in one study (Berlyne *et al.*, 1972). The trigger for susceptibility seems to be hypophosphatemia (Thurston *et al.*, 1972).

The compound that provides the aluminum is important in determining toxicity. Soluble salts (acetate, chloride, nitrate, and sulfate) were toxic to chicks, while insoluble salts (oxide and phosphate) were not (Storer and Nelson, 1968). The lack of toxicity of the phosphate could have been due to prevention of a secondary phosphate deficiency. Street (1942) concluded that aluminum from a soluble salt (sulfate) fed in amounts equal to the dietary phosphorus content on a molar basis caused nearly complete precipitation of the phosphorus in the gut. When the aluminum was from the insoluble hydroxide, about one-fourth to one-third of it precipitated with phosphorus.

A high fluoride content in the diet decreases the retention of ingested aluminum (LSRO, 1975). Conversely, it has been suggested that aluminum be used to prevent or treat fluorine toxicosis in ruminants (Thompson *et al.*, 1959) and poultry (Cakir *et al.*, 1978).

## TISSUE LEVELS

There is not good agreement on tissue aluminum levels, as shown in Table 3. Dietary aluminum increases the concentration of aluminum in several tissues in rats (Ondreicka *et al.*, 1966; Berlyne *et al.*, 1972), sheep (Valdivia-Rodriguez, 1977), and cattle (Valdivia *et al.*, 1978).

There appear to be marked differences in measured levels among species and between investigators working with the same species.

## MAXIMUM TOLERABLE LEVELS

The main factors that can either increase or decrease the severity of aluminum toxicity are the phosphorus level in the diet and the solubility of the aluminum source. Using soluble salts of aluminum, Bailey (1977) and Valdivia *et al.* (1978) found no adverse effects from feeding dietary levels of 1,200 ppm aluminum to calves. Studies with sheep showed no adverse effects from dietary levels of 1,215 ppm aluminum (Bailey, 1977). Based on these results, the maximum tolerable level of aluminum for cattle and sheep is about 1,000 ppm.

No untoward effects on chicks were found with dietary aluminum levels of 486 ppm (Cakir *et al.*, 1978), but Storer and Nelson (1968) showed that growth and efficiency of feed conversion were significantly reduced when chicks were fed 500 ppm aluminum. Feeding aluminum levels of 486 ppm had no effect on the performance of turkeys (Cakir *et al.*, 1978). Although no studies show the effects of feeding various levels of aluminum to pigs, it would seem desirable to limit aluminum from soluble salts to about 200 ppm for nonruminants.

## SUMMARY

Aluminum is the third most abundant element in the earth's crust but is found in only trace amounts in plants and animals. It has not been proved to be biologically essential, but there are indirect indications that it may be. Its toxic effects are mostly exerted through interference with phosphorus absorption and metabolism. Levels that produce toxicosis are determined by several factors.

TABLE 2 Effects of Aluminum Administration in Animals

| Class and Number of Animals[a] | Age or Weight | Administration | | | | Effect(s) | Reference |
|---|---|---|---|---|---|---|---|
| | | Quantity of Element[b] | Source | Duration | Route | | |
| Cattle—6 | 226 kg | 1,200 ppm | $AlCl_3 \cdot 6H_2O$ | 84 d | Diet | No adverse effect | Valdivia *et al.*, 1978 |
| Cattle—4 | Mature | 1,215 ppm | Aluminum sulfate | 14 d | Diet | No adverse effect | Bailey, 1977 |
| Cattle | Newborn | 1,215 ppm | Aluminum sulfate | | | Refusal of diet after weaning | |
| Sheep—4 | 37 kg | 900 ppm | Aluminum sulfate | 14 d | Diet | No adverse effect | Thompson *et al.*, 1959 |
| Sheep—4 | | 1,215 ppm | Aluminum sulfate | 20 d | Diet | No adverse effect | Bailey, 1977 |
| Sheep—10 | 36.7 kg | 2,200 ppm | $AlCl_3 \cdot 6H_2O$ | 77 d | Diet | Reduced growth and P absorption, especially with low dietary P | Valdivia-Rodriguez, 1977 |
| Chicken—24 | | 30 ppm | $Al(NH_4)(SO_4)_2 \cdot 12H_2O$ | 56 d | Diet | Increased incidence of perosis | Insko *et al.*, 1938 |
| Chicken—30 | 1 d | 324 ppm | $Al_2(SO_4)_3 \cdot 18H_2O$ | 28 d | Diet | No adverse effect | Cakir *et al.*, 1978 |
| | | 486 ppm | | | | No adverse effect | |
| Chicken—20 | 1 d | 500 ppm | Aluminum chloride | 2 wk | Diet | Reduced growth rate, feed efficiency, and bone ash | Storer and Nelson, 1968 |
| | | 1,000 ppm | Aluminum sulfate | | | Reduced growth rate, feed efficiency, and bone ash | |
| Chicken—20 | | 2,000 ppm | Aluminum chloride | | | Reduced growth rate, feed efficiency, and bone ash | |

| | | | | | | | |
|---|---|---|---|---|---|---|---|
| Chicken—20 | 1 d | 2,000 ppm | Aluminum sulfate | 2 wk | Diet | Reduced growth rate, feed efficiency, and bone ash | Storer and Nelson, 1968 |
| | | 4,000 ppm | Aluminum chloride | | | Death; reduced growth rate, feed efficiency, and bone ash | |
| | | 4,000 ppm | Aluminum sulfate | | | Death; reduced growth rate, feed efficiency, and bone ash | |
| | | 5,000 ppm | Aluminum chloride | 3 wk | | Death | |
| | | 5,000 ppm | Aluminum sulfate | | | Death | |
| | | 5,000 ppm | Aluminum nitrate | | | Death | |
| | | 5,000 ppm | Aluminum acetate | | | Death; reduced growth rate and bone ash | |
| | | 5,000 ppm | Aluminum phosphate | | | No adverse effect | |
| | | 16,000 ppm | $Al_2O_3$ | | | No adverse effect | |
| Chicken | 1 d | 2,200 ppm | $Al_2(SO_4)_3 \cdot 18H_2O$ | 5 wk | Diet | Reduced bone ash and serum | Deobald and Elvehjem, 1935 |
| | | 3,300 ppm | | | | Reduced bone ash and serum P | |
| | | 4,400 ppm | | | | Severe rickets and death | |
| Chicken—108 | 1 d | 3,500 ppm | $Al(OH)_3$ | 19 d | Diet | No adverse effect | Miller and Kifer, 1970 |
| Chicken—12 | 10 d | 4,900 ppm | Alum | 8 d | Diet | No adverse effect | Grau and Klein, 1957 |
| | | 8,000 ppm | Alum | | | Reduced growth rate | |
| Chicken—30 | 1 d | 8,000 ppm | Reactive aluminum hydroxide gel | 35 d | Diet | No adverse effect | Williams and Rodbard, 1957 |
| Chicken—45 | | 25,000 ppm | Reactive aluminum hydroxide gel | 35 d | | Weakness and death | |
| Chicken—60 | | | Nonreactive aluminum hydroxide gel | | | Occasional transient weakness | |

TABLE 2 *Continued*

| Class and Number of Animals[a] | Age or Weight | Administration: Quantity of Element[b] | Administration: Source | Administration: Duration | Administration: Route | Effect(s) | Reference |
|---|---|---|---|---|---|---|---|
| Chicken—15 | | | Aluminum phosphate gel | 6 wk | | Transient weakness, occasional death | |
| Turkey—30 | 1 d | 324 ppm | $Al_2(SO_4)_3 \cdot 18H_2O$ | 28 d | Diet | No adverse effect | Cakir *et al.*, 1978 |
| | | 486 ppm | | | | No adverse effect | |
| Rabbit | | 1 mg/kg | $Al(NO_3)_3$ | 6 mo | Oral | No adverse effect | Nekipelov, M. K., 1966 |
| | | 10 mg/kg | | 6 mo | | Decreased working capacity | |
| Dog—8 (Beagles) | | 600 ppm | Basic sodium aluminum phosphate | 90 d | Diet | No adverse effect | Industrial Bio-Test Laboratories, Inc., 1972a |
| Dog—8 (Beagles) | | 1,700 ppm | | 90 d | Diet | Occasional microconcretions in renal tubules | Industrial Bio-Test Laboratories, Inc., 1972a |
| Dog—8 (Beagles) | | 2,600 ppm | $NaAl_3H_{14}(PO_4)_8 \cdot 4H_2O$ | 90 d | | No adverse effect | Industrial Bio-Test Laboratories, Inc., 1972b |
| Dog (Beagles) | | 2,600 ppm | Acidic sodium aluminum phosphate | 90 d | | No adverse effect | Industrial Bio-Test Laboratories, Inc., 1972c |
| Rat—104 | Weanling | 5 ppm | Aluminum potassium sulfate | Lifetime | Water | Increased incidence of tumors in males but not in females | Schroeder and Mitchener, 1975 |
| Rat—8 | 200 g | 160–180 ppm | Aluminum sulfate | 24 d | Diet | No adverse effect | Ondreicka *et al.*, 1966 |
| | | 2,835 ppm | | | | Increased Al concentration in liver, testis, and bone | |

| | | | | | | |
|---|---|---|---|---|---|---|
| Rat—6 | 36.5 mg/kg | Aluminum chloride | 52 d | Gavage | Decreased incorporation of $^{32}P$ into phospholipids, RNA, and DNA | |
| | 240 mg/kg | | Single dose | | Decreased incorporation of $^{32}P$ into phospholipids, RNA, and DNA | |
| Rat—10 | 36.5 mg/kg | | 55 d | | Decreased ATP:ADP ratio | |
| | 223 mg/kg | | Single dose | | Decreased ATP:ADP ratio | |
| Rat | 69 mg/kg | | Single dose | | Decreased liver and muscle glycogen | |
| | 122 mg/kg | Aluminum chloride | Single dose | Gavage | Disturbed carbohydrate metabolism | Ondreicka *et al.*, 1966 |
| | 200 mg/kg | | | | Drastically reduced liver glycogen and coenzyme A; reduced muscle glycogen; increased liver and muscle lactate; and liver and serum pyruvate | |
| Rat—30 | 170 ppm | Basic sodium aluminum phosphate | 90 d | Diet | Microconcretions in renal tubules of females, not males | Industrial Bio-Test Laboratories, Inc., 1972d |
| | 1,700 ppm | | | | Microconcretions in renal tubules of females, not males | |
| | 260 ppm | $NaAl_3H_{14}(PO_4)_8 \cdot 4H_2O$ | | Diet | Microconcretions in renal tubules of females, not males | Industrial Bio-Test Laboratories, Inc., 1972e |

TABLE 2 *Continued*

| Class and Number of Animals[a] | Age or Weight | Administration: Quantity of Element[b] | Source | Duration | Route | Effect(s) | Reference |
|---|---|---|---|---|---|---|---|
| | | 2,600 ppm | | | | Microconcretions in renal tubules of females, not males | |
| Rat—6 | Weanling | 1,100 ppm | $Al(OH)_3$ | 4 wk | Diet | Reduced growth rate, increased bone Al concentration, rachitic bone changes | Thurston *et al.*, 1972 |
| Rat—5 | 200 g | 1,579 ppm | $Al_2(SO_4)_3$ | 3 wk | Water | No adverse effect | Berlyne *et al.*, 1972 |
| Rat | 200 g | 3,158 ppm | $Al_2(SO_4)_3$ | 3 wk | Water | Periorbital bleeding | |
| | | 150 mg/kg | $Al(OH)_3$ | | Gavage | No adverse effect | |
| Rat—10 | 40–50 g | 1,800 ppm | | 6 wk | Diet | Decreased weight gain and serum P (Al:P atomic ratio 1:1) | Street, 1942 |
| | | 2,160 ppm | | | | Decreased weight gain and serum P (Al:P atomic ratio 1:1) | |
| Rat—20 | | 2,160 ppm | $Al_2(SO_4)_3 \cdot 18H_2O$ | | | Severely decreased weight gain and serum P (Al:P atomic ratio 1:1) | |
| Rat—10 | | 2,160 ppm | | | | No adverse effect (Al:P atomic ratio 2:5) | |

| | | | | | | | |
|---|---|---|---|---|---|---|---|
| Rat—20 | | 3,600 ppm | $Al(OH)_3$ | | | Decreased weight gain and serum P (Al:P atomic ratio 1:1) | |
| | | 3,600 ppm | | | | No adverse effect (Al:P atomic ratio 2:3) | |
| Mouse—10 | 200 g | 160–180 ppm | Aluminum chloride | 40 d | Diet | No adverse effect | Ondreicka *et al.*, 1966 |
| | | 355 ppm | | | | Decreased P retention | |
| | | 19.3 mg/kg | Aluminum chloride | 180–300 d (3 generations) | Water | Decreased growth rate beginning with 2nd litter from original mice | |
| | | 770 mg/kg | Aluminum chloride | Single dose | Gavage | $LD_{50}$ | |
| | | 980 mg/kg | Aluminum sulfate | | | $LD_{50}$ | |
| Mouse—40 | | 2,070 ppm | Alum—phosphate baking powder | 4 mo | Diet | Gastrointestinal and ovarian lesions; impaired reproduction | Schaeffer *et al.*, 1928 |
| | | 4,100 ppm | Alum baking powder | | | Gastrointestinal and severe ovarian lesions; severely impaired reproduction | |
| Mouse—20 | | 13,000 ppm | Alum—phosphate baking powder | | | Impaired reproduction | |
| | | 44,000 ppm | | | | Impaired reproduction; increased mortality of offspring; ovarian lesions | |

[a] Number of animals per treatment.
[b] Quantity expressed in parts per million or as milligrams per kilogram of body weight.

TABLE 3 Tissue Levels of Aluminum (ppm fresh tissue)

| | Species and Treatment | | | | | | | | |
|---|---|---|---|---|---|---|---|---|---|
| | Rats[a] | | Rats[b] | | | Cattle[c] | | Sheep[d] | |
| Tissue | Control | A1 | Control | A1 | A1–NX | Control | A1 | Control | A1 |
| Blood | 6.5 | 10.8 | — | — | — | — | — | — | — |
| Plasma or serum | — | — | 0.24 | 1.46 | 2.25 | 0.10 | 0.12 | — | — |
| Liver | 15.5 | 29.9 | 0.12 | 0.46 | 2.34 | 7.6 | 11.2 | 2.79 | 6.52 |
| Kidney | 8.6 | 9.0 | — | — | — | 4.5 | 5.4 | 2.86 | 4.92 |
| Skeletal muscle | 0.39 | 0.44 | 0.56 | 0.67 | 9.69 | 3.8 | 4.7 | 2.57 | 3.98 |
| Heart | 10.8 | 12.7 | 0.34 | 0.53 | 1.41 | — | — | — | — |
| Brain | 7.1 | 10.8 | — | — | — | 6.4 | 7.7 | 4.54 | 4.54 |
| Bone | 702.0 | 912.0 | 2.93 | 2.76 | 9.38 | — | — | — | — |
| Whole body | — | — | 0.28 | 0.76 | 3.52 | — | — | — | — |

[a] Ondreicka *et al.*, 1966. A1 treatment is 160–180 ppm A1 in the diet for 8 days, followed by 2,835 ppm for 8 days, as aluminum sulfate.
[b] Berlyne *et al.*, 1972. A1 treatment is 1,579 ppm A1 in drinking water, as aluminum sulfate. A1–NX treatment is the same with partial nephrectomy. Values presented assume same tissue dry matter concentration in A1 and A1–NX as in controls.
[c] Valdivia *et al.*, 1978. A1 treatment is 1,200 ppm in the diet as aluminum chloride.
[d] Valdivia-Rodriguez, 1977. A1 treatment is 2,200 ppm in the diet as aluminum chloride.

## REFERENCES

Alfrey, A. C., G. R. Le Gendre, and W. D. Kaehny. 1976. The dialysis encephalopathy syndrome. Possible aluminum intoxication. N. Engl. J. Med. 294:184.

Ammerman, C. B., S. M. Miller, K. R. Fick, and S. L. Hansard II. 1977. Contaminating elements in mineral supplements and their potential toxicity: A review. J. Anim. Sci. 44:485.

Bailey, C. B. 1977. Influence of aluminum hydroxide on the solubility of silicic acid in rumen fluid and the absorption of silicic acid from the digestive tract of ruminants. Can. J. Anim. Sci. 57:239.

Belous, A. M. 1961. Content of some trace elements in healing of femur fracture. Ukr. Biokhim. Zh. 23:856; cited in Chem. Abstr. 56:10756, 1962.

Benett, R. W., T. V. N. Persaud, and K. L. Moore. 1975. Experimental studies on the effects of aluminum on pregnancy and fetal development. Anat. Anz. Bd. 138:365.

Berlyne, G. M., J. B. Ari, E. Knopf, R. Yagil, G. Weinberger, and G. M. Danovitch. 1972. Aluminum toxicity in rats. The Lancet 1:564.

Butt, E. M., R. E. Nusbaum, T. C. Gilmour, and S. L. Didio. 1956. Trace metal patterns in disease states. I. Hemochromatosis and refractory anemia. Am. J. Clin. Pathol. 26:225.

Cakir, A., T. W. Sullivan, and F. B. Mather. 1978. Alleviation of fluorine toxicity in starting turkeys and chicks with aluminum. Poult. Sci. 57:498.

Crapper, D. R., and G. T. Tomko. 1975. Neuronal correlates of an encephalopathy associated with aluminum neurofibrillary degeneration. Brain Res. 97:253.

Crapper, D. R., S. S. Krishnan, and A. J. Dalton. 1973. Brain aluminum distribution in Alzheimer's disease and experimental neurofibrillary degeneration. Science 180:511.

Deobald, H. J., and C. A. Elvehjem. 1935. The effects of feeding high amounts of soluble iron and aluminum salts. Am. J. Physiol. 111:118.

Field, A. C., and D. Purves. 1964. The intake of soil by grazing sheep. Proc. Nutr. Soc. 23:XXIV.

Grau, C. R., and N. W. Klein. 1957. Sewage-grown algae as a feedstuff for chicks. Poult. Sci. 36:1046.

Harmon, B. G., J. Simon, D. E. Becker, A. H. Jensen, and D. H. Baker. 1970. Effect of source and level of dietary phosphorus on structure and composition of turbinate and long bones. J. Anim. Sci. 30:742.

Hutchinson, G. E. 1943. The geochemistry of aluminum and certain related elements. Q. Rev. Biol. 18:242.

Industrial Bio-Test Laboratories, Inc. 1972a. 90-Day Subacute Oral Toxicity Study with Kasal in Beagle Dogs. Report IBT No. J749. Prepared for Stauffer Chemical Company. Industrial Bio-Test Laboratories, Inc., Northbrook, Ill. 55 pp. Cited in LSRO, 1975.

Industrial Bio-Test Laboratories, Inc. 1972b. 90-Day Subacute Oral Toxicity Study with Levair in Beagle Dogs. Report IBT No. J749. Prepared for Stauffer Chemical Company. Industrial Bio-Test Laboratories, Inc., Northbrook, Ill. 55 pp. Cited in LSRO, 1975.

Industrial Bio-Test Laboratories, Inc. 1972c. 90-Day Subacute Oral Toxicity Study with Levn-Lite in Beagle Dogs. Report BTL 71-49; IBT No. J749. Prepared for Monsanto Company. Industrial Bio-Test Laboratories, Inc., Northbrook, Ill. 54 pp. Cited in LSRO, 1975.

Industrial Bio-Test Laboratories, Inc. 1972d. 90-Day Subacute Oral Toxicity Study with Kasal in Albino Rats. Report IBT No. B747. Prepared for Stauffer Chemical Company. Industrial Bio-Test Laboratories, Inc., Northbrook, Ill. 31 pp. Cited in LSRO, 1975.

Industrial Bio-Test Laboratories, Inc. 1972e. 90-Day Subacute Oral Toxicity Study with Levair in Albino Rats. Report IBT No. B747. Prepared for Stauffer Chemical Company. Industrial Bio-Test Laboratories, Inc., Northbrook, Ill. 31 pp. Cited in LSRO, 1975.

Insko, W. M., M. Lyons, and J. H. Martin. 1938. The effect of manganese, zinc, aluminum, and iron salts on the incidence of perosis in chicks. Poult. Sci. 17:264.

Kortus, J. 1966. Effect of increased doses of aluminum and of fluoride ion on the aluminum level in organs. Biologia (Bratislava) 21:33; cited in Chem. Abstr. 64:20501, 1966.

Kotter, L., C. Herrmann, C. Ring, and G. Lorsico. 1966. Increasing the formation of antibodies against heated protein with the help of metallic compounds. Zentralbl. Veterinaermed. Reihe B 13:613; cited in Chem. Abstr. 66:45035.

Lauro, V., and C. Giornelli. 1963. Indagini sperimentali sui rapporti tra carenza nella dieta di alcuni metalli (alluminio, ferro, manganese) ed attivita riproduttiva. Arch. Ostet. Ginecol. 68:121; cited in Sorenson *et al.,* 1974.

Litton Bionetics, Inc. 1975. Mutagenic Evaluation of Sodium Aluminum Phosphate (Acidic) (Compound 007785-88-8). Report prepared under DHEW Contract No. FDA 75-1. Kensington, Md. Cited in LSRO, 1975.

LSRO. 1975. Evaluation of the Health Aspects of Aluminum Compounds as Food Ingredients. Life Sciences Research Office, Federation of American Societies for Experimental Biology, Bethesda, Md. SCOGS-43. Contract No. FDA 223-75-2004.

Mayor, G. H., J. A. Keiser, D. Makdani, and P. K. Ku. 1977. Aluminum absorption and distribution: Effect of parathyroid hormone. Science 197:1187.

Miller, D., and R. R. Kifer. 1970. Effect of glutamic acid and antiacids on chick bioassay of protein quality of fish meals. Poult. Sci. 49:1327.

Nekipelov, M. K. 1966. Hygienic standard for aluminum nitrate in water basins. Hyg. Sanit. (USSR) 31:204; cited in Sorenson *et al.,* 1974.

O'Gara, R. W., and J. M. Brown. 1967. Comparison of the carcinogenic actions of subcutaneous implants of iron and aluminum in rodents. J. Natl. Cancer Inst. 38:947; cited in Sorenson *et al.,* 1974.

Ondreicka, R., E. Ginter, and J. Kortus. 1966. Chronic toxicity of aluminum in rats and mice and its effects on phosphorus metabolism. Br. J. Ind. Med. 23:305; cited in Tracor-Jitco, Inc., 1973.

Pragay, D. A. 1962. Muscular dystrophy in chicks caused by dietary aluminum hydroxide gel. Fed. Proc. 21:388. (Abstr.)

Schaeffer, G., G. Fontes, E. Le Breton, C. Oberling, and L. Thivolle. 1928. The dangers of certain mineral baking powders based on alum, when used for human nutrition. J. Hyg. 28:92.

Schroeder, H. A., and M. Mitchener. 1975. Life-term studies in rats: Effects of aluminum, barium, beryllium and tungsten. J. Nutr. 105:421.

Schroeder, H. A., and A. P. Nason. 1971. Trace-element analysis in clinical chemistry. Clin. Chem. 17:461.

Sorenson, J. R. J., I. R. Campbell, L. B. Tepper, and R. D. Lingg. 1974. Aluminum in the environment and human health. Environ. Health Perspect. 8:3.

Storer, N. L., and T. S. Nelson. 1968. The effect of various aluminum compounds on chick performance. Poult. Sci. 47:244.

Street, H. R. 1942. Influence of aluminum sulfate and aluminum hydroxide upon the absorption of dietary phosphorus by the rat. J. Nutr. 24:111.

Struwe, F. J., and T. W. Sullivan. 1975. Fluorine and phosphorus relationships in diets for starting and growing–finishing turkeys. Poult. Sci. 54:1821 (Abstr.).

Thompson, A., S. L. Hansard, and M. C. Bell. 1959. The influence of aluminum and zinc upon the absorption and retention of calcium and phosphorus in lambs. J. Anim. Sci. 18:187.

Thurston, H., G. R. Gilmore, and J. D. Swales. 1972. Aluminum retention and toxicity in chronic renal failure. The Lancet 1:881.

Tomimatsu, Y., and J. W. Donovan. 1976. Spectroscopic evidence of perturnation of tryptophan in aluminum (III) and gallium (III) binding to ovotransferrin and human serum transferrin. FEBS Lett. 71:229; cited in Valdivia-Rodriguez, 1977.

Tracor-Jitco, Inc. 1973. Scientific Literature Reviews on Generally Recognized as Safe (GRAS) Food Ingredients—Aluminum Compounds. Prepared for Food and Drug Administration. NTIS No. PB-223 862.

Valdivia, R., C. B. Ammerman, C. J. Wilcox, and P. R. Henry. 1978. Effect of dietary aluminum on animal performance and tissue mineral levels in growing steers. J. Anim. Sci. 47:1351.

Valdivia-Rodriguez, R. J. 1977. Effect of dietary aluminum on phosphorus utilization by ruminants. Ph. D. thesis. University of Florida, Gainesville.

Williams, C., and S. Rodbard. 1957. Weakness in young chicks on a diet supplemented with aluminum hydroxide gel. Poult. Sci. 36:602.

# Antimony

Antimony (Sb) is a lustrous, silver-white metal with a bluish tinge that is classified with arsenic and bismuth as a Group VA metal in the periodic table. Very little antimony occurs free in nature, and most is derived from the principal antimony ore, stibnite ($Sb_2S_3$), which contains 71 to 75 percent of this element when nearly pure. Abundant deposits are found in China, Mexico, Bolivia, Algeria, Portugal, and France. Other valuable antimony ores include cervantite ($Sb_2O_4$), valentinite ($Sb_2O_3$), and kermesite ($Sb_2S_2O$). Mean antimony concentration in the earth's crust has been estimated to be 0.2 ppm (Schroeder, 1973).

Metallic antimony was used in the manufacture and plating of vases and household vessels as early as 4000 B.C. (Mellor, 1939) and as a constituent of ancient bronzes in the fifth or sixth Egyptian Dynasty (Fink and Kopp, 1933). Use of stibnite as a cosmetic (Kohl) is referred to in the Old Testament and in early Chinese and Arabic writings. Industrial uses are diverse, but a major portion serves as a constituent of alloys with lead, tin, and copper. Lead–antimony alloys are used in storage battery grids, pewter, printer's type, lead shot, lead electrodes, cable coverings, foil, and solder. The trisulfide ($Sb_2S_3$) and pentasulfide ($Sb_2S_5$) are used in the compounding of rubber. Antimony trioxide ($Sb_2O_3$) is used as a textile dye, and lead antimonate [$Pb_3(SbO_4)_2$] as a paint pigment. Oxides are used as opacifiers in enamels and as decolorizing and refining agents in glass maufacture. An important pharmaceutical compound is tartar emetic, potassium antimonyl tartrate

($KSbC_4H_4O_7 \cdot \frac{1}{2}H_2O$), which has been used for years in the treatment of schistosomiasis.

## ESSENTIALITY

Antimony has no known essential metabolic function in living organisms, and Liebscher and Smith (1968) have used this mineral as a model for nonessential elements. Thiol-containing enzymes are inhibited *in vitro* by antimony salts. The possibility of similar enzyme inhibitions *in vivo* and the affinity of trivalent antimony for erythrocytes may be significant for the effectiveness of antimony tartrates in treatment of schistosomiasis.

## METABOLISM

Soluble antimony compounds, such as antimonites and tartrates, are slowly absorbed from the alimentary tract. Waitz *et al.* (1965) reported that ingestion of potassium antimonyl tartrate by monkeys, rats, and mice led to greater fecal than urinary excretion, and these workers concluded that gastrointestinal absorption was poor. Felicetti *et al.* (1974) found that very little of either trivalent or pentavalent antimony tartrate was absorbed by hamsters from a gavage. About 2 percent of the initial body burden was present 4 days later, and nearly two-thirds of this was found in the gastrointestinal tract. Antimony halides are hydrolyzed to oxides in the alimentary tract and are apparently not absorbed.

The metabolic behavior of antimony is affected by its valence state. Trivalent antimony concentrates in the liver of all species studied (Brady *et al.*, 1945; Djuric *et al.*, 1962; Gellhorn *et al.*, 1946; Otto *et al.*, 1947; Otto and Maren, 1950; Rowland, 1968; Tarrant *et al.*, 1971; Thomas *et al.*, 1973), in the thyroid and parathyroid of dogs (Brady *et al.*, 1945), and in the erythrocytes of many species, including man (Otto and Maren, 1950; Otto *et al.*, 1947). Hair, skin, and skeletal accumulations of trivalent antimony have been reported in mice (Molokhia and Smith, 1969; Thomas *et al.*, 1973) and hamsters (Felicetti *et al.*, 1974). Pentavalent antimony has a lesser affinity for the liver than does trivalent antimony and accumulates more in the spleen (Gellhorn *et al.*, 1946). Human erythrocytes do not concentrate antimony in the pentavalent state (Otto *et al.*, 1947). In most rodents, trivalent antimony is excreted primarily in the feces and pentavalent primarily in the urine

(Otto and Maren, 1950), but in humans, both valence states of antimony are excreted in the urine (Otto *et al.*, 1947).

## SOURCES

Potentially toxic exposures occur as a consequence of industrial processing and use of antimony, preparation, or storage of food in containers improperly glazed with an antimony enamel and accidental ingestion or parenteral administration of excessive doses of antimony compounds. The most common industrial exposures to antimony are during the mining, smelting, and refining of the ore; in the production of alloys; in the manufacture of abrasives; and in type-setting (Browning, 1969). Inhalation of antimony-containing dusts or fumes constitute the main hazard, and the American Conference of Industrial Hygienists has set a threshold limit of 0.5 mg/m$^3$ of air. Stibine ($SbH_3$) is a colorless gas evolved when certain antimony alloys are treated with acid and subjected to electrolysis (e.g., during the charging of storage batteries). Stibine is also released when some antimony compounds are treated with steam, or when nascent hydrogen comes into contact with metallic antimony or a soluble antimony compound. The threshold limit for stibine in air for an 8-hour working day has been set at 0.1 mg per liter. Antimony trioxide is frequently used as an opacifier in vitreous coatings. Enamel glazes of this type, particularly if low in silica, are readily attacked by acids of foodstuffs. Monier-Williams (1925, 1934) found that exposure of enamelled containers to a 1 percent solution of citric acid dissolved 0.01 g of antimony per liter.

Published studies of antimony levels in foods or animal feedstuffs are few. Murthy *et al.* (1971) analyzed the total diets for 7 days of institutionalized children in 28 localities in the United States. Intakes varied from 0.25 to 1.28 mg antimony per day, and dietary concentrations ranged from 0.21 to 0.69 ppm. Hamilton and Minski (1972/1973) analyzed adult English diets and found a mean intake of 34 $\mu$g/day. These workers also noted that the proportion of refined foods in the diet may influence such values, since brown sugar contained 0.08 ppm antimony and refined white sugar contained $<$0.002 ppm.

# TOXICOSIS

## LOW LEVELS

### *Oral*

The dietary concentration of antimony that induces toxicosis is dependent upon valence state, pentavalent compounds being less irritating than trivalent compounds (Goodwin, 1944). Bradley and Fredrick (1941) reported that rats fed potassium antimonyl tartrate in increasing daily doses over 12 months grew normally on amounts up to 100 mg/kg body weight (36.5 mg/kg antimony), but cardiac pathology was evident. Antimony metal up to 1 g/kg body weight produced similar effects. Pribyl (1927) fed rabbits 15 mg of potassium antimonyl tartrate per kilogram body weight (5.5 mg/kg antimony) and found increased concentrations of nonprotein nitrogen in blood and urine. After 5 to 20 days, icterus was noted and some individuals showed fatty degeneration and parenchymal necrosis of the liver. Wieland (1937) found no pathology in rabbits fed 2 to 6 mg potassium antimonyl tartrate per kg body weight (0.7 to 2.2 mg/kg antimony). Lifetime studies (Schroeder *et al.*, 1968) with mice showed that 5 ppm antimony as potassium antimonyl tartrate in drinking water reduced mean life span of females slightly. No evidence of carcinogenesis or tumorigenesis was obtained.

### *Inhalation*

Guinea pigs were exposed by Dernehl *et al.* (1945) to a dust concentration of antimony trioxide of 45.4 mg/m$^3$ of air (19 mg/m$^3$ antimony) for 2 hours daily, 7 days a week for 3 weeks, followed by 3 hours daily exposure. All animals showed extensive interstitial pneumonitis, and 4 died. No electrocardiogram changes were evident, but 11 out of 15 guinea pigs having 138 or more hours of exposure showed fatty degeneration of the liver. There was also a leucocytopenia, a polymorphonuclear neutrophilopenia, and a relative lymphocytosis. Gross *et al.* (1951) exposed rats to antimony trioxide for periods up to 14 months and induced a chronic lipoid pneumonia. The lipoid deposits were intra-alveolar with some perifocal fibrosis.

Brieger *et al.* (1954) exposed rabbits to the dust of antimony trisulfide in a concentration of 5.6 mg/m$^3$ of air (2 mg/m$^3$ antimony), 7 hours per day, 5 days per week for 6 weeks. The lungs showed a mild degree of venous congestion and areas of focal hemorrhage. Higher concentrations of 27.8 mg/m$^3$ of air (10 mg/m$^3$ antimony) induced lung inflammation. Slight to moderate myocardial damage produced changes in the T waves. In some rabbits the myocardium was flabby and dilated, with swelling of the fibers and granular cytoplasmic inclusions. Dogs exposed by Brieger *et al.* (1954) to 5.3 and 5.6 mg antimony trisulfide per cubic meter of air (1.9 to 2.0 mg/m$^3$ antimony) were not affected adversely.

Industrial poisoning of humans by antimony may not be clear-cut, since industrial antimony usually contains other substances, such as arsenic, and toxic signs and symptoms are somewhat similar. Renes (1953), in a study of mining, concentrating, and smelting stibnite, found air concentrations of antimony to be 4.7 to 11.8 mg/m$^3$, while arsenic concentrations were 0.4 to 1.1 mg/m$^3$. Early signs of arsenical poisoning were absent, and he concluded that the illness observed was due to the antimony. Renes (1953) suggested that apparent differences in the toxicity of antimony in different industrial settings may be related to particle size. In his study, particle size was usually less than 1 $\mu$m. Bulmer and Johnston (1948) found no ill health in humans working in a laboratory where antimony trisulfide was crushed and ground. Particles were small but not as minute as in a fume. Two men were exposed for a year to air antimony concentrations of 52 mg/m$^3$. Schrumpf and Zabel (1910) described a variety of symptoms in 15 to 20 percent of typesetters, including irritability, fatigue, pains in the limbs, anorexia, and gastrointestinal complaints. Gocher (1945) noted muscular pains, headache, dizziness, and oppression in the chest. Feil (1939) and Renes (1953) recorded laryngitis and tracheitis in antimony smelters. The latter researcher also observed abdominal cramps, diarrhea, vomiting, dizziness, nerve tenderness and tingling, severe headaches, and prostration. Acutely ill individuals showed pneumonitis but no peripheral parenchymal pulmonary damage. Removal from exposure and treatment with penicillin aerosols rapidly alleviated the symptoms. Feil (1939) found a characteristic skin eruption in foundry workers in which the eruption was pustular, sometimes covered with a crust (like the lesions of chickenpox) and present on limbs, face, and chest. There was also some irritation of eyes and throat, some gingivitis, and clinical appearance of anemia. Brieger *et al.* (1954) studied men in a plant where resinoid grinding wheels were manufactured and where antimony trisulfide had replaced lead for the preceding 2 years. Air concentrations of antimony varied between 0.6 and 5.5 mg/m$^3$, with most above 3.0. Several men had died of heart attacks, and 35 out of 75 men examined electrocardiographically showed abnormalities, mostly in the T waves.

## HIGH LEVELS

### *Oral*

Acute toxicosis in rabbits was studied by Oelkers (1937), who found that a single oral dose of 125 mg of potassium antimonyl tartrate per kilogram of body weight (46 mg/kg antimony) was fatal in all cases; 120 mg/kg of body weight (44 mg/kg antimony) was almost certainly fatal in 24 to 36 hours; and 115 mg/kg of body weight (42 mg/kg antimony) killed 50 percent of the animals. In oral studies with rats, Bradley and Fredrick (1941) found the minimum lethal dose of potassium antimonyl tartrate expressed as antimony was 300 mg/kg of body weight. Franz (1937) described centrolobular fatty degeneration of the liver and degenerative changes in the kidneys of rats.

Monier-Williams (1934) reported that in 1928 there was an outbreak of sickness in Newcastle-on-Tyne in 56 individuals who had drunk lemonade made in white enamelled buckets. A drinking glass of lemonade was found to contain the equivalent of about 100 mg potassium antimonyl tartrate (36.5 mg antimony). Symptoms reported included a burning sensation in the stomach, colic, nausea, vomiting, and collapse. The usual emetic dose of potassium antimonyl tartrate is about 30 to 60 mg (11 to 22 mg antimony).

### *Injection*

Bradley and Fredrick (1941) established the $LD_{50}$ for antimony metal and five antimony compounds administered intraperitoneally to the rat. These values (per kilogram of body weight) were 100 mg of antimony metal, 3,250 mg of antimony trioxide (1,360 mg antimony), 4,000 mg of antimony pentoxide (1,500 mg antimony), 1,000 mg of antimony trisulfide (360 mg antimony), 1,500 mg of antimony pentasulfide (450 mg antimony), and 11 mg of potassium antimonyl tartrate (4 mg antimony). Animals dying within a few days showed loss of weight, general weakness, loss of hair, dyspnea, and myocardial insufficiency. At necropsy, myocardial congestion and dilatation of the right heart were prominent. There was little change in the lungs. The livers were congested and showed some degeneration and polymorphonuclear infiltration. Toxic glomerular nephritis was present most markedly in rats receiving the tartrate, while metallic antimony produced moderate splenic hyperplasia with some eosinophilia.

Seitz (1924) reported that subcutaneous injections of rabbits and guinea pigs with 53 mg of metallic antimony per kilogram of body weight produced a mild polycythemia and eosinophilia. The $LD_{50}$ of a single subcutaneous injection of antimony compounds in mice was found (Ercoli, 1971) to be (per kilogram of body weight) 48 mg for sodium antimonyl tartrate (19 mg antimony), 55 mg for potassium antimonyl tartrate (20 mg antimony), 390 mg for a chelate of sodium antimonyl tartrate and 3-mercaptovaline (penicillamine) (57 mg antimony), 2,000 mg for stibocaptate (500 mg antimony), and 670 mg for stibophen (110 mg antimony). Cotton and Logan (1966) noted that potassium antimonyl tartrate injected intravenously into dogs at the rate of 64 mg/kg of body weight (23.3 mg/kg antimony) depressed cardiac contractile force and induced bradycardia.

Potassium antimonyl tartrate has been administered frequently intravenously for treatment of human schistosomiasis and leishmaniasis. Usual dosage has been 20 to 25 mg/kg of body weight (7 to 9 mg/kg antimony) for 20 days. Symptoms of acute intoxication following therapeutic intravenous injection have been occasionally reported. They include headache, giddiness, sore throat, a metallic taste in the mouth, cough, nausea, diarrhea, tachycardia, muscular stiffness and debility (Strong, 1942). Cardiac arrhythmias resulting in death have been reported numerous times (Ming-Hsin *et al.*, 1958). Khalil (1931) noted bradycardia (both acute and of short duration, and chronic) during a course of injections, returning to normal when they were discontinued. Mainzer and Krause (1940) concluded that this is not due to vagal stimulation, but to a direct toxic action of antimony on the heart muscle. Ming-Hsin *et al.* (1958) suggested that the cardiac disturbance is a combined effect of autonomic system dysfunction caused by antimony inhibiting the cerebral cortex and inducing hyperexcitability of the myocardium. Gastrointestinal disturbances and mild jaundice commonly appear 2 to 3 weeks after a course of injections (Chopra, 1927). A frequent sequel of antimony administration is pneumonia (Cushny, 1941). Papular skin eruptions are also sometimes seen following repeated injections.

### *Inhalation*

Stibine ($SbH_3$) toxicity closely resembles that of arsine, death occurring rapidly on exposure to concentrations of 1 percent in air. It is a powerful hemolytic poison. The lethal dose for white mice was estimated to be 100 mg/l (98 mg/l antimony) for 1 hour and 40 minutes (Stock and Guttman, 1904), and for week-old chicks, 25 to 30 mg/l (Steele *et al.*,

1944). The hemolytic effect has been described by Webster (1946) and Dunn and Webster (1945). Webster (1946) found a single exposure to 40 to 45 mg/1 for 1 hour was dangerous for dogs and cats, death occurring within a few hours to a day. These concentrations produced marked hemoconcentration but no hemoglobinuria. In guinea pigs, erythrocytes underwent morphological change (crenation) within a few minutes after exposure. Exposure of guinea pigs to 65 mg/1 stibine (63 mg/1 antimony) for 1 hour produced hemoglobinuria, and anemia followed in a few days. With 92 mg/1 (90 mg/1 antimony), animals died on the second to sixth days with casts in the renal tubules.

Although the threshold limit of stibine in air for man has been set at 0.1 mg/1, no chronic stibine poisoning has been reported. Several cases of toxicity have been reported (Dernehl *et al.*, 1944; Nau *et al.*, 1944) from a mixture of gases including arsine, hydrogen sulfide, and stibine. These were liberated when water was added to a hot dross containing aluminum, compounds of arsenic and antimony, and an excess of free sulfur. Affected individuals experienced severe headache, nausea, weakness, abdominal and lumbar pain, hematuria, and profound hemolytic anemia. Recovery followed hospitalization and treatment by transfusion and intravenous injection of glucose.

## FACTORS INFLUENCING TOXICITY

The severity of antimony toxicosis is influenced by the form in which it occurs, the animal species affected, and the route of administration. In general, trivalent antimony is more toxic than pentavalent antimony. Small particles ($<1$ $\mu$m) of antimony-containing dust are more toxic than larger particles when inhaled. A decreasing order of toxicity of antimony in several forms ingested by the rat would be potassium antimonyl tartrate, antimony metal, and antimony trioxide.

Removal from exposure is important. The nausea and emesis induced by oral ingestion of potassium antimonyl tartrate tends to be self-protecting. Men developing pneumonitis, from inhalation of antimony compounds produced in the smelting of ores, recovered when exposure ceased and they were treated with penicillin aerosols. Humans made ill by exposure to a mixture of arsine, hydrogen sulfide, and stibine recovered following hospitalization, blood transfusion, intravenous injection of glucose, and oral administration of ferrous sulfate. British antilewisite has been administered intramuscularly in acute toxicosis, followed by a daily course of sodium thiosulfate intravenously for 2 weeks, depending on the clinical course of the disease.

## TISSUE LEVELS

Using neutron activation, Smith (1967) found median antimony concentrations in human tissues between 0.05 and 0.15 ppm (dry basis). Highest levels were found in lungs (0.28 ppm) and hair (0.34 ppm). Liebscher and Smith (1968), also using neutron activation, reported antimony concentrations in 23 different human tissues. These were taken from healthy adults who died as a result of violence and who had no known direct industrial exposure to antimony dust or fumes. They lived and worked in the area of Glascow, Scotland. On a dry basis, highest mean concentrations were found in hair (0.69 ppm), lung (0.48 ppm), and prostate (0.42 ppm). Lowest mean concentrations were found in spleen (0.07 ppm) and ovary (0.07 ppm). However, great variation was evident, e.g., antimony concentration in teeth ranged from 0.005 to 0.665 ppm and in hair from 0.080 to 6.58 ppm. Molokhia and Smith (1967) determined that the apex of the lungs had the highest concentrations and the base of the lungs the lowest. These workers also found that the lymph glands had relatively high antimony concentrations (0.34–0.43 ppm wet weight). Hamilton *et al.* (1972/1973) reported generally lower antimony concentrations in human tissues than did Smith (1967) and Liebscher and Smith (1968) but similar to those for mouse tissues obtained by Molokhia and Smith (1969). Mean human blood antimony concentrations were 0.005 ppm. Rib antimony was approximately 1.5 ppm in ash. Nixon *et al.* (1967) reported a mean of 0.034 ppm antimony in dental enamel of Scottish subjects and 0.070 ppm for Egyptian subjects treated with antimony for bilharzia. Smith (1947), studying the excretion rate of potassium antimonyl tartrate in man, found the 50 percent excretion time to be approximately 500 hours.

## MAXIMUM TOLERABLE LEVELS

Data are insufficient to set a dietary maximum tolerable level for antimony with precision. Based on limited evidence, a level of 70–150 ppm in the dry diet is suggested for the rabbit.

## SUMMARY

Antimony is a lustrous, white metal classed with arsenic and bismuth (Group VA) in the periodic table. It occurs in nature primarily as stibnite ($Sb_2S_3$) and is extensively used in alloys with lead in storage battery grids and printer's type, as the sulfide in compounding of rubber, and as oxides in flame-retarding textiles and in enamel and glass manufacture. Humans may be poisoned by industrial dusts and fumes or by consuming acid food or drink prepared or stored in vessels coated with antimony-containing enamel or glaze. Food and feeds seem generally low in antimony. Trivalent antimony appears to be more toxic than pentavalent antimony, and toxicity from dusts is greater when particle size is small ($<1\ \mu m$). Stibine ($SbH_3$), released by treatment of antimony with acid, is an extremely toxic hemolytic gas. Animal toxicity is primarily an experimental phenomenon.

Symptoms of illness in humans include anorexia, nausea, abdominal cramps, diarrhea, emesis, muscular pains, nerve tenderness, headaches, oppression in the chest, pneumonitis, and pustular skin eruptions. Stibine induces hemolysis and anemia.

Antimony has no known essential metabolic function and inhibits thiol-containing enzymes *in vitro*. Potassium antimonyl tartrate has been used extensively in treatment of human schistosomiasis.

TABLE 4 Effects of Antimony Administration in Animals

| Class and Number of Animals[a] | Age or Weight | Administration: Quantity of Element[b] | Source | Duration | Route | Effect(s) | Reference |
|---|---|---|---|---|---|---|---|
| Chicken | 1 wk | 24 mg/l | Stibine ($SbH_3$) | | Air | Erythrocyte hemolysis; death | Steele *et al.*, 1944 |
| Rabbit | | 0.7–2.2 mg/kg | Potassium antimonyl tartrate | | Diet | Nopathology | Wieland, 1937 |
| Rabbit—4 | | 5.5 mg/kg | Potassium antimonyl tartrate | In several doses over 7–22 d | Drench | Increased NPN in blood and urine; icterus; fatty degeneration and parenchymal necrosis of liver; gastrointestinal hemorrhage; renal hemorrhage; death | Pribyl, 1927 |
| Rabbit | | 46 mg/kg | Potassium antimonyl tartrate | Single | Drench | Death | Oelkers, 1937 |
| Rabbit—6 | | 2 mg/m³ | Antimony trisulfide | 7 h/d 5 d/wk, 6 wk | Dust in air | Mild venous congestion and focal hemorrhage in lungs; myocardial damage (flattened T waves in EKG, cloudy swelling and cytoplasmic granules) | Brieger *et al.*, 1954 |
| Rabbit—5 | | 10 mg/m³ | | 5 d | | Lung inflammation; myocardial damage; parenchymatous degeneration of liver and renal tubular epithelium | |
| Rabbit | | 53 mg | Metallic antimony | Single | Subcutaneous injection | Mild polycythemia and eosinophilia | Seitz, 1924 |

| | | | | | | |
|---|---|---|---|---|---|---|
| Dog—20 | 0.4–23 mg/kg | Potassium antimonyl tartrate | Geometrically increasing doses at 15 min intervals | Intravenous injection | Depressed cardiac contractile force and bradycardia; death | Cotton and Logan, 1966 |
| Dog and cat | 39–44 mg/1 | Stibine | 1 h | Air | Death in a few hours to a day; pulmonary congestion and edema; hemoconcentration | Webster, 1946 |
| Rat | 100 mg/kg | Potassium antimonyl tartrate | 1 yr | Diet | Normal growth but cardiac pathology | Bradley and Fredrick, 1941 |
| | 300 mg/kg | | | | Minimum lethal dose | |
| Rat | 4 mg/kg | Potassium antimonyl tartrate | Single | Intraperitoneal injection | $LD_{50}$; alopecia; dyspnea; myocardial insufficiency; congestion of lungs | Bradley and Fredrick, 1941 |
| | 100 mg/kg | Antimony metal | Single | Intraperitoneal injection | $LD_{50}$; alopecia; dyspnea; myocardial insufficiency; congestion of lungs | |
| | 360 mg/kg | Antimony trisulfide | Single | Intraperitoneal injection | $LD_{50}$; alopecia; dyspnea; myocardial insufficiency; congestion of lungs | |
| | 450 mg/kg | Antimony pentasulfide | Single | Intraperitoneal injection | $LD_{50}$; alopecia; dyspnea; myocardial insufficiency; congestion of lungs | |
| | 1,360 mg/kg | Antimony trioxide | Single | Intraperitoneal injection | $LD_{50}$; alopecia; dyspnea; myocardial insufficiency; congestion of lungs | |
| | 1,500 mg/kg | Antimony pentoxide | Single | Intraperitoneal injection | $LD_{50}$; alopecia; dyspnea; myocardial insufficiency; congestion of lungs | |
| Mouse | 98 mg/1 | Stibine | 1 h 40 min | Air | Lethal dose | Stock and Guttman, 1904 |

TABLE 4 *Continued*

| Class and Number of Animals[a] | Age or Weight | Administration: Quantity of Element[b] | Source | Duration | Route | Effect(s) | Reference |
|---|---|---|---|---|---|---|---|
| Mouse—105 | 20–25g | 19 mg/kg | Sodium antimonyl tartrate | Single | Subcutaneous injection | $LD_{50}$ | Ercoli, 1971 |
| Mouse—95 | | 20 mg/kg | Potassium antimonyl tartrate | Single | Subcutaneous injection | $LD_{50}$ | |
| Mouse—270 | | 57 mg/kg | Chelate of sodium antimonyl tartrate and 3-mercap-tovaline | Single | Subcutaneous injection | $LD_{50}$ | |
| Mouse—80 | | 110 mg/kg | Stibophen | | | | |
| Mouse—80 | | 500 mg/kg | Stibocaptate | | | | |
| Guinea pig—15 | | 19 mg/m³ | Antimony trioxide | 2 h/d, 7 d/wk, 3 wk, followed by 3 h/d | Dust in air | With 138 h or more exposure, fatty degeneration of liver, leucocytopenia, and relative lymphocytosis; four deaths | Dernehl *et al.*, 1945 |
| Guineapig | | 63 mg/l | Stibine | 1 h | Air | Hemoglobinuria and anemia in a few days | Webster, 1946 |
| | | 90 mg/l | | | | Death in 2–6 d | |

[a] Number of animals per treatment.

[b] Quantity expressed in parts per million or as milligrams per kilogram of body weight.

## REFERENCES


Bradley, W. R., and W. G. Fredrick. 1941. Toxicity of antimony-animal studies. Ind. Med. 2:15.

Brady, F. J., A. H. Lawton, D. B. Corvie, H. L. Andrews, A. T. Ness, and G. E. Ogden. 1945. Localisation of trivalent radioactive Sb following intravenous administration to dogs infected with *Dirofilaria immitis*. Am. J. Trop. Med. 25:103.

Brieger, H., C. W. Semisch, J. Stasney, and D. A. Piatnek. 1954. Industrial antimony poisoning. Ind. Med. Surg. 23:521.

Browning, E. 1969. Toxicity of Industrial Metals, 2nd ed. Butterworth and Co., Ltd., London. 383 pp.

Bulmer, F. M. R., and J. H. Johnston. 1948. Antimony trisulphide. J. Ind. Hyg. 30:26.

Chopra, R. N. 1927. Experimental investigation into the action of organic compounds of antimony. Indian J. Med. Res. 15:41.

Cotton, M. D., and M. E. Logan. 1966. Effects of Sb on the cardiovascular system and intestinal smooth muscle. J. Pharmacol. 151:7.

Cushny, A. R. 1941. Antimony, Pharmacology and Therapeutics, p. 81. Churchill, London.

Dernehl, C. U., F. M. Stead, and C. A. Nau. 1944. Arsine, stibine and $H_2S$. Accidental generation in a metal refinery. Ind. Med. Surg. 13:361.

Dernehl, C. U., C. A. Nau, and H. H. Sweets. 1945. Animal studies on the toxicity of inhaled antimony trioxide. J. Ind. Hyg. Toxicol. 27:256.

Djuric, D., R. G. Thomas, and R. Lie. 1962. The distribution and excretion of trivalent antimony in the rat following inhalation. Int. Arch. Gewerbepathol. Gewerbehyg. 19:529.

Dunn, R. C., and S. H. Webster. 1945. Haemoglobinuria, crystals, casts and globules in renal tubules of guinea-pigs following chemical analyses. Am. J. Pathol. 23:967.

Ercoli, N. 1971. Significance of the chemotherapeutic index in the treatment of schistosomiasis with antimony compounds. Bull. WHO 45:371.

Feil, A. 1939. Le role de l'antimonine en pathologie professionelle. Pr. Med. 47:1133.

Felicetti, S. A., R. G. Thomas, and R. O. McClellan. 1974. Metabolism of two valence states of inhaled antimony in hamsters. Am. Ind. Hyg. Assoc. J. 35:292.

Fink, C. G., and A. H. Kopp. 1933. A rediscovered ancient Egyptian craft. Metrop. Mus. Stud. 4:163.

Franz, G. 1937. Zur pathologischen Anatomie der Antimonvergiftung. Arch. Exp. Pathol. Pharmakol. 186:661.

Gellhorn, A., N. A. Tupikova, and H. B. Van Dyke. 1946. Tissue distribution and excretion of four organic antimonials after single or repeated administration to normal hamsters. J. Pharmacol. 87:169.

Gocher, T. E. P. 1945. Antimony intoxication. Northwest Med., Seattle 44:92.

Goodwin, L. G. 1944. The toxicity and trypanocidal activity of some organic antimonials. J. Pharmacol. 81:224.

Gross, P., J. H. Brown, and T. F. Hatch. 1951. Experimental endogenous lipoid pneumonia. Am. J. Pathol. 57:690.

Hamilton, E. I., and M. J. Minski. 1972/1973. Abundance of the chemical elements in man's diet and possible relations with environmental factors. Sci. Total Environ. 1:375.

Hamilton, E. I., M. J. Minski, and J. J. Cleary. 1972/1973. The concentration and distribution of some stable elements in healthy human tissue from the United Kingdom (An environmental study). Sci. Total Environ. 1:341.

Khalil, B. M. 1931. The specific treatment of human schistosomiasis. Beih. Arch. Schiffs-u. Trop. Hyg. 35:225.

Liebscher, K., and H. Smith. 1968. Essential and nonessential trace elements. A method of determining whether an element is essential or nonessential in human tissue. Arch. Environ. Health 17:881.

Mainzer, F., and M. Krause. 1940. Changes of the electrocardiogram during antimony treatment. Trans. R. Soc. Trop. Med. Hyg. 33:405.

Mellor, J. W. 1939. Treatise on Inorganic and Theoretical Chemistry, vol. 9, p. 339. Longmans Green, New York.

Ming-Hsin, H., C. Shaoh-Chi, P. Yu-Siu, and Y. Kuo-Yuei. 1958. Mechanism and treatment of cardiac arrhythmias in tartar emetic intoxication. Chim. Med. J. 76:103.

Molokhia, M. M., and H. Smith. 1967. Trace elements in the lung. Arch. Environ. Health 15:745.

Molokhia, M. M., and H. Smith. 1969. Tissue distribution of trivalent antimony in mice infected with *Schistosoma mansoni*. Bull. WHO 40:123.

Monier-Williams, G. W. 1925. The solubility of glazes and enamels used in cooking utensils. Analyst 50:133.

Monier-Williams, G. W. 1934. Antimony in Enamelled Hollow-Ware. Min. Health Rep. Publ. Health Med. Subj. No. 73. H. M. Stationery Office, London.

Murthy, G. K., U. Rhea, and J. T. Peeler. 1971. Levels of antimony, cadmium, chromium, cobalt, manganese, and zinc in institutional total diets. Environ. Sci. Technol. 5:436.

Nau, C. A., W. Andersen, and R. E. Cone. 1944. Arsine, stibine and $H_2S$. Accidental industrial poisoning by a mixture. Ind. Med. Surg. 13:308.

Nixon, G. S., H. D. Livingstone, and H. Smith. 1967. Estimation of antimony in human enamel by activation analysis. Caries Res. 1:327.

Oelkers, H. A. 1937. Zur Pharmakologie des Antimons. Arch. Exp. Pathol. Pharmakol. 187:56.

Otto, G. F., T. H. Maren, and H. W. Brown. 1947. Blood levels and excretion rates of antimony in persons receiving trivalent and pentavalent antimonials. Am. J. Hyg. 46:193.

Otto, G. F., and T. H. Maren. 1950. Studies on the chemotherapy of filariasis. VI. Studies on the excretion and concentration of antimony in blood and other tissues following the injection of trivalent and pentavalent antimonials into experimental animals. Am. J. Hyg. 51:370.

Pribyl, E. 1927. Nitrogen metabolism in experimental subacute arsenic and antimony poisoning. J. Biol. Chem. 74:775.

Renes, L. E. 1953. Antimony poisoning in industry. Arch. Ind. Hyg. 7:99.

Rowland, H. A. K. 1968. Stilbokinetics I: Studies on mice with $^{124}$Sb-labelled sodium antimony dimercaptosuccinate (astiban). Trans. R. Soc. Trop. Med. Hyg. 62:632.

Schroeder, H. A. 1973. Recondite toxicity of trace elements, pp. 107–199. *In* W. J. Hayes, Jr., ed. Essays in Toxicology, vol. 4. Academic Press, New York.

Schroeder, H. A., M. Mitchener, J. J. Balassa, M. Kanisawa, and A. P. Nason. 1968. Zirconium, niobium, antimony and fluorine in mice: Effects on growth, survival and tissue levels. J. Nutr. 95:95.

Schrumpf, P., and B. Zabel. 1910. Klinische und experimentelle Untersuchungen über die Antimonvergiftung der Schriftsetzer. Arch. Exp. Pathol. Pharmakol. 63:242.

Seitz, A. 1924. Die Hygiene in Schrifftgiessereigewerbe. Arch. Hyg. 94:284.

Smith, H. 1967. The distribution of antimony, arsenic, copper and zinc in human tissue. J. Forensic Sci. Soc. 7:97.

Smith, R. E. 1947. Studies on the biological exchange of radio-antimony in animals and man. Fed. Proc. 6:205.

Steele, J. M., F. J. Gerrard, and D. R. Matheson. 1944. Gases as antimalarials. Effect of stibine. U.S. Nav. Med. Res. Inst. Res. Proj. 10:150.

Stock, A., and O. Guttman. 1904. Über den Antimonwasserstoff und das gelbe Antimon. Ber. Dtsch. Chem. Ges. 37:885.

Strong, R. P. 1942. *In* Still's Diagnosis, Prevention and Treatment of Tropical Diseases. Blakiston, Philadelphia.

Tarrant, M. E., S. Wadley, and J. J. Woodage. 1971. The effect of penicillamine on the treatment of experimental schistosomiasis with tartar emetic. Ann. Trop. Med. Parasitol. 65:233.

Thomas, R. G., S. W. Felicetti, R. V. Lucchino, and R. O. McClellan. 1973. Retention patterns of antimony in mice following inhalation of particles formed at different temperatures. Proc. Soc. Exp. Biol. Med. 144:544.

Waitz, J. A., R. E. Ober, J. E. Musenhelder, and P. E. Thompson. 1965. Physiological disposition of antimony after administration of $^{124}$Sb labelled tartar emetic to rats, mice and monkeys and the effects of tris (p-amino phenyl) carbonium palmoate on this distribution. Bull. WHO 33:537.

Webster, S. H. 1946. Volatile hydrides of toxicological importance. J. Ind. Hyg. 28:167.

Wieland, M. 1937. Zur Pharmakologie des Antimons. Cited by Browning. 1969.

# Arsenic

Arsenic (As) is a solid, brittle nonmetal, tin-white to steel-gray in color, with a metallic luster. It is found usually combined in the minerals orpiment and realgar (natural sulfides), arsenolite, arsenopyrite, cobaltite, and niccolite. Arsenic occurs in tri- or pentavalent states, with arsenic trioxide ($As_2O_3$) the most common compound. In the United States arsenic trioxide or white arsenic is produced principally as a by-product of copper and lead smelting and in the recovery of other metals such as gold and silver. White arsenic is used for manufacturing calcium and lead arsenate insecticides, wood preservatives, and herbicides and is the starting material for almost all organic and inorganic arsenic compounds (Lansche, 1965). Arsenic is widely distributed in the biosphere. In areas near smelters and refineries, arsenic contamination of soil and herbage is quite common (Lillie, 1970). Addition of 0.01 percent organic arsenic compounds to diets for growing swine has been shown to increase average daily gain and improve feed efficiency (Carpenter, 1951). Several reviews on arsenic are available (Frost, 1953; NRC, 1977; Underwood, 1977).

## ESSENTIALITY

Schroeder and Balassa (1966) fed arsenic to rats and mice at 0.053 ppm over long periods and noted normal growth and development. They concluded that if arsenic is an essential trace element for these animals, requirements are of the order of 1.0 mg per rat daily or less. There have been reports on the improved appearance of skin and hair of mice, rats, and horses supplemented with arsenic (Sollman, 1953; Schroeder and Balassa, 1966). Nielsen *et al.* (1975) maintained rats on diets with 0.030 ppm (air dry basis) and found a rough hair coat, decreased growth, decreased hematocrits, and enlarged spleens when compared to controls receiving 4.5 ppm arsenic. Anke *et al.* (1977) observed deficiency signs in goats and minipigs fed less than 50 ppb arsenic, including impaired reproductive performance, decreased birth weights, increased neonatal mortality, and lower weight gains in second-generation animals.

Chicks fed 15 to 25 ppb arsenic for 4 weeks weighed less than controls supplemented with 1 ppm additional arsenic (Nielsen and Shuler, 1978). Neonatal mortality in rats was decreased with arsenic supplementation (Nielsen *et al.*, 1977).

The beneficial effects of various organic arsenicals on growth, health, and feed efficiency of poultry and swine have been reviewed by Frost (1953, 1967) and Frost *et al.* (1955). Arsanilic acid, 4-nitrophenylarsonic acid, 3-nitro-4-hydroxyphenylarsonic acid, and phenylarsenoxide have been found to be valuable in animal production. The phenylarsenoxides are more potent as coccidiostats than arsonic acids, but only the arsonic acids are recognized as growth stimulants for swine and poultry (Frost *et al.*, 1955). The precise mechanism of action is unknown but closely resembles that of antibiotics and is to some extent complementary to them (Underwood, 1977). Addition of 0.01 percent 3-nitro-4-hydroxyphenylarsonic acid to the diet of growing swine (12.7 kg) increased the average daily gain from 0.20 to 0.33 kg and improved feed efficiency by 6.3 percent (Carpenter, 1951).

## METABOLISM

Arsenic in the forms in which it usually occurs in foods is readily absorbed and rapidly excreted, mainly via the urine (Coulson *et al.*, 1935). Less than 10 percent of the usual soluble forms of arsenic appear in the feces. Inorganic arsenic ingested as arsenic trioxide ($As_2O_3$) is also well absorbed but has a longer retention period in tissues and is excreted almost equally in the feces and urine in man and in the rat (Coulson *et al.*, 1935; Overby and Frederickson, 1963). Organic arsenic compounds (arsanilic acid) are well absorbed and deposited in tissues of swine and chickens at levels proportional to the dietary level. Organic arsenic is removed rapidly from tissues and excreted mostly in the feces (Frost *et al.*, 1955; Hanson *et al.*, 1955; Overby and Frost, 1960; Overby and Frederickson, 1963).

The toxicity of arsenic varies with the species of animal, valence of the compound, solubility, route of exposure, and absorption and excretion rates. No forms of the element accumulate in tissues; some are simply excreted more rapidly than others (Frost, 1967).

A recent report (Lakso and Peoples, 1975) showed that ingested inorganic arsenate ($Na_2HAsO_4 \cdot 2H_2O$) and arsenite ($HAsO_2$) could be methylated *in vivo* by both the ruminant (cow) and nonruminant (dog) and that the methylated arsenic found in the urine is not necessarily due to its ingestion as such in plant material.

An "organic" form of arsenic was found in muscle and liver of fish following oral or parenteral administration of radioactive arsenic as sodium arsenate. The conversion appeared to be endogenous and a result of action by intestinal microflora (Penrose, 1975). Biomethylation of metal ions capable of redox activity can occur by way of methyl vitamin $B_{12}$ or methyl iodine (Wood, 1975). Arsenicals appear to be oxidized *in vivo* from trivalent to pentavalent, but the reverse does not occur (Overby and Frederickson, 1963).

Studies on tissue accumulation and excretion of $^{76}As$ injected as sodium arsenite have shown wide species differences (Hunter *et al.*, 1942; Ducoff *et al.*, 1948). Greatest arsenic concentrations in all animals were found generally in liver, kidney, spleen, and lung (Frost, 1953). In rats, unlike other species studied, arsenic was concentrated in the blood and retained apparently in bound form.

## SOURCES

Arsenic occurs in normal soils at levels ranging from 1 to 40 ppm, but higher levels can result from the extensive use of arsenical sprays for control of insects and weeds. Most fruits, vegetables, cereal grains, meat, and dairy products contain less than 0.5 ppm and rarely exceed 1 ppm (fresh basis), but this level can increase due to contamination (Underwood, 1977).

The arsenic concentration in seawater ranges from 2 to 5 ppb (Schroeder and Balassa, 1966). The arsenic content of commercial fishmeals used in livestock production ranged from 2.6 to 9.1 ppm (air dry basis) (Lunde, 1968). The arsenic contents of fish and crustaceans on a fresh weight basis are as follows: freshwater fish, 0.75 ppm (average for 15 species) (Ellis *et al.*, 1941); cod, eel, and mackerel, 1.5–4.1 ppm

(Holmes and Remington, 1934); shellfish and crustaceans, 3–174 ppm (Chapman, 1926).

Sheep and cattle do not find arsenic distasteful and, in fact, may develop a taste for it (Clarke and Clarke, 1975). Ruminants were reported to graze selectively contaminated forage.

## TOXICOSIS

Many reviews on arsenic toxicosis are available (Vallee *et al.*, 1960; Buchanan, 1962; Albert, 1965; Hammond, 1965; Schroeder and Balassa, 1966; Fowler and Weissberg, 1974; Clarke and Clarke, 1975; Ammerman *et al.*, 1977). Arsenicals differ widely in their toxicity. Trivalent arsenicals, which specifically block lipoate-dependent enzymes, are generally more toxic than pentavalent arsenicals (Frost, 1967).

Arsenic also appears to exert a toxic action by attachment to sulfhydryl groups of protein. The attachment is loose enough that compounds with sulfhydryl groups with greater affinity for arsenic can withdraw the tissue arsenic for urinary excretion.

Generally, inorganic arsenicals are more toxic than the organic forms. Factors such as absorption into cells, rate of oxidation, rate of elimination, etc., vary with circumstances, and most generalizations concerning arsenicals have important exceptions (Frost, 1967). Degree of toxicity in ruminants is variable and may depend on route of exposure, animal age, nutritional status, and duration of exposure (Case, 1974; Selby *et al.*, 1974).

### LOW LEVELS

Chronic toxicosis due to arsenic is seldom reported, but Selby *et al.* (1974) noted that chronic arsenic administration resulted in an improvement in the appearance of the hair of cattle. With arsenic withdrawal, the improvement in hair coat was lost and animals appeared unthrifty, lost weight, and had inflamed eyes. Other signs of arsenic intoxication include inflamed mucous membranes of the upper respiratory tract, diarrhea, cachexia, eczema, and incoordination of gait.

The recommended growth promotant level of arsanilic acid in the diet for swine is about 100 ppm of the compound. When 10 times this level was fed to swine for 20 days (Ledet *et al.*, 1973), severe posterior paresis or quadriplegia was observed in several animals by day 15. The tissue arsenic levels were correlated with time on the diet, and peripheral nerve tissue had an affinity for arsenic. Carpenter (1951) reported 3-nitro-4-hydroxyphenylarsonic acid to be toxic to swine when fed at a level of 0.02 percent of the total ration for 2–4 weeks. This toxicosis was characterized by stiffness of the hind legs.

In chronic studies with rats, Franke and Moxon (1937) found 50 ppm of dietary arsenic, as $Na_2HAsO_3$, were slightly toxic. Arsanilic acid at 70 ppm arsenic was well tolerated by growing rats (Frost, 1953). Rats had increased levels of tissue arsenic when fed diets containing 16 ppm arsenic, as $As_2O_3$, or protein-bound arsenic in turkey liver derived from an organic pentavalent arsenical (Morgareidge, 1963).

### HIGH LEVELS

Arsenic poisoning is commonly an acute clinical syndrome and death usually occurs so rapidly that preceding illness, if observed, is of a few days' duration. The signs of inorganic arsenic toxicosis vary with the quantity and method of administration. The usual signs displayed by cattle that have been dipped in solutions containing excessive arsenic include colicky pain, vomiting, diarrhea, marked depression, and dermatitis usually due to increased capillary permeability and cellular necrosis. The time until onset and severity are governed by the amount of arsenic absorbed through the skin (Kinsley, 1929). Signs of acute ingestion toxicosis are similar; however, skin lesions are rarely present. The lesions at necropsy of acute cases include gastroenteritis, glomerular nephritis, and, frequently, dermatitis. Acute signs in horses are cerebral involvement and signs of intense pain with head banging (Lillie, 1970).

Animals may survive a high single oral dose that can be toxic with a short period of repeated exposures, although it has been suggested that tolerances of low oral doses may be increased by repeated dosage (Clarke and Clarke, 1975). Following a single oral dose, almost all of the administered arsenic is excreted within a few days. Ruminants that survive arsenic intoxication by a single dose should be withheld from market for 2 weeks and for 6 weeks following multiple dosage (Selby *et al.*, 1974). Swine and poultry that have received arsenic as a growth promotant must be held 5 days following withdrawal before slaughter for market (AAFCO, 1978).

Most nonruminants are more susceptible to intoxication than are ruminants or horses. Arsanilic acid is the least toxic of the arsenicals investigated. The fatal dose for horses and cows was 300 grains (approximately 40 mg/kg of body weight) arsenic per day, as contrasted with 4–8 grains (approximately 6–12 mg/kg of body weight) for sheep. Horses and cattle could ingest 20–30 grains (2.66 to 4 mg/kg of body weight) arsenic daily continuously with no apparent ill effects (Reives, 1925). The oral $LD_{50}$ of arsanilic acid and 3-nitro-4-hydroxyphenylarsonic acid in rats was about 800 and 100 mg of arsenic per kilogram of body weight, respectively (Frost, 1953).

### FACTORS INFLUENCING TOXICITY

Arsenic salts were effective in counteracting selenium toxicity (see section on selenium) in poultry (Thapar *et al.*, 1969) and in rats (Hendrick *et al.*, 1953; Olson *et al.*, 1963). The latter authors measured exhaled selenide and concluded that arsenic had no effect, however, on selenide excretion with low levels of selenium. Muth *et al.* (1971) reported that 1 ppm dietary arsenic as sodium arsenate in selenium-deficient diets for ewes reduced incidence of myopathy in the lambs but that 0.1 ppm arsenic was not protective. Arsenic decreased the retention of selenium in the liver and also decreased the elimination of selenium via the volatilization pathway thus increasing the excretion of selenium into the gut (Levander, 1971).

## TISSUE LEVELS

A period of about 6 weeks is necessary to deplete body tissues of arsenic when animals have been exposed repeatedly to excessive levels of the element over a long period of time (Selby *et al.*, 1974). Arsenic levels are higher in skin, hair, liver, kidney, and spleen than in other tissues of intoxicated ruminants.

In cattle poisoned under field conditions (Reagor, 1973), liver levels were 3.5 to 60.4 ppm (dry weight). Liver arsenic levels were 27 ppm (dry weight) in dead steers that had consumed about 300 mg arsenic per day as MSMA (monosodium acid methanearsonate) for 7 days (Dickinson, 1972). Muscle arsenic averaged 8.8 ppm (dry weight) in these acutely poisoned cattle. Cattle that had consumed 1.25 ppm dietary arsenic for 8 weeks had muscle arsenic levels of 0.2 ppm (dry weight) and all tissue concentrations tested were 1 ppm arsenic or less (Peoples, 1964).

Cows consuming 18 ppm arsenic (40 mg per head daily) as dried manure from poultry that had received 3-nitro-4-hydroxyphenylarsonic acid had no arsenic residue in milk after 5 days on the diet (Calvert and Smith, 1972). When arsanilic acid or the above arsonic compound was administered by gelatin capsule in amounts of 0, 1.6, or 3.2 mg arsenic per kilogram of body weight to cows for 5 days, milk arsenic increased from 0.015 to 0.026 ppm for arsanilic acid only at the highest level. No changes in milk were found with the other treatments. Jersey cows fed 0, 0.026, 0.015, or 0.103 mg arsenic per kilogram of body weight as lead arsenate for 126 days all had milk arsenic levels below 0.05 ppm (Marshall *et al.*, 1963).

Overby and Frederickson (1963) reported that muscle arsenic levels were lower for animals that had received organic arsenic than for the inorganic forms. The arsenic content of spleen, kidneys, and muscle of cattle ranged between 2.4 to 9.3 ppm (dry weight) when exposed to contaminated pasture containing 3 to 227 ppm arsenic on a dry weight basis (Lillie, 1970).

## MAXIMUM TOLERABLE LEVELS

Arsenic-containing compounds differ widely in their toxicity with the inorganic forms being more toxic than the organic forms (Table 5). This has been observed more from the standpoint of acute arsenic toxicosis, since chronic cases are seldom seen. Arsenic, as lead arsenate, at a level of 4.68 mg/kg of body weight (approximately 200 ppm dietary arsenic) was fed to cows without adverse effect. Potassium arsenite was tolerated at a level of 285 ppm arsenic by swine but caused decreased feed consumption and weight loss at a level of 570 ppm.

Arsanilic acid has been fed to swine and poultry at 100 ppm to increase animal performance, and 1,000 ppm of this compound have been tolerated in several studies without adverse effect.

Although there are suggested differences in tolerance to arsenic among species, the maximum tolerable dietary levels are set at 50 ppm for inorganic forms and 100 ppm for organic forms of arsenic for domestic animals.

## SUMMARY

Arsenic is widely distributed in the biosphere and can be a major source of contamination for livestock in areas surrounding smelters and where arsenicals are used to control weeds and insects. Fish and crustaceans have unusually high levels of arsenic and may represent a source of increased intake of the element when fishmeal products are fed. Several organic arsenicals are recognized as growth stimulants for swine and poultry, and they also act as coccidiostats. Their mechanism of action as a growth stimulant resembles that of the antibiotics. Organic and inorganic arsenicals differ greatly in their toxic level, metabolism, and excretion. Trivalent arsenicals specifically block lipoate-dependent enzymes and are more toxic than pentavalent forms. Arsenic also appears to exert a toxic action by attachment to sulfhydryl groups of protein. Arsenic poisoning is frequently an acute clinical syndrome, and death usually occurs so rapidly that preceding illness is of only a few days duration. The signs vary with quantity and method of administration but usually include colicky pain, diarrhea, depression, glomerular nephritis, and dermatitis usually due to increased capillary permeability and cellular necrosis.

TABLE 5 Largest Single Oral Dose of Organic Arsenic Compounds Tolerated by Different Species[a]

| Source | Rat, mg/kg[b,c] | Chicken, mg/kg[b,d] | Duck, mg[e] |
|---|---|---|---|
| Phenylarsonic acid | 10 | 35 | — |
| Arsenosoaniline | 25 | 35 | — |
| Arsanilic acid | 400 | 300–400 | 1,000 |
| Dodecylamine *p*-chlorophenylarsonate | <100 | 100 | — |
| 3-Nitro-4-hydroxyphenylarsonic acid | 20 | 100 | <100 |
| 4-Nitrophenylarsonic acid | 75 | <100 | — |

[a] Frost *et al.*, 1955; dose expressed as milligrams of arsenic-containing compound.

[b] Doses resulting in no mortality or less than 10 percent mortality.

[c] Adult rats weighing 100 to 150 g were used.

[d] Chickens weighing 1,130 to 1,360 g were used.

[e] Maximum tolerated single dose for ducks weighing approximately 1,130 g.

TABLE 6 Effects of Arsenic Administration in Animals

| Class and Number of Animals[a] | Age or Weight | Administration: Quantity of Element[b] | Source | Duration | Route | Effect(s) | Reference |
|---|---|---|---|---|---|---|---|
| Cattle—2 | 400 kg | 1.17 mg/kg | Lead arsenate | 126 d | Diet | No adverse effect | Marshall *et al.*, 1963 |
| | | 2.34 mg/kg | | | | No adverse effect | |
| | | 4.68 mg/kg | | | | No adverse effect | |
| Cattle—1 | Mature | 343 mg | $As_2O_3$ | 3 d | Drench | No adverse effect | Fitch *et al.*, 1939 |
| | | 1.367 g | | | | Increased milk arsenic | |
| Cattle—4 | Mature | 40 mg/head/day | Dried poultry manure | 18 d | Diet | No adverse effect | Calvert and Smith, 1972 |
| Cattle—2 | Mature | 1.6 mg/kg | Arsonic acid[c] | 5 d | Capsule | No adverse effect | Calvert and Smith, 1972 |
| | | 1.6 mg/kg | Arsanilic acid | | | No adverse effect | |
| | | 3.2 mg/kg | Arsonic acid[c] | | | No adverse effect | |
| | | 3.2 mg/kg | Arsanilic acid | | | Increased arsenic in milk from 0.015 to 0.026 ppm | |
| Sheep—3 | 45 kg | 5.7 ppm | Potassium arsenite | 84 d | Diet | No adverse effect | Bucy *et al.*, 1954 |
| | | 11.7 ppm | | | | No adverse effect | |
| | | 17.1 ppm | | | | No adverse effect | |
| | | 5.7 ppm | Arsanilic acid | | | No adverse effect | |
| | | 11.4 ppm | | | | No adverse effect | |
| | | 17.1 ppm | | | | No adverse effect | |
| | | 5.7 ppm | Arsonic acid[c] | | | No adverse effect | |
| | | 11.4 ppm | | | | No adverse effect | |
| | | 17.1 ppm | | | | No adverse effect | |
| Sheep—2 | | 17.1 ppm | Arsonic acid[c] | 63 d | Diet | No adverse effect | Bucy *et al.*, 1954 |
| | | 34.2 ppm | | | | No adverse effect | |
| | | 68.4 ppm | | | | No adverse effect | |

| | | | | | | | |
|---|---|---|---|---|---|---|---|
| | | 17.1 ppm | Arsanilic acid | | | No adverse effect | |
| | | 34.2 ppm | | | | No adverse effect | |
| | | 68.4 ppm | | | | No adverse effect | |
| | | 17.1 ppm | Potassium arsenite | | | No adverse effect | |
| | | 34.2 ppm | | | | No adverse effect | |
| | | 68.4 ppm | | | | No adverse effect | |
| Sheep—32 | 32 kg | 68.4 ppm | Arsonic acid[c] | 56 d | Diet | No adverse effect | Bucy *et al.*, 1954 |
| Sheep—1 | 35 kg | 142.4 ppm | Arsonic acid[c] | 56 d | Diet | No adverse effect | Bucy *et al.*, 1955 |
| | 30 kg | 284.8 ppm | | | | No adverse effect | |
| | 28 kg | 570 ppm | | | | No adverse effect | |
| | 30 kg | 1,139 ppm | | | | No adverse effect | |
| | 32 kg | 1,424 ppm | Arsanilic acid | | | No adverse effect | |
| | 25 kg | 284.8 ppm | | | | No adverse effect | |
| | 34 kg | 570 ppm | | | | Convulsions by 56 d, weight loss | |
| | 30 kg | 1,139 ppm | | | | Convulsions by 28 d | |
| | 28 kg | 142 ppm | Potassium arsenite | | | No adverse effects | |
| | 29 kg | 285 ppm | | | | No adverse effects | |
| | 30 kg | 570 ppm | | | | Weight loss, decreased feed consumption | |
| | 32 kg | 1,139 ppm | Potassium arsenite | 56 d | Diet | Weight loss, decreased feed consumption | |
| Swine—27 | 17 kg | 1,000 ppm | Arsanilic acid | 18 d | Diet | Quadriplegia | Ledet *et al.*, 1973 |
| Swine—6 | 18 kg | 10,000 ppm | | 19 d | | Increased arsenic in tissues | |
| Chicken | 1.2 kg | 400 mg/kg | Arsanilic acid | Single dose | Oral | Maximum tolerated single oral dose | Frost *et al.*, 1955 |
| Chicken—4 | 22 wk | 1 ppm | Arsenic pentoxide | 56 d | Diet | No adverse effects | Hermayer *et al.*, 1977 |
| | | 10 ppm | | | | No adverse effects | |
| | | 100 ppm | | | | Decreased body weight, feed intake, and egg production | |

TABLE 6 *Continued*

| Class and Number of Animals[a] | Age or Weight | Administration: Quantity of Element[b] | Source | Duration | Route | Effect(s) | Reference |
|---|---|---|---|---|---|---|---|
| | | 1,000 ppm | | | | Decreased body weight, feed intake, and egg production | |
| Turkey | Young | 250 ppm | Arsonic acid[c] | 42 d | Diet | Two-fold increase in liver arsenic | Frost *et al.*, 1955 |
| Duck | 1.1 kg | 1,000 mg/kg | Arsanilic acid | Single dose | Oral | Maximum tolerated single oral dose | Frost *et al.*, 1955 |
| Rat—15 | 80 g | 16 ppm | $As_2O_3$ | 28 d | Diet | 25% increase in tissue arsenic | Morgareidge, 1963 |
| | 80 g | 16 ppm | Protein bound arsenic from turkey liver | 28 d | | No adverse effect | |
| Rat—106 | 100–150 g | 400 mg/kg | Arsanilic acid | Single dose | Oral | 0–10% death rate | Frost *et al.*, 1955 |

[a] Number of animals per treatment group.
[b] Quantity expressed in parts per million or as milligrams per kilogram of body weight.
[c] 3-Nitro-4-hydroxyphenylarsonic acid.

## REFERENCES

AAFCO (Association of American Feed Control Officials). 1978. Official Publication, p. 147.

Albert, A. 1965. Selective Toxicity. Wiley, New York.

Ammerman, C. B., S. M. Miller, K. R. Fick, and S. L. Hansard II. 1977. Contaminating elements in mineral supplements and their potential toxicity: A review. J. Anim. Sci. 44:485.

Anke, M., M. Gunn, and M. Partshefeld. 1977. The essentiality of arsenic for animals. *In* D. D. Hemphill, ed. Trace Substances in Environmental Health—X. University of Missouri, Columbia.

Buchanan, W. D. 1962. Toxicity of Arsenic Compounds. Elsevier, Amsterdam.

Bucy, L. L., U. S. Garrigus, R. M. Forbes, H. W. Norton, and M. F. James. 1954. Arsenical supplements in lamb fattening rations. J. Anim. Sci. 13:668.

Bucy, L. L., U. S. Garrigus, R. M. Forbes, H. W. Norton, and W. W. Moore. 1955. Toxicity of some arsenicals to growing-fattening lambs. J. Anim. Sci. 14:435.

Calvert, C. C., and L. W. Smith. 1972. Arsenic in milk and blood of cows fed organic arsenic compounds. J. Dairy Sci. 55:706.

Carpenter, L. E. 1951. The effect of 3-nitro-4-hydroxyphenyl arsonic acid on the growth of swine. Arch. Biochem. Biophys. 32:181.

Case, A. A. 1974. Toxicity of various chemical agents to sheep. J. Am. Vet. Med. Assoc. 164:277.

Chapman, C. A. 1926. On the presence of arsenic in marine crustaceans and shell fish. Analyst 51:548.

Clarke, E. G. C., and M. L. Clarke. 1975. Veterinary Toxicology, 3rd ed., p. 477. Williams & Wilkins Co., Baltimore, Md.

Coulson, E. J., R. E. Remington, and K. M. Lynch. 1935. Toxicity of naturally occurring arsenic in foods. J. Nutr. 10:255.

Dickinson, J. O. 1972. Toxicity of the arsenical herbicide monosodium acid methane-arsonate in cattle. Am. J. Vet. Res. 33:1889.

Ducoff, H. S., W. B. Neal, R. L. Straube, L. O. Jacobsen, and A. M. Bruess. 1948. Biological studies with arsenic 76. Excretion and tissue localization. Proc. Soc. Exp. Biol. Med. 69:548.

Ellis, M. M., B. A. Westfall, and M. D. Ellis. 1941. Arsenic in fresh water fish. Ind. Eng. Chem. 33:1331.

Fitch, L. W. N., E. R. Grimmett, and E. M. Wall. 1939. Occurrence of arsenic in soils and waters of Waiotapu Valley and its relation to stock health. II. Feeding experiments at Wallaceville. N.Z. J. Sci. Technol. 21:146a.

Fowler, B. A., and J. B. Weissberg. 1974. Arsine poisoning. N. Engl. J. Med. 291:1171.

Franke, K. W., and A. L. Moxon. 1937. The toxicity of orally ingested arsenic, selenium, tellurium, vanadium and molybdenum. J. Pharm. Exp. Therp. 61:89.

Frost, D. V. 1953. Considerations on the safety of arsanilic acid for use in poultry feeds. Poult. Sci. 32:217.

Frost, D. V. 1967. Arsenicals in biology—Retrospect and prospect. Fed. Proc. 26:194.

Frost, D. V., L. R. Overby, and H. C. Spruth. 1955. Studies with arsanilic acid and related compounds. J. Agric. Food Chem. 3:235.

Hammond, P. B. 1965. Toxic Minerals: Arsenic, fluorine, selenium, lead, thallium, nitrates, ch. 56, p. 960. *In* L. M. Jones (ed.) Veterinary Pharmacology and Therapeutics, 3rd ed. Iowa State University Press, Ames.

Hanson, L. E., L. E. Carpenter, W. J. Anunan, and E. F. Ferrin. 1955. The use of arsanilic acid in the production of market pigs. J. Anim. Sci. 14:513.

Hendrick, C., H. L. Klug, and O. E. Olson. 1953. Effect of 3-nitro-4-hydroxy-phenylarsonic acid and arsanilic acid on selenium poisoning in the rat. J. Nutr. 51:131.

Hermayer, K. L., P. E. Stake, and R. L. Shippe. 1977. Evaluation of dietary zinc, cadmium, tin, lead, bismuth and arsenic toxicity in hens. Poult. Sci. 56:1721 (Abstr.).

Holmes, A. D., and R. E. Remington. 1934. Arsenic content of American cod liver oil. Ind. Eng. Chem. 26:573.

Hunter, F. T., A. F. Kip, and J. W. Irvine, Jr. 1942. Radioactive tracer studies on arsenic injected as potassium arsenite. I. Excretion and localization in tissues. J. Pharm. Exp. Ther. 76:207.

Kinsley, A. T. 1929. Arsenical poisoning. Vet. Med. 24:445.

Lakso, J. V., and S. A. Peoples. 1975. Methylation of inorganic arsenic by mammals. J. Agric. Food Chem. 23:674.

Lansche, A. M. 1965. Arsenic. Bureau of Mines Bull. 640. Washington, D.C.

Ledet, A. E., J. R. Duncan, W. B. Buck, and F. K. Ramsey. 1973. Clinical, toxicological, and pathological aspects of arsanilic acid poisoning in swine. Clin. Toxicol. 6:439.

Levander, O. A. 1971. Factors that modify the toxicity of selenium. *In* W. Mertz and W. E. Cornatzer (eds.). Newer Trace Elements in Nutrition. Marcel Dekker, Inc., New York.

Lillie, R. J. 1970. Arsenic. *In* Air Pollutants Affecting the Performance of Domestic Animals. Agric. Handb. No. 380. U.S. Department of Agriculture, Washington, D.C.

Lunde, G. 1968. Activation analysis of trace elements in fishmeal. J. Sci. Food Agric. 19:432.

Marshall, S. P., F. W. Hayward, and W. R. Meagher. 1963. Effects of feeding arsenic and lead upon their secretion in milk. J. Dairy Sci. 46:580.

Morgareidge, K. 1963. Metabolism of two forms of dietary arsenic by the rat. J. Agric. Food Chem. 11:377.

Muth, O. H., P. D. Whanger, P. H. Weswig, and J. E. Oldfield. 1971. Occurrence of myopathy in lambs of ewes fed added arsenic in a selenium-deficient ration. Am. J. Vet. Res. 32:1621.

Nielsen, F. H., and T. R. Shuler. 1978. Arsenic deprivation studies in chicks. Fed. Proc. 37:893 (Abstr.).

Nielsen, F. H., S. H. Givand, and D. R. Myron. 1975. Evidence of a possible requirement for arsenic by the rat. Fed. Proc. 34:923 (Abstr.).

Nielsen, F. H., D. R. Myron, and E. O. Uthus. 1977. Newer trace elements—Vanadium (V) and arsenic (As) deficiency signs and possible metabolic roles. *In* M. Kirchgessner, ed. Trace Element Metabolism in Man and Animals—3, p. 244. Technical University, Munich.

NRC. 1977. Arsenic. National Academy of Sciences–National Research Council, Washington, D.C.

Olson, O. E., B. M. Schulte, E. I. Whitehead, and A. W. Halverson. 1963. Effect of arsenic on selenium metabolism in rats. J. Agric. Food Chem. 11:531.

Overby, L. R., and R. L. Frederickson. 1963. Metabolic stability of arsanilic acid in chickens. J. Agric. Food Chem. 11:378.

Overby, L. R., and D. V. Frost. 1960. Excretion studies in swine fed arsanilic acid. J. Anim. Sci. 19:140.

Penrose, W. L. 1975. Organic arsenic compounds in aquatic organisms, p. C-20. *In* Int. Conf. Heavy Metals Environ., Toronto, Canada.

Peoples, S. A. 1964. Arsenic toxicity in cattle. Ann. N.Y. Acad. Sci. 111:644.

Reagor, J. C. 1973. Arsenic poisoning in cattle. Southwest Vet. 26:295.
Reives, G. I. 1925. The arsenical poisoning of livestock. J. Econ. Entomol. 18:83.
Schroeder, H. A., and J. J. Balassa. 1966. Abnormal trace metals in man: Arsenic. J. Chron. Dis. 19:85.
Selby, L. A., A. A. Case, C. R. Dorn, and D. J. Wagstaff. 1974. Public health hazards associated with arsenic poisoning in cattle. J. Am. Vet. Med. Assoc. 165:1010.
Sollman, T. 1953. Manual of Pharmacology. Saunders, Philadelphia.
Thapar, N. T., E. Guenthner, C. W. Carlson, and O. E. Olson. 1969. Dietary selenium and arsenic additions to diets for chickens over a life cycle. Poult. Sci. 48:1988.
Underwood, E. J. 1977. Trace Elements in Human and Animal Nutrition, 4th ed. Academic Press, New York.
Vallee, B. L., D. D. Ulmer, and W. E. C. Wacker. 1960. Arsenic toxicology and biochemistry. AMA Arch. Ind. Health 23:132.
Wood, J. M. 1975. Metabolic cycles for toxic elements, p. A-5. *In* Int. Conf. Heavy Metals Environ., Toronto, Canada.

# Barium

Barium (Ba) is one of the alkaline earth metals, and, because of its high chemical activity, it never occurs free in nature. It is found chiefly as barite or heavy spar, $BaSO_4$, and witherite, $BaCO_3$ (Pidgeon and Preisman, 1964). The element is of little importance in animal nutrition. Barium sulfate is used in taking X-ray photographs of the intestinal tract. Even though the barium ion, $Ba^{+2}$, is extremely toxic when absorbed, barium sulfate ($BaSO_4$) is so slightly soluble that it is nontoxic.

## ESSENTIALITY

Evidence is not available to show that barium performs any essential function in vertebrate animals, although there is a report of depressed growth in rats and guinea pigs when barium was omitted from the mineral mixture of the specially purified diets fed to these animals (Rygh, 1949). It is interesting that barium sulfate ($BaSO_4$) forms the exoskeleton of the rhizopod *Xenophyophora* (Vinogradov, 1953).

## METABOLISM

No study appears to have been made of the absorption of natural barium, but a study of the metabolism of $^{140}Ba$ in young (2 months) rats showed 24-hour urinary and fecal excretions to be 7 and 20 percent, respectively (Bauer *et al.*, 1956). The $^{140}Ba$ was administered by intraperitoneal injection. The study also demonstrated that barium was deposited in the skeleton.

## SOURCES

Barium is present in all soils and in all plants. Only small quantities are found in plants in most cases. The barium content of different plant species growing on different soils ranged from 0.5 to 40 ppm with a mean value of 10 ppm (Underwood, 1977). Certain plant species can accumulate high concentrations of barium from barium-rich soils. For example, *Juglans regia* and *Fraxinus pennsylvanica* were reported to contain 2,600 and 1,700 ppm barium, respectively. This ability to accumulate barium is unusual considering the fact that barium is toxic to most plants even at low concentrations. With the exception of plants that accumulate barium, the quantities of barium taken up by plants, even from a soil high in barium, are so low that there is little likelihood of animals suffering toxic effects by eating the plants.

## TOXICOSIS

Barium is extremely toxic when absorbed (Sollmann, 1957). The characteristic systemic action of barium is a marked stimulation of muscles of all types, regardless of innervation. In poisoning by soluble barium salts, the major signs and symptoms are related to this powerful muscular stimulation. The action of barium on the gastrointestinal musculature causes vomiting, severe colic and diarrhea, and hemorrhage. The cardiovascular effects consist of a marked hypertension due to a spasm of the arteriolar musculature, intense myocardial stimulation, and ultimately death from systolic cardiac arrest. In addition, barium causes tremors of skeletal muscles. Paralysis of the central nervous system may develop late in the course of barium intoxication. Death usually occurs within an hour, but may be delayed for some time. The fatal dose for the human is about 0.8 to 0.9 g of barium chloride. Barium carbonate and sulfide are also toxic but act more slowly and require larger doses.

The treatment of barium poisoning consists in the precipitation of the unabsorbed salt remaining in the intestinal tract by the ingestion of a solution of sodium or magnesium sulfate, which forms the insoluble barium sulfate and also acts as a cathartic to hasten the elimination of barium from the bowel. Therapy is otherwise purely symptomatic and supportive.

There is little information published on feeding varying levels of soluble barium compounds to livestock. Most reports (Spector, 1956), where these compounds have been administered, indicate that the dose level was lethal (Table 7). It should be pointed out that these are not necessarily minimum lethal dose levels. Maximum tolerance levels have not been stated.

Martinez and Church (1970) reported on the effect of barium in *in vitro* studies with washed suspensions of rumen microorganisms. Thirty ppm of barium as barium chloride ($BaCl_2 \cdot 2H_2O$) significantly depressed cellulose digestion by 15 percent. Subsequent additions of barium above 30 ppm progressively depressed digestion of cellulose. An inhibition of slightly over 50 percent resulted with additions of 200 ppm barium.

## TISSUE LEVELS

There are very limited data on the level of barium in animal tissue. It has been reported that humans contain 22 mg of barium, of which 93 percent is present in the bone (Underwood, 1977). The remainder is widely distributed throughout the soft tissues in very low concentrations (0.02 and 0.1 ppm wet weight). Garner (1959) indicated that barium is concentrated by the choroid of the eye and reported a value of 3.5 ppm wet weight for the rabbit.

## MAXIMUM TOLERABLE LEVEL

Based upon extrapolations from toxicity data and *in vitro* rumen work, the level of soluble barium in a diet probably should not exceed 20 ppm. It should be pointed out that the measurement of total barium in a feed or feed ingredient is not a measure of biologically available barium. A much better guide for safety purposes would be a procedure whereby the acid or water soluble barium is determined. Much of the barium present in most feedstuffs is not available to the animal and is of little concern. Barium is a potential problem in the feed industry only if ingredients or mixed feeds inadvertently become contaminated with soluble forms of the element.

## SUMMARY

Barium is one of the alkaline earth metals, and, because of its high chemical activity, it never occurs free in nature. Conclusive evidence is not available to show that barium performs any essential function in vertebrate animals. Barium is extremely toxic when absorbed with the characteristic systemic action marked by stimulation of muscles of all types, regardless of innervation. Based upon extrapolation of available data, the level of soluble barium in a diet probably should not exceed 20 ppm.

TABLE 7 Effects of Barium Administration in Animals

| Class of Animals | Age or Weight[a] | Administration: Quantity of Element[b] | Source | Duration[a] | Route | Effect(s) | Reference |
|---|---|---|---|---|---|---|---|
| Swine | | 733 mg/kg | $BaCO_3$ | | | Death | Esser, 1935 |
| Chicken | | 330 ppm of compound | Barium chloride and carbonate | | Diet | Depressed weight gain | Taucins *et al.*, 1969 |
| Chicken | | 6,000 ppm of compound | Barium chloride and carbonate | | | Death | |
| Horse | | 450–675 mg/animal | $BaCl_2 \cdot 2H_2O$ | | Oral | Death | Esser, 1935 |
| Rabbit | | 95 mg/kg | $BaCl_2 \cdot 2H_2O$ | | Oral | Death | Schwartze, 1920 |
| Rabbit | | 124 mg/kg | $BaCO_3$ | | | Death | |
| Rabbit | | 118 mg/kg | $Ba(C_2H_3O_2)_2 \cdot H_2O$ | | Oral | Death | Crawford, 1908 |
| Dog | | 50 mg/kg | $BaCl_2 \cdot 2H_2O$ | | Oral | Death | Schwartze, 1920 |

[a] Not available.

[b] Quantity expressed in parts per million or as milligrams per kilogram of body weight.

## REFERENCES


Bauer, G. C. H., A. Carlsson, and B. Lindquist. 1956. A comparative study on the metabolism of $^{140}Ba$ and $^{45}Ca$ in rats. Biochem. J. 63:535.

Crawford, A. C. 1908. Barium, a Cause of the Loco-Weed Disease. U.S. Dept. Agric. Bur. Plant Ind. Bull. 129.

Esser, A. 1935. Clinical, anatomical and spectrographic investigation of the central nervous system in acute metal poisoning with particular consideration of their importance for forensic medicine and tissue pathology. I. Strontium, barium, magnesium, aluminum, thorium (radioactive material), thallium, zinc, cadmium, mercury. Dtsch. Z. Ger. Med. 25:239.

Garner, R. J. 1959. Distribution of radioactive barium in eye tissues. Nature (Lond.) 184:733.

Martinez, A., and D. C. Church. 1970. Effect of various mineral elements on *in vitro* rumen cellulose digestion. J. Anim. Sci. 31:982.

Pidgeon, L. M., and L. Preisman. 1964. Kirk-Othmer Encyclopedia of Chemical Technology, vol. 3, 2nd ed. John Wiley & Sons, New York.

Rygh, O. 1949. Research on trace elements. 1. Importance of strontium, barium and zinc. Bull. Soc. Chim. Biol. 31:1052.

Schwartze, E. W. 1920. Toxicity of Barium Carbonate to Rats. U.S. Dept. Agric. Bull. 915.

Sollmann, T. 1957. A Manual of Pharmacology, 8th ed. W. B. Saunders Co., Philadelphia.

Spector, W. S., ed. 1956. Handbook of Toxicology, vol. 1. W. B. Saunders Co., Philadelphia.

Taucins, E., A. Svilane, A. Valdmanis, A. Buike, R. Zarina, and E. Ya. Fedorova. 1969. Barium, strontium, and copper salts in chick nutrition. Fiziol. Akt. Komponenty Pitan. Zhivotn. 199.

Underwood, E. J. 1977. Trace Elements in Human and Animal Nutrition, 4th ed. Academic Press, New York.

Vinogradov, A. P. 1953. The Elementary Chemical Composition of Marine Organisms. Sears Foundation, New Haven, Conn.

# Bismuth

Bismuth (Bi) is a pinkish-white, lustrous, crystalline, brittle metal widely distributed in small quantities throughout the world. Bismuth occurs naturally as a metal, oxide, sulfide, and carbonate and until about 1775 was often confused with tin and lead. Most bismuth destined for use in the Western Hemisphere today is obtained as a by-product of copper-, lead-, and tin-refining processes. The most significant physical property of bismuth is its expansion upon solidification (Hempel and Hawley, 1973). Accordingly, bismuth alloys are especially suited to use in safety devices such as plugs in compressed gas cylinders, automatic fire sprinkler systems, and firedoor releases. While bismuth has been used in the treatment of venereal disease, contemporarily, bismuth is used as a coloring agent in decorative cosmetics (Beaver and Burr, 1963), in ointments for burns, to delineate viscous surfaces in X-ray analyses, as a fungicide, in the treatment of warts, and to regulate stool odor and consistency in colostomy patients (Burns *et al.*, 1974).

## ESSENTIALITY

Despite rather extensive experience with bismuth as a therapeutic agent for gastrointestinal disturbances, no evidence exists to indicate that bismuth is an essential nutrient.

## METABOLISM

Metallic bismuth is poorly absorbed; however, several organic forms of bismuth are absorbed from parenteral administration sites as well as through skin and mucous membranes. A dark brown to black pigment frequently occurs in the skin, gum line, and mucosa of bismuth-treated individuals (Wachstein and Zak, 1946). This pigment is believed to represent precipitation and accumulation of bismuth sulfide (Urizar and Vernier, 1966). Some aspect of bismuth metabolism, in man at least, results in a reversible myoclonic encephalopathy following administration of bismuth subnitrate (Cambier *et al.*, 1974; Lhermitte *et al.*, 1975). This indicates the blood brain barrier is permeable to bismuth. Another bismuth-related phenomenon is the development of intranuclear inclusion bodies within proximal renal tubule cells in humans and animals exposed to bismuth (Wachstein, 1949). These inclusions are spherical, eosinophilic, slightly acid-fast, electron dense, homogenous structures that, by X-ray microanalysis, have been demonstrated to contain bismuth (Fowler and Goyer, 1975).

## SOURCES

The potential for animal exposure to parenteral compounds has greatly diminished since the advent of penicillin and other antibiotics that replaced basic bismuth chloride and other bismuth compounds as antisyphilitic agents. Additional bismuth compounds, bismuth aluminate, bismuth oxide, bismuth subnitrate, bismuth subgallate, and bismuth tannate have been and/or continue to be used as gastrointestinal protectives at levels of 0.3 to 2 g for dogs and up to 15 to 30 g for horses and cattle. Some of the above preparations and bismuth iodosubgallate, bismuth iodide oxide, bismuth sodium tartrate, and bismuth subcarbonate are still used as external astringents, on buccal warts (Urizar and Vernier, 1966) as topical antiseptics, and in cosmetics. The industrial or manufacturing uses for bismuth include the manufacture of dry cell cathodes, industrial catalysts, disinfectants, magnets, semiconductors, glazes, and carriers for $^{235}U$ (Hempel and Hawley, 1973).

## TOXICOSIS

### LOW LEVELS

There are numerous reports of iatrogenic bismuth toxicosis in humans, but data from natural or experimental bismuth toxicoses in animals are very limited.

Lechat *et al.* (1968) orally administered bismuth subnitrate to rabbits at a rate equivalent to 70 to 74 mg of bismuth per kilogram of body weight for up to 34 weeks without any noticeable effect regardless of whether the vehicle used for diluting and carrying the bismuth was saccharose, mannitol, or sorbitol. Elevations in urinary and renal concentrations of bismuth were noted during the experiment but disappeared soon after cessation of the bismuth administration. Of major significance in this experiment was the absence of any inclusion bodies in the kidneys. Similar studies conducted by Lechat *et al.* (1968) with rats yielded comparable results.

The chronic studies with bismuth conducted by Steinfeld and Meyer (1886) involved rabbits, cats, and dogs. Rabbits weighing 0.75 to 1 kg were subcutaneously injected with 15 mg of bismuth oxide per day on 4 different days. Cats weighing 2.3 to 3.4 kg were injected subcutaneously with 10 to 20 mg bismuth oxide daily for 4 consecutive days. In each of these species, bismuth toxicosis was manifested by lassitude, stomatitis, salivation, anorexia, weight loss, diarrhea, fever, albuminuria, and tetanic convulsions. At necropsy, the bismuth-intoxicated animals had ulceration and necrosis of the mucosa of the large intestine.

Low-level bismuth toxicity studies have been conducted with laboratory rodents for periods of up to 2 years. Wilson (1975a) observed no effect of 1 to 2 ml tripotassium dicitrato bismuthate (TDB) administered daily for 30 days to rats weighing 190 to 215 g. In similar studies with TDB in 100- to 120-g rats, Wilson (1975b) revealed that administering 4 to 32 mg TDB/kg per day by gavage for 40 days caused no toxic effects, while the highest level was effective in decreasing the healing time of experimentally induced ulcers. Preussmann and Ivankovic (1975) found no carcinogenic or other toxic effects of bismuth oxychloride administered at the rates of 1 to 5 percent in the diets of rats for a period of 2 years.

Haddow and Horning (1960) were concerned about the carcinogenic effect of bismuth and reported that bismuth dextran, administered to mice in weekly subcutaneous injections at the rate of 1 mg per week for 23 weeks was not carcinogenic. The mice were observed for a total period of 10 months. The same results were obtained with hamsters administered the above dosage of bismuth dextran intramuscularly for the same period. Innes *et al.* (1969) orally dosed mice with bismuth dimethylthiocarbonate at a rate equivalent to 34 ppm bismuth without producing toxic effects.

### HIGH LEVELS

Acute studies of bismuth toxicity were conducted by Steinfeld and Meyer (1886) using rabbits, cats, dogs, and frogs. Rabbits weighing about 1 kg were injected subcutaneously with 72 mg bismuth as $Bi_2O_3$, and cats weighing 2.5 to 3.5 kg were similarly given 36–900 mg bismuth as $Bi_2O_3$. The effects in both species included increased respiration rates, slowed heart rate, convulsions, and death. One dog weighing 6.2 kg was also given two 40 mg subcutaneous injections of bismuth oxide. In addition to many of the above signs, the dog also exhibited vomition and tenesmus. Frogs injected in the dorsal lymph sack with 7 to 2,700 mg of bismuth as $Bi_2O_3$ exhibited the same effects as the rabbits and cats. The effects of parenterally administered bismuth subnitrate in rabbits have been studied by Langhans (1886). He was the first to report on the "wismuthcellen," the bismuth-induced intranuclear and intracytoplasmic inclusion bodies in the renal tubular epithelium.

The acute toxic effects of orally administered bismuth have been observed in cats by Novak and Gutig (1908). These workers found 10 g of bismuth subnitrate (7.0 to 7.4 g Bi) administered in a single oral dose cause cyanosis, methemoglobinemia, and death. This response is typical of nitrate poisoning. The same dosage of bismuth subnitrate administered orally to dogs was reported to be without effect (Dalche and Villegean, 1887, as cited by Mayer and Baehr, 1912).

Acute bismuth toxicity studies in rodents have been conducted by Wilson (1975a) using TDB. Rats weighing about 200 g were not affected by oral administration of 1 to 5 ml of TDB per 100 g of body weight in a 24-hour period. Similarly, mice weighing 24 to 32 g were not affected by 0.2 to 0.9 ml orally administered TDB per 100 g of body weight during a 24-hour period.

The acute toxic effects of bismuth cymol administered parenterally to rats have been studied by Pappenheimer and Maechling (1934). They gave 77 to 3,930 mg/kg in one to four intramuscular injections. This caused renal tubular degeneration and the formation of the inclusion bodies previously mentioned. The $LD_{50}$ for a single oral dose of bismuth as $Bi_2O_3$ in mice has been calculated from unpublished data by Preussmann and Ivankovic (1975) to be 19.3 g/kg. These workers never found a high enough level of bismuth oxide to effect an $LD_{50}$ for rats. Vu Ngoc and Garcet (1967) have indicated the rat gastric mucosa is hypersensitive to bismuth.

### FACTORS AFFECTING TOXICITY

The limited data from the rather wide variety of bismuth compounds used in the reviewed toxicity experiments herein preclude meaningful statements on factors influencing toxicity. The data, however, suggest that rodents are rather refractive to several bismuth compounds but do develop bismuth nephrotoxicity. Of the larger species, dogs seem less susceptible to bismuth toxicity than cats, but this species difference in bismuth toxicity may be a simple difference in the rates of bismuth administration to these two species.

## TISSUE LEVELS

Data on tissue levels of bismuth are fragmentary at best. While specific concentrations are lacking, evidence for the persistence of bismuth in tissue comes from experience with chronic use of bismuth medicaments in humans (Mayer and Baehr, 1912; Burr *et al.* 1965). Randall *et al.* (1972) demonstrated that cases of bismuth nephrotoxicity and inclusion body formation have followed the topical use of bismuth preparations and that the bismuth inclusion bodies are very persistent, being present in 86 percent of the kidneys of humans treated with bismuth preparations 30 years previously (Beaver and Burr, 1963). Fowler and Goyer (1975) demonstrated that bismuth is concentrated in these inclusions but gave no quantitative estimates.

The blood levels of bismuth are more transient than bismuth renal inclusion bodies. Kruger *et al.* (1976) noted levels of 10 to 16 mg bismuth/dl in the venous blood of patients using bismuth skin creams and found no detectable bismuth in blood within 3 weeks after the bismuth creams were discontinued. Similar levels have been found by Buge *et al.* (1974) in serum, blood, and plasma of bismuth-toxic humans.

## MAXIMUM TOLERABLE LEVELS

There seems to be insufficient continuity or similarity among the various bismuth toxicity experiments conducted to make any definitive statements on the maximum tolerable levels. Extrapolating from the bismuth toxicity data from rabbits, one can calculate that 2,310–3,360 ppm of bismuth as bismuth subnitrate are below the MTL. If one uses a 100-fold margin-of-safety factor for cross-species extrapolation in conjunction with the work of Preussmann and Ivankovic (1975) indicating 40,000 ppm of bismuth was a no-effect level in rats, the MTL for the domestic animals may approximate 400 ppm of bismuth.

## SUMMARY

Overt signs of experimental bismuth toxicosis (lassitude, salivation, anorexia, diarrhea, fever, and convulsions) have been demonstrated in rabbits, cats, and dogs, with the latter species being perhaps the least susceptible of the three. Rodents appear to be quite resistant to bismuth toxicosis as indicated by the fact that the calculated $LD_{50}$ for bismuth oxychloride in mice is 21.5 g/kg and that no overt signs of bismuth toxicosis in rodents were reported at lesser levels of administration. Bismuth gradually induces nephrotoxicity characterized by the formation of intranuclear inclusion bodies in renal tubular epithelium and the persistence of these inclusions for perhaps the life of the host.

TABLE 8 Effects of Bismuth Administration in Animals

| Class and Number of Animals[a] | Age or Weight | Administration | | | | Effect(s) | Reference |
|---|---|---|---|---|---|---|---|
| | | Quantity of Element[b] | Source | Duration | Route | | |
| Rabbit—2 | 0.75–1.0 kg | 13.5 mg | $Bi_2O_3$ | Once/d for 4 d | Subcutaneous | Lassitude; stomatitis; salivation; anorexia; weight loss; fever; diarrhea; convulsions; and intestinal ulcers | Steinfeld and Meyer, 1886 |
| | 1 kg | 72 mg | | Single injection | | Rapid respiration; bradycardia; convulsions and death in 12 h | |
| Rabbit—9 | 2.3–3.0 kg | 70 mg/kg | Bismuth subnitrate | 5 d per wk for 34 wk | Drench | No adverse effects | Lechat *et al.*, 1968 |
| Dog—1 | 6.2 kg | 72 mg | $Bi_2O_3$ | Two injections 2 d apart | Subcutaneous | Diarrhea; vomition; tetanic convulsions; tenesmus; death in 4 d | Steinfeld and Meyer, 1886 |
| Dog | | 7.2 g | Bismuth subnitrate | Once/d | Drench | No adverse effects in 30 d | Dalche and Villegean, 1887 (Cited by Mayer and Baehr, 1912) |
| Cat—1 | 3 kg | 9 mg | $Bi_2O_3$ | Once/d for 4 d | Subcutaneous | Lassitude; stomatitis; salivation; anorexia; weight loss; fever; diarrhea; convulsions; and intestinal ulcers | Steinfeld and Meyer, 1886 |

| | | | | | | | |
|---|---|---|---|---|---|---|---|
| | 3 kg | 18 mg | | | | Same as above | |
| | 2.5 kg | 72 mg | | Single injection | Intravenous | Rapid respiration; bradycardia; convulsions; death in 13 h | |
| | 3.1–3.6 kg | 90 mg | | | Subcutaneous | Death in 12 h | |
| Cat | | 7.2 g | Bismuth subnitrate | Single dose | Drench | Cyanosis, methemaglobinuria; death | Novak and Gutig, 1908 |
| Rat—8 | | 77 mg/kg | Bismuth cymol | 1–4 injections | Intramuscular | Renal tubular degeneration; renal inclusion bodies | Pappenheimer and Maechling, 1934 |
| Rat | | 3,930 mg/kg | | | | Same as above | |
| Rat—40 | 100 d | 19.3 g/kg | Bismuth oxychloride | Single dose | Gavage | $LD_{50}$ | Preussmann and Ivankovic, 1975 |
| Rat—144 | 100–120 g | 2.8 mg/kg | Tripotassium dicitrato bismuthate | 40 d | Drench | No adverse effects | Wilson, 1975b |
| | | 22 mg/kg | | | | | |
| | | | | 40 d | | No adverse effects | |
| | 190–215 g | 700 mg | | 30 d | | No adverse effects | |
| | | 1,400 mg | | | | No adverse effects | |
| Rat—144 | 190–215 g | 7 g/kg | | 24 h | | No adverse effects | Wilson, 1975a |
| | | 35 g/kg | | 24 h | | No adverse effects | |
| Rat—40 | Weanling | 8,000 ppm | Bismuth oxychloride | 2 yr | Diet | No effect on growth; noncargiogenic | Preussmann and Ivankovic, 1975 |
| | | 40,000 ppm | | 2 yr | | No effect on growth; noncarcinogenic | |
| Mouse | 24–32 g | 14 mg/g | Tripotassium dicitrato bismuthate | 24 h | Gavage | No adverse effects | Wilson, 1975a |
| | | 63 mg/g | | 24 h | | No adverse effects | |
| Mouse—20 | | 1 mg | Bismuth dextran | Once/wk for 23 wk | Subcutaneous | No neoplasia (10 mo) | Haddow and Horning, 1960 |

TABLE 8 *Continued*

| Class and Number of Animals[a] | Age or Weight | Administration: Quantity of Element[b] | Source | Duration | Route | Effect(s) | Reference |
|---|---|---|---|---|---|---|---|
| Mouse—72 | Weanling | 34 ppm | Bismuth dimethylthiocarbonate | | Diet | Nonneoplastic | Innes *et al.*, 1969 |
| Hamster | | 1 mg | Bismuth dextran | Once/wk for 23 wk | Subcutaneous | No neoplasia (10 mo) | Haddow and Horning, 1960 |
| Frog—2 | Medium to large | 7 mg | $Bi_2O_3$ | Single injection | Dorsal lymph sack | Rapid respiration; bradycardia convulsions; death in 105 h | Steinfeld and Meyer, 1886 |
| | | 135 mg | | | | Death in 82 h | |

[a] Number of animals per treatment.
[b] Quantity expressed in parts per million or as milligrams per kilogram of body weight or milligram per day.

## REFERENCES


Beaver, D. L., and R. E. Burr. 1963. Bismuth inclusions in the human kidney. Arch. Pathol. 76:89.

Buge, A., G. Rancurel, M. Poisson, and J. Dechy. 1974. Encephalopathies myocloniques par les sels de bismuth. Nouv. Presse Med. 3:2315.

Burns, R., D. W. Thomas, and V. J. Barron. 1974. Reversible encephalopathy possibly associated with bismuth subgallate ingestion. Br. Med. J. 1:220.

Burr, R. E., A. M. Gotto, and D. L. Beaver. 1965. Isolation and analysis of bismuth inclusions. Toxicol. Appl. Pharmacol. 7:588.

Cambier, J., M. Masson, and R. Dairou. 1974. Encephalopathie myoclonique et intoxication par les sels de bismuth. Nouv. Presse Med. 3:2662.

Fowler, B. A., and R. A. Goyer. 1975. Bismuth localization within nuclear inclusions by X-ray microanalysis. Effects of accelerating voltage. J. Histochem. Cytochem. 23:722.

Haddow, A., and E. S. Horning. 1960. On the carcinogenicity of an iron-dextran complex. J. Natl. Cancer Inst. 24:109.

Hempel, C. A., and G. G. Hawley. 1973. The Encyclopedia of Chemistry, 3rd ed. Van Nostrand Reinhold Co., New York.

Innes, J. R. M., B. M. Ulland, M. G. Valerio, L. Petrucelli, L. Fishbein, E. R. Hart, L. Pallotta, R. R. Bates, H. L. Falk, J. J. Gart, M. Klein, I. Mitchell, and J. Peters. 1969. Bioassay of pesticides and industrial chemicals for tumorigenicity in mice. A preliminary note. J. Natl. Cancer Inst. 42:1101.

Kruger, G., D. J. Thomas, F. Weinhardt, and S. Hoyer. 1976. Disturbed oxidative metabolism in organic brain syndrome caused by bismuth in skin creams. Lancet 176:485.

Langhans, T. 1886. Pathologisch-anatomische Befunde Bei mit Bismuth um subnitricum vergifteten thieren 2. Chirurgie 13:263.

Lechat, P., L. Morel-Maroger, R. Cluzan, F. Flouvat, and J. Fontagne. 1968. Etude experimentale des effets de l'ingestion prolongee de sous-nitrate de bismuth associe a des doses. Therapie 23:445.

Lhermitte, F., C. F. Degos, and J. L. Signoret. 1975. Encephalopathies reversibles par les sels insolubles de bismuth, cinq nouveaux cas. Nouv. Presse Med. 4:419.

Mayer, L., and G. Baehr. 1912. Bismuth poisoning. A clinical and pathological report. Surg. Gynecol. Obstet. 15:309.

Novak, J., and C. Gutig. 1908. Nitritvergiftung durch Bismutun subnitricum. Ber. Klin. Wochenschr. 45:1764.

Pappenheimer, A. M., and E. H. Maechling. 1934. Inclusions in renal epithelial cells following the use of certain bismuth preparations. Am. J. Pathol. 10:577.

Preussmann, R., and S. Ivankovic. 1975. Absence of carcinogenic activity in BD rats after oral administration of high doses of bismuth oxychloride. Food Cosmet. Toxicol. 13:543.

Randall, R. W., R. J. Osheroff, S. Bakerman, and J. G. Setter. 1972. Bismuth nephrotoxicity. Ann. Intern. Med. 77:481.

Steinfeld, W., and H. Meyer. 1886. Untersuchungen uber die toxischen und therapeutischen Wirkungen des Bismuths. Arch. Exp. Pathol. Pharmakol. 20:40.

Urizar, R., and R. L. Vernier. 1966. Bismuth nephropathy. J. Am. Med. Assoc. 198:187.

Vu Ngoc, H., and S. Garcet. 1967. Hypersensitivity of the rat gastric mucosa to repeated administration of bismuth. Sem. Ther. 43:34.

Wachstein, M., and G. F. Zak. 1946. Bismuth pigmentation. Its histochemical identification. Am. J. Pathol. 22:603.

Wachstein, M. 1949. Studies on inclusion bodies. I. Acid-fastness of nuclear inclusion bodies that are induced by ingestion of lead and bismuth. Am. J. Clin. Pathol. 19:608.

Wilson, T. R. 1975a. The pharmacology of tri-potassium di-citrato bismuthate (TDB). Postgrad. Med. J. 51:18.

Wilson, T. R. 1975b. Effect of tri-potassium di-citrato bismuthate (TDB) on the healing of experimental gastric ulcers in rats. Postgrad. Med. J. 51(Suppl. 5):22.

# Boron

Boron (B) is an amorphous, dark brown nonmetal. A native sodium tetraborate is found in California, Nevada, Oregon, and Asia Minor from which borax ($Na_2B_4O_7 \cdot 10H_2O$) is prepared and from which boron can be isolated. Boron also occurs in nature as colemanite ($Ca_2B_6O_{11} \cdot 5H_2O$), boronatrocalcite ($CaB_4O_7NaBO_2 \cdot 8H_2O$), and boracite ($Mg_7Cl_2B_{16}O_{30}$). A small amount (0.003 percent) of boron in steel increases its hardness and improves its mechanical properties when quenched or drawn. Boron also strongly absorbs neutrons, and boron steel is used in shielding and in controlling the operating rate of atomic power plants. Calcium boride is used in deoxidation of copper–brass bronze. Borax is used in the manufacture of glass, enamels, soap, sizing for paper, and as a preservative for wood and meats. Boron carbide is used as an abrasive, and boron hydrides (boranes) are used as high-energy fuels. Borax and boric acid solutions have long been used as mild antiseptics, and fused borax containing colored metal oxides may be used for artificial gems or, when ground, as pigments.

## ESSENTIALITY

Boron has been established as essential for higher plants for over 50 years, and this element is added frequently to fertilizers for plants with high requirements such as alfalfa, apples, and certain root and cruciferous crops (rutabaga, turnips, red beets, sugar beets, cabbage, and cauliflower). Boron deficiencies in plants most often occur on light-colored sands and silt loams in humid regions. Liming may somewhat decrease boron bioavailability in soils, but this decrease is probably related to the need by plant tissues of a specific calcium–boron ratio.

Purified diets containing 0.15 to 0.16 ppm boron have been used in an attempt to induce boron deficiency in rats, but they grew and reproduced as well as those receiving supplemental boron (Hove *et al.*, 1939; Orent-Keiles, 1941; Teresi *et al.*, 1944). Earlier suggestions that potassium-deficient rats would benefit from boron supplementation were not supported by later investigations (Follis, 1947). If boron is required by rats, dietary need must be below 0.15 ppm. In rats fed a diet containing not more than 0.001 ppm boron, hepatic RNA synthesis was stimulated by an intraperitoneal injection of 20 $\mu$M borate as boric acid (Weser, 1967).

## METABOLISM

Boron in food or administered as soluble borate or boric acid is rapidly and almost completely absorbed from the gastrointestinal tract. Ochsner (1917) and Kahlenberg (1924) indicated that boron can be absorbed through intact skin, but the amounts are apparently too slight to produce systemic toxicosis. However, toxic amounts can be absorbed through damaged skin (Pfeiffer *et al.*, 1945). The application of a 5 percent solution of boric acid to normal skin of human subjects produced barely detectable levels of boron in the urine, while application to granulated wounds or burns produced prompt urinary boron excretion. Cope (1943) found 2.0 to 2.5 g of boron at necropsy in some patients treated for severe burns with saturated boric acid solution, and there have been reports of toxicosis and death following absorption from open skin lesions (Witthaus, 1911). The boron hydrides (boranes) may be absorbed from the lungs and are highly toxic.

The main excretory pathway for boron is via the urine (Kent and McCance, 1941; Owen, 1944; Tipton *et al.*, 1966).

## SOURCES

The average boron concentration in the earth's crust is about 10 ppm, with most soils ranging from 7 to 80 ppm (Krauskopf, 1972). Among plants, monocotyledons generally contain less boron than dicotyledons. Boron deficiency is seen in a wide variety of plants when vegetative dry matter boron concentrations are less than 15 ppm. Adequate but not excessive boron concentrations range from 20 to 100 ppm in the dry matter. Boron toxicosis in plants occurs usually when dry tissue concentrations of boron exceed 200 ppm. Neubert *et al.* (1969) have published boron values for 24 crops and Bradford (1966) for 55 crops. Zook and Lehmann (1968) analyzed tropical and subtropical fruits and found that avocados had the highest boron concentration (7–10 ppm, fresh basis), followed by stone fruits (1.4–3.5 ppm), and pome and citrus fruits and berries (0.3–2.4 ppm). Cereal grains contain about 1 to 5 ppm boron (Beeson, 1941). Animal muscle and organ concentrations range mostly between 0.5 and 1.5 ppm (dry basis). Bone concentrations are several times higher (Underwood, 1977). Cow's milk normally contains 0.5 to 1.0 ppm boron when fed diets containing 16 to 34 ppm boron with little variation associated with breed or stage of lactation (Hove *et al.*, 1939; Owen, 1944). The World Health Organization (1973) has calculated daily boron intakes of Australian infants (0–6 mo) from cow's milk to be 0.4 to 0.85 mg. A representative U.S. diet for adults was calculated to provide 3 mg of boron per day, although estimated boron intakes of adult humans around the world vary from 0.3 to 41.0 mg due to geographical differences (Schlettwein-Gsell and Mommsen-Straub, 1973).

The U.S. Department of the Interior has established an upper limit for boron in public water supplies at 1 ppm (Bradford, 1971), and the Environmental Protection Agency (1973) has proposed 5 ppm boron as the allowable maximum in water for livestock. However, Green and Weeth (1977) have found water boron concentrations in Nevada ranging from 0.2 ppm in the Humboldt River to 80 ppm at Borax Flat. Most water samples were above 1 ppm, and a number exceeded 5 ppm.

A child consuming milk containing approximately 0.7 g boric acid per liter developed coeliac diseaselike symptoms (Forsyth, 1919). Human burn patients who were treated topically with saturated boric acid solution accumulated up to 2.0–2.5 g boron in their bodies (Cope, 1943). Industrial poisoning from the "oxygen-bonded" salts of boron has not been reported. However, the boron hydrides (boranes) are highly toxic and constitute a significant industrial hazard. Diborane, decaborane, and pentaborane are most frequently encountered. These are used chiefly as high-energy fuels, and decaborane has also been used as a vulcanizer of rubber instead of sulfur (Browning, 1969). Diborane ($B_2H_6$) is a gas with a nauseating odor; decaborane ($B_{10}H_{14}$) is a solid with an intense, bitter, chocolatelike odor; and pentaborane ($B_5H_9$) is a volatile liquid with a sweetish odor.

## TOXICOSIS

### LOW LEVELS

#### *Oral*

Little quantitative information defining chronic intake limits for boron is available. Browning (1969) stated that 0.25 percent boric acid (437 ppm boron) in drinking water inhibited growth in animals but produced no detected changes in the blood nor observable lesions at necropsy. Green and Weeth (1977) conducted a study with Hereford heifers fed grass hay (38 ppm boron) and water to which boron (as borax) had been added at various concentrations. They suggested that water containing less than 29 ppm boron would not be discriminated against, while in the range of 29 to 95 ppm, cattle would show preference for water with lower boron concentrations. Consumption of 150 or 300 ppm boron in water for 30 days produced inflammation and edema in the legs and around the dew claws and decreased hay consumption, gain, hematocrit, and hemoglobin concentrations. Lethargy and occasional diarrhea was seen in heifers consuming the higher level. Plasma boron concentrations were 0.53 ppm on control water (0.8 ppm boron) and 11.2 and 18.9 ppm on water containing 150 and 300 ppm boron, respectively. Thirty, 67, and 69 percent of the total daily boron intake on these three respective treatments was excreted in the urine. Glomerular filtration rate and osmolal clearance were unaffected by the high boron levels, but a relative diuresis was indicated by modifications in free water clearance. Urinary phosphate excretion was decreased.

Green *et al.* (1973) concluded that 75 ppm boron in drinking water did not affect growth or reproduction in rats. When boron concentrations exceeded 150 ppm, these workers reported reduced body size, continued prepubescent fur, lack of incisor pigmentation, aspermia, and impaired ovarian development.

Prolonged consumption of small amounts of boric acid by human beings has been reported by Browning (1969) to lead to mild gastrointestinal irritation, anorexia, disturbed digestion, nausea, emesis, and an erythematous rash. As in the case reported by Sanders (1912), this rash may be firm to the touch with a tendency to become purpuric. One case of coeliac diseaselike symptoms has been reported (Forsyth, 1919) in a child fed milk containing about 0.7 g of boric acid per liter. Medicinal use of boric acid and borax for babies has resulted in anorexia, nausea, emesis, diarrhea, marked cardiac weakness, and a red papular eruption over the entire body (British Medical Association, 1966).

### *Inhalation*

Repeated exposure of rats to 20 ppm of decaborane in air for 6 hours a day, 5 days a week, produced nervousness, restlessness, loss of weight, unsteadiness, a tendency to belligerency, tremors of the head, and convulsions (Svirbely, 1954b). Repeated exposure of rats and mice to about 3 ppm of pentaborane in air produced a similar but more marked neurological effect, particularly with respect to belligerency.

The symptoms of human exposure to diborane (Lowe and Freeman, 1957; Cardasco *et al.,* 1962) included pulmonary irritation, chest tightness, dyspnea, nonproductive cough and wheezing, pneumonia, headache, vertigo, chills, fatigue, and muscular weakness. Increased blood nonprotein nitrogen concentrations and positive cephalin flocculation tests were also seen.

Mild exposure to decaborane resulted in headache, nausea, dizziness, and drowsiness (Lowe and Freeman, 1957). Exposure to pentaborane produced headache, dizziness, often hiccups, and nervousness. Drowsiness and nausea were sometimes experienced initially. Muscular pain and cramps were common. Liver and kidney damage were evident as illustrated by abnormal liver function tests and elevated blood nonprotein nitrogen and urea levels.

Maximum allowable air concentrations have been set at 0.1 ppm for diborane, 0.5 ppm for decaborane, and 0.01 ppm for pentaborane (Browning, 1969).

## HIGH LEVELS

### *Oral*

The lethal dose of boric acid varies according to the species. In animals it has been reported (Pfeiffer *et al.,* 1945) to range from 1.2 to 3.45 g (210 to 603 mg boron) per kilogram of body weight, and death is due to central nervous paralysis and gastrointestinal irritation (Buzzo and Ceratola, 1932). In human adults, the single toxic dose of boric acid has been reported to vary from 20 to 45 g (Potter, 1921). Infant deaths have been reported from single feedings of saturated boric acid solutions containing 1 to 6 g (McNally and Rust, 1928; Young *et al.,* 1949). Initial symptoms include nausea, emesis (sometimes with blood), abdominal pain, and diarrhea. A generalized erythematous rash, or even exfoliation, may follow. In severe cases, shock with low blood pressure, tachycardia, and cyanosis may be seen. Death appears due to central nervous system depression. Necropsy signs include cloudy swelling of the kidneys, centrolobular hepatic necrosis, and hemorrhagic enteritis.

### *Inhalation*

The $LD_{50}$ of diborane for rats is about 50 ppm (Wills, 1953). Earliest signs of nonlethal exposure include respiratory embarrassment, followed by a slight fall in blood pressure, increased intestinal contraction, initial stimulation, and subsequent depression of the cerebral cortex, bradycardia, and ventricular fibrillation. Stumpe (1960) exposed golden hamsters to 50–600 ppm of diborane and found mean survival time decreased with increasing concentration.

The $LD_{50}$ of decaborane for mice (exposed 4 hours) was about 36 ppm (Svirbely, 1954a). Rats were initially resistant to this level, but upon further exposure the $LD_{50}$ was found to range from 32 to 84 ppm. Principal signs included restlessness, depressed respiration, incoordination, weakness, spasmodic movements, convulsions, and corneal opacities. Decaborane inhaled by dogs led to bradycardia and periods of moderate hypertension preceding the terminal fall in blood pressure (Walton *et al.*, 1955).

The $LD_{50}$ of pentaborane for mice (exposed 4 hours) was about 11 ppm and about 18 ppm for rats. Acute neurotoxicosis was evident, and signs included restlessness, tremors, spasms, and convulsions. Corneal opacities were seen at necropsy.

Acute human intoxication following exposure to boron hydrides was first reported by Rozendaal (1951). At that time it was not known which compounds were responsible, and some affected individuals were exposed to at least diborane and pentaborane. Symptoms reported included generalized muscular cramps, mental confusion, disorientation, loss of memory, exhaustion, shortness of breath, chills, fever, and spasmodic seizures.

### *Injection*

Walton *et al.* (1955) administered decaborane to dogs by subcutaneous, intraperitoneal, or intravenous injection and produced toxicosis. Intravenous injection produced an epinephrinelike response, while all routes of administration resulted in signs like those from high-level inhalation. Decaborane (30 mg/kg of body weight) in corn oil was injected intraperitoneally into rabbits by Merritt *et al.* (1964). Increased irritability, then lethargy, and finally loss of response to sensory stimuli developed over a 3- to 6-hour interval. All rabbits died in less than 24 hours. Cole *et al.* (1954) found that mice injected with suspensions of decaborane in alcohol or gelatin, or with aqueous sodium bicarbonate solutions of decaborane, were most severely affected by the alcohol suspension.

### TOPICAL ABSORPTION

Pfeiffer *et al.* (1945) reported that the application of a 5 percent solution of boric acid to damaged skin can result in systemic toxicosis. As much as 2.0 to 2.5 g were found at necropsy by Cope (1943) in patients treated for severe burns with saturated boric acid solution.

### FACTORS INFLUENCING TOXICITY

Boric acid applied to intact skin will not be absorbed sufficiently to cause systemic toxicosis, while such application to damaged skin may result in intoxication and death. The boranes are appreciably more toxic than boric acid or soluble borates, and pentaborane is the most hazardous of all. The central nervous system toxicosis of decaborane was most pronounced in rabbits, intermediate in rats, and least in dogs.

### THERAPY

Removal from exposure is important. Hill and Svirbely (1954) found that protection from decaborane vapors could be sustained for several hours by a conventional chemical cartridge respirator filled with silica gel. Acute diborane intoxication has been treated by Cardasco *et al.* (1962) by oxygen or intermittent positive pressure breathing. Chronic cases were also treated with bronco-dilators and expectorants. Merritt (1965) has recommended methylene blue for decaborane poisoning on the basis of its oxidizing effect, which counters the reducing potential of boron hydrides. Cole *et al.* (1954) reported a favorable response to methylene blue when a bicarbonate solution of decaborane had been administered intravenously to mice. However, the toxicosis resulting from intravenous administration of an alcoholic suspension of decaborane yielded best to a combination of atrolactamide and sodium lactate.

## TISSUE LEVELS

Boron concentrations in most soft animal tissues range from 0.5 to 1.5 ppm (dry basis) and severalfold higher in bones. Hamilton *et al.* (1972/1973) reported the following mean boron concentrations (parts per million wet basis) in human tissues: blood, 0.4; liver, 0.2; kidney, 0.6; muscle, 0.1; brain, 0.06; testis, 0.09; lung, 0.6; and lymph nodes, 0.6. Human ribs from hard water areas in England contained 10.2 ppm boron in the ash, while ribs from soft water areas contained 6.2 ppm boron in the ash. Human dental enamel contained 0.5 to 69.0 ppm boron (dry basis) with a mean of 18.2 ppm (Losee *et al.*, 1973). Ingestion of large amounts of boric acid increased tissue boron levels, particularly those in the brain (Pfeiffer *et al.*, 1945).

The boron concentration of cow's milk normally ranges from 0.5 to 1.0 ppm (Hove *et al.*, 1939; Owen, 1944), but these values can be altered by dietary boron intake. Owen (1944) increased milk boron concentration from 0.7 to over 3.0 ppm by adding 20.0 g of borax (3.5 g boron) daily to the cow's diet.

Green and Weeth (1977) fed Hereford heifers grass hay containing 38.3 ppm boron and water containing 0.8 ppm boron, or water containing 150 or 300 ppm boron (added as borax). The three levels of boron in water produced the following respective levels of boron in plasma (milligrams per deciliter): 0.05, 1.12, or 1.89. Total daily boron intakes were 0.7, 15.3, or 26.0 mg/kg of body weight, respectively. Daily urinary boron excretion was 0.2, 10.3, or 17.9 mg/kg of body weight, respectively.

## MAXIMUM TOLERABLE LEVELS

Boron has been added to lactating dairy cattle diets at 145 to 157 ppm (in the form of borax) with no adverse effects (Owen, 1944). Additions of 150 ppm to water consumed by yearling cattle decreased feed consumption and produced weight loss, edema, and inflammation of the legs (Green and Weeth, 1977). Both studies were short term (42 and 30 d, respectively). Based on these limited experimental data with cattle and field experience with high-boron water, a maximum tolerable level of 150 ppm boron (as borax) in the dry diet of cattle is suggested. Extrapolation of this level to other species seems reasonable, based on data with laboratory animals.

## SUMMARY

Boron is a dark brown, nonmetal that occurs in nature as borax, colemanite, boronatrocalcite, and boracite. It is used (in a variety of forms) to harden steel, to absorb neutrons in atomic energy plants, in deoxidation of bronze, in the manufacture of glass and porcelain enamels, as a fire-proofing agent, in pharmaceuticals, and as high-energy fuel. Animal toxicosis is primarily an experimental phenomenon, although livestock in certain regions may be exposed to high-boron water (up to 80 ppm) that has not been shown to be toxic. Toxicosis in humans has resulted from ingestion of boric acid or borax solutions, topical application of boric acid solutions to burn-damaged skin, and inhalation of boranes. Symptoms of illness include anorexia, nausea, emesis, diarrhea, cardiac weakness, and an erythematous rash when the toxicosis results from boric acid or borax. Borane toxicosis by inhalation causes pulmonary irritation, dyspnea, pneumonia, headache, vertigo, nausea, muscular pain, impaired cardiac function, and central nervous system depression.

Boron is required by plants, but it has no known function in animals.

TABLE 9 Effects of Boron Administration in Animals

| Class and Number of Animals[a] | Age or Weight | Administration | | | | Effect(s) | Reference |
|---|---|---|---|---|---|---|---|
| | | Quantity of Element[b] | Source | Duration | Route | | |
| Cattle—2 | Adult in mid-lactation | 145–157 ppm | Borax | 42 d | Diet | No adverse effect | Owen, 1944 |
| Cattle—12 | Yearling, 290 kg | 150 ppm | Borax | 30 d | Water | Decreased feed consumption and weight loss; edema and inflammation of legs | Green and Weeth, 1977 |
| | | 300 ppm | | | | Decreased feed consumption and weight loss; edema and inflammation of legs; lethargy; slight diarrhea | |
| Rat—6 | Young, 30 d | 75 ppm | Borax | 45 d | Water | No adverse effect | Green *et al.*, 1973 |
| | | 150 ppm | | | | Reduced testis size; irreversible pathology | |
| Rat—11 | Weanlings, 21 d | 300 ppm | Borax | 49 d | Water | Reduced body size; continued prepubescent fur; lack of incisor pigmentation; aspermia; and impaired ovarian development | Green *et al.*, 1973 |
| Rat—6 | Young, 30 d | 300 ppm | Borax | 45 d | Water | Same as above | Green *et al.*, 1973 |

| | | | | | | | |
|---|---|---|---|---|---|---|---|
| Rat—15 | Adult | 18 ppm | Decaborane | 6 h/d, 5 d/wk for 21 exposures | Air | Nervousness; restlessness; weight loss; unsteadiness; tendency to belligerency; head tremors; convulsions | Svirbely, 1954b |
| | | 3 ppm | Pentaborane | 5 h/d, 5 d/wk up to 10 exposures | Air | As above, but neurological effects more marked especially belligerency | |
| Rat | | 50 ppm | Diborane | | Air | Respiratory embarrassment; slight fall in blood pressure; increased intestinal contraction; initial stimulation and subsequent depression of cerebral cortex; bradycardia; ventricular fibrillation; death of 50% | Wills, 1953 |
| Guinea pig—6 | Adult | 18 ppm | Decaborane | 6 h/d, 5 to 6 exposures | Air | Dyspnea; eye inflammation and exudate; listlessness; convulsions; emaciation | Svirbely, 1954b |

[a] Number of animals per treatment.
[b] Quantity expressed in parts per million.

## REFERENCES

Beeson, K. C. 1941. The Mineral Composition of Crops with Particular Reference to the Soil. U.S. Dept. Agric. Misc. Publ. 369.

Bradford, G. R. 1966. Boron, pp. 33–61. *In* H. D. Chapman, ed. Diagnostic Criteria for Plants and Soils. University of California, Berkeley.

Bradford, G. R. 1971. Trace elements in the water resources of California. Hilgardia 41:45.

British Medical Association Public Health Committee. 1966. Borax and boric acid. Br. Med. J. 2:188.

Browning, E. 1969. Boron, pp. 90–97. *In* Toxicity of Industrial Metals, 2nd ed. Butterworth & Co., Ltd., London. 383 pp.

Buzzo, A., and R. D. Ceratola. 1932. The toxicity of boric acid and of borates used as preservatives and antiseptics. Rev. Assoc. Med. Argent. 46:1493.

Cardasco, A. M., R. W. Cooper, C. Anderson, and J. V. Murphy. 1962. Pulmonary aspects of some toxic experimental space fluids. Dis. Chest 41:68.

Cole, V. V., D. L. Hill, and A. H. Oikemos. 1954. Problems in study of decaborane and possible therapy of its poisoning. Arch. Ind. Hyg. 10:158.

Cope, O. 1943. Care of victims of the Cocoanut Grove fire at the Massachusetts General Hospital. N. Engl. J. Med. 229:138.

Environmental Protection Agency. 1973. Proposed Criteria for Water Quality, vol. 1. U.S. Environmental Protection Agency, Washington, D.C.

Follis, R. H., Jr. 1947. Effect of adding boron to potassium-deficient diet in rat. Am. J. Physiol. 150:520.

Forsyth, D. 1919. Coeliac disease or boric acid poisoning. Lancet 2:728.

Green, G. H., and H. J. Weeth. 1977. Responses of heifers ingesting boron in water. J. Anim. Sci. 46:812.

Green, G. H., M. D. Lott, and H. J. Weeth. 1973. Effects of boron-water on rats. Proc. West. Sec. Am. Soc. Anim. Sci. 24:254.

Hamilton, E. I., M. J. Minski, and J. J. Cleary. 1972/1973. The concentration and distribution of some stable elements in healthy human tissues from the United Kingdom. Sci. Total Environ. 1:341.

Hill, W. H., and J. L. Svirbely. 1954. Gas mask protection against decaborane. Arch. Ind. Hyg. 10:69.

Hove, E., C. A. Elvehjem, and E. B. Hart. 1939. Boron in animal nutrition. Am. J. Physiol. 127:689.

Kahlenberg, L. 1924. On the passage of boric acid through the skin by osmosis. J. Biol. Chem. 62:149.

Kent, N. L., and R. A. McCance. 1941. The absorption and excretion of "minor" elements by man. I. Silver, gold, boron and vanadium. Biochem. J. 35: 837.

Krauskopf, K. B. 1972. Geochemistry of micronutrients. *In* J. J. Mortvedt, P. M. Giordano, and W. L. Lindsay (eds.). Micronutrients in Agriculture. Soil Science Society of America, Inc., Madison, Wis.

Losee, F., T. E. Cutress, and R. Brown. 1973. Trace elements in human dental enamel. *In* D. D. Hemphill (ed.). Trace Substances in Environmental Health—VII. University of Missouri, Columbia.

Lowe, H. J., and G. Freeman. 1957. Boron hydride (borane) intoxication in man. Arch. Ind. Health 16:523.

McNally, W. D., and C. A. Rust. 1928. The distribution of boric acid in human organs in 6 deaths due to boric acid poisoning. J. Am. Med. Assoc. 90:382.

Merritt, J. H., E. J. Schultz, and A. A. Wykes. 1964. Effect of decaborane on the norepinephrine content of rat brain. Biochem. Pharmacol. 13:1364.

Merritt, J. A., Jr. 1965. Methylene blue in treatment of decaborane toxicity. Arch. Environ. Health 10:452.

Neubert, P., W. Wrazidlo, H. P. Vielemeyer, I. Hundt, F. Gullmick, and W. Bergmann. 1969. Tabellen zur Pflanzenanalyzre-Erste orientierende Ubersicht. Institut fur Pflanzenernährung Jena, Berlin. 30 pp.

Ochsner, E. H. 1917. Biochemistry of topical application. J. Am. Med. Assoc. 68:220.

Orent-Keiles, E. 1941. The role of boron in the diet of the rat. Proc. Soc. Exp. Biol. Med. 44:199.

Owen, E. C. 1944. The excretion of borate by the dairy cow. J. Dairy Res. 13:243.

Pfeiffer, C. C., L. F. Hallman, and I. Gersh. 1945. Boric acid ointment: Possible intoxication in treatment of burns. J. Am. Med. Assoc. 128:266.

Potter, C. 1921. A case of borax poisoning. J. Am. Med. Assoc. 76:378.

Rozendaal, H. M. 1951. Clinical observations on the toxicology of boron hydrates. Arch. Ind. Hyg. 4:257.

Sanders, J. H. 1912. Boracic acid poisoning. Br. Med. J. 1:605.

Schlettwein-Gsell, D., and S. Mommsen-Straub. 1973. Übersicht Spurenelemente in Lebensmitteln. IX. Bor. Int. J. Vit. Nutr. Res. 43:93.

Stumpe, A. R. 1960. Toxicity of diborane in high concentration. Arch. Ind. Health 21:519.

Svirbely, J. L. 1954a. Acute toxicity studies of decaborane and pentaborane by inhalation. Arch. Ind. Hyg. 10:298.

Svirbely, J. L. 1954b. Subacute toxicity of decaborane and pentaborane. Arch. Ind. Hyg. 10:305.

Teresi, J. D., E. Hove, C. A. Elvehjem, and E. B. Hart. 1944. Further study of boron in the nutrition of the rat. Am. J. Physiol. 140:513.

Tipton, I. H., P. L. Stewart, and P. G. Martin. 1966. Trace elements in diets and excreta. Health Phys. 12:1683.

Underwood, E. J. 1977. Trace Elements in Human and Animal Nutrition, 4th ed. Academic Press, New York.

Walton, R. P., J. A. Richardson, and O. J. Brodie. 1955. Cardiovascular actions of decaborane. J. Pharmacol. Exp. Ther. 114:367.

Weser, U. 1967. Stimulation of rat liver RNA synthesis by borate. Proc. Soc. Exp. Biol. Med. 126:669.

Wills, J. H. 1953. Toxicity and Pharmacology of Boron Hydrides. Spec. Rep. No. 15. U.S. Army Chemical Center Chem. Corps Med. Labs.

Witthaus, R. A. 1911. Manual of Toxicology. William Wood and Co., New York.

World Health Organization Expert Committee. 1973. Trace Elements in Human Nutrition. WHO. Tech. Rep. Ser. 532.

Young, E. G., R. P. Smith, and O. C. Mackintosh. 1949. Boric acid as a poison. Can. Med. Assoc. J. 61:44.

Zook, E. G., and J. Lehmann. 1968. Mineral composition of fruits. J. Am. Diet. Assoc. 52:225.

# Bromine

Bromine (Br) is the only nonmetallic element that is a liquid at ambient temperature and pressure. As summarized by Standen (1964), bromine is widely distributed in nature but in relatively small proportions. Thus, the earth's crust contains an average of 1.6 ppm bromine. Natural minerals containing this element consist of certain rare silver halides such as bromyrite, embolite, and iodobromite. The bromine available for commercial exploitation occurs in the oceans, in water of closed basins (salt lakes), and in brines or salt deposits. The bromine content of ocean water averages 65 ppm, whereas the bromine content of the other aqueous sources, such as salt lakes and brine wells, can be as high as 5,600 ppm. In 1977, the domestic production of bromine amounted to about 200 million kg (U.S. Department of the Interior, 1977). Bromine derivatives occur in gasoline, fumigation compounds, fire-retardant materials, pharmaceutical preparations, dyes, and other chemicals. Biological impacts of bromine could arise from its occurrence as a natural constituent of soils, plants, and animals and from its use in fumigants and pharmaceutical preparations.

## ESSENTIALITY

Literature evidence for the essentiality of bromine in animals is conflicting. Winnek and Smith (1937) did not observe a bromine deficiency in rats fed purified diets containing about 0.5 ppm bromine. Bromine supplementation, as KBr, at levels of 20 ppm did not significantly affect the rate of growth, feed intake, or the general appearance of the treated rats. In a similarly conducted gestation study, no effects on reproductive parameters were observed that could be attributed to the various dietary bromine treatments. Conversely, a nutritional requirement for bromine in mice and chickens, respectively, was postulated by Huff *et al.* (1956) and Bosshardt *et al.* (1956). In the mice studies, bromine, as KBr, at 3.75 ppm was effective in reversing the growth inhibition caused by the dietary inclusion of 2 percent iodinated casein. In the poultry studies, day-old chicks were reared for 31 days on diets supplemented with bromine, as NaBr, at 8 and 15 ppm. Improvements in growth attributed to the supplemental bromine ranged from 8 to 10 percent. In swine, although the study was not designed to assess essentiality, Barber *et al.* (1971) did not observe any improvement in productive performance when bromine (200 ppm as an equimolar mixture of $NH_4Br$, KBr, and NaBr) was fed throughout the growing–finishing period.

The status of bromine in animal nutrition is summarized best by Mertz (1970) in that: "growth responses to bromine supplementation have been observed in chickens and mice, but the essentiality, biological function, or mode of action of the element have not yet been unequivocally proved."

## METABOLISM

Excellent reviews on the metabolism of bromine in animals have been published by Gross (1962) and Underwood (1977).

Winnek and Smith (1937) measured the distribution of bromine in the bodies of rats reared on diets containing either 0.5 or 20.0 ppm of bromine, as KBr. At the end of the 120- to 200-day test period, the tissues of the rats fed the 0.5 ppm diet contained bromine levels, expressed on a dry weight basis, that varied from 3.5 ppm for liver to 32.0 ppm for spleen. Tissues from the rats treated with 20 ppm contained more bromine than those from the rats fed the low-bromine diet with corresponding values for liver and spleen being 120 and 190 ppm, respectively. Cole and Patrick (1958) examined the uptake and excretion of a radioactive isotope of bromine, as $K^{82}Br$, in young rats that had been reared for 2 weeks on a diet low in bromine. The radiotracer was given intraperitoneally, and animals were sacrificed for tissue analysis at various time intervals postinjection. After 24 hours, the pancreas contained the highest relative amount of the radiotracer and the brain contained the lowest. Overall differences were small with the percentages of administered dose per gram of tissue (wet weight) varying from 0.5 for brain to 1.4 for spleen. All tissues were essentially depleted of radiotracer by 72 hours postinjection. Bromine excretion, as evidenced by the quantity of radiotracer in the urine, was found to occur principally through the kidney. Rauws (1975) has summarized the pharmacokinetics of the bromide ion in animal systems by concluding that it is completely absorbed in the gastrointestinal tract, it is distributed primarily in the extracellular fluid (like chloride), it penetrates the blood–brain barrier, and it is excreted mainly via the kidney.

## SOURCES

Bromine is ubiquitous in nature, and consequently it is an ingredient of all feedstuffs. Bowen (1966) has stated that the average bromine content of soils is 5 ppm and that this figure for land plants is 15 ppm. Other dietary sources of bromine would include salt that has been prepared from brines containing relatively high levels of bromine.

## TOXICOSIS

### LOW LEVELS

Bromine toxicity per se has not been studied in large animals. In a study of lactating dairy cows, Lynn *et al.* (1963) did not observe an adverse effect on milk production when bromine, as NaBr, was fed in sequential dosages of 9.5, 19.0, and 38.0 ppm for a 72-day period. Likewise in swine, Barber *et al.* (1971) observed that 200 ppm of bromine, as a mixture of inorganic bromides, did not adversely affect rate of gain, feed conversion, dressing percentage, or several other carcass quality parameters.

### HIGH LEVELS

In chicks, Doberenz *et al.* (1965) noted that a dietary concentration of bromine, as NaBr, of 20,000 ppm caused death by 2 weeks of age. Dosages of 5,000 and 10,000 ppm resulted in a reduced rate of gain, while a dosage of 2,500 ppm was without effect on production parameters. In a series of studies in rats, Van Logten *et al.* (1973, 1974) noted that at dietary levels of bromine, as NaBr, of 19,200 ppm the treated rats did not groom themselves normally and showed signs of motor incoordination of the hind legs. At this level, relative kidney weights were increased. No clear effects on growth rate, food intake, or histology were observed.

### FACTORS INFLUENCING TOXICITY

The primary factor that can influence bromine toxicity is chloride. Winnek and Smith (1937) demonstrated, in rats, that the Br:Cl ratio of the diet influenced the amount of bromine deposited in the tissues. Czerwinski (1958) showed that chloride administration to rabbits that had induced acute and chronic bromism caused a threefold increase in the rate of bromine excretion. In a similar manner, Rauws and Van Logten (1975) showed that the biological half-life of bromine (assessed by measuring blood levels) in rats pretreated with diets containing 2,000 ppm bromine and then treated with sodium chloride in the drinking water approximated 25.1, 12.0, 6.9, and 2.5 days. The levels of salt causing these decreases in half-life were, correspondingly, 0, 80, 200, and 600 ppm.

## TISSUE LEVELS

Lynn *et al.* (1963) found that dairy cows fed 9.5 to 38.0 ppm of bromine, as NaBr, resulted in milk residues that ranged from 1 to 12 ppm on a wet weight basis.

In regard to humans, FAO/WHO (1966) has established an acceptable daily intake of 1 mg bromine, as inorganic bromides, per kg of body weight. The Food and Drug Administration (1976) has established a tolerance of 125 ppm in or on processed foods with certain exceptions and when it occurs by use of certain specified chemicals.

## MAXIMUM TOLERABLE LEVELS

Data on which to base a maximum tolerable level for bromine in animal feeds are sparse. Growing pigs tolerated 200 ppm, growing chickens tolerated 5,000 ppm, and growing rats tolerated 4,800 ppm without adverse effects. Based on these data and the interrelationships of bromine and chlorine, no adverse effects would be expected in poultry consuming 2,500 ppm of bromine and in other animals consuming 200 ppm of bromine.

## SUMMARY

Bromine is widely distributed in nature with land plants having an average content of about 15 ppm. Evidence to support its nutritional essentiality is inconclusive, although a few workers have noted growth effects from dietary supplementation. Generally, bromine competes with chlorine in metabolism, and ingestion of inorganic bromides causes a generalized distribution in the tissue extracellular fluid. It is rapidly excreted, primarily through the kidney, so that tissue concentrations soon decline to normal upon cessation of administration.

TABLE 10 Effects of Bromine Administration in Animals

| Class and Number of Animals[a] | Age or Weight | Administration: Quantity of Element[b] | Source | Duration | Route | Effect(s) | Reference |
|---|---|---|---|---|---|---|---|
| Cattle—4, lactating, dairy | Unspecified | Up to 43 ppm, 9.5 ppm 22 d, 19 ppm 20 d, 38 ppm 30 d | NaBr | 72 d | Diet | No adverse effects | Lynn *et al.*, 1963 |
| Swine—8 | 20 kg | 200 ppm | Equimolar mixture of NaBr $NH_4Br$ KBr | To 90 kg body weight | Diet | No adverse effects | Barber *et al.*, 1971 |
| Chicken—14 | 1 d | 8 ppm | NaBr | 31 d | Diet | No adverse effects | Bosshardt *et al.*, 1956 |
| Chicken—29 | | 8 ppm | | | | No adverse effects | |
| Chicken—14 | | 15 ppm | | | | No adverse effects | |
| Chicken—26 | | 15 ppm | | | | No adverse effects | |
| Chicken—29 | | 15 ppm | | | | No adverse effects | |
| Chicken—20 | 1 d | 2,500 ppm | NaBr | 28 d | Diet | No adverse effects | Doberenz *et al.*, 1965 |
| | | 5,000 ppm | | | | No adverse effects | |

TABLE 10 *Continued*

| Class and Number of Animals[a] | Age or Weight | Administration | | | | Effect(s) | Reference |
|---|---|---|---|---|---|---|---|
| | | Quantity of Element[b] | Source | Duration | Route | | |
| Chicken—30 | | 5,000 ppm | | | Diet (low fat) | Reduced rate of weight gain | |
| Chicken—90 | | 10,000 ppm | | | Diet | Reduced rate of weight gain and increased death rate | |
| Chicken—20 | | 20,000 ppm | | | | 100 percent death rate | |
| Rat—4 | 110–130 g | 300 ppm | NaBr | 28 d | Diet | No adverse effects | Van Logten *et al.*, 1973 |
| | | 1,200 ppm | | | | No adverse effects | |
| | | 4,800 ppm | | | | No adverse effects | |
| | | 19,200 ppm | | | | Increased relative kidney weight; poor grooming | |

| | | | | | | | |
|---|---|---|---|---|---|---|---|
| | | | | | | behavior; motor incoordination in hind legs | |
| Rat—20 | 50–60 g | 75 ppm | NaBr | 90 d | Diet | No adverse effects | Van Logten *et al.*, 1974 |
| | | 300 ppm | | | | No adverse effects | |
| | | 1,200 ppm | | | | Increased relative thyroid weight | |
| | | 4,800 ppm | | | | Increased relative thyroid weight | |
| | | 19,200 ppm | | | | Increased relative thyroid, spleen, and adrenal weights; poor grooming behavior; motor incoordination in hind legs; growth retardation; feed conversion decrease; some histopathology | |

[a] Number of animals per treatment.
[b] Quantity expressed in parts per million.

## REFERENCES

Barber, R. S., R. Braude, and K. G. Mitchell. 1971. Arsanilic acid, sodium salicylate and bromide salts as potential growth stimulants for pigs receiving diets with and without copper sulfate. Br. J. Nutr. 25:381.

Bosshardt, D. K., J. W. Huff, and R. H. Barnes. 1956. Effect of bromine on chick growth. Proc. Soc. Exp. Biol. Med. 92:219.

Bowen, H. J. M. 1966. Trace Elements in Biochemistry. Academic Press, New York and London.

Cole, B. T., and H. Patrick. 1958. Tissue uptake and excretion of bromine-82 by rats. Arch. Biochem. Biophys. 74:357.

Czerwinski, A. L. 1958. Bromide excretion as affected by chloride administration. J. Am. Pharm. Assoc. 47:467.

Doberenz, A. R., A. A. Kurnick, B. J. Hulett, and B. L. Reid. 1965. Bromide and fluoride toxicities in the chick. Poult. Sci. 44:1500.

FAO/WHO. 1966. Evaluation of Some Pesticide Residues in Food, p. 115. FAO, PL: CP/15 WHO/Food Add. 67.32.

Food and Drug Administration. 1976. Inorganic Bromide, Ch. I, p. 608. Title 21, Code of Federal Regulations.

Gross, J. 1962. Iodine and Bromine. *In* C. L. Comar and F. Bronner (eds.). Mineral Metabolism, vol. II, part B, pp. 221–285. Academic Press, New York and London.

Huff, J. W., D. K. Bosshardt, O. P. Miller, and R. H. Barnes. 1956. A nutritional requirement for bromine. Proc. Soc. Exp. Biol. Med. 92:216.

Lynn, G. E., S. A. Shrader, O. H. Hammer, and C. A. Lassiter. 1963. Occurrence of bromides in the milk of cows fed sodium bromide and grain fumigated with methyl bromide. J. Agric. Food Chem. 11:87.

Mertz, W. 1970. Some aspects of nutritional trace element research. Fed. Proc. 29:1482.

Rauws, A. G. 1975. Bromide pharmacokinetics: A model for residue accumulation in animals. Toxicology 4:195.

Rauws, A. G., and M. J. Van Logten. 1975. The influence of dietary chloride on bromide excretion in the rat. Toxicology 3:29.

Standen, A. (ed.). 1964. Kirk-Othmer Encyclopedia of Chemical Technology, vol. 3. John Wiley & Sons, New York, London, Sydney.

Underwood, E. J. 1977. Trace Elements in Human and Animal Nutrition. Academic Press, New York and London.

U.S. Department of the Interior. 1977. Bureau of Mines Minerals Yearbook, Bromine chapter.

Van Logten, M. J., M. Wolthius, A. G. Rauws, and R. Kroes. 1973. Short-term toxicity study on sodium bromide in rats. Toxicology 1:321.

Van Logten, M. J., M. Wolthius, A. G. Rauws, R. Kroes, E. M. Don Tonkelaar, H. Berkvens, and G. J. Van Esch. 1974. Semichronic toxicity study of sodium bromide in rats. Toxicology 2:257.

Winnek, P. S., and A. H. Smith. 1937. Studies on the role of bromide in nutrition. J. Biol. Chem. 121:345.

# Cadmium

Cadmium (Cd) is a silvery white metal that resembles aluminum. Cadmium constitutes about 0.000011 percent of the earth's crust. It occurs geologically in zinc ores and is a by-product of zinc production, either as a vapor from roasting zinc ores or as a sludge from zinc sulfate purification. Several cadmium salts possess commercially desirable colors, ranging from yellow through oranges and reds. Their heat resistance makes them useful for ceramics, enamelware, and plastics. Although the metal oxidizes readily, it is highly resistant to corrosion and is widely used to plate iron and steel. Cadmium is also used in solders, nickel–cadmium batteries, and in stabilizers for polyvinyl chloride. The U.S. industrial uses (in metric tons) of cadmium in 1974 were as follows: batteries, 544; pigments, 997; stabilizers, 906; plating, 2,718; and others (including alloys), 441 (Stubbs, 1978). Mining and smelting operations represent point sources of environmental pollution; however, the widespread sources are corrosion of metal-plated iron, discarded cadmium-containing consumer products, and losses from industrial operations such as plating baths. Inorganic fertilizers contain some cadmium; however, the use of urban sewage sludges to fertilize pastures or food croplands represents a potentially serious source of cadmium access to animal and human foods. Various aspects of cadmium flow in the United States were described by Yost (1979).

Similarities in chemical reactivity of zinc and cadmium lead to similar metabolic pathways in biological systems. Whereas zinc is an important essential element, cadmium is best known for its toxicity and metabolic antagonisms of zinc and other essential elements. Anemia, bone demineralization, and kidney damage are the principal adverse effects of cadmium ingested in "moderate" amounts. Higher levels can lead to death. As evidence of long-term adverse effects of cadmium in man has accumulated, any cadmium pollution of the environment is recognized as a potentially serious health hazard to man.

The biological importance of cadmium has been considered from a variety of aspects in several reviews (Schroeder and Balassa, 1961; Schroeder *et al.,* 1967; Flick *et al.,* 1971; Friberg *et al.,* 1971, 1974; Fassett, 1972; Neathery and Miller, 1975; Fleischer *et al.,* 1974; Underwood, 1977).

## ESSENTIALITY

Very limited data suggest that cadmium may be an essential element. In rats fed a highly purified diet containing <0.004 ppm cadmium, a growth depression was observed when they were maintained in a metal-free environment but not under conventional laboratory conditions (Schwarz and Spallholz, 1978). The dietary concentrations that increased growth are less than those of present human diets.

In studies of hypertension, Perry *et al.* (1977) maintained their rats in stainless steel cages in a room designed to exclude airborne contaminants. Their specially prepared rye-based diet contained 0.0137 ± 0.0019 (SD) ppm cadmium. During the first 18 months, rats receiving 1 ppm cadmium in the drinking water consistently weighed at least 5 percent more than rats fed the control diet; however, a statistical analysis was not presented.

## METABOLISM

In this review, primary consideration will be given to the metabolism and effects of cadmium taken orally. An extensive literature deals with injected cadmium; however, it is not clear how applicable many of these findings are to the effects of dietary cadmium.

There appears to be no homeostatic control mechanism to limit cadmium absorption and retention below a nontoxic threshold. With $^{109}$Cd, absorption of cadmium in mice occurred irrespective of the body burden, and cadmium was not cleared from the body by subsequent cadmium dosing (Cotzias *et al.,* 1961).

The intestinal tract limits cadmium absorption as discussed below.

There was little transfer of cadmium across the placenta in mice (Berlin and Ullberg, 1963), rats (Lucis *et al.,* 1972; Pietrzak-Flis *et al.,* 1978), and cows (Neathery *et al.,* 1974). Sonawane *et al.* (1975) found a higher proportion of an injected dose of $^{109}$Cd, as the chloride, was transported across the placenta of pregnant rats late in pregnancy. The mammary gland markedly limited cadmium transport into the milk of rats (Lucis *et al.,* 1972) and cows (Miller *et al.,* 1967; Neathery *et al.,* 1974; Sharma *et al.,* 1979). Very little cadmium was transported into avian eggs (Sell, 1975; Leach *et al.,* 1979).

The duodenum of the young Japanese quail was shown to take up a high proportion of ingested cadmium (Jacobs *et al.,* 1974). The cadmium retained by the gastrointestinal tract appears to represent primarily the fraction that is most rapidly cleared from the body. This phase usually takes 4 to 12 days in rats (Moore *et al.,* 1973a; Kello and Kostial, 1977), cows (Miller *et al.,* 1968), and goats (Miller *et al.,* 1969). Small amounts of injected doses of cadmium were excreted via bile and the intestinal tract wall (Berlin and Ullberg, 1963; Cikrt and Tichý, 1974; Havrdová *et al.,* 1974; Horner and Smith, 1975; Stowe, 1976). Urinary excretion of cadmium is typically very small; however, data in humans suggest that it may be an index of body burden (Kjellström, 1979).

Absorption of cadmium was approximately 0.3 to 2.5 percent of a single oral dose in rats (Moore *et al.,* 1973b; Kello and Kostial, 1977) and 5 percent of cadmium fed in the diet for 1 week to Japanese quail (Jacobs *et al.,* 1978b). Growing sheep fed 60 ppm cadmium excreted 95 percent of their cadmium intake in the feces (Doyle *et al.,* 1974). Miller *et al.* (1969) estimated that 0.3 to 0.4 percent was retained by young goats 14 days after an oral dose of $^{109}CdCl_2$. The intestinal tract still retained a significant amount of cadmium. Similar results were obtained for dairy cows (Neathery *et al.,* 1974).

The biological half-life for whole-body retention of orally administered cadmium was reported to be 206 days for rats (Moore *et al.,* 1973b), less than 200 days for mice (Richmond *et al.,* 1966), 99 days for Chipping sparrows (Anderson and Von Hook, Jr., 1973), and 116 days for young Japanese quail (Jacobs *et al.,* 1978b). Perry *et al.* (1977) gave rats 0, 1.0, 2.5, 5.0, 10.0, 25.0, or 50.0 ppm cadmium in the drinking water for 2 years; subsets of animals were killed at 6-month intervals. Assuming that the liver and kidneys accounted for half of the absorbed cadmium in the body (as estimated by Friberg *et al.,* 1974), the rats retained 1 percent of the ingested cadmium at the end of the first year and 0.6 to 0.9 percent at the end of the second year. The proportions retained were similar for all dose levels.

Berlin and Ullberg (1963) injected $^{109}CdCl_2$ intravenously into mice and made sagittal whole-body autoradiographs of animals killed at intervals from 5 minutes to 16 days after injection. Cadmium left the blood rapidly and accumulated in the liver, kidney, gastrointestinal mucosa, salivary glands, pancreas, hypophysis, adrenal, thyroid, spleen, lymph glands, testes, hair follicles, heart, and major blood vessels. Nordberg and Nishiyama (1972) also prepared autoradiographs from whole-body sections of mice. They observed high cadmium concentrations in the liver, kidney, salivary glands, testicle, and pancreas 112 days after injection. Horner and Smith (1975) injected $^{109}CdCl_2$ into the femoral vein of rats and measured $^{109}Cd$ in 27 tissues and body fluids at intervals between 5 minutes and 60 days after injection. After 24 hours, the concentrations of cadmium in most tissues remained relatively constant; however, the $^{109}Cd$ content of the kidney gradually increased throughout the experimental period. Cadmium per gram of three muscles amounted to 0.02 to 0.03 percent of the total body burden between 1 and 60 days.

Young Japanese quail were fed a diet containing $^{115m}Cd$ as the chloride (total 1 ppm cadmium) for 1 week followed by basal diet for 50 days (Jacobs *et al.,* 1978b). At the end of the experiment, 3.58 percent of the cadmium remained in the whole bird. This was distributed as follows: liver, 24.8; kidney, 24.5; intestinal tract, 11.9; skin and feathers, 6.8; and carcass, 33.7 percent. Miller *et al.* (1969) found that 14 days after a single oral dose of $^{109}CdCl_2$ in young goats, the tissue concentrations of $^{109}Cd$ were in decreasing order as follows: kidney, liver, duodenum, and abomasum. Under similar conditions for cows, Neathery *et al.* (1974) found the following sequence: kidney, liver, and small intestine. The bovine fetal tissues with the highest concentrations were kidney, tibia, and liver. Numerous investigators have demonstrated the rapid concentrating effect of cadmium in the liver and kidney and the gradual shift of cadmium from other tissues to the kidney. Cross-sectional studies of human populations in the United States, Sweden, and Japan showed 2.7- to 5.6-fold increases of cadmium in muscle between adolescence and middle age (Kjellström, 1979). The biological half-time of cadmium-109 in muscle, 77 days, was longer than that for liver and kidney, 65 days, in adult male Japanese quail (A. O. L. Jones, FDA, Washington, D.C., 1979, personal communication).

Margoshes and Vallee (1957) were the first to isolate an unusual protein with a high metal content from equine renal cortex. Isolation procedures and characterization of the protein, metallothionein, have been studied by several workers (Kägi and Vallee, 1960; Pulido *et al.,* 1966; Shaikh and Lucis, 1971; Nordberg *et al.,* 1972). In 1974 Kägi *et al* (1974) described the isolation and properties of equine hepatic and renal metallothioneins. The molecular weight of each was about 6,600. Both were high in cysteine (approximately one-third of the amino acid residues), whereas phenylalanine, tyrosine, tryptophan, and histidine were absent. Both proteins exhibited considerable microheterogeneity even within tissues from the same animal. The total metal content (cadmium, zinc, and copper) for each protein was 6 g atoms per mole, or one metal atom per three cysteinyl residues. Zinc was the predominant metal in the liver protein, and cadmium predominated in the kidney protein. Bremner (1978) reviewed information on the distribution of metallothionein in various tissues and its involvement in cadmium metabolism.

## SOURCES

The industrial uses of cadmium provide the largest sources of environmentally hazardous amounts of cadmium. The points at which pollution is most apt to occur begin with mining and smelting, followed by manufacturing, loss from manufactured products during use and when discarded, and the reclamation and use of waste products contaminated with cadmium. The air, water, and soil provide pathways by which cadmium may be dissipated and enter animals and man either directly or via the food chain. Numerous aspects of these problems have been reviewed (Fassett, 1972; Fleischer *et al.*, 1974; and Friberg *et al.*, 1971, 1974).

Underwood (1977) reviewed the effects of pollution on the cadmium content of animal feeds. Pollution has been shown to increase the cadmium content of mixed pasture herbage by more than 40-fold. A considerable portion of the cadmium taken up is retained by the root system. Some plants, such as clover, have a special capacity for concentrating cadmium from the soil. Parts of the seed containing highest concentrations of cadmium may be removed during milling to produce somewhat lower cadmium concentrations in products for human use; however, many of the cadmium-rich by-products may be used in animal feeds. Urban sewage sludges contain significant amounts of cadmium. Use of high-cadmium sludges for fertilizing either animal or human food croplands has been shown to increase substantially the cadmium content of animal and human foods (Council for Agricultural Science and Technology, 1976). Superphosphate fertilizers contain some cadmium, but they do not supply nearly as much cadmium as equivalent fertilizer application of most sludges. There are great variations in the capacity of different plant species, varieties, and tissues to take up cadmium. In general, the leaves contain the highest amounts. Leafy vegetables can contain over 100 ppm cadmium (dry weight) without evidence of toxicosis in the plant (Chaney and Hornick, 1978). It was shown that cadmium taken up by Swiss chard and romaine lettuce from sludge was available for absorption by guinea pigs (Furr *et al.,* 1976) and mice (Chaney *et al.,* 1978b), respectively. The seeds of some plants, such as corn, contain less cadmium than the leaves, whereas for others, such as wheat or soybeans, the quantities are similar. Cadmium uptake is greater at lower soil pH, at lower soil organic matter, and at higher soil temperature. Most forages and plant materials fed to animals contain levels of cadmium well below 0.5 ppm on a dry weight basis (Table 12; Underwood, 1977; Baker *et al.,* 1979; Chaney *et al.,* 1978a).

Feed phosphates from Florida deposits typically contain 6 to 7 ppm cadmium (D. J. Thompson, International Minerals and Chemical Corporation, 1979, personal communication). Since these phosphates are present at levels of approximately 1 percent in finished feeds, they contribute 0.06 to 0.07 ppm cadmium to the diet.

Kopp and Kroner (1968) reported relatively low values for cadmium in 1,577 samples of surface waters collected throughout the United States. The mean for the 40 positive samples was 9.5 $\mu$g/1.

Information on tissue levels of cadmium is summarized in Table 12. The cadmium content of the unpurified basal diets, which was similar to commercial diets for domestic animals, ranged from 0.18 to 0.32 ppm. These were for cattle, swine, and chickens. The highest level, 0.7 ppm, was in a grass diet for sheep in Scotland and the lowest level was in a purified soy isolate diet for chickens which contained 0.07 ppm.

## TOXICOSIS

The effects of various oral intake levels of cadmium in animals are summarized in Table 11.

### LOW LEVELS

Most studies of cadmium have been designed to define the effects of high intakes, generally in a short period of time, rather than to establish no-effect levels. With an adequate diet, 5 ppm dietary cadmium is the level at which gross adverse effects are most apt to begin.

Mills and Dalgarno (1972) fed sheep diets with 3.5 ppm cadmium during the latter part of pregnancy and for 7.5 to 8.0 weeks following parturition. The lambs had normal copper and zinc levels in the liver at birth, but these elements were markedly depressed at the end of the experiment.

Hansen and Hinesly (1979) observed decreased liver iron and kidney manganese in swine fed 0.47 ppm cadmium in the form of corn grown on sludge-fertilized land. Controls received 0.1 ppm cadmium, the background dietary level. Hepatocyte microsomal protein and *o*-dealkylation of *p*-nitrophenitole were increased. Decreased egg production was observed by Leach *et al.* (1979) in hens fed 3 ppm cadmium in a soy isolate diet. A similar effect did not occur with a nonpurified diet. As little as 1 ppm cadmium in the diet of young Japanese quail produced acute degenerative damage to the absorptive cells of the intestinal villi (Mason *et al.*, 1977). These changes became less marked after continuous exposure for 28 or 49 days.

With 0.5 or 2.5 ppm cadmium in the drinking water, dogs ate normally and had no adverse effects during 4 years; pathological changes in the kidney occurred with 5 ppm (Anwar *et al.*, 1961). Shortened life span, kidney damage, arteriosclerosis, and ventricular hypertrophy occurred in rats receiving 5 ppm cadmium in their drinking water (Schroeder *et al.*, 1965; Kanisawa and Schroeder, 1969). Increased sodium and water retention occurred in rats receiving 5 ppm cadmium in their drinking water (Doyle *et al.*, 1975).

With 1 ppm of cadmium in the drinking water of rats, Perry *et al.* (1977) showed that the systolic blood pressure was elevated at 12 and 18 months of exposure, but not at earlier or later periods. With 2.5 ppm cadmium in the drinking water, blood pressure was elevated at 6, 12, and 18 months. Ohanian *et al.* (1978) reported that rats genetically sensitive to hypertension developed increases in blood pressure, cardiac hypertrophy, and kidney damage with 1 ppm cadmium in the drinking water. This hypertension was most pronounced with high salt intake. The liver and kidney cadmium concentrations were highest in those rats with highest blood pressure. Similar effects of cadmium and salt either did not occur or were much smaller in a line of rats resistant to hypertension. The two lines were derived from the same pool of Sprague-Dawley rats.

### HIGH LEVELS

Details of conditions that caused cadmium toxicity are presented in Table 11.

The higher levels of cadmium produced a wide range of changes in metabolic measurements that have been observed in one or more species. These include decreased serum ceruloplasmin, decreased renal leucine aminopeptidase, decreased bone ash, decreased serum albumin, increased serum $\alpha_2$-, $\beta_2$-, and $\gamma$-globulins, and increased transferrin. Tissue mineral level changes include decreased iron and copper in liver, increased zinc in the liver and kidney, increased copper in the kidney, and decreased zinc and iron in the tibia. Decreased tissue levels of zinc, copper, and iron have also been reported for the newborn.

In the young animal, cadmium can reduce growth rate. Cadmium can cause anemia (decreased hemoglobin, hematocrit, and red blood cell counts), neutrophilia, lymphocytopenia, enteropathy, renal tubular damage, bone marrow hypoplasia, decreased granulation of the adrenal medulla, hypertrophy of the heart ventricles, hypertension, and splenomegaly. Sheep fed cadmium have lost the crimp in their wool, a characteristic of copper deficiency. Shortened life span was reported for animals that were otherwise reasonably healthy.

Reproductive problems related to ingested cadmium have been produced in cattle, sheep, goats, and mice. These included abortions, deformed young, atrophy of ovaries, testicular hypoplasia, decreased egg production, decreased egg weight, and infertility (Table 11). At very high levels cadmium can cause death. The oral $LD_{50}$ for four species (dogs, rats, mice, guinea pigs) ranged from 39 to 107 mg/kg of body weight.

## FACTORS INFLUENCING TOXICITY

Since cadmium is retained in the body so long, factors that influence the absorption of small amounts of cadmium are important as well as factors that modify overt cadmium toxicity. Several physiological considerations have been shown to be important. Male rats retained less cadmium than females (Kello *et al.*, 1979). Newborn rats had a very high absorption of cadmium (Kello and Kostial, 1977; Sasser and Jarboe, 1977). The palatability of diets containing 40 ppm cadmium was not involved in toxicosis in cattle (Powell *et al.*, 1964a). Cousins *et al.* (1977) found that inanition was not a factor in most of a large number of toxic changes produced in rats by dietary cadmium.

Terhaar *et al.* (1965) reported that doses of cadmium as low as 0.01 mg/kg protected against 100 mg/kg given 24 hours later. Protection was observed with pretreatments administered between 2 weeks and 7 hours prior to the large dose. The effects of cadmium and zinc on induction of metallothionein synthesis have been studied by many investigators, including the relation of the preexisting metallothionein in protecting against acute cadmium toxicity (Leber and Miya, 1976;

Webb and Verschoyle, 1976; Probst *et al.*, 1977). The relationships of zinc and cadmium to metallothionein synthesis and degradation have been reviewed (Cousins, 1979).

Valberg *et al.* (1977) found that an oral dose of cadmium bound in thionein was taken up by the intestinal mucosa to the same extent as cadmium as the chloride. Less of the cadmium in thionein was transported out of the mucosa, and the cadmium–thionein caused severe mucosal lesions, whereas cadmium as the chloride caused little damage. The bioavailability of cadmium in animal and plant products as incorporated into feeds needs investigation.

With respect to interactions between cadmium and other nutrients, many of these result from the effects of cadmium in markedly altering tissue concentrations of several minerals. The relationships of specific nutrients to the effects of cadmium have been reviewed (Fox, 1974, 1979a; Bremner, 1978). In animals receiving marginal or deficient levels of zinc, cadmium caused either the appearance or exacerbation of the signs of zinc deficiency, including lower tissue levels of zinc. With higher levels of dietary zinc, the effects of cadmium were lessened or entirely counteracted. These studies have been carried out in turkeys (Supplee, 1961), chicks (Supplee, 1963), calves (Powell *et al.*, 1964b, 1967), goats (Powell *et al.*, 1967), and rats (Petering *et al.*, 1971).

Bunn and Matrone (1966) fed cadmium to copper-deficient rats and mice. The cadmium caused decreased weight gains and lowered hemoglobin; these effects were almost completely overcome by supplements of copper and zinc. When Campbell and Mills (1974) fed only the copper level required by the rat for growth and hemoglobin formation, as little as 1.5 ppm of cadmium in the diet caused decreased plasma ceruloplasmin. Under similar experimental conditions, Davies and Campbell (1977) found that 4.4, 8.8, and 17.6 ppm of cadmium markedly increased the uptake of $^{64}Cu$ by the duodenal mucosa; however, 17.6 ppm of cadmium decreased the absorption of the $^{64}Cu$ dose. As dietary cadmium increased, there was an inverse relation between the copper and cadmium concentrations in a low-molecular-weight protein fraction isolated from the intestinal mucosa.

Van Campen (1966) demonstrated that cadmium could interfere with copper absorption from ligated intestinal loops in rats. Starcher (1969) showed that oral administration of a very large dose of cadmium displaced copper from a low-molecular-weight mucosal protein. He suggested that the protein was important in copper absorption. Evans *et al.* (1970) reached similar conclusions from *in vitro* studies of the effect of cadmium on copper binding to bovine low-molecular-weight proteins isolated from the duodenum and liver.

Hill *et al.* (1963) fed 25 to 400 ppm of cadmium to chicks in a copper- and iron-deficient diet. Supplements of copper, zinc, and iron each partially corrected the adverse effects of cadmium. Stowe *et al.* (1974) found pyridoxine deficiency to protect against the anemia caused by cadmium in rats. Banis *et al.* (1969) showed that supplements of iron and zinc had marked protection against cadmium in the rat. Fox *et al.* (1971) showed that iron (II) was much more protective against cadmium than iron (III) and that the cadmium antagonism was primarily that for iron (III). Pond and Walker (1972) reported that injected iron prevented the anemia associated with feeding cadmium to rats. Freeland and Cousins (1973) found that cadmium decreased iron absorption in chicks. Hamilton and Valberg (1974) found that low amounts of cadmium inhibited only mucosal iron uptake, whereas larger amounts diminished both uptake and transfer of iron. During high iron absorption by iron-deficient mice, the absorption of cadmium was greater than in iron-replete animals absorbing less iron.

Vitamin C has been shown to decrease markedly the toxicoses produced by dietary cadmium in young Japanese quail as shown by improvements in the following responses: depressed growth, anemia, decreased bone mineralization, depression of spermatogenesis, bone marrow hyperplasia, and decreased granulation of the adrenal (Fox and Fry, 1970; Fox *et al.,* 1971; Richardson *et al.,* 1974). Ascorbic acid supplements normalized most of the changes in tissue mineral concentrations produced by cadmium. It appears that the primary effect of ascorbic acid was to improve iron absorption. Tissue cadmium concentrations were not affected.

Decreased bone mineralization resulting from oral cadmium intake is noted in Table 11. Itokawa *et al.* (1973) observed a marked curvature of the spinal column in rats fed cadmium in diets that were low in zinc, calcium, and protein. In a subsequent experiment with rats fed calcium-deficient diets, dietary cadmium caused renal hypertrophy, degeneration of glomeruli and tubules, thinning of cortical osseous tissue, decreased osteocytes, and decreased acid mucopolysaccharides in epiphyseal cartilage (Itokawa *et al.,* 1974).

Increased kidney and liver levels of cadmium have been found in mice fed diets low in calcium, with cadmium supplied by polluted rice (Kobayashi *et al.,* 1971). Similar effects of a low-calcium diet on cadmium levels in the kidney cortex were obtained by Larsson and Piscator (1971) and Pond and Walker (1975) in rats fed purified diets. Ando *et al.* (1977) found a marked increase in the fecal and urinary excretion of either an oral or an intravenous $^{47}Ca$ dose by rats that had received 10 mg cadmium per day by gavage. The uptake of $^{47}Ca$ by bone was decreased in cadmium-exposed rats irrespective of the route by which $^{47}Ca$ was administered.

Washko and Cousins (1976) gave a dose of $^{109}Cd$ by stomach tube to calcium-deficient rats and controls. The calcium-deficient rats had greater amounts of the cadmium dose in the intestinal mucosa, serum, lungs, liver, and urine, with less in the feces. Cadmium bound to a low-molecular-weight protein in the intestinal mucosa was present in greater amounts in calcium-deficient rats. Except for the lungs, they found no effects of calcium deficiency on the distribution of an injected dose of $^{109}Cd$. Subsequently, Washko and Cousins (1977) found a similar distribution pattern in calcium-deficient and control rats that had received cadmium in drinking water for 8 weeks. Rats fed a low-calcium diet had a greater capacity to absorb either calcium or cadmium. There was enhanced binding of $^{45}Ca$ and $^{115m}Cd$ to intestinal calcium-binding protein, and it was concluded that this was the factor responsible for increased cadmium absorption in calcium deficiency.

The relationships between calcium and cadmium are especially important because the painful cadmium-induced disease, Itai-Itai Byo, occurred in human beings consuming a low-calcium diet (Friberg *et al.*, 1971). The patients were postmenopausal women who had borne many children. The primary features of the disease were kidney damage and loss of bone mineral, which resulted in multiple fractures.

Because of its role in calcium metabolism, the effects of vitamin D on cadmium metabolism and toxicosis have also been investigated. When rachitic chicks were treated with vitamin D, Worker and Migicovsky (1961) observed an increased cadmium uptake by the tibia. Cousins and Feldman (1973) found no effect of vitamin D upon cadmium uptake in liver and kidney when the vitamin was administered to vitamin D-deficient chicks. In other experiments, they observed an *in vitro* effect of cadmium in decreasing conversion of 25-hydroxycholecalciferol to 1,25-dihydroxycholecalciferol by kidney homogenates and isolated mitochondria from vitamin D-deficient chicks. They also showed decreased conversion by kidney mitochondria from chicks that had received 50 ppm cadmium in their drinking water (Feldman and Cousins, 1973).

Suzuki *et al.* (1969) produced a rapid effect on cadmium absorption by feeding a low-protein diet to mice for 24 hours before and 24 hours after an oral dose of $^{115m}Cd$. This treatment caused a significant increase in the uptake of cadmium by the liver, kidneys, and whole body. In Japanese quail receiving 75 ppm cadmium, toxicosis was markedly less with dried egg white as the protein source than with either soy isolate or casein plus gelatin (Fox *et al.*, 1973). With very low levels of dietary cadmium, the uptake of cadmium by the small intestinal tract tissue was approximately threefold greater with the soy isolate than with the casein–gelatin diet (Fox *et al.,* 1979a). Kello and Kostial (1977) obtained greater retention of an oral dose of $^{115m}CdCl_2$ in rats fed a milk diet as compared with a nonpurified diet.

Most of the above studies were carried out with toxic levels of cadmium. Recently, Jacobs *et al.* (1978a) studied the effect in young Japanese quail of a combined supplement of zinc, copper, and manganese (30, 5, and 12 mg/kg, respectively) on uptake of dietary cadmium beginning at levels similar to those in the diet of man and domestic animals—0.02, 0.082, 0.145, 0.270, 0.520, and 1.020 ppm. Cadmium-109, as the chloride, was used as a tracer. The combined supplement (each element equal to that in the basal diet) caused the amount of cadmium retained by the liver, kidney, and whole body to be less than that by birds fed the basal diet. Figure 1 shows the marked effect of the supplement on cadmium concentration in the kidney at each level of dietary cadmium. In a longer-term study with cadmium-115m, the zinc, copper, and manganese supplement decreased the amount of cadmium retained (Jacobs *et al.,* 1978b). The effect was most marked during the initial phases of cadmium uptake (the rapid clearance phase); however, there was also a beneficial effect of the supplement on the long-term turnover of cadmium.

Jacobs *et al.* (1977) showed that zinc was the principal element having a protective effect against uptake of low dietary cadmium in the short-term experiments. With graded levels of zinc, cadmium was markedly affected in the intestinal tract tissues and decreased in the liver by the highest level of zinc (Fox *et al.,* 1979b). Wide variations in response were observed between control birds fed a casein–gelatin diet and those fed a soy isolate diet. A marginally deficient level of iron appeared to accelerate movement of low dietary cadmium to the kidney (Fox *et al.,* 1979b). Flanagan *et al.* (1978) found that a dose of 25 $\mu$g of cadmium with a tracer of $^{115m}Cd$ as the chloride given in a cereal-based breakfast to human subjects was absorbed in greater amounts in subjects with low serum ferritin levels than in those with higher levels (7.5 vs. 2.6 percent of the dose). Men and women were in each range and the hemoglobin levels were 14 and 16 g/dl, respectively. The range in absorption for all subjects was from less than 1 to greater than 26 percent of the dose.

The above studies with the young Japanese quail and with human beings show that zinc and iron not only decrease cadmium toxicoses, but also affect the absorption and tissue retention of very low levels of cadmium, typical of those in the diet of humans and domestic animals.

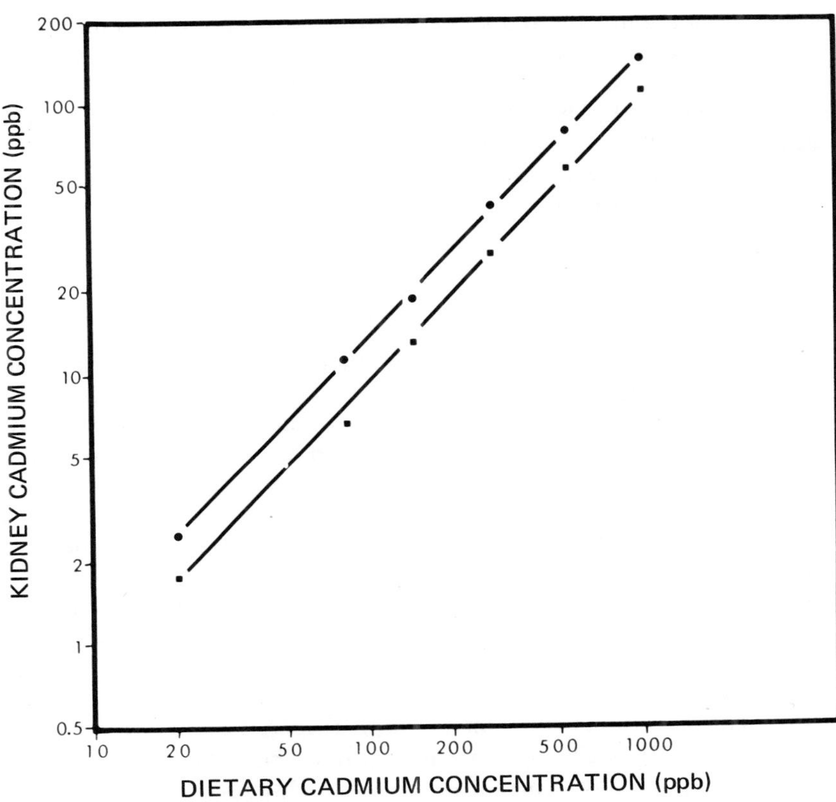


FIGURE 1 Effects of doubling dietary levels of zinc, copper, and manganese on concentrations of cadmium in the kidney of young Japanese quail fed cadmium at dietary levels bracketing typical dietary intakes of humans and domestic animals. The control and supplemented diets contained 30 and 60 ppm zinc, 5 and 10 ppm copper, and 12 and 24 ppm manganese, respectively. Differences between values for controls (circles, upper line) and supplemented birds (squares, lower line) were statistically significant ($P < 0.05$ for lowest cadmium intake and $P < 0.01$ for all others) (Jacobs *et al.*, 1978a).

The intake of deficient or marginally adequate levels of zinc and iron are clearly important determinants of cadmium absorption and retention. The same is probably true for other nutrients known to affect toxicity of high levels of cadmium.

Once discovered, it is important to remove a source of cadmium exposure. Exon *et al.* (1977) gave mice drinking water containing 300 ppm cadmium for 10 weeks. The cadmium was removed and cadmium was determined in liver and kidneys periodically at intervals up to 180 days postexposure. The cadmium concentrations remained remarkably constant in the liver for 20 days; at 40 and 180 days the cadmium in the liver had declined. There appeared to be a slight increase in kidney cadmium by 10 days that did not change by 180 days. Thus, most absorbed cadmium will remain in the body; however, tissue concentrations at a given exposure are lower in the growing animal.

## TISSUE LEVELS

Cadmium is a toxic element for human beings. The biological halftime of cadmium in the kidney, the target organ for damage, has been estimated to be 18 to 33 years in humans. Although body burden of cadmium can be determined *in vivo* by neutron activation analysis of the liver and one kidney, this is presently only a research tool. There is no therapeutic procedure for removing cadmium from the body that does not cause kidney damage. There is considerable evidence that when the concentration of cadmium in the renal cortex reaches 200 ppm, cadmium is excreted in the urine and damage to the proximal tubule occurs. Recent evidence in humans (Kjellström *et al.,* 1977; Nogawa *et al.,* 1978) and in horses (Elinder *et al.,* in press) shows graded pathological kidney damage beginning at lower levels of cadmium intake or renal cortex concentrations. A Joint FAO/WHO Expert Committee on Food Additives (1972) established provisional tolerable intakes of cadmium at 57 to 71 $\mu$g per day. Current estimates of cadmium in the U.S. food supply show intakes near the FAO/WHO values (Mahaffey *et al.,* 1975). Considerations in evaluating safe cadmium intakes for humans have been reviewed (Fox, 1979b). It is generally considered undesirable to permit cadmium levels in the human food supply to increase.

Data on the cadmium concentrations in liver, kidney, and muscle of animals fed 15 ppm or less dietary cadmium are summarized in Table 12. The concentrations in livers and kidneys of control animals were mostly below 1 ppm wet weight. Exceptions were 1 ppm in the kidney of sheep (Sharma *et al.,* 1979) and 1.6, 3.2, and 3.0 ppm in kidneys of laying hens (Leach *et al.,* 1979; Sharma *et al.,* 1979). These values were higher than the cadmium levels in the protein-source food groups of the market basket survey of the Food and Drug Administration (Mahaffey *et al.,* 1975). The latter values were as follows: dairy products, 0.005 ppm; meat, fish, and poultry, 0.0093 ppm; and grains and cereal products, 0.028 ppm. The values for muscle cadmium of control animals in

Table 12 are more nearly similar to the food composite groups. As cadmium exposure level and/or time of exposure increased, the concentrations of cadmium in liver, kidney, and muscle increased. Although concentrations of cadmium are higher in liver and kidney than in muscle, increases of cadmium may be of greater importance in muscle because of the greater consumption of muscle meats by humans.

## MAXIMUM TOLERABLE LEVELS

Concentrations of cadmium as low as 1 ppm in the diet or drinking water have produced adverse effects in monogastric animals. These changes included hypertension, decreased kidney manganese, and acute degenerative damage to the intestinal villi. More severe effects have been observed at 5 ppm or higher levels of dietary cadmium in all species. The profound influences of mild deficiencies of several nutrients in increasing cadmium retention at low levels and toxicity at high levels suggest that more severe effects than the above might sometimes occur under practical conditions with 1 ppm of cadmium in the diet. The maximum tolerable level of dietary cadmium had been set at 0.5 ppm for reasons of cadmium content of animal tissues used for human food and the need to avoid increases of cadmium in the U.S. food supply. Somewhat higher levels of consumption for a very short period of time would not be expected to create a hazard to the animal's health and probably not to human food use, particularly if the animal were to be slaughtered at a young age.

## SUMMARY

Cadmium is industrially valuable for plating iron and steel and in batteries, plastics, ceramics, and solders. Cadmium pollution is associated with its production and use and with the discard of products containing cadmium. Although cadmium may be an essential nutrient at very low levels, most information about its biological behavior shows it to be toxic and a potent antagonist of several essential minerals, notably zinc, iron, copper, and calcium. Diets low in these minerals and in protein permit greater absorption/toxicity of cadmium. Almost nothing is known of interactions between cadmium and other nonessential toxic elements. A small proportion of dietary cadmium is absorbed; it accumulates in the liver and kidney where it has a very long biological half-life and can eventually cause renal tubular damage. High intakes of cadmium in the diet can cause reduced growth rates, anemia, enteropathy, kidney damage, infertility, deformed fetuses, abortions, and hypertension. Concentrations of 5 ppm are almost always associated with some adverse health effect; levels as low as 1 ppm have had undesirable effects. A dietary concentration of 0.5 ppm is the maximum tolerable level.

TABLE 11 Effects of Cadmium Administration in Animals

| Class and Number of Animals[a] | Age or Weight | Administration | | | | Effect(s) | Reference |
|---|---|---|---|---|---|---|---|
| | | Quantity of Element[b] | Source | Duration | Route | | |
| Cattle—8 | 2.5–4.5 mo | 40 ppm | $CdCl_2$ | 5 d | Diet | No adverse effect | Powell *et al.*, 1964a |
| | | 160 ppm | | | | Decreased food intake | |
| | | 640 ppm | | | | Marked decrease in food intake | |
| | | 2,560 ppm | | | | Almost no diet eaten | |
| Cattle—2 | 61 kg | 15 mg/kg, 3 times/wk (ca. 600 mg/d) | $CdCl_2$ | 63 d | Capsule | Decreased body weight, food intake, and RBC–ALAD | Lynch *et al.*, 1976 |
| Cattle—2 | Adult | 50 ppm | Cadminate® (cadmium succinate) | 343 d | Diet | 1 abnormal calf; 1 dead calf; low body weight gain | Wright *et al.*, 1977 |
| | | 100 ppm | | | | 2 dead calves | |
| | | 200 ppm | | | | Decreased RBC, hematocrit, hemoglobin; 1 premature calf; 1 aborted fetus; | |
| | | 300 ppm | | | | 1 dead immature calf; cow died 14 d later; 1 aborted fetus | |
| Cattle—2 | Adult | 2.40 ppm[c] | $CdCl_2$ | 3 mo | Diet | Increased liver and kidney Cd | Sharma *et al.*, 1979 |
| | | 2.40 ppm | | 6 mo | | No change by 6 mo in liver Cd from 3 mo exposure | |

## TABLE 11 *Continued*

| Class and Number of Animals[a] | Age or Weight | Administration: Quantity of Element[b] | Source | Duration | Route | Effect(s) | Reference |
|---|---|---|---|---|---|---|---|
| | | 11.29 ppm | | 3 mo | | Increased liver and kidney Cd | |
| | | 11.29 ppm | | 6 mo | | Cd in liver high by 6 mo | |
| Sheep—6 | 4 mo | 5 ppm | $CdCl_2$ | 191 d | Diet | Decreased liver Fe and Cu | Doyle *et al.*, 1974; Doyle and Pfander, 1975 |
| | | 15 ppm | | | | Increased liver and kidney Zn | |
| | | 30 ppm | | | | Decreased gain; increased kidney Cu | |
| | | 60 ppm | | | | Decreased hematocrit and liver Mn; increased with increasing Cd dose; rumen, abomasum, ileum, liver, heart, spleen, lung, testis, and kidney Cd | |
| Sheep—6 | Adult, 13–14 wk preg. | 3.5 ppm | $CdSO_4$ | To parturition | Diet, grass | No adverse effects | Mills and Dalgarno, 1972 |
| | | | | As above + 7.5–8 wk | | Decreased lamb's liver and kidney Zn | |
| | | 7.1 ppm | | To parturition | | No adverse effects | |

| | | | | | | | |
|---|---|---|---|---|---|---|---|
| | | | | As above + 7.5–8 wk | | Decreased lamb's liver and kidney Zn | |
| | | 12.3 ppm | | To parturition | | Decreased lamb's liver Cu | |
| | | | | As above + 7.5–8 wk | | Decreased lamb's plasma Zn, Cu and ceruloplasmin; ewe's liver Cu and crimp in wool | |
| Sheep—2 | Adult | 50 ppm | Cadminate® | 287 d | Diet | 1 normal lamb, difficult delivery, ewe died; 1 ewe infertile | Wright *et al.*, 1977 |
| | | 100 ppm | | | | Decreased RBC, hemoglobin, hematocrit; 1 normal lamb, died after 15 wk (urinary calculi); 1 viable immature malformed lamb | |
| | | 200 ppm | | | | 1 viable immature lamb with malformation; 1 dead lamb with corneal opacity | |
| | | 300 ppm | | | | 1 aborted fetus with keratitis; 1 ewe infertile | |
| | | 500 ppm | | | | 2 aborted fetuses with keratitis; 1 ewe died later | |

TABLE 11 *Continued*

| Class and Number of Animals[a] | Age or Weight | Administration: Quantity of Element[b] | Source | Duration | Route | Effect(s) | Reference |
|---|---|---|---|---|---|---|---|
| Goat | 2 yr | 75 ppm | | 19 mo | Diet | Fed to death; 50% abortions; no normal young, high liver and kidney Cd in adults; low Cd in newborn and milk; increased liver and kidney Zn (adult), lower in young; decreased milk Zn, liver Cu in adult and young | Anke *et al.*, 1970 |
| Swine—4 | 55 d | 50 ppm | $CdCl_2$ | 42 d | Diet | Decreased hematocrit | Cousins *et al.*, 1973 |
| | | 150 ppm | | | | Decreased weight gain and renal leucine aminopeptidase | |
| | | 450 ppm | | | | Decreased weight gain and renal leucine aminopeptidase | |
| Swine—3 | | 1,350 ppm | | | | Minimal gain; decreased serum P; increased liver and kidney Zn; Cd in liver, spleen, lung, and heart increased in relation to dose; kidney Cd near maximum at 450 ppm; high Cd in teeth, little in bone | |

| | | | | | | | |
|---|---|---|---|---|---|---|---|
| Swine—3 | 11 wk | ca. 0.47 ppm[d] | Corn | 2 wk | Diet | Increased liver Cd, RBC, MCV, and MCH; decreased kidney Mn by 6 wk | Hansen and Hinesly, 1979 |
| | | ca. 0.47 ppm | Corn[e] | 8 wk | Diet | Increased liver and kidney Cd, RBC, MCB, MCH, liver microsomal protein, and *o*-dealkylation of *p*-nitrophenitole; decreased liver Fe and kidney Mn | |
| Swine—2 | 9–11 wk | 2.41 ppm[f] | $CdCl_2$ | 3 mo | Diet | Increased liver and kidney Cd | Sharma *et al.*, 1979 |
| | | 2.41 ppm | | 6 mo | | Cd in liver by 6 mo similar to 2.41 ppm Cd for 3 mo | |
| | | 10.12 ppm | | 3 mo | | Increased liver and kidney Cd | |
| | | 10.12 ppm | | 3 mo | | Liver and kidney Cd by 6 mo increased from 10 ppm for 3 mo | |
| Chicken—5–20 | 1 d | 75 ppm | $CdCl_2$ | 21 d | Diet | Decreased body weight, hematocrit, hemoglobin, liver, and kidney Fe and serum Zn; increased serum TIBC and kidney Zn and Cu | Freeland and Cousins, 1973 |
| Chicken—6 | 1 d | 3 ppm | $CdSO_4 \cdot 8H_2O$ | 6 wk | Diet | Increased kidney Cd | Leach *et al.*, 1979 |
| | | 12 ppm | | | | Increased liver, kidney, and breast muscle Cd | |
| | | 48 ppm | | | | Increased liver, kidney, and breast muscle Cd; | |

TABLE 11 *Continued*

| Class and Number of Animals[a] | Age or Weight | Administration: Quantity of Element[b] | Source | Duration | Route | Effect(s) | Reference |
|---|---|---|---|---|---|---|---|
| | | | | | | Cd concentrations were dose-related | |
| Chicken—20 | 14 d | 400 ppm | $CdCO_3$ | 24 d | Diet | Minimal weight gain; 10% mortality | Pritzl *et al.*, 1974 |
| | | 600 ppm | | | | 55% mortality | |
| | | 800 ppm | | | | 100% mortality | |
| | | 1,000 ppm | | | | 100% mortality | |
| Chicken—10 | Adult | 3 ppm | $CdSO_4$ | 12 wk | Diet | Decreased egg production; increased kidney Cd | Leach *et al.*, 1979 |
| | | 12 ppm | | | | Decreased egg production; increased liver and kidney Cd | |
| | | 48 ppm | | | | Decreased egg production; increased liver and kidney Cd and kidney Zn | |
| Chicken—15 | | 3 ppm | | 48 wk | Diet | No adverse effect | |
| | | 12 ppm | | | | Increased kidney Cd; decreased eggshell thickness | |
| Chicken—12 | | 48 ppm | | | | Decreased egg production and shell thickness; increased egg, breast muscle, liver, and kidney Cd | |

| | | | | | | | |
|---|---|---|---|---|---|---|---|
| Chicken—ca.22 | 1 yr | 1.88 ppm[g] | $CdCl_2$ | 1–24 wk | Diet | Increased Cd in muscle, liver, and kidney | Sharma *et al.*, 1979 |
| | | 13.06 ppm | | 1–24 wk | | Increased muscle, liver, and kidney Cd | |
| Chicken | Adult | 50 ppm | | 6 d | Diet | Decreased egg production and egg weight | Anke *et al.*, 1970 |
| | | 100 ppm | | | | Decreased egg production and egg weight | |
| | | 100 ppm | | | | Atrophy of ovary by 28 d | |
| | | 200 ppm | | | | Stopped egg production | |
| | | 200 ppm | | | | Marked atrophy of ovary by 28 d; gizzard appeared Zn deficient; increased tissue Cd and liver Zn; decreased muscle, femur feather Zn in 5 tissues and egg; Fe in 3 tissues | |
| Japanese quail—20 | 7 d | 0.062 ppm | $CdCl_2$ + $^{109}CdCl_2$ | 7 d | Diet | Dose-related increases in duodenum, liver, and kidney Cd | Jacobs *et al.*, 1978a |
| Japanese quail—10 | | 0.125 ppm | | | | | |
| Japanese quail—20 | | 0.250 ppm | | | | | |
| Japanese quail—10 | | 0.500 ppm | | | | | |
| Japanese quail—20 | | 1.020 ppm | | | | | |
| Japanese quail—10 | 1 d | 1 ppm | $CdCl_2$ | 14 d | Diet | Half of birds had accelerated degeneration | Mason *et al.*, 1977; M. R. S. Fox, |

TABLE 11 *Continued*

| Class and Number of Animals[a] | Age or Weight | Administration: Quantity of Element[b] | Source | Duration | Route | Effect(s) | Reference |
|---|---|---|---|---|---|---|---|
| | | | | | | duodenal absorptive cells at villous tips | FDA, 1979, perseronal communication |
| | 12 d | | | 2 d | | All birds affected as above | |
| | 1 d | | | 28 or 49 d | | Minimal changes | |
| | | 10 ppm | | 14 d | | Shortened villi; cellular infiltration of lamina propria; aggregation of goblet cells at villous tips | |
| | 12 d | | | 2 d | | As 10 ppm for 14 d | |
| | 1 d | | | 28 or 49 d | | Lesser changes than with 10 ppm at 14 d | |
| Japanese quail—80 | 1 d | 75 ppm | $CdCl_2$ | 28 d | Diet | Decreased body weight, hematocrit, total plasma proteins, and albumin; increased transferrin and mortality | Jacobs *et al.*, 1969 |
| Japanese quail—80 | 1 d | 75 ppm | $CdCl_2$ | 28 d | Diet | Decreased weight gain, hematocrit | Fox *et al.*, 1971 |
| Japanese quail—11 | | | | | | Decreased RBC, liver, kidney, and tibia Fe, | |

| | | | | | | | |
|---|---|---|---|---|---|---|---|
| | | | | | | tibia Zn, and tibia total ash; increased RBC Zn, RBC, and kidney Cu and liver Ca | |
| Japanese quail—53 | 1 d | 75 ppm | $CdCl_2$ | 28 or 42 d | Diet | Testicular hypoplasia; bone marrow hyperplasia; hypertrophy of heart ventricles (42 d); enteropathy, decreased granules in adrenal medullary cells | Richardson *et al.*, 1974 |
| Japanese quail—10 | 1 d | 75 ppm | $CdCl_2$ | 28 d | Diet | Small intestine: dilated, thin wall, short villi and microvilli, hyperplasia goblet cells, atrophic absorptive cells, degeneration of some nerve pleuses in the muscular propria | Richardson and Fox, 1974 |
| Rabbit—10 | 3 mo | 160 ppm | $CdCl_2$ | 200 d | Water[h] | Decreased growth, hematocrit, hemoglobin; neutrophilia, lymphopenia, hypoalbuminemia; increased $\alpha_2$-, $\beta_2$-, and $\gamma$-globulins, splenomegaly, cardiomegaly, and renal enlargement; interlobular hepatic and interstitial renal | Stowe *et al.*, 1972 |

TABLE 11 *Continued*

| Class and Number of Animals[a] | Age or Weight | Administration: Quantity of Element[b] | Source | Duration | Route | Effect(s) | Reference |
|---|---|---|---|---|---|---|---|
| | | | | | | fibrosis, biliary hyperplasia; renal cortex: coagulation necrosis, glomerular and interstitial fibrosis; proximal tubule epithelium contained apical cysts, dilated endoplasmic reticulum, and nuclear degeneration | |
| Rabbit | | 43 mg/kg | $CdCl_2$ | 1 dose | Oral | $LDL_0$[i] | Fairchild *et al.*, 1977 |
| Dog—2 | 6 wk | 0.5 ppm | $CdCl_2$ | 4 yr | Water[h] | Liver, kidney, and pancreas Cd levels dose-related | Anwar *et al.*, 1961 |
| | | 2.5 ppm | | | | No adverse effects | |
| | | 5 ppm | | | | Some fat droplets in glomeruli; some tubular atrophy and inflammatory cells | |
| | | 10 ppm | | | | Some fat droplets in glomeruli; some tubular atrophy and inflammatory cells | |

| | | | | | | | |
|---|---|---|---|---|---|---|---|
| Dog | | 105 mg/kg | $CdSO_4$ | 1 dose | Oral | $LD_{50}$ | Fairchild *et al.*, 1977 |
| Rat—46[i] | Weanling | 1 ppm | Cadmium acetate | 30 mo | Water[h] | Slightly higher weight gain 3–20 mo; elevated systolic blood pressure at 12 and 18 mo; increased renal Cd at 18 mo | Perry *et al.*, 1977 |
| | | 2.5 ppm | | | | Increased blood pressure at 6, 12, and 18 mo | |
| | | 5 ppm | | | | Increased blood pressure at 6, 12, and 18 mo | |
| | | 10 ppm | | | | Increased blood pressure at 6 mo, liver Zn at 18 mo | |
| | | 25 ppm | | | | Increased blood pressure at 6 mo | |
| | | 50 ppm | | | | Decreased gain, weight plateau by 10 mo; no change in blood pressure; general: whole blood and serum Cd increased with dose (little time effect); liver Cd increased with dose and time; kidney Cd increased with dose and time to 18 mo; no mortality trends related to dose; prior to 18 mo, increases in blood pressure were as- | |

TABLE 11 *Continued*

| Class and Number of Animals[a] | Age or Weight | Administration: Quantity of Element[b] | Source | Duration | Route | Effect(s) | Reference |
|---|---|---|---|---|---|---|---|
| | | | | | | sociated with kidney Cd of ca. 5–50 ppm | |
| Rat—20 | 55 g | 5 ppm | | 330 d | Water[h] | By 294 d greater $^{24}Na$ retention in females; by 320 d water retention greater in males and females | Doyle *et al.*, 1975 |
| Rat—100 | Weanling | 5 ppm | Cadmium acetate | Lifetime | Water[h] | Shortened life span; arteriolar sclerosis of kidneys; arteriosclerosis and ventricular hypertrophy of heart | Kanisawa and Schroeder, 1969; Schroeder *et al.*, 1965 |
| Rat | | 63 mg/kg | CdO | 1 dose | Oral | $LD_{50}$ | Fairchild *et al.*, 1977 |
| Mouse—10 | Weanling | 10 ppm | Water soluble | 2 generations | Water[h] | $F_1$ through $F_2$ generations: dead litters; | Schroeder and Mitchener, 1971 |

| | | | | | | |
|---|---|---|---|---|---|---|
| | | | | | young deaths; runts; decreased number of offspring; sharp angulation in the distal third of the tail; three of the five pairs failed to breed in the $F_2$ generation | |
| Mouse | 107 mg/kg | $CdCl_2$ | 1 dose | Oral | $LD_{50}$ | Fairchild *et al.*, 1977 |
| Guinea pig | 39 mg/kg | | | | $LD_{50}$ | |

[a]Number of animals per treatment.
[b]Quantity expressed as parts per million (concentration in diet) or as milligrams per kilogram of body weight; amounts are those added to the diet unless indicated otherwise.
[c]Total dietary cadmium, control 0.18 ppm.
[d]Total dietary cadmium, control 0.10 ppm.
[e]Sewage sludge fertilized.
[f]Total dietary cadmium, control 0.23 ppm.
[g]Total dietary cadmium, control 0.32 ppm.
[h]Drinking water.
[i]Lowest published lethal dose.
[j]Sixteen rats were used for blood pressure measurements and five rats were killed at 6, 12, 18, and 24 months for organ analysis.

TABLE 12 Effects of Graded Levels of Total Dietary Cadmium upon Cadmium in Liver, Kidney, and Muscle

| Class and Number of Animals | Age or Weight | Quantity of Element | Principle Source | Duration | Tissue Cd, ppm fresh weight[a] | | | Reference |
|---|---|---|---|---|---|---|---|---|
| | | | | | Liver | Kideny | Muscle | |
| Cattle—2 | Adult, lactating | 0.18 ppm | Diet | 3 mo | 0.6 | — | —[b] | Sharma *et al.*, 1979 |
| | | 2.40 ppm | $CdCl_2$ | | 0.7 | — | — | |
| | | 11.29 ppm | | | | 2.2 | — | — |
| Sheep—6 | 4 mo | 0.2 ppm | Diet | 6.4 mo | 0.5 | 1.0 | 0:025 | |
| | | 5 ppm | $CdCl_2$ | | 4.5 | 13.5 | 0.047 | |
| | | 15 ppm | | | 15.5 | 43 | 0.091 | |
| Sheep—6 | Adult | 0.7 ppm | Grass | ca. 15–16 wk[c] | 0.29 | — | — | Mills and Dalgarno, 1972 |
| | | 3.5 ppm | $CdSO_4$ | — | 0.60 | — | — | |
| | | 7.1 ppm | | — | 1.05 | — | — | |
| | | 12.3 ppm | | — | 3.36 | — | — | |
| Swine—2 | 9–11 wk | 0.23 ppm | Diet | 3 mo | 0.1 | 0.09, 0.06[d] | —[b] | Sharma *et al.*, 1979 |
| | | 0.23 ppm | | 6 mo | 0.1 | — | — | |
| | | 2.41 ppm | $CdCl_2$ | 3 mo | 1.0 | 1.14, 6.52 | — | |
| | | 2.41 ppm | | 6 mo | 1.3 | 1.11, 10.97 | — | |
| | | 10.12 ppm | | 3 mo | 4.8 | 5.15, 28.12 | — | |
| | | 10.12 ppm | | 6 mo | 10.2 | 3.08, 42.30 | — | |

| | | | | | | | | |
|---|---|---|---|---|---|---|---|---|
| Swine—3 | 11 wk | ca. 0.10 ppm | Diet | 8 wk | 0.06 | 0.15 | — | Hansen and Hinesly, 1979 |
| | | ca. 0.47 ppm | Corn[e], diet | | 0.10 | 0.32 | — | |
| Chicken—6 | 1 d | 0.21 ppm | Diet | 6 wk | 0.06 | 0.07 | 0.02 | Leach *et al.*, 1979 |
| | | 3.2 ppm | $CdSO_4 \cdot 8H_2O$ | | 1.14 | 1.76 | 0.03 | |
| | | 12.1 ppm | | | 1.23 | 9.45 | 0.06 | |
| Chicken—10 | Adult, laying | 0.07 ppm | Diet | 12 wk | 0.4 | 1.6 | — | |
| | | 3.07 ppm | $CdSO_4 \cdot 8H_2O$ | | 2.2 | 5.9 | — | |
| | | 12.07 ppm | | | 6.4 | 17.6 | — | |
| Chicken—15 | Adult, laying | 0.22 ppm | Diet | 48 wk | 0.7 | 3.2 | 0.029 | |
| | | 3.22 ppm | $CdSO_4 \cdot 8H_2O$ | | 8.0 | 52 | 0.137 | |
| | | 12.22 ppm | | | 10.0 | 135 | 0.398 | |
| Chicken—15 | 1 yr, laying | 0.32 ppm | Diet | 6 wk | 0.2 | 3 | 0.063 | Sharma *et al.*, 1979 |
| | | 1.88 ppm | $CdCl_2$ | | 6.2 | 45 | 0.140 | |
| | | 13.06 ppm | | | 28.0 | 160 | 0.263 | |

[a] When values were reported on dry weight, a moisture content of 70 percent was assumed for liver and 77 percent for kidney; for values that were reported on fat-free dry weight, moisture plus fat content were assumed as follows: liver, 76 percent; kidney, 81 percent; chicken breast muscle (6 wk), 79 percent; and hens, 76 percent.

[b] It was stated in the text that dietary Cd did not affect muscle Cd.

[c] Cd was begun 13–14 wk of pregnancy and continued 7.5–8 wk into lactation.

[d] Values for medulla, cortex.

[e] Corn grown on sewage sludge-amended soil.

## REFERENCES

Anderson, S. H., and R. I. Van Hook, Jr. 1973. Uptake and biological turnover of $^{109}$Cd in chipping sparrows, *Spizella passerina*. Environ. Physiol. Biochem. 3:243.

Ando, M., Y. Sayato, M. Tonomuro, and T. Osawa. 1977. Studies on excretion and uptake of calcium by rats after continuous oral administration of cadmium. Toxicol. Appl. Pharmacol. 39:321.

Anke, M., A. Hennig, H. J. Schneider, H. Ludke, W. von Gargen, and H. Schlegel. 1970. The interrelations between cadmium, zinc, copper and iron in metabolism of hens, ruminants and man. *In* C. F. Mills, ed. Trace Element Metabolism in Animals, p. 317. E. S. Livingstone, Edinburgh.

Anwar, R. A., R. F. Langham, C. A. Hoppert, B. V. Alfredson, and R. U. Byerrum. 1961. Chronic toxicity studies. II. Chronic toxicity of cadmium and chromium in dogs. Arch. Environ. Health 3:92.

Baker, D. E., M. C. Amacher, and R. M. Leach. 1979. Sewage sludge as a source of cadmium in soil–plant–animal systems. Environ. Health Perspect. 28:45.

Banis, R. J., W. G. Pond, E. F. Walker, Jr., and J. R. O'Connor. 1969. Dietary cadmium, iron and zinc interactions in the growing rat. Proc. Soc. Exp. Biol. Med. 130:802.

Berlin, M., and S. Ullberg. 1963. The fate of $Cd^{109}$ in the mouse. Arch. Environ. Health 7:686.

Bremner, I. 1978. Cadmium toxicity. Nutritional influences and the role of metallothionein. World Rev. Nutr. Dietet. 32:165.

Bunn, C. R., and G. Matrone. 1966. *In vivo* interactions of cadmium, copper, zinc and iron in the mouse and rat. J. Nutr. 90:395.

Campbell, J. K., and C. F. Mills. 1974. Effects of dietary cadmium and zinc on rats maintained on diets low in copper. Proc. Nutr. Soc. 33:15A.

Chaney, R. L., and S. B. Hornick. 1978. Accumulation and effects of cadmium on crops, p. 125. *In* Cadmium 77—Edited Proceedings, First International Cadmium Conference, San Francisco. Met. Bull. London.

Chaney, R. L., P. T. Hundemann, W. T. Palmer, R. J. Small, M. C. White, and A. M. Decker. 1978a. Plant accumulation of heavy metals and phytotoxicity resulting from utilization of sewage sludge and sludge composts on cropland, p. 86. *In* Proceedings 1977 National Conference on Composting of Municipal Residues and Sludges. Information Transfer, Inc., Rockville, Md.

Chaney, R. L., G. S. Stoewsand, C. A. Bache, and D. J. Lisk. 1978b. Cadmium deposition and hepatic microsomal induction in mice fed lettuce grown on municipal sewage sludge-amended soil. J. Agric. Food Chem. 26:992.

Cikrt, M., and M. Tichý. 1974. Excretion of cadmium through bile and intestinal wall in rats. Br. J. Ind. Med. 31:134.

Cotzias, G. C., D. C. Borg, and B. Selleck. 1961. Virtual absence of turnover in cadmium metabolism: $Cd^{109}$ studies in the mouse. Am. J. Physiol. 201:927.

Council for Agricultural Science and Technology. 1976. Application of Sewage Sludge to Cropland: Appraisal of Potential Hazards of Heavy Metals to Plants and Animals, p. 29. U.S. Environmental Protection Agency Publ. No. MCD-33, Washington, D.C.

Cousins, R. J. 1979. Metallothionein synthesis and degradation: Relationship to cadmium metabolism. Environ. Health Perspect. 28:131.

Cousins, R. J., and S. L. Feldman. 1973. Effect of cholecalciferol on cadmium uptake in the chick. Nutr. Rep. Int. 8:363.

Cousins, R. J., A. K. Barber, and J. R. Trout. 1973. Cadmium toxicity in growing swine. J. Nutr. 103:964.

Cousins, R. J., K. S. Squibb, S. L. Feldman, A. deBari, and B. L. Silbon. 1977. Biomedical responses of rats to chronic exposure to dietary cadmium fed in *ad libitum* and equalized regimes. J. Toxicol. Environ. Health 2:929.

Davies, N. T., and J. K. Campbell. 1977. The effect of cadmium on intestinal copper absorption and binding in the rat. Life Sci. 20:955.

Doyle, J. J., and W. H. Pfander. 1975. Interactions of cadmium with copper, iron, zinc, and manganese in ovine tissues. J. Nutr. 105:599.

Doyle, J. J., W. H. Pfander, S. E. Grebing, and J. O. Pierce II. 1974. Effect of dietary cadmium on growth, cadmium absorption and cadmium tissue levels in growing lambs. J. Nutr. 104:160.

Doyle, J. J., R. A. Bernhoft, and H. H. Sandstead. 1975. The effects of a low level of dietary cadmium on blood pressure, $^{24}Na$, $^{42}K$, and water retention in growing rats. J. Lab. Clin. Med. 86:57.

Elinder, C.-G., L. Jönsson, M. Piscator, and B. Palmster. In press. Histopathological changes in relation to cadmium concentration in horse kidneys. Environ. Res.

Evans, G. W., P. F. Majors, and W. E. Cornatzer. 1970. Mechanism for cadmium and zinc antagonism of copper metabolism. Biochem. Biophys. Res. Comm. 40:1142.

Exon, J. H., J. G. Lamberton, and L. D. Koller. 1977. Effect of chronic oral cadmium exposure and withdrawal on cadmium residues in organs of mice. Bull. Environ. Contamin. Toxicol. 18:74.

Fairchild, E. J., R. J. Lewis, and R. L. Tatkin (eds.). 1977. Registry of Toxic Effects of Chemical Substances, vol. 2, p. 526. DHEW Publ. No. (NIOSH) 78-104-B.

Fassett, D. W. 1972. Cadmium. *In* D. H. K. Lee, ed. Metallic Contaminants and Human Health, p. 97. Academic Press, New York.

Feldman, S. L., and R. J. Cousins. 1973. Influence of cadmium on the metabolism of 25-hydroxycholecalciferol in chicks. Nutr. Rep. Int. 8:251.

Flanagan, P. R., J. S. McLellan, J. Haist, G. Cherian, M. J. Chamberlain, and L. S. Valberg. 1978. Increased dietary cadmium absorption in mice and human subjects with iron deficiency. Gastroenterology 74:841.

Fleischer, M., A. F. Sarofim, D. W. Fassett, P. Hammond, H. T. Schacklette, I. C. T. Nisbet, and S. Epstein. 1974. Environmental impact of cadmium: A review by the panel on hazardous trace substances. Environ. Health. Perspect., Exp. Issue No. 253.

Flick, D. F., H. F. Kraybill, and J. M. Dimitroff. 1971. Toxic effects of cadmium: A review. Environ. Res. 4:71.

Fox, M. R. S. 1974. Effect of essential minerals on cadmium toxicity. A review. J. Food Sci. 39:321.

Fox, M. R. S. 1979a. Nutritional influences on metal toxicity—Cadmium as a model toxic element. Environ. Health Perspect. 29:95.

Fox, M. R. S. 1979b. Safe levels of cadmium in intravenous diets. Am. J. Clin. Nutr. 32:725.

Fox, M. R. S., and B. E. Fry, Jr. 1970. Cadmium toxicity decreased by dietary ascorbic acid supplements. Science 169:989.

Fox, M. R. S., B. E. Fry, Jr., B. F. Harland, M. E. Schertel, and C. E. Weeks. 1971. Effect of ascorbic acid on cadmium toxicity in the young coturnix. J. Nutr. 101:1295.

Fox, M. R. S., R. M. Jacobs, B. E. Fry, Jr., and B. F. Harland. 1973. Effect of protein source on response to cadmium. Fed. Proc. 32:924 (Abstr.).

Fox, M. R. S., R. M. Jacobs, A. O. L. Jones, and B. E. Fry, Jr. 1979a. Effects of nutritional factors on metabolism of dietary cadmium at levels similar to those in man. Environ. Health Perspect. 28:107.

Fox, M. R. S., R. M. Jacobs, A. O. L. Jones, and B. E. Fry, Jr. 1979b. Tissue cadmium levels affected by dietary ascorbic acid and iron. Fed. Proc. 38:553 (Abstr.).

Freeland, J. H., and R. J. Cousins. 1973. Effect of dietary cadmium on anemia, iron absorption, and cadmium binding protein in the chick. Nutr. Rep. Int. 8:337.

Friberg, L., M. Piscator, and G. Nordberg. 1971. Cadmium in the Environment. CRC Press, Cleveland.

Friberg, L., M. Piscator, G. F. Nordberg, and T. Kjellström. 1974. Cadmium in the Environment, 2nd ed. CRC Press, Cleveland.

Furr, A. K., G. J. Stoewsand, C. A. Bache, and D. J. Lisk. 1976. Study of guinea pigs fed swiss chard grown on municipal sludge-amended soil. Multi-element content of tissues. Arch. Environ. Health 31:87.

Hamilton, D. L., and L. S. Valberg. 1974. Relationship between cadmium and iron absorption. Am. J. Physiol. 227:1033.

Hansen, L. G., and T. D. Hinesly. 1979. Cadmium from soil amended with sewage sludge: Effects and residues in swine. Environ. Health Perspect. 28:51.

Havrdová, J., M. Cikrt, and M. Tichý. 1974. Binding of cadmium and mercury in the rat bile: Studies using gel filtration. Acta Pharmacol. Toxicol. 34:246.

Hill, C. H., G. Matrone, W. L. Payne, and C. W. Barber. 1963. *In vivo* interactions of cadmium with copper, zinc and iron. J. Nutr. 80:227.

Horner, D. B., and J. C. Smith. 1975. The distribution of tracer doses of cadmium in the normal rat. Arch. Environ. Contam. Toxicol. 3:307.

Itokawa, Y., T. Abe, and S. Tanaka. 1973. Bone changes in experimental chronic cadmium poisoning. Arch. Environ. Health 26:241.

Itokawa, Y., T. Abe, R. Tabei, and S. Tanaka. 1974. Renal and skeletal lesions in experimental cadmium poisoning. Arch. Environ. Health 28:149.

Jacobs, R. M., M. R. S. Fox, and M. H. Aldridge. 1969. Changes in plasma proteins associated with the anemia produced by dietary cadmium in Japanese quail. J. Nutr. 99:119.

Jacobs, R. M., M. R. S. Fox, B. E. Fry, Jr., and B. F. Harland. 1974. The effect of a two-day exposure to dietary cadmium in the concentration of elements in duodenal tissue of Japanese quail. *In* W. G. Hoekstra, J. W. Suttie, H. E. Ganther, and W. Mertz, eds. Trace Element Metabolism in Animals—2, p. 684. University Park Press, Baltimore, Md.

Jacobs, R. M., M. R. S. Fox, A. O. L. Jones, R. P. Hamilton, and J. Lener. 1977. Cadmium metabolism: Individual effects of Zn, Cu and Mn. Fed. Proc. 36:1152 (Abstr.).

Jacobs, R. M., A. O. L. Jones, M. R. S. Fox, and B. E. Fry, Jr. 1978a. Retention of dietary cadmium and the ameliorative effect of zinc, copper and manganese in Japanese quail. J. Nutr. 108:22.

Jacobs, R. M., A. O. L. Jones, B. E. Fry, Jr., and M. R. S. Fox. 1978b. Decreased long-term retention of $^{115m}$Cd in Japanese quail produced by a combined supplement of zinc, copper and manganese. J. Nutr. 108:901.

Joint FAO/WHO Expert Committee on Food Additives. 1972. Evaluation of Certain Food Additives and the Contaminants Mercury, Lead and Cadmium. WHO Tech. Rep. Ser. No. 505:20, 32.

Kägi, J. H. R., and B. L. Vallee. 1960. Metallothionein: a cadmium- and zinc-containing protein from equine renal cortex. J. Biol. Chem. 235:3460.

Kägi, J. H. R., S. R. Himmelhoch, P. D. Whanger, J. L. Bethune, and B. L. Vallee. 1974. Equine hepatic and renal metallothioneins. Purification, molecular weight, amino acid composition and metal content. J. Biol. Chem. 249:3537.

Kanisawa, M., and H. A. Schroeder. 1969. Life term studies on the effect of trace elements on spontaneous tumors in mice and rats. Cancer Res. 29:892.

Kello, D., D. Dkanić, and K. Kostial. 1979. Influence of sex and dietary calcium on intestinal absorption in rats. Arch. Environ. Health 34:30.

Kello, D., and K. Kostial. 1977. Influence of age and milk diet on cadmium absorption from the gut. Toxicol. Appl. Pharmacol. 40:277.

Kjellström, T. 1979. Exposure and accumulation of cadmium in populations from Japan, the United States, and Sweden. Environ. Health Perspect. 28:169.

Kjellström, T., K. Shiroishi, and P.-E. Evrin. 1977. Urinary $B_2$-microglobulin excretion among people exposed to cadmium in the general environment. An epidemiological study in cooperation between Japan and Sweden. Environ. Res. 13:318.

Kobayashi, J., H. Nakahara, and T. Hasegawa. 1971. Accumulation of cadmium in organs of mice fed on cadmium-polluted rice. Jap. J. Hyg. 26:401 (see Friberg *et al.*, 1974, p. 28).

Kopp, J. F., and R. C. Kroner. 1968. Trace metals in waters of the United States. U.S. Department of the Interior, Cincinnati.

Larsson, S.-E., and M. Piscator. 1971. Effect of cadmium on skeletal tissue in normal and calcium-deficient rats. Isr. J. Med. Sci. 7:495.

Leach, R. M., Jr., K. W.-L. Wang, and D. E. Baker. 1979. Cadmium and the food chain: The effect of dietary cadmium on tissue composition in chicks and laying hens. J. Nutr. 109:437.

Leber, A. P., and T. S. Miya. 1976. A mechanism for cadmium- and zinc-induced tolerance to cadmium toxicity: Involvement of metallothionein. Toxicol. Appl. Pharmacol. 37:403.

Lucis, O. J., R. Lucis, and Z. A. Shaikh. 1972. Cadmium and zinc in pregnancy and lactation. Arch. Environ. Health 25:14.

Lynch, G. P., D. F. Smith, M. Fisher, T. L. Pike, and B. T. Weinland. 1976. Physiological responses of calves to cadmium and lead. J. Anim. Sci. 42:410.

Mahaffey, K. R., P. E. Corneliussen, C. F. Jelinek, and J. S. Fiorino. 1975. Heavy metal exposure from foods. Environ. Health Perspect. 12:63.

Margoshes, M., and B. L. Vallee. 1957. A cadmium containing protein from equine kidney cortex. J. Am. Chem. Soc. 79:4813.

Mason, K. E., M. E. Richardson, and M. R. S. Fox. 1977. Enteropathy caused by 1 or 10 ppm dietary Cd. Fed. Proc. 36:1152 (Abstr.).

Miller, W. J., B. Lampp, G. W. Powell, C. S. Salotti, and D. M. Blackmon. 1967. Influence of a high level of dietary cadmium on cadmium content in milk, excretion and cow performance. J. Dairy Sci. 50:1404.

Miller, W. J., D. M. Blackmon, and Y. G. Martin. 1968. $^{109}Cd$ absorption, excretion, and tissue distribution following single tracer oral and intravenous doses in young goats. J. Dairy Sci. 51:1836.

Miller, W. J., D. M. Blackmon, R. P. Gentry, and F. M. Pate. 1969. Effect of dietary cadmium on tissue distribution of $^{109}Cd$ following a single oral dose in young goats. J. Dairy Sci. 52:2029.

Mills, C. F., and A. C. Dalgarno. 1972. Copper and zinc status of ewes and lambs receiving increased dietary concentrations of cadmium. Nature 239:171.

Moore, W., Jr., J. F. Stara, and W. C. Crocker. 1973a. Gastrointestinal absorption of different compounds of $^{115m}$cadmium and the effect of different concentrations in the rat. Environ. Res. 6:159.

Moore, W., Jr., J. F. Stara, W. C. Crocker, M. Malachuk, and R. Iltis. 1973b. Comparison of $^{115m}$cadmium retention in rats following different routes of administration. Environ. Res. 6:473.

Neathery, M. W., and W. J. Miller. 1975. Metabolism and toxicity of cadmium, mercury and lead in animals: A review. J. Dairy Sci. 58:1767.

Neathery, M. W., W. J. Miller, R. P. Gentry, P. E. Stake, and D. M. Blackmon. 1974. Cadmium-109 and methyl mercury-203 metabolism, tissue distribution, and secretion into milk of cows. J. Dairy Sci. 57:1177.

Nogawa, K., A. Ishizaki, and S. Kawano. 1978. Statistical observations of the dose-response relationships of cadmium based on epidemiological studies in the Kakehashi River basin. Environ. Res. 15:185.

Nordberg, G. F., and K. Nishiyama. 1972. Whole-body and hair retention of cadmium in mice. Including an autoradiographic study on organ distribution. Arch. Environ. Health 24:209.

Nordberg, G. F., M. Nordberg, M. Piscator, and O. Vesterberg. 1972. Separation of two forms of rabbit metallothionein by isoelectric focusing. Biochem. J. 126:491.

Ohanian, E. V., J. Iwai, G. Leitl, and R. Tuthill. 1978. Genetic influence on cadmium-induced hypertension. Am. J. Physiol. 235:H385.

Perry, H. M., Jr., M. Erlanger, and E. F. Perry. 1977. Elevated systolic pressure following chronic low-level cadmium feeding. Am. J. Physiol. 232:H114.

Petering, H. G., M. A. Johnson, and K. L. Stemmer. 1971. Studies of zinc metabolism in the rat. I. Dose-response effects of cadmium. Arch. Environ. Health 23:93.

Pietrzak-Flis, Z., G. L. Rehnberg, J. J. Favor, D. F. Cahill, and J. W. Laskey. 1978. Chronic ingestion of cadmium and/or tritium in rats. Environ. Res. 16:9.

Pond, W. G., and E. F. Walker, Jr. 1972. Cadmium-induced anemia in growing rats: Prevention by oral or parenteral iron. Nutr. Rep. Int. 5:365.

Pond, W. G., and E. F. Walker, Jr. 1975. Effect of dietary Ca and Cd level of pregnant rats on reproduction and on dam and progeny tissue mineral concentrations. Proc. Soc. Exp. Biol. Med. 148:665.

Powell, G. W., W. J. Miller, and D. M. Blackmon. 1967. Effects of dietary EDTA and cadmium on absorption, excretion and retention of orally administered $^{65}Zn$ in various tissues of zinc deficient and normal goats and calves. J. Nutr. 93:203.

Powell, G. W., W. J. Miller, and C. M. Clifton. 1964a. Effect of cadmium on the palatability of calf starters. J. Dairy Sci. 47:1017.

Powell, G. W., W. J. Miller, J. D. Morton, and C. M. Clifton. 1964b. Influence of dietary cadmium level and supplemental zinc on cadmium toxicity in the bovine. J. Nutr. 84: 205.

Pritzl, M. C., Y. H. Lie, E. W. Kienholz, and C. E. Whiteman. 1974. The effect of dietary cadmium on development of young chickens. Poult. Sci. 53:2026.

Probst, G. S., W. F. Bousquet, and T. S. Miya. 1977. Correlation of hepatic metallothionein concentrations with acute cadmium toxicity in the mouse. Toxicol. Appl. Pharmacol. 39:61.

Pulido, P., J. H. R. Kägi, and B. L. Vallee. 1966. Isolation and some properties of human metallothionein. Biochemistry 5:1768.

Richardson, M. E., and M. R. S. Fox. 1974. Dietary cadmium and enteropathy in the Japanese quail. Histochemical and ultrastructural studies. Lab. Invest. 31:722.

Richardson, M. E., M. R. S. Fox, and B. E. Fry, Jr. 1974. Pathological changes produced in Japanese quail by ingestion of cadmium. J. Nutr. 104:323.

Richmond, C. R., J. S. Findlay, and J. E. London. 1966. Whole Body Retention of Cadmium-109 by Mice Following Oral, Intraperitoneal, and Intravenous Administration. Atomic Energy Commission, Los Alamos Scientific Laboratory, University of California. LA-3610-MS, 195.

Sasser, L. B., and G. E. Jarboe. 1977. Intestinal absorption and retention of cadmium in neonatal rat. Toxicol. Appl. Pharmacol. 41:423.

Schroeder, H. A., and J. J. Balassa. 1961. Abnormal trace metals in man: Cadmium. J. Chron. Dis. 14:236.

Schroeder, H. A., and M. Mitchener. 1971. Toxic effects of trace elements on the reproduction of mice and rats. Arch. Environ. Health 23:102.

Schroeder, H. A., J. J. Balassa, and W. H. Vinton, Jr. 1965. Chromium, cadmium, and lead in rats: Effects on life span, tumors and tissue levels. J. Nutr. 86:51.

Schroeder, H. A., A. P. Nason, E. H. Tipton, and J. J. Balassa. 1967. Essential trace metals in man: Zinc. Relation to environmental cadmium. J. Chron. Dis. 20:179.

Schwarz, K., and J. E. Spallholz. 1978. Growth effects of small cadmium supplements in rats maintained under trace-element controlled conditions, p. 105. *In* Cadmium 77—Edited Proceedings, First International Cadmium Conference, San Francisco, Met. Bull., London.

Sell, J. L. 1975. Cadmium and the laying hen: Apparent absorption, tissue distribution and virtual absence of transfer into eggs. Poult. Sci. 54:1674.

Shaikh, Z. A., and O. J. Lucis. 1971. Isolation of cadmium-binding proteins. Experientia 27/9, 1024.

Sharma, R. P., J. C. Street, M. P. Verma, and J. L. Shupe. 1979. Cadmium uptake from feed and its distribution to food products of livestock. Environ. Health Perspect. 28:59.

Sonawane, B. R., M. Nordberg, G. F. Nordberg, and G. W. Lucier. 1975. Placental transfer of cadmium in rats: Influence of dose and gestational age. Environ. Health Perspect. 12:97.

Starcher, B. C. 1969. Studies on the mechanism of copper absorption in the chick. J. Nutr. 97:321.

Stowe, H. D. 1976. Biliary excretion of cadmium by rats: Effects of zinc, cadmium, and selenium pretreatments. J. Toxicol. Environ. Health 2:45.

Stowe, H. D., M. Wilson, and R. A. Goyer. 1972. Clinical and morphologic effects of oral cadmium toxicity in rabbits. Arch. Pathol. 94:389.

Stowe, H. D., R. A. Goyer, P. Medley, and M. Cates. 1974. Influence of dietary pyridoxine on cadmium toxicity in rats. Arch. Environ. Health 28:209.

Stubbs, R. L. 1978. Cadmium—The metal of benign neglect. *In* Cadmium 77—Edited Proceedings, First International Cadmium Conference, San Francisco. Met. Bull., London.

Supplee, W. C. 1961. Production of zinc deficiency in turkey poults by dietary cadmium. Poult. Sci. 40:827.

Supplee, W. C. 1963. Antagonistic relationship between dietary cadmium and zinc. Science 139:119.

Suzuki, S., T. Taguchi, and G. Yokahashi. 1969. Dietary factors influencing upon the retention rate of orally administered $^{115m}CdCl_2$ in mice with special reference to calcium and protein concentrations in diet. Ind. Health 7:155.

Terhaar, C. J., E. Vis, R. L. Roudabush, and D. W. Fassett. 1965. Protective effects of low doses of cadmium chloride against subsequent high oral doses in the rat. Toxicol. Appl. Pharmacol. 7:500 (Abstr.).

Underwood, E. J. 1977. Trace Elements in Human and Animal Nutrition, 4th ed. Academic Press, New York.

Valberg, L. S., J. Haist, M. G. Cherian, L. Delaquerriere-Richardson, and R. A. Goyer. 1977. Cadmium-induced enteropathy: Comparative toxicity of cadmium chloride and cadmium–thionein. J. Toxicol. Environ. Health 2:963.

Van Campen, D. R. 1966. Effects of zinc, cadmium, silver and mercury on the absorption and distribution of copper-64 in rats. J. Nutr. 88:125.

Washko, P. W., and R. J. Cousins. 1976. Metabolism of $^{109}Cd$ in rats fed normal and low-calcium diets. J. Toxicol. Environ. Health 1:1055.

Washko, P. W., and R. J. Cousins. 1977. Role of dietary calcium and calcium-binding protein in cadmium toxicity in rats. J. Nutr. 107:920.

Webb, M., and R. D. Verschoyle. 1976. An investigation of the role of metallothioneins in protection against the acute toxicity of the cadmium ion. Biochem. Pharmacol. 25:673.

Worker, N. A., and B. B. Migicovsky. 1961. Effect of vitamin D on the utilization of zinc. J. Nutr. 75:222.

Wright, F. C., J. S. Palmer, J. C. Riner, M. Haufler, J. A. Miller, and C. A. McBeth. 1977. Effects of dietary feeding of organocadmium to cattle and sheep. J. Agric. Food Chem. 25:293.

Yost, K. J. 1979. Some aspects of cadmium flow in the U.S. Environ. Health Perspect. 28:5.

# Calcium

Calcium (Ca), a white, silvery, alkaline earth metal, was discovered in 1808 by Sir Humphrey Davey. Its name was derived from the Latin word *calx,* a name by which the oxide of the element was known to early Romans. Calcium does not occur in the free state, but its compounds are widely distributed in nature. It is the fifth most abundant element in the earth's crust and the most abundant cation in the animal body, comprising 1 to 2 percent of the total weight. Approximately 99 percent of the calcium in the animal body is found in the bones and teeth, with the remaining 1 percent widely distributed in various soft tissues. Calcium has a very close interrelationship with phosphorus and vitamin D (Hegsted, 1973).

## ESSENTIALITY

Calcium has been recognized as an essential element in animal nutrition for at least 100 years. Quantitatively, the participation of calcium in the formation of bone is its most important function. Bone acts not only as a supporting or structural component of the body, but also as a vital physiological tissue serving to provide a readily available source of calcium for maintenance of homeostasis. The 1 percent of the body's calcium outside of the bone functions in a number of essential processes and is found in extracellular fluid, soft tissue, and as a component of various membrane structures (Bronner, 1964). Calcium is involved in blood coagulation; it is necessary for muscle contractility, myocardial function, normal neuromuscular irritability, and integrity of intracellular cement substances and various membranes. Calcium is also an activator of numerous enzymes.

## METABOLISM

There are many factors that influence the absorption, distribution, and excretion of calcium. Calcium is absorbed from the intestine by an active transport mechanism. Vitamin D is required for this active transport, and there is strong evidence that a vitamin D-induced, calcium-binding protein plays a role (Wasserman and Corradino, 1973). There is evidence that the active transport of calcium is regulated to meet the calcium needs of the body and is most active when dietary calcium is restricted or when the needs are great such as during pregnancy, lactation, and egg production (Wills, 1973). Calcium is also absorbed by passive ionic diffusion, a process that may be vitamin D dependent. The active process is believed to occur mainly in the proximal duodenum and the passive process in the remainder of the small intestine (Avioli, 1972).

The level of calcium in serum, maintained remarkably constant at a concentration of about 10 mg/dl in most species, is regulated by parathyroid hormone (PTH), calcitonin, and metabolically active vitamin D. It is believed that a slight decrease in serum calcium causes an increased secretion of PTH. PTH stimulates the biosynthesis of 1,25-dihydroxy vitamin $D_3$, which causes an increased absorption of calcium from the intestine and an increased resorption of calcium from bone. On the other hand, a slight increase in serum calcium results in a decrease in PTH secretion and an increase in calcitonin release. These changes effect a decreased production of 1,25-dihydroxy vitamin $D_3$, causing a reduction in intestinal absorption and bone resorption of calcium (Tanaka *et al.*, 1973).

Approximately 60 percent of the serum calcium is ionized and physiologically active. The remaining calcium is nonionized and physiologically inert; 35 percent of the serum calcium is bound to protein and 5 percent is complexed with citrate, bicarbonate, and phosphate (White *et al.*, 1968). A significant decrease in serum ionized calcium results in tetany, while an increase can cause cardiac or respiratory failure through an impairment of muscle function (Bronner, 1964; Goodman and Gilman, 1975).

Calcium metabolism is influenced by a large number of factors.

Several reviews have presented detailed discussions (Nicolaysen *et al.*, 1953; Rasmussen and DeLuca, 1963; Bronner, 1964; Hegsted, 1973; Borle, 1973, 1974; Mueller *et al.*, 1973; Schraer *et al.*, 1973).

## SOURCES

Calcium is present in variable amounts in almost all feedstuffs (National Research Council, 1971). The calcium content of natural feeds varies widely, depending upon the species of plant, portion of plant fed, and stage of maturity. In general, grains such as barley, corn, milo, oats, and wheat are very low in calcium, having levels ranging from 0.02 to 0.10 percent. Nonlegume roughages such as grass hay and mature range forages are intermediate in calcium content (0.31 to 0.36 percent), and legume forages such as alfalfa and clover hay contain 1.2 to 1.7 percent calcium.

Several supplemental sources of calcium are used in animal diets, with the most common being ground limestone or calcium carbonate. Other common sources include oyster shell, calcium sulfate, calcium chloride, calcium phosphates, and bone meal. These range in calcium content from 16 to 38 percent.

## TOXICOSIS

Calcium compounds ingested orally are usually not considered toxic. The homeostatic mechanism of the animal tends to protect against the absorption of excessive quantities of the element; however, because of the interrelationship with other nutrients, especially phosphorus, the feeding of high or excessive levels of calcium for extended periods of time can have a detrimental effect on animal performance. Optimum animal performance is linked very closely with calcium and phosphorus levels in the diet. Most animals require a fairly narrow calcium-to-phosphorus ratio—usually no wider than 2:1; however, ruminants can tolerate wider ratios than monogastric animals providing the phosphorus level is adequate. The quantitative aspects of the dietary calcium-to-phosphorus ratios and subsequent animal performance are discussed in several recent reviews (Beeson *et al.*, 1975; Harms *et al.*, 1976; Peo, 1976).

Parturient paresis (milk fever) resulting from calcium deficiency is a metabolic problem, primarily affecting high-producing dairy cows, which is probably related to endocrine function (Beeson *et al.*, 1975).

Both the parathyroid gland and calcitonin-secreting cells of the thyroid may be involved. Dietary calcium intake prior to and during early lactation is important in influencing the capability of these glands to respond appropriately to the sudden change in metabolic demands for calcium imposed by lactation. Black *et al.* (1973) conducted a study with dairy cows in which they compared a high- and a low-calcium diet. The high-calcium diet provided 150 g of calcium per cow daily, whereas the low-calcium diet provided 25 g. The phosphorus intake was 25 g per cow daily for each calcium level. They concluded that cows fed high-calcium diets depend primarily upon intestinal absorption of calcium, whereas in cows receiving less calcium, the significance of the parathyroid hormones in maintaining proper blood calcium levels is greater. They stated that the long-term feeding of a high-calcium diet appears to suppress the chief cells of the parathyroids, so that they are less able to respond quickly to a hypocalcemic condition.

Dowe *et al.* (1957) studied the effect of excessive dietary calcium on growing beef calves. The animals were fed a diet that provided 12 g of phosphorus per day. Calcium-to-phosphorus ratios of 1.3:1, 4.3:1, 9.1:1, and 13.7:1 were compared by the addition of ground limestone. The two higher ratios depressed gain. Ammerman *et al.* (1963) reported that a level of 4.4 percent dietary calcium caused a significant depression in protein and energy digestion by beef steers. Lewis *et al.* (1951) found that excess calcium and a ration borderline in phosphorus content reduced feed consumption and rate of gain of steer calves. The long-term nutritional effects of excessive dietary calcium on bulls was studied by Krook *et al.* (1971). Relative intake varied from 3.5 times recommended requirements in young bulls to 5.9 times in old bulls. It was suggested that the excessive calcium intake resulted in hypersecretion of calcitonin with the result that normal bone resorption was inhibited and bone mass increased. With adequate phosphorus, ruminant animals have been observed to perform satisfactorily with dietary calcium-to-phosphorus ratios between 1:1 and 7:1. Depressed performance has been observed with ratios above 7:1, but these effects were not as extreme as those below 1:1. Most investigators, however, suggest that calcium-to-phosphorus ratios for ruminants should be in the range of 1.5:1 to 2:1 (Beeson *et al.,* 1975).

Fontenot *et al.* (1964) fed calcium-to-phosphorus ratios of 1:1, 2:1, 4:1, and 8:1 (0.3 percent phosphorus) to lambs with and without supplemental zinc (100 ppm). When no supplemental zinc was fed, rate of gain tended to be decreased at the higher calcium levels (4:1, 8:1). Feed efficiency was depressed 9 percent when these high-calcium levels were fed (compared to 1:1). In the zinc-supplemented groups, calcium level had no such effect on rate and efficiency of gain.

Zimmerman *et al.* (1963) conducted a series of experiments to determine optimum calcium and phosphorus levels for maximum body weight gain, feed conversion, and metatarsal calcification of swine 2 to 7 weeks of age. High calcium levels in the diets reduced rate of body weight gain. In general, calcium-to-phosphorus ratios of 1.6:1 or wider adversely influenced gains. High calcium levels in the diet (above 0.8 percent) reduced feed efficiency. The results of this work suggested a maximum dietary calcium level of 0.8 percent and a minimum phosphorus level of 0.6 percent for optimum performance and adequate skeletal development of young swine. Combs *et al.* (1966) observed that when the level of calcium was increased in the diet of young pigs from 0.48 percent to 0.88 percent and to 1.32 percent, average daily gain and feed conversion decreased significantly. The highest level of calcium decreased apparent digestibility of calcium and dry matter. The skin disease, parakeratosis, can be induced by high levels of calcium. Luecke *et al.* (1956) produced a 100 percent incidence of parakeratosis in pigs on a diet with 1.5 percent calcium, 0.8 percent phosphorus, and 31 ppm zinc. The addition of 20 ppm of zinc to the diet essentially eliminated the disease. Newland *et al.* (1958) stated that pigs fed high-calcium diets had an accelerated rate of zinc metabolism, thereby increasing the requirement for zinc. A more complete review of the effect of dietary calcium levels in swine is presented by Peo (1976).

Urbanyi (1960) determined that levels of calcium carbonate above 3 percent of the diet (1.2 percent calcium) had an adverse effect on the body weight gains from a 4-week chick experiment. In a large-scale broiler trial by Smith and Taylor (1961), significantly different growth rate and feed conversion values were detected between groups fed either 0.83 or 1.35 percent calcium with a phosphorus level of 0.52 percent for a 10-week feeding period, with the 0.83 percent treatment being superior. Fangauf *et al.* (1961) fed chicks with diets containing 1.2, 2.0, 3.0, 4.5, and 6.5 percent calcium and concluded that broiler diets should not contain over 2 percent calcium. Levels above this significantly depressed feed intake and body weight gain and increased mortality. Shane *et al.* (1969) reported on a growing pullet study showing that levels of calcium above 2.5 percent fed between 8 and 20 weeks of age would cause nephrosis, visceral gout, calcium urate deposits, and high mortality. Parathyroid size and activity were reduced, along with feed consumption and weight gains. They recommended that diets should not contain more than 1.2 percent calcium for pullets under 18 to 20 weeks of age. There are conflicting reports on the effect of excess calcium in the diet of laying hens. Gutowska and Parkhurst (1942) reported that a high level (3.95 percent) of calcium in the diet decreased egg production and feed efficiency. There was no significant difference in eggshell breaking strength or egg weight. MacIntyre *et al.* (1963) reported that levels of dietary calcium up to 6 percent did not depress egg production, feed efficiency, or egg weight. Harms and Waldroup (1971) also found that calcium levels up to 5 percent did not significantly affect the rate of egg production, egg weight, eggshell thickness, or feed consumption per hen per day during the 4-month feeding period.

There is evidence that levels of calcium greater than 1 percent in the diet may have an adverse effect on the growth of laboratory animals (Davis, 1959). Goto and Sawamura (1973) reported that rats fed high-calcium diets (2.83 percent added calcium) had significant weight loss and decreased absorption and retention of nitrogen when compared to the control group (0.72 percent added calcium). Increasing the calcium level in the diet (above 1 percent) can have a depressing effect upon the utilization of other nutrients in the diet, including protein, fats, vitamins, the macromineral elements—phosphorus and magnesium—and the trace mineral elements, especially iron, iodine, zinc, and manganese (Davis, 1959). In those instances where the intake of these nutrients is just adequate, increasing the calcium in the diet may have a markedly adverse effect and may produce a deficiency of the other nutrients.

There are many factors that influence the absorption, distribution, and excretion of calcium (Church and Pond, 1974). The level of dietary phosphorus, the ratio of calcium to phosphorus, and the level of dietary vitamin D are the most important.

## MAXIMUM TOLERABLE LEVELS

Many sources of calcium are inexpensive, and animals have a tolerance for widely differing levels in the diet. This has often led to the neglect of dietary calcium levels and their balance with other nutrients, primarily mineral elements. Except for the laying hen, which has a high requirement for calcium, it is inadvisable to maintain dietary calcium levels much above 1 percent. Experiments with most species have shown that optimum animal performance is obtained when the ratio of calcium to phosphorus in the diet is in the range of approximately 1.5:1 to 2:1. If adequate amounts of phosphorus are included in the diet, then a wider ratio of calcium to phosphorus can be tolerated. In ruminant animals, diets with a ratio of calcium to phosphorus as wide as 7:1 have performed satisfactorily; however, the phosphorus has been present at levels well above that which is considered a requirement level.

Assuming the presence of adequate levels of dietary phosphorus, the following calcium levels can be tolerated; however, depending on the age and production status of the animal, even these levels may reduce performance: cattle, 2 percent; sheep, 2 percent; swine, 1 percent; poultry, 1.2 percent; laying hen, 4 percent; horse, 2 percent; rabbit, 2 percent.

## SUMMARY

Calcium does not occur in the free state, but its compounds are widely distributed and it is the most abundant cation in the animal body. It is an essential element in animal nutrition and serves a principal role in bone formation, as well as a number of other physiological processes. Primary factors that influence the metabolism of calcium include phosphorus, vitamin D, hormonal systems, and the age of the animal. Calcium compounds ingested orally are usually not considered toxic. The homeostatic mechanism of the animal tends to protect against the absorption of excessive quantities of the element; however, because of the interrelationship with other nutrients, especially phosphorus, the feeding of high or excessive levels of calcium for extended periods of time can have a detrimental effect on animal performance.

The addition of extra calcium to an otherwise adequate diet may precipitate a deficiency of other essential elements, i.e., phosphorus, magnesium, iron, iodine, zinc, and manganese. In every case, it appears that the injurious effect of the calcium is due to an interaction rather than to a harmful effect of the calcium itself. Additional quantities of the respective elements will overcome the adverse effect of the calcium similar to that observed by reducing the level of calcium in the diet. A relationship also exists between high dietary calcium and the utilization of protein and fat. Because of the intricate interrelations between calcium metabolism and that of other nutrients, caution must be observed when calcium supplements are being added to the diet.

TABLE 13 Effects of Calcium Administration in Animals

| Class and Number of Animals[a] | Age or Weight | Administration: Quantity of Element | Source | Duration | Route | Effect(s) | Reference |
|---|---|---|---|---|---|---|---|
| Cattle—19 | 8 mo (213 kg) | 16, 51, 106, and 160 g/d (12 g/d of P resulting in Ca:P ratios of 1.4:1, 4.4:1, 9.1:1, and 13.7:1) | Limestone ($CaCO_3$) | 140 d | Diet | Gains decreased as the calcium in the diet was increased | Dowe *et al.*, 1957 |
| Cattle—2 | 2 yr | 1.84% Ca, 0.18% P<br>1.84% Ca, 0.36% P<br>4.40% Ca, 0.36% P | $CaCO_3$ | 28 d | Diet | Protein and energy digestion significantly reduced at 4.4% dietary calcium | Ammerman *et al.*, 1963 |
| Cattle | 0.5–17 yr | 67, 72, and 88 g/d | $CaHPO_4 \cdot 2H_2O$ Natural feedstuffs | Up to 13 yr | Diet | Hypersecretion of calcitonin; arrested bone resorption and osteopetrosis | Krook *et al.*, 1971 |
| Sheep—40 | Fattening lambs | 0.3, 0.6, 1.2, and 2.4% (0.3% P) | | Growing and finishing period | Diet | Rate of gain and feed efficiency decreased at the higher calcium levels; this was overcome by zinc supplementation | Fontenot *et al.*, 1964 |
| Swine—30 | Baby pigs, 2–3 wk | 0.5, 0.7, 0.9, and 1.0% (0.5 or 0.7% P) | $CaCO_3$ | 28 d | Diet | High calcium levels in the diet reduced rate of gain and feed effi- | Zimmerman *et al.*, 1963 |

| | | | | | | | |
|---|---|---|---|---|---|---|---|
| | | | | | | ciency; results suggested a maximum calcium level of 0.8% with a minimum phosphorus level of 0.6% | |
| Swine—24 | Baby pigs, 2 wk | 0.48, 0.88, and 1.32% (Ca:P ratio 1:1, 2:1, and 3:1) | Limestone ($CaCO_3$) | 6 wk | Diet | Significant decrease in gain and feed efficiency as dietary calcium level increased | Combs *et al.*, 1966 |
| Chicken | 1 d | 0.15, 0.5, 0.9, 1.2, 2.1, and 4.1% (0.65% P, Ca:P ratio from 0.2:1 to 6.5:1) | $CaCO_3$ | 4 wk | Diet | Body weight gain and feed efficiency reduced when dietary calcium increased beyond 1.2% | Urbanyi, 1960 |
| Chicken—1,715 | 1 d | 0.83 and 1.35% (0.52% P) | $CaCO_3$ | 10 wk | Diet | Significantly reduced body weight and feed efficiency at the 1.35% level as compared to the 0.83% level | Smith and Taylor, 1961 |
| Chicken—50 | 8 wk | 0.6 and 3.0% (0.4% P) | $CaCO_3$ | 12 wk | Diet | Visceral urate deposits, mortality, nephrosis, and smaller parathyroid glands observed on the high-calcium treatment | Shane *et al.*, 1969 |
| Rat | 60 g | 0.72 and 2.83% (0.17 and 1.6% P) | Ca lactate | 30 d | Diet | Significant loss and decreased absorption and retention of nitrogen when fed the high-calcium diet | Goto and Sawamura, 1973 |

[a] Number of animals per treatment.

## REFERENCES

Ammerman, C. B., L. R. Arrington, M. C. Jayaswal, R. L. Shirley, and G. K. Davis. 1963. Effect of dietary calcium and phosphorus levels on nutrient digestibility by steers. J. Anim. Sci. 22:248.

Avioli, L. V. 1972. Intestinal absorption of calcium. Arch. Int. Med. 129:345.

Beeson, W. M., T. W. Perry, N. L. Jacobson, K. D. Wiggers, and G. N. Jacobson. 1975. Calcium in Beef and Dairy Nutrition. National Feed Ingredients Association, Des Moines, Iowa.

Black, H. E., C. C. Capen, and C. D. Arnaud. 1973. Ultrastructure of parathyroid glands and plasma immunoreactive parathyroid hormone in pregnant cows fed normal and high calcium diets. Lab. Invest. 29:173.

Borle, A. B. 1973. Calcium metabolism at the cellular level. Fed. Proc. 32:1944.

Borle, A. B. 1974. Calcium and phosphate metabolism. Ann. Rev. Physiol. 36:361.

Bronner, F. 1964. Dynamics and function of calcium, pp. 341–444. *In* C. L. Comar and F. Bronner, eds., Mineral Metabolism, vol. 2, part A. Academic Press, New York.

Church, D. C., and W. G. Pond. 1974. Basic Animal Nutrition and Feeding. D. C. Church, Corvallis, Oreg.

Combs, G. E., T. H. Berry, H. D. Wallace, and R. C. Crum, Jr. 1966. Levels and sources of vitamin D for pigs fed diets containing varying quantities of calcium. J. Anim. Sci. 25:827.

Davis, G. K. 1959. Effects of high calcium intakes on the absorption of other nutrients. Fed. Proc. 18:1119.

Dowe, T. W., J. Matsushima, and V. H. Arthaud. 1957. The effects of adequate and excessive calcium when fed with adequate phosphorus in growing rations for beef calves. J. Anim. Sci. 16:811.

Fangauf, R., H. Vogt, and W. Penner. 1961. Studies of calcium tolerance in chickens. Arch. Geflugelk. 25:82.

Fontenot, J. P., R. F. Miller, and N. O. Price. 1964. Effects of calcium level and zinc supplementation of fattening lamb rations. J. Anim. Sci. 23:874.

Goodman, L. S., and A. Gilman. 1975. The Pharmacological Basis of Therapeutics, 5th ed. The Macmillan Company, New York.

Goto, S., and T. Sawamura. 1973. Effect of excess calcium intake on absorption of nitrogen, fat, phosphorus, and calcium in young rats. J. Nutr. Sci. Vitaminal. 19:355.

Gutowska, M. S., and R. T. Parkhurst. 1942. Studies in mineral nutrition of laying hens. 11. Excess of calcium in the diet. Poult. Sci. 21:321.

Harms, R. H., and P. W. Waldroup. 1971. The effect of high dietary calcium on the performance of laying hens. (Research Notes) Poult. Sci. 50:967.

Harms, R. H., B. L. Damron, D. A. Roland, and L. M. Potter. 1976. Calcium in Broiler, Layer and Turkey Nutrition. National Feed Ingredients Association, Des Moines, Iowa.

Hegsted, D. M. 1973. Modern Nutrition in Health and Disease, 5th ed. Lea and Febiger, Philadelphia.

Krook, L., L. Lutwak, K. McEntee, P. A. Hendrickson, K. Braun, and S. Roberts. 1971. Nutritional hypercalcitonism in bulls. Cornell Vet. 61:625.

Lewis, J. K., W. H. Burkitt, and F. S. Willson. 1951. The effect of excess calcium with borderline and deficient phosphorus in rations of steer calves. J. Anim. Sci. 10:1053.

Luecke, R. W., J. A. Hoefer, W. S. Brammell, and F. Thorp, Jr. 1956. Mineral interrelationships of parakeratosis of swine. J. Anim. Sci. 15:347.

MacIntyre, T. M., H. W. R. Chancey, and E. E. Gardiner. 1963. Effect of dietary energy and calcium level on egg production and egg quality. Can. J. Anim. Sci. 43:337.

Mueller, W. J., R. L. Brubaker, C. V. Gay, and J. N. Boelkins. 1973. Mechanisms of bone resorption in laying hens. Fed. Proc. 32:1951.

National Research Council. 1971. Atlas of Nutritional Data on United States and Canadian Feeds. National Academy of Sciences, Washington, D.C.

Newland, H. W., D. E. Ullrey, J. A. Hoefer, and R. W. Luecke. 1958. The relationship of dietary calcium to zinc metabolism in pigs. J. Anim. Sci. 17:886.

Nicolaysen, R., N. Eeg-Larsen, and O. J. Malm. 1953. Physiology of calcium metabolism. Physiol. Rev. 33:424.

Peo, E. R., Jr. 1976. Calcium in Swine Nutrition. National Feed Ingredients Association, Des Moines, Iowa.

Rasmussen, H., and H. F. DeLuca. 1963. Calcium homeostasis. Ergeb. Physiol. 53:109.

Schraer, R., J. A. Elder, and H. Schraer. 1973. Aspects of mitochondrial function in calcium movement and calcification. Fed. Proc. 32:1938.

Shane, S. M., R. J. Young, and L. Krook. 1969. Renal and parathyroid changes produced by high calcium intake in growing pullets. Avian Dis. 13:558.

Smith, H., and J. H. Taylor. 1961. Effect of feeding two levels of dietary calcium on the growth of broiler chickens. Nature 190:1200.

Tanaka, Y., H. Frank, and H. F. DeLuca. 1973. Role of 1,25-dihydroxy-cholecalciferol in calcification of bone and maintenance of serum calcium concentration in the rat. J. Nutr. 102:1569.

Urbanyi, L. 1960. Chicken feeding trials with diets containing sufficient phosphorus and increasing amounts of calcium carbonate. Nutr. Abstr. 30:691.

Wasserman, R. H., and R. A. Corradino. 1973. Vitamin D, calcium and protein synthesis. Vitam. Horm. 31:43.

White, A., P. Handler, and E. L. Smith. 1968. Principles of Biochemistry, 4th ed. McGraw-Hill, New York.

Wills, M. R. 1973. Intestinal absorption of calcium. Lancet 1:820.

Zimmerman, D. R., V. C. Speer, V. W. Hays, and D. V. Catron. 1963. Effect of calcium and phosphorus levels on baby pig performance. J. Anim. Sci. 22:658.

# Chromium

The name for the element chromium (Cr) was derived from the Greek designation for color because all chromium compounds have color. Chromium is a hard metal that takes a high polish. It is used in steel and other alloys, in numerous industrial and chemical procedures, in paints and dyes, and in the tanning industry. When electroplated, it produces the hard, noncorrosive, lustrous surface commonly seen as "chrome" on many consumer items. Chromium is derived from the ore, chromite ($FeOCr_2O_3$), by the reduction of the oxide with aluminum (Chemical Rubber Co., 1971/1972). Use of this element in the United States increased by 50 percent in the years 1948–1968 and was estimated at 1⅓ million tons in 1968 (Knapp, 1971). Reviews on chromium include Underwood (1977) and the National Research Council (1974).

## ESSENTIALITY

Mertz (1967) and Schroeder (1968) have investigated the role of chromium as an essential element for animals and concluded that chromium (III) is required for normal carbohydrate and lipid metabolism, as first suggested by Curran (1954). In rats fed low-chromium diets in a nearly chromium-free environment, severe impairment of glucose tolerance was observed. The condition was particularly evident in older breeding rats and was improved gradually by chromium supplementation (Mertz *et al.,* 1965a,b). There was a higher requirement for chromium in human subjects with an impaired glucose tolerance test (Mertz *et al.*, 1977).

Life-term studies with mice and rats given 5 ppm chromium (III) in drinking water showed increased growth over controls for both sexes and decreased mortality of males (Schroeder *et al.*, 1963a,b). Subsequent life-term studies with mice, given 5 ppm chromium (VI) in drinking water, showed slight weight decreases compared with controls.

The inorganic salts of chromium also improve glucose tolerance (Gurson and Saner, 1973). More information is needed on the biologically active chemical state(s) and physiological role(s) of chromium. It is possible that a deficiency of this element may exist for species ingesting highly purified diets and for animals in which stress has depleted body stores. Supplementing laying hens with 10 ppm chromium as $CrCl_3$ for 28 days improved interior egg quality as measured by Haugh units (Jensen *et al.*, 1978).

## METABOLISM

Chromium (III) is required for utilization of glucose in peripheral tissues, acting in conjunction with insulin. The biologically active form of chromium is called glucose tolerance factor (GTF). It is a small organic molecule containing nicotinic acid, glycine, glutamic acid, cysteine, and chromium (Mertz *et al.*, 1974; Mertz, 1975), but its exact structure is not yet known. The content of chromium in the human body is known to decrease throughout life (Schroeder *et al.*, 1962), and evidence of chromium (GTF) deficiency in adults has been obtained (Freund *et al.*, 1979). The extent to which GTF deficiency occurs is unknown; however, it is thought to be common in the elderly.

Absorption of orally administered chromium (III) is very low regardless of nutritional status and dosage (Mertz, 1967); the major excretory route of absorbed chromium is the urinary tract, although feces contain some $^{51}Cr$ activity following intravenous dosage with $^{51}CrCl_3$. Oral administration of $^{51}CrCl_3$ to rats resulted in 5–10 percent absorption within 5 minutes of stomach tubing, but chromium retained by the animals decreased to less than 1 percent at 1 hour (Polansky and Anderson, 1978).

When Mertz *et al.* (1969) administered chromium (III) up to 250 $\mu$Ci as $^{51}CrCl_3 \cdot 6H_2O$ to pregnant rats, either by stomach tube or intravenously, no activity was found in the young; but, with intragastric administration of labeled chromic chloride incorporated into brewer's yeast, the isotope was transported to the young at an average level of 20 percent of the dose administered to the mother. Diets high in natural chromium also increased chromium levels in the young, while 2 ppm chromium (III), as the acetate, in drinking water had no effect.

The greatest proportion of chromium (III) in tissues of rats 96 hours after having been injected intravenously with $^{51}CrCl_3$ was found in kidney and spleen. Chromium (III) was excreted mainly via urine (Hopkins, 1965; Mathur and Doisy, 1972). When Kraintz and Talmage (1952) injected rats and rabbits with $^{51}CrCl_3$, the greatest concentration was found in bone marrow. The chromium (III) was associated closely with all serum proteins in rabbits 24 hours following intravenous injection. Sukhacheva *et al.* (1978) injected 0.05 mg of radioactive chromium per rat as $Na_2Cr_2O_7$ (chromium VI) and $CrCl_3$ (chromium III). Chromium (III) and (VI) were accumulated in spleen but only the trivalent form accumulated in liver. The blood clearance of chromium (VI) was more rapid than that of chromium (III). Chromium (III) associated closely with serum proteins, while chromium (VI) was bound to red blood cells (Gray and Sterling, 1950).

Chromium (III) as chromic oxide ($Cr_2O_3$) has been used for several decades as a fecal marker in digestibility and absorption studies in many species: chicks (Dansky and Hill, 1952; Hill and Anderson, 1958), rats (Schurch *et al.*, 1950), sheep (Reid *et al.*, 1950; Woolfolk *et al.*, 1950; Lassiter *et al.*, 1966), and humans (Irwin and Crampton, 1951; Whitby and Lang, 1960). Fecal chromium recoveries have been variable, generally ranging between 90 and 100 percent. Radioactive chromium has been a useful tagging agent in tracing the fate of river waters to the sea, because it is not appreciably concentrated by river or oceanic biota (Osterberg *et al.*, 1965).

## SOURCES

Chromium is ubiquitous in water, soil, and living matter. Wide variation among concentrations reported may be due to differences in analytical procedures, standards, and geographical locations. Values for chromium in foods, flora, forage, and soil were given by Schroeder *et al.* (1962), most of which were less than 100 ppb. The level of chromium in feed grade phosphates ranged from 39 ppm in defluorinated phosphate to 128 ppm in dicalcium phosphate (unpublished data, International Minerals and Chemical Corp., Libertyville, Ill.). Summarization of water analysis (National Research Council, 1974) indicated a range of 1 to 112 ppb chromium with a mean of 8 ppb.

Definition of the chemistry of naturally occurring chromium complexes is incomplete, and information on the biological availability of these compounds is limited. Naturally occurring complexed chromium, however, appears to be better utilized than the inorganic salts (Mertz *et al.*, 1969). The trivalent and hexavalent forms of chromium are the most stable and are encountered more commonly than is the divalent state. Most of the chromic and dichromate compounds are soluble, while chromites are highly insoluble. Of the three common valency states—II, III, and VI—only the trivalent chromium forms octahedral coordination complexes and polynuclear olated complexes by hydrolysis in aqueous solution, decreasing solubility of the ion. Hexavalent chromium has acidic properties, does not form coordination compounds, and is easily reduced to chromium (III); but it has also been found in natural materials (Mertz, 1967). Chromium (II) is also rapidly converted to chromium (III).

## TOXICOSIS

Because of their protein-precipitating and oxidizing properties, chromium trioxide, chromates, and bichromates are potent protoplasmic poisons, but chromic oxide, trivalent chromium salts, and metallic chromium are much less toxic (Pascale *et al.*, 1952). Chromium toxicity has been reviewed by MacKenzie *et al.* (1958).

### LOW LEVELS

Signs of chronic oral toxicosis of chromium (III) and chromium (VI) differ among species. They consist primarily of skin contact dermatitis and sores, irritation of respiratory passages, ulceration and perforation of the nasal septum, and lung cancer.

The toxicity of chromium (III) when administered by the oral route has been studied very little. No adverse effects (Table 14) were observed in either mice or rats given 5 ppm chromium (III) as chromium acetate in drinking water throughout their life (Schroeder *et al.*, 1964, 1965). MacKenzie *et al.* (1958) gave both forms to rats in drinking water for 1 year at levels up to 25 ppm. Tissue deposition of chromium increased in liver, kidney, bone, and spleen, with the spleen considerably higher in chromium content than the other three tissues when chromium (VI) was administered. Physiological effects were not different at 0 to 11 ppm chromium. With an intake of 25 ppm, animals receiving chromium (VI) showed tissue chromium contents 9 times greater than those ingesting chromium (III) but with no adverse effects on weight gain, food consumption, or pathology from either form of chromium. Table 14 shows effects on young chicks of chromium in poultry feeds. Chromium (III) at levels up to 500 ppm in water did not adversely affect growth of rats and mice (Table 14), while 25 ppm chromium (VI) decreased water consumption in rats. Growing rats fed varying levels of Cr-nicotinic acid complex up to 276 ppm chromium exhibited no abnormalities after 20 weeks (Mertz, 1975; Mertz and Roginski, 1975).

### HIGH LEVELS

Few cases of acute systemic intoxication by chromium ingestion in man or animals have been reported. The $LD_{50}$ was 18 and 60 mg chromium per kilogram of body weight, respectively, for chromium as chromalum [$KCr(SO_4)_2 \cdot 12H_2O$] and a chromium (III) nicotinic acid complex with high GTF activity in rats injected intravenously (Mertz, 1975; Mertz and Roginski, 1975). The lethal single oral dose for chromium (VI) in young rats was 130 mg/kg, while as much as 650 mg/kg body weight of chromium (III) produced no overt toxicosis (Samitz *et al.*, 1962). Garner's *Veterinary Toxicology* (1967) has given the acute lethal dose of chromate (VI) for mature cattle at around 700 mg/kg, while 30–40 mg/kg of body weight produced chromium poisoning in young calves. Inflammation and congestion of the stomach, ulceration of the rumen and abomasum, and high blood and liver chromium levels were characteristic findings. Chromium levels of 30 ppm in liver and 4 ppm in blood, compared with normal levels of less than 2 ppm, were suggested as indicative of chromium poisoning.

### FACTORS AFFECTING TOXICITY

Hill and Matrone (1970) investigated the influence of $CrCl_3 \cdot 6H_2O$, chromium (III), in alleviating adverse effects of ammonium vanadate fed to chicks at 0, 10, and 20 ppm added vanadium. At 20 ppm supplementary vanadium, growth was depressed and mortality was high. Additions of chromium (III) at dietary levels of 500, 1,000, and 2,000 ppm significantly alleviated both problems but did not completely overcome vanadium toxicosis. Without added vanadium, 2,000 ppm chromium (III) depressed growth, but mortality was unaffected. Levels of 500 and 1,000 ppm supplementary chromium (III) without added vanadium produced no significant differences in chicks from those receiving no added chromium or vanadium.

## TISSUE LEVELS

Generally, mammalian tissue concentrations range from 10 to several hundred ppb (Mertz, 1967). Chromium (VI) rarely occurs in biological tissues. Schroeder *et al.* (1962) listed chromium concentrations on a fresh weight basis of various animal and human tissues as 100 ppb or less and found that tissue concentrations decrease with age except in the lung. An intake of 25 ppm chromium (VI) increased tissue chromium levels 9 times that which was found when chromium (III) was fed.

## MAXIMUM TOLERABLE LEVELS

Chromic oxide ($Cr_2O_3$) (III) has been used as a fecal marker in cattle and sheep for periods of several weeks at levels as high as 3,000 ppm chromium with no evidence of adverse effects. Chicks were fed 1,000 ppm chromium as chromic chloride ($CrCl_3$) (III) without effect, but 2,000 ppm resulted in reduced growth.

Potassium chromate ($K_2CrO_4$) (VI) and sodium chromate ($Na_2CrO_4$) (VI) have been fed to chicks at levels of 100 ppm with no adverse effects. Tissue levels of chromium were increased in rats offered 7.7 ppm chromium in the water as potassium chromate, and decreased water intake occurred with 25 ppm of the element in the water. Chromic chloride had no effect on rats when offered as 25 ppm chromium in the water.

Maximum tolerable dietary levels are set at 3,000 ppm chromium as the oxide and 1,000 ppm as the chloride for domestic animals.

## SUMMARY

Chromium (III) is essential in animal nutrition as a constituent of glucose tolerance factor (GTF). The GTF organic complex is 50 times more active biologically than inorganic chromium (III), while chromium (VI) is rarely found in living systems. Chromium trioxide, chromates, and bichromates are toxic due to their protein-precipitating and oxidizing properties. Chronic chromium toxicosis results in skin contact dermatitis, irritation of respiratory passages, ulceration and perforation of the nasal septum, and lung cancer. Chromium (VI) compounds appear to be more toxic than chromium (III) compounds. Acute systemic chromium intoxication is rare but was produced with a single oral dose of 700 mg/kg of body weight chromium (VI) in mature cattle and 30–40 mg/kg of body weight chromium (VI) in young calves. Signs of acute toxicosis included inflammation and congestion of the stomach, ulceration of the rumen and abomasum, and increased concentration of blood and liver chromium.

TABLE 14 Effects of Chromium Administration in Animals

| Class and Number of Animals[a] | Age or Weight | Administration: Quantity of Element[b] | Source | Duration | Route | Effect(s) | Reference |
|---|---|---|---|---|---|---|---|
| Chicken—24 | Immature | 3 ppm | $CrCl_3$ | 27 d | Diet | No adverse effect | Baker and Molitoris, 1975 |
| Chicken—52 | Immature | 30 ppm | $Na_2CrO_4$ | 32 d | Diet | No adverse effect | Romoser *et al.*, 1961 |
| | | 100 ppm | | | | No adverse effect | |
| Chicken | Immature | 100 ppm | $K_2CrO_4$ | 21 d | Diet | No adverse effect | Mertz and Roginski, 1975; Mertz, 1975 |
| Chicken—15 | Immature | 500 ppm | $CrCl_3$ | 21 d | Diet | No adverse effect | Hill and Matrone, 1970 |
| | | 1,000 ppm | | | | No adverse effect | |
| | | 2,000 ppm | | | | Reduced growth | |
| Rat—50 | Immature–mature | 5 ppm | $Cr(CH_3COO)_3$ | Life-term | Water | No adverse effect | Schroeder *et al.*, 1965 |
| Rat | Immature | 300 ppm | $K_2CrO_4$ | 180 d | Water | No adverse effect | Gross and Heller, 1946 |
| | | 500 ppm | | | | No adverse effect | |
| Rat—16 | 34 d | 0.45 ppm | $K_2CrO_4$ | 1 yr | Water | No adverse effect | MacKenzie *et al.*, 1958 |
| | | 2.2 ppm | | | | No adverse effect | |
| | | 4.5 ppm | | | | No adverse effect | |
| | | 7.7 ppm | | | | Increased chromium in tissues | |

TABLE 14 *Continued*

| Class and Number of Animals[a] | Age or Weight | Administration: Quantity of Element[b] | Source | Duration | Route | Effect(s) | Reference |
|---|---|---|---|---|---|---|---|
| | | 11 ppm | | | | Increased chromium in tissues | |
| Rat—20 | 32 d | 25 ppm | $K_2CrO_4$ | 1 yr | Water | Decreased water intake; concentrated in tissues 9 times more than $Cr^{3+}$ but no toxic signs | MacKenzie *et al.*, 1958 |
| | | 25 ppm | $CrCl_3$ | | | No adverse effect | |
| Rat | Immature | 276 ppm | Chromium (III) nicotinic acid complex | 140 d | Diet | No adverse effect | Mertz and Roginski, 1975; Mertz, 1975 |
| Mouse—50 | Immature–mature | 5 ppm | $Cr(CH_3COO)_3$ | Life-term | Water | No adverse effect | Schroeder *et al.*, 1964 |

| | | | | | | | |
|---|---|---|---|---|---|---|---|
| Mouse—54 | Weanling | 5 ppm | Chromium (VI) | Life-term | Water | Reduced growth | Schroeder and Mitchener, 1971 |
| Mouse—62 | Weanling | 5 ppm | $Cr(CH_3COO)_3$ | Life-term | Water | No adverse effect | Schroeder *et al.*, 1963a,b, |
| Mouse | Weanling | 10 ppm | Chromium (VI) | Life-term | Water | Increased chromium in tissues | Schroeder and Nason, 1976 |
| Mouse | Mature | 100 ppm | $K_2CrO_4$ | 180 d | Water | No adverse effect | Gross and Heller, 1946 |
| | | 200 ppm | | | | No adverse effect | |
| | | 300 ppm | | | | No adverse effect | |
| | | 400 ppm | | | | No adverse effect | |
| | | 500 ppm | | | | No adverse effect | |
| Guinea pig—12 | Adult | 5 ppm | $CrCl_3$ | 147 d | Diet | Increased glucose tolerance | Mertz and Roginski, 1975; Mertz, 1975 |
| | | 50 ppm | | | | Increased glucose tolerance | |

[a] Number of animals per treatment group.
[b] Quantity expressed in parts per million.

## REFERENCES

Baker, D. H., and B. A. Molitoris. 1975. Lack of response to supplemental tin, vanadium, chromium and nickel when added to a purified crystalline amino acid diet for chicks. Poult. Sci. 54:925.

Chemical Rubber Co. 1971/1972. Handbook of Chemistry and Physics, 52nd ed. The Chemical Rubber Co., Cleveland, Ohio.

Curran, G. L. 1954. Effect of certain transition group elements on hepatic synthesis of cholesterol in the rat. J. Biol. Chem. 210:765.

Dansky, L. M., and F. W. Hill. 1952. Application of the chromic oxide indicator method to balance studies with growing chickens. J. Nutr. 47:449.

Freund, H., S. Atamian, and J. E. Fischer. 1979. Chromium deficiency during total parenteral nutrition. J. Am. Med. Assoc. 241:496.

Garner, R. J. 1967. Veterinary Toxicology, 3rd ed. Balilliere, Tindal & Cassell, London.

Gray, S. J., and K. Sterling. 1950. The tagging of red cells and plasma proteins with radioactive chromium. J. Clin. Invest. 29:1604.

Gross, W. G., and V. G. Heller, 1946. Chromates in animal nutrition. J. Ind. Hyg. Toxicol. 28:52.

Gurson, C. T., and G. Saner. 1973. Effects of chromium supplementation on growth in marasmic protein–calorie malnutrition. Am. J. Clin. Nutr. 26:988.

Hill, C. H., and G. Matrone. 1970. Chemical parameters in the study of *in vivo* and *in vitro* interactions of transition elements. Fed. Proc. 29:1474.

Hill, F. W., and D. L. Anderson. 1958. Comparison of metabolizable energy and productive energy determinations with growing chicks. J. Nutr. 64:587.

Hopkins, L. L., Jr. 1965. Distribution in the rat of physiological amounts of injected $Cr^{51}$ (III) with time. Am. J. Physiol. 209:731.

Irwin, M. I., and E. W. Crampton. 1951. The use of chromic oxide as an index material in digestion trials with human subjects. J. Nutr. 43:77.

Jensen, L. S., D. V. Maurice, and M. W. Murray. 1978. Evidence for a new biological function of chromium. Fed. Proc. 37:404.

Knapp, C. E. 1971. Beryllium-hazardous air pollutant. Environ. Sci. Technol. 5:584.

Kraintz, L., and R. V. Talmage. 1952. Distribution of radioactivity following intravenous administration of trivalent chromium 51 in the rat and rabbit. Proc. Soc. Exp. Biol. Med. 81:490.

Lassiter, J. W., V. Alligood, and C. H. McGaughey. 1966. Chromic oxide as an index of digestibility of all-concentrate rations for sheep. J. Anim. Sci. 25:44.

MacKenzie, R. D., R. U. Byerrum, C. F. Decker, C. A. Hoppert, and R. F. Langham. 1958. Chronic toxicity studies. II. Hexavalent and trivalent chromium administered in drinking water to rats. AMA Arch. Ind. Health 18:232.

Mathur, R. K., and R. J. Doisy. 1972. Effect of diabetes and diet on the distribution of tracer doses of chromium in rats. Proc. Soc. Exp. Biol. Med. 139:836.

Mertz, W. 1967. Biological role of chromium. Fed. Proc. 26:186.

Mertz, W. 1975. Effects and metabolism of glucose tolerance factor. Nutr. Rev. 33:129.

Mertz, W., and E. E. Roginski. 1975. Some biological properties of chromium (Cr)–nicotinic acid (NA) complexes. Fed. Proc. 34:922 (Abstr.).

Mertz, W., E. E. Roginski, and H. A. Schroeder. 1965a. Some aspects of glucose metabolism of chromium-deficient rats raised in a strictly controlled environment. J. Nutr. 86:107.

Mertz, W., E. E. Roginski, and R. C. Reba. 1965b. Biological activity and fate of trace quantities of intravenous chromium (III) in the rat. Am. J. Physiol. 209:489.

Mertz, W., E. E. Roginski, F. J. Feldman, and D. E. Thurman. 1969. Dependence of chromium transfer into the rat embryo on the chemical form. J. Nutr. 99:363.

Mertz, W., E. W. Toepfer, E. E. Roginski, and M. M. Polansky. 1974. Present knowledge of the role of chromium. Fed. Proc. 33:2277.

Mertz, W., W. R. Wolf, and E. E. Roginski. 1977. Relation of chromium excretion to glucose metabolism in human subjects. Fed. Proc. 36:1152 (Abstr.).

National Research Council. 1974. Chromium. National Academy of Sciences–National Research Council, Washington, D.C.

Osterberg, C., N. Cutshall, and J. Cronin. 1965. Chromium-51 as a radioactive tracer of Columbia River water at sea. Science 150:1585.

Pascale, L. R., S. S. Waldstein, G. Engbring, and A. Dubin. 1952. Chromium intoxication with special reference to hepatic injury. J. Am. Med. Assoc. 149:1385.

Polansky, M. M., and R. A. Anderson. 1978. Rapid absorption of chromium. Fed. Proc. 37:895 (Abstr.).

Reid, J. T., P. G. Woolfolk, C. R. Richards, R. W. Kaufmann, J. K. Loosli, K. L. Turk, J. I. Miller, and R. E. Blaser. 1950. A new indicator method for the determination of digestibility and consumption of forages by ruminants. J. Dairy Sci. 33:60.

Romoser, G. L., W. A. Dudley, L. J. Machlin, and L. Loveless. 1961. Toxicity of vanadium and chromium for the growing chick. Poult. Sci. 40:1171.

Samitz, M. H., J. Shrager, and S. Katz. 1962. Studies on the prevention of injurious effects of chromates in industry. Ind. Med. Surg. 31:427.

Schroeder, H. A. 1968. The role of chromium in mammalian nutrition. Am. J. Clin. Nutr. 21:230.

Schroeder, H. A., and M. Mitchener. 1971. Scandium, chromium (VI), gallium, yttrium, rhodium, palladium, indium in mice: Effects on growth and life span. J. Nutr. 101:1431.

Schroeder, H. A., and A. P. Nason. 1976. Interactions of trace metals in mouse and rat tissues; zinc, chromium, copper and manganese with 13 other elements. J. Nutr. 106:198.

Schroeder, H. A., J. J. Balassa, and I. H. Tipton. 1962. Abnormal trace metals in man—chromium. J. Chron. Dis. 15:941.

Schroeder, H. A., W. H. Vinton, Jr., and J. J. Balassa. 1963a. Effect of chromium, cadmium and other trace metals on the growth and survival of mice. J. Nutr. 80:39.

Schroeder, H. A., W. H. Vinton, Jr., and J. J. Balassa. 1963b. Effects of chromium, cadmium and lead on the growth and survival of rats. J. Nutr. 80:48.

Schroeder, H. A., J. J. Balassa, and W. H. Vinton, Jr. 1964. Chromium, lead, cadmium, nickel and titanium in mice: Effect on mortality, tumors and tissue levels. J. Nutr. 83:239.

Schroeder, H. A., J. J. Balassa, and W. H. Vinton, Jr. 1965. Chromium, cadmium and lead in rats: Effects on life span, tumors and tissue levels. J. Nutr. 86:51.

Schurch, A. F., L. E. Lloyd, and E. W. Crampton. 1950. The use of chromic oxide as an index for determining the digestibility of a diet. J. Nutr. 41:629.

Sukhacheva, E. I., T. P. Archipova, and V. M. Masha. 1978. The peculiarities of behavior of trivalent and hexavalent chromium in rats. *In* M. Kirchgessner (ed.). Trace Element Metabolism in Man and Animal—3, p. 268. Proc. 3rd. Int. Symp.

Underwood, E. J. 1977. Trace Elements in Human and Animal Nutrition, 4th ed. Academic Press, New York.

Whitby, L. G., and D. Lang. 1960. Experience with the chromic oxide method of fecal marking in metabolic balance investigations on humans. J. Clin. Invest. 39:854.

Woolfolk, P. G., C. R. Richards, R. W. Kaufman, C. M. Martin, and J. T. Reid. 1950. A comparison of fecal nitrogen excretion rate, chromium oxide and "chromagen(s)" methods for evaluating forages and roughages. J. Dairy Sci. 33:385 (Abstr.).

# Cobalt

## INTRODUCTION

Cobalt (Co) is in the same family in the periodic table as nickel and iron. It makes up only 0.0023 percent of the earth's crust, where it usually occurs with nickel or sulfide and arsenic ores. Industrially, cobalt is used in alloy steel.

Although cobalt had been shown to be present in plant and animal tissues, the first clear evidence that it was a dietary essential was in 1935 from Australian research into the cause of certain diseases of cattle and sheep known as "coast disease" and "wasting disease" (Underwood, 1977). There are cobalt-deficiency areas in a number of different parts of the world. Cobalt is distributed widely in the animal body, with the highest concentration in liver, bone, and kidney (Underwood, 1977). The only known function of the element in the animal is its role as a component of vitamin $B_{12}$. Microbes use cobalt to synthesize $B_{12}$.

## ESSENTIALITY

Cobalt per se is a dietary essential for ruminants and horses, in which it is incorporated into vitamin $B_{12}$ by gastrointestinal microbes. It is essential in other animals as a component of vitamin $B_{12}$, but they have more limited ability to synthesize vitamin $B_{12}$. In nonruminants vitamin $B_{12}$ is absorbed from the lower digestive tract. The animals may also ingest vitamin $B_{12}$ via coprophagy. Signs of cobalt deficiency in cattle and sheep are loss of appetite, body weight loss, emaciation, and anemia. The appearance is that of a starved animal. The dietary requirement of cobalt for ruminants is 0.10 ppm, dry basis.

## METABOLISM

Cobalt is not absorbed to a high degree by ruminants. In rats 80 percent of orally administered cobalt appeared in the feces (Comar *et al.*, 1946). The element is used by microorganisms to synthesize vitamin $B_{12}$ in the rumen of ruminants or in the cecum and colon of nonruminants. When cobalt is injected or given orally in very large amounts, it accumulates in the liver. Cobalt can replace zinc in certain proteolytic enzymes such as pancreatic carboxypeptidase A (Vallee, 1974).

## SOURCES

Cobalt is found in variable quantities in plants. Although most feeds are adequate in cobalt, those grown in cobalt-deficient soils may be deficient. Supplemental sources include cobalt oxide and salts such as cobalt sulfate and cobalt chloride. The supplemental sources can be administered by incorporating in feeds or mineral mixes, by drenching with cobalt solutions, or use of cobalt "bullets" composed of cobalt oxide and finely divided iron. The bullets, administered orally, lodge against the ruminoreticular fold to supply a steady amount of cobalt to the ruminal digesta. Two problems have arisen from use of the bullets, namely, regurgitation by the animal and coating of the pellets inside the rumen with calcium phosphate (Underwood, 1977), which prevents bioavailability of cobalt.

## TOXICOSIS

Under practical conditions, cobalt deficiency in ruminants is more likely than cobalt toxicosis. Nevertheless, in supplementation to prevent a deficiency, accidental oversupplementation is possible, which will produce deleterious effects.

### LOW LEVELS

Characteristic signs of chronic toxicosis for most species are reduced feed intake and body weight, emaciation, anemia, hyperchromemia, debility, and increased liver cobalt (Ely *et al.,* 1948; Keener *et al.,* 1949; Becker and Smith, 1951; Turk and Kratzer, 1960). The signs are similar to those of cobalt deficiency except the elevated liver cobalt levels. Addition of 2.88 ppm of cobalt to a diet of young pigs did not affect performance (Kline *et al.,* 1954).

### HIGH LEVELS

Administration of large amounts of cobalt induces polycythemia in simple-stomached animals (Underwood, 1977). The disturbance does not occur in functional ruminants, but affects calves prior to rumen development. The polycythemia is accompanied by hyperplasia of the bone marrow, reticulocytosis, and increased blood volume. Toxic signs from intravenous injection of 0.81 to 1.80 mg/kg of body weight in calves were lacrimation, salivation, dypsnea, incoordination, defecation, and urination (Dunn *et al.,* 1952). Oral administration of excessive cobalt in cattle resulted in lack of appetite, decreased water consumption, increased hemoglobin, red cell count and packed red cell volume, and incoordination (Keener *et al.,* 1949). Fatty infiltration of the liver, slight pulmonary edema, and congestion and petechial to ecchymotic hemorrhages in small intestine were reported in sheep that died of cobalt toxicosis (Becker and Smith, 1951).

Adding cobalt in the form of cobalt chloride ($CoCl_2 \cdot 6H_2O$) to the diet at levels up to 200 ppm did not result in toxicosis in pigs fed a diet adequate in iron (Huck and Clawson, 1976). The addition of 400 or 600 ppm cobalt caused anorexia, growth depression, stiff-leggedness, humped back, incoordination, and extreme muscular tremors. Serum cobalt was increased and iron in serum was decreased by added cobalt.

### FACTORS INFLUENCING TOXICITY

Increasing the protein level of the diet by feeding casein resulted in less depression in growth rate of calves fed 200 mg cobalt per day, and the calves fed the high-protein diet returned to normal gains in body weight earlier (Ely *et al.,* 1948). However, in calves fed 100 mg cobalt per day, changing the protein content of the diet by addition of casein to the grain mixture had no effect on cobalt toxicity. Administration of 500 mg methionine intravenously to calves prior to injection with 50 or 75 mg cobalt (0.64–1.85 mg/kg) prevented or decreased the severity of the signs (Ely *et al.,* 1953). Ethylenediaminetetraacetate (EDTA), cystine, or cysteine alleviated toxicosis in chicks fed 50 ppm cobalt (Turk and Kratzer, 1960).

Addition of 200 ppm iron, 400 ppm manganese, and 400 ppm zinc alleviated the growth depression in swine caused by adding 400 ppm cobalt and partially restored feed intake and growth from adding 600 ppm cobalt (Huck and Clawson, 1976). Feeding 0.5 or 1.0 percent methionine alleviated the toxicosis caused by feeding 600 ppm cobalt.

## TISSUE LEVELS

There are only limited data concerning the effect of feeding excessive cobalt on tissue levels of the mineral. Oral administration of cobalt to cattle increased cobalt in liver and kidney up to over 10-fold (Keener *et al.*, 1949). The levels were 2.1 to 15.4 ppm in liver and 1.9 to 5.4 ppm in kidney, dry basis. Levels for control animals were 0.44 to 0.85 ppm for liver and 0.26 to 0.41 ppm, dry basis, for kidney. Liver cobalt in swine was increased by all levels of cobalt supplementation, the level being generally proportional to the supplemental level (Huck and Clawson, 1976). The effect of supplementing 200, 400, and 600 ppm cobalt on cobalt in liver, spleen, kidney, and heart was studied by these workers. Cobalt was increased in these tissues by supplementation.

## MAXIMUM TOLERABLE LEVELS

Cattle tolerated cobalt at a level of 66 mg/100 kg (Keener *et al.*, 1949) and sheep tolerated up to 352 mg/100 kg (Becker and Smith, 1951). The 66-mg/100-kg level would mean 26 ppm if the dry matter intake is assumed to be 2.5 percent of body weight. Thus, 10 ppm appears safe. No signs of toxicosis were observed in chicks fed diets with 4.7 ppm cobalt, and severe toxicosis was observed at 50 ppm (Turk and Kratzer, 1960). In swine 200 ppm cobalt produced no adverse effect (Huck and Clawson, 1976). It appears that swine and poultry should be able to tolerate 10 ppm.

## SUMMARY

Cobalt is a dietary essential for ruminants, which use it for the synthesis of vitamin $B_{12}$. Nonruminants are usually fed vitamin $B_{12}$ rather than cobalt, since they possess only limited capacity to use the mineral to advantage in meeting their vitamin $B_{12}$ requirement. Sources of cobalt in addition to amounts in feedstuffs are cobalt oxide and salts of cobalt.

A deficiency is more likely to occur than toxicosis. Signs of toxicosis are polycythemia in simple-stomached animals, and reduced feed intake and body weight, emaciation, anemia, debility, increased hemoglobin and packed cell volume, and elevated liver cobalt in ruminants. Toxic levels appear to be at least 300 times the requirement, so the likelihood of a problem appears remote. Also, increased protein or methionine administration appears to help in protecting against cobalt toxicosis.

TABLE 15 Effects of Cobalt Administration in Animals

| Class and Number of Animals[a] | Age or Weight | Administration: Quantity of Element[b] | Administration: Source | Administration: Duration | Administration: Route | Effect(s) | Reference |
|---|---|---|---|---|---|---|---|
| Cattle—5 | 1–14 wk | 55–66 mg/100 kg | Cobaltous sulfate | Up to 28 wk | Water | No adverse effect | Keener *et al.*, 1949 |
| Cattle—7 | | 110–880 mg/100 kg | | Up to 13 wk | | Hyperchromemia | |
| Cattle—2 | | 2,200–4,400 mg/100 kg | | 4–6 wk | | Loss of appetite, decreased water consumption, rough hair coat; incoordination; increase in hemoglobin and packed red cell volume | |
| Cattle | 84–148 kg | 77–86 mg/100 kg | Cobalt sulfate | 21 and 13 d | Diet | No adverse effect | Ely *et al.*, 1948 |
| | | 92–238 mg/100 kg | | 4–16 d | | Loss of appetite | |
| Cattle—1 | 176 d | 31 mg/100 kg | Cobalt sulfate | | IV | No adverse effect | Dunn *et al.*, 1952 |
| Cattle—18 | 73–185 d | 81–180 mg/100 kg | | | | Lacrimation; salivation; dypsnea; incoordination; defecation; urination | |
| Sheep—17 | 1 yr | 22–352 mg/100 kg | $CoCl_2 \cdot 6H_2O$ | 8 wk | Diet | No adverse effect | Becker and Smith, 1951 |
| Sheep—10 | | 440 and 1,100 mg/100 kg | | | | Depressed appetite; body weight losses; anemia; death of 1 of 5 at lower and 3 of 5 at higher level; fatty liver and pul- | |

## TABLE 15 *Continued*

| Class and Number of Animals[a] | Age or Weight | Administration: Quantity of Element[b] | Source | Duration | Route | Effect(s) | Reference |
|---|---|---|---|---|---|---|---|
| | | | | | | monary edema on necropsy (fatalities) | |
| Sheep—6 | 1 yr+ | 165 mg/100 kg | $CoCl_2 \cdot 6H_2O$ | 8 d | IV | Death (2 of 6) | |
| Sheep—5 | | 10.7 g/100 kg[c] | $CoSO_4 \cdot 7H_2O$ | Single dose | Oral | Death (1 of 5) | Andrews, 1965 |
| | | 21.4 g/100 kg[c] | | | | Death (3 of 5) | |
| | | 64.3 g/100 kg[c] | | | | Death (5 of 5) | |
| Swine—6 | 5.2 kg | 2.88 ppm | | Up to 100 kg | Diet | No adverse effect | Kline *et al.*, 1954 |
| Swine—16 | | 25 ppm | $CoCl_2 \cdot 6H_2O$ | | Diet | No adverse effect | Huck and Clawson, 1976 |
| | | 50 ppm | | | | No adverse effect | |
| Swine—32 | | 100 ppm | | | | No adverse effect | |
| Swine—24 | | 200 ppm | | | | No adverse effect | |
| Swine—32 | | 400 ppm | | | | Anorexia; decreased gain; stiff legs; humped back; incoordination; muscular tremors | |
| Swine—12 | | 600 ppm | | | | Same as 400 ppm | |
| Chicken | Chicks | 4.7 ppm | | | Diet | No adverse effect | Turk and Kratzer, 1960 |
| | | 50 ppm | | | | Emaciation; debility; inanition; death | |

[a] Number of animals per treatment.
[b] Quantity expressed in parts per million or as milligrams per kilogram of body weight.
[c] Estimated weights of sheep.

## REFERENCES

Andrews, E. D. 1965. Cobalt poisoning in sheep. N.Z. Vet. J. 13:101.

Becker, D. E., and S. E. Smith. 1951. The level of cobalt tolerance in yearling sheep. J. Anim. Sci. 10:266.

Comar, C. L., G. K. Davis, and R. F. Taylor. 1946. Cobalt metabolism studies: Radioactive cobalt procedures with rats and cattle. Arch. Biochem. 9:149.

Dunn, K. M., R. E. Ely, and C. F. Huffman. 1952. Alleviation of cobalt toxicity in calves by methionine administration. J. Anim. Sci. 11:326.

Ely, R. E., K. M. Dunn, and C. F. Huffman. 1948. Cobalt toxicity in calves resulting from high oral administration. J. Anim. Sci. 7:239.

Ely, R. E., K. M. Dunn, C. F. Huffman, C. L. Comar, and G. K. Davis. 1953. The effect of methionine on the tissue distribution of radioactive cobalt injected intravenously into dairy calves. J. Anim. Sci. 12:394.

Huck, D. W., and A. J. Clawson. 1976. Excess dietary cobalt in pigs. J. Anim. Sci. 43:1231.

Keener, H. A., G. P. Percival, and K. S. Marrow. 1949. Cobalt tolerance in young dairy cattle. J. Dairy Sci. 32:527.

Kline, E. A., J. Kostelic, G. C. Ashton, P. G. Homeyer, L. Quinn, and D. V. Catron. 1954. The effect of the growth performance of young pigs of adding cobalt, vitamin $B_{12}$ and antibiotics to semipurified rations. J. Nutr. 53:543.

Turk, J. L., Jr., and F. H. Kratzer. 1960. The effects of cobalt in the diet of the chicks. Poult. Sci. 39:1302. (Abstr.)

Underwood, E. J. 1977. Trace Elements in Human and Animal Nutrition, 4th ed. Academic Press, New York.

Vallee, B. L. 1974. The entatic properties of cobalt carboxypeptidase and cobalt procarboxypeptidase. *In* W. G. Hoekstra, J. W. Suttie, H. E. Ganther, and W. Mertz, eds. Trace Element Metabolism in Animals—2, p. 5. University Park Press, Baltimore, Md.

# Copper

The attractive, ductile, and conductive metal copper (Cu) has played a significant role in civilizations since the Stone Age. From the crude hammered artifacts dating to about 6000 B.C. to the electrical use for copper by Western man, one appreciates the past and present reliance upon this metal. Its presence in biological systems was well established by the nineteenth century, but not until the early twentieth century did copper become recognized as an essential trace element.

The literature on biological aspects of copper is voluminous. Several comprehensive reviews on the subject exist, including those of the National Research Council (1977), Schroeder *et al.* (1966), Scheinberg and Sternlieb (1960), and Underwood (1977). Specific reviews on the toxicologic aspects of copper include those of Buck *et al.* (1973) and Clarke and Clarke (1975).

## ESSENTIALITY

The essentiality of copper was suggested by McHargue (1925) in the early 1920's; however, conclusive evidence of the biological requirement for copper was actually provided by Hart *et al.* (1928) working with anemic, milk-fed rats. While their anemia was not corrected by either iron supplementation alone or by a liver extract alone, feeding iron and liver together caused a marked elevation in the hemoglobin and packed cell volume within approximately 2 weeks. A bluish tinge of the ashed liver preparation was a clue to its copper content and prompted simultaneous copper and iron supplementation of the milk-fed anemic rats. Their dramatic response in hemoglobin formation was a milestone in the history of nutrition research. Since then the roles of copper in ovine enzootic ataxia (swayback), bovine falling disease, aortic rupture in swine and turkeys, wool and hair depigmentation, and anemia have been elucidated. Numerous copper dependent enzymes, including lysyl oxidase, cytochrome C oxidase, ferroxidase, and tyrosinase, have been recognized (O'Dell, 1976).

The level of dietary copper required for health is somewhat species-dependent and is usually positively correlated with dietary levels of molybdenum (Mo) and inorganic sulfur. Various data suggest that the copper requirements for specific biological processes increase in the rat, for instance, in order as follows: hemoglobin formation, growth, hair pigmentation, and lactation. When dietary conditions are optimal for utilization of copper, 4 to 5 ppm copper in swine and poultry rations and 8 to 10 ppm copper in ruminant rations appear adequate (Underwood, 1977).

## METABOLISM

Copper absorption for most species appears to take place in the duodenum and jejunum. Absorption is affected significantly by the chemical form of the ingested copper. In domestic species, the absorption rate may be as low as 10 percent (Comar, 1950). In general, copper carbonate ($CuCO_3$) and the water soluble forms, copper sulfate, nitrate, and chloride, are absorbed to a greater extent than copper oxide ($CuO$). Metallic copper is very poorly absorbed.

Absorbed copper appears first in plasma as cupric ion loosely bound to albumin. During hepatic synthesis of ceruloplasmin, copper is tightly bound to this metalloprotein, which is then released to the general circulation (Scheinberg and Sternlieb, 1960). Ultimately cuproprotein is present in brain, erythrocytes, and liver as cerebrocuprein, erythrocuprein, and hepatocuprein, respectively. The biliary system is the major excretory pathway for absorbed copper in most species studied (Underwood, 1977). Copper is also excreted during perspiration and lactation. Large quantities of copper are excreted by the urinary system in cases of biliary obstruction or Wilson's disease.

## SOURCES

There is a great geographical variability in the copper content of soils as reflected in the natural incidence of copper deficiency in livestock in various parts of the world. Essentially all plant materials contain copper, which has an affinity for the plant lipids (Schroeder *et al.,* 1966). Of the animal products, oysters have the highest concentration of copper, approximately 137 ppm on a dry weight basis.

Numerous copper-containing compounds used in agriculture and veterinary medicine such as plant and animal fungicides, molluscacides, and foot baths for the control of foot rot in cattle and sheep have provided sources of copper in some instances of copper toxicosis. The copper residues in litter from copper-supplemented swine and poultry have become a significant dietary copper source when this litter is recycled in livestock diets (Fontenot, 1972; Davis, 1974).

Elevations in hepatic, serum, and urine copper levels have occurred in Australian livestock from the consumption of lupin containing toxic alkaloids, although dietary levels of copper are in the low to normal range. Additionally, the plant *Heliotropium europium* contains hepatotoxic alkaloids (heliotrine and lasiocarpine), the ingestion of which impairs hepatic capacity to metabolize copper and results in toxic elevations of liver copper in ruminants (Bull *et al.,* 1956; Underwood, 1977).

## TOXICOSIS

### LOW LEVELS

A significant time period (weeks to months) is usually required for the development of chronic copper toxicosis signs; however, their ultimate expression is so rapid that the fatal course appears to be caused by an acute process. Calves fed copper, as copper sulfate, at 115 ppm (Shand and Lewis, 1957) and 300 ppm (Weiss and Baur, 1968) for up to 129 days exhibited thirst, apathy, hemolytic crises, icterus, hepatic necrosis, and death. Adult cattle are believed to be more resistant to copper toxicosis than younger cattle. Felsman *et al.* (1973) found growing calves were not affected adversely by supplemental copper at levels of up to 900 ppm as cupric sulfate over a 90-day period. Ferguson (1943) and Cunningham *et al.* (1959) have fed 1.2 to 5 g copper sulfate (40 to 500 ppm of copper) daily for up to 16 months to cattle older than 7 months without apparent effects, even in pregnant animals. Kidder (1949), however, observed copper toxicosis in a 227-kg steer fed 5 g of copper sulfate per day for 122 days.

Sheep are more sensitive to copper toxicity than the other mammals studied. The pathogenesis of chronic ovine copper toxicosis has two stages, i.e., a phase of copper accumulation in the tissues and a phase of acute illness including a hemolytic crisis (Gopinath *et al.,* 1974). The phase of copper accumulation or the prehemolytic phase may take place over a period of 6 to 10 weeks or longer. During this time liver copper levels, on a dry weight basis, increase gradually from a normal range of 6 to 279 ppm to a range of 1,000 to 3,000 ppm (McCosker, 1968). Whole blood copper levels usually remain within normal limits of about 70 to 120 $\mu$g/dl for about the first half of the prehemolytic period. During the second half of the prehemolytic period, blood copper levels increase to at least twice their normal value and then suddenly become 8 to 10 times normal coincident with the hemolytic crisis. Serum glutamic oxalacetic transaminase (SGOT) values begin to rise during the period that blood copper levels are rising (Thompson and Todd, 1974). At the start of the hemolytic crisis, the creatine phosphokinase (CPK) levels markedly increase.

The development of copper toxicosis in sheep during the prehemolytic period may go unobserved unless growth retardation of the copper toxicosis is recognized. In this regard, Luke and Marquering (1972) have calculated that for every 100 ppm increase in liver copper in sheep there is an 8.5 g reduction in the daily rate of gain.

During the hemolytic crisis, sheep become icteric and have swollen, partially cirrhotic livers and very dark, hemoglobin-stained kidneys (Adamson *et al.,* 1969). Histopathology of these copper-toxic sheep reveals cytoplasmic lipofuscin granules (indicative of a lysosome response) in the renal tubular epithelium and hepatic parenchyma (Gopinath *et al.,* 1974). In addition there are renal tubular necrosis and spongy lesions in the white matter of the brain, pons, and cerebellum (Morgan, 1973). Occasionally death occurs in copper-toxic sheep without a hemolytic crisis taking place.

Hill and Williams (1965) found that the hepatic copper levels in sheep fed 26.6 ppm copper approached 1,000 ppm, a level that has been associated with chronic copper toxicosis in sheep. Adamson *et al.* (1969) reported that 38 ppm copper fed for 16 to 20 weeks produced icterus, hepatitis, myocardial hemorrhage, hemolysis, and death among housed lambs that were also being fed 0.12 to 1.12 ppm molybdenum and 0.06 to 0.24 percent sulfur. Doherty *et al.* (1969) observed nephropathy and encephalopathy in sheep fed 80 ppm copper for extended periods. Dietary copper supplementation as copper sulfate at rates of 20 to 30 ppm (Gopinath *et al.,* 1974; Gopinath and Howell, 1975), 114 to 228 ppm (Todd *et al.,* 1962), 384 to 576 ppm (McCosker, 1968), and an estimated 398 ppm (Morgan, 1973) for sheep of various ages and weights have all resulted in the induction of hemolytic crises and variable rates of mortality within 45 to 115 days.

Swine appear more tolerant of dietary copper than ruminants. In fact, 250 ppm of copper as $CuSO_4$ have been used routinely for its antimicrobial effect and growth promotion in swine. This level of copper fed to pigs 3 to 7 weeks old has been associated with decreased growth, hemoglobin, and liver iron levels and increased liver copper and zinc levels (Ritchie *et al.*, 1963; Gipp *et al.*, 1973a,b, 1974). Allcroft *et al.* (1961) found that 400 ppm copper in swine rations were nontoxic, but liver copper levels rose sharply under this regimen. Dietary copper at 500 ppm (as $CuSO_4$) caused reduced gains, anemia, and death among swine despite molybdenum supplementation (Combs *et al.*, 1966; DeGoey *et al.*, 1971). Suttle and Mills (1966) reported that feeding swine 425 to 750 ppm dietary copper caused reduced feed intake, retarded growth, anemia, jaundice, increased liver and serum copper, and elevated aspartic transaminase levels. The toxic effects of lower levels of dietary copper in that experiment were eliminated by supplementing with 150 ppm zinc and 150 ppm iron. At the highest level of copper supplementation, the toxicosis was eliminated by 500 ppm zinc and 750 ppm iron. Rations containing 1,000 ppm copper are reported to be lethal in pigs (Allcroft *et al.*, 1961).

Copper has been used as an antimicrobial and/or growth-promoting agent in poultry, i.e., growth rates in young ducks are increased by feeding 100 ppm oral copper as cupric sulfate during an 8-week period (King, 1975). Mehring *et al.* (1960) reported that 500 ppm copper in standard diets was the minimal toxic level for copper in growing chickens, although Mayo *et al.* (1956) found 324 ppm copper caused growth retardation and muscular dystrophy in growing chickens fed a corn–soy diet. Dietary copper at 1,176 ppm fed to growing chickens for a 10-week period resulted in a 51 percent weight loss in these birds (Mehring *et al.*, 1960). In adult hens, Goldberg *et al.* (1956) found 800 to 1,600 ppm copper as copper acetate caused weight loss, anemia, and a 33 percent mortality. The minimal toxic effect level of copper for young turkeys appears to be in the range of 300 to 400 ppm (Supplee, 1964; Vohra and Kratzer, 1968). Their studies indicated that 800 to 900 ppm copper in normal turkey rations caused reduced growth, while 3,240 ppm copper caused death in 21 days. Copper in purified turkey diets appears much more toxic than in standard diets inasmuch as 100 ppm copper as $CuSO_4$ or $CuCO_3$ caused decreased growth and mortality in young turkeys fed a purified diet for a 3-week period (Waibel *et al.*, 1964).

Horses appear to be more resistant to copper toxicosis than either cattle, swine, sheep, or poultry. Smith *et al.* (1975a) fed ponies diets containing 791 ppm of copper as cupric carbonate ($CuCO_3$) for a period of 6 months without ill effects in the experimental animals or their offspring. This level of copper resulted in liver copper levels between 3,445 and 4,294 ppm, dry basis. Despite high liver copper values, hemolytic crises were not induced, and no copper was present in the urine, albeit fecal copper increased steadily during the course of the experiment.

Dietary copper at 200 ppm as cupric sulfate has proven to improve growth rates in rabbits (King, 1975).

The toxic effects of copper in tank water for gilled fish include congestion of the respiratory lamellae (P'equignot and Moga, 1975) and gradual reduction leading to total ablation of mucous cells in the respiratory lamellae (P'equignot *et al.,* 1975), inhibition of growth (Hubschman, 1967), fatty degeneration of liver, renal necrosis, decreased hematopoietic centers (Baker, 1969), blockage of spawning (Mount, 1968), increased mortality (Hazel and Meith, 1970), and decreased hatchability of eggs (Brungs *et al.,* 1976). The maximal acceptable toxicant concentration (MATC) for copper in continuously flowing water for minnows has been calculated at between 66 and 118 ppb (Brungs *et al.,* 1976). Mount and Stephan (1969) have shown that the softer the tank water for the fish, the lower the MATC for copper will be. Young fish (fry) are inhibited in growth and have an increased mortality at approximately one-fourth the levels of copper that are required to reduce hatchability of fish eggs. Crayfish and lobsters seem to be less tolerant of copper than finned fish, i.e., copper concentrations as low as 15 ppb retard growth in young crayfish (Hubschman, 1967), while 56 ppb is the lethal threshold ($LT_{50}$) for lobsters in 20 to 30 percent saline water (McLeese, 1974).

Laboratory rats appear very resistant to dietary copper, as indicated by Boyden *et al.* (1938), who found the minimal toxic effect level for copper as $CuSO_4$ in growing rat diets to be approximately 1,000 ppm; 2,000 ppm (11.8 mg Cu/rat/day) caused weight loss, and 4,000 ppm caused death within 1 week. Cho (1973) reported copper-toxic rats developed hyperplasia of adrenal gland and anterior hypophysis.

### HIGH LEVELS

Considering the quantities of copper compounds that have been used in agriculture and veterinary medicine, there are relatively few veterinary examples of acute copper toxicoses. Cases of acute copper toxicosis have occurred in accidental overdosing or the accidental consumption of copper-containing anthelmintics, foot baths, and fungicides. Copper

EDTA and copper glycinate are parenteral copper supplements that could produce copper toxicosis (Buck *et al.,* 1973). The cellulitis and abscess formation, which often accompany the subcutaneous injection of copper glycinate, might be considered an acute local copper toxicosis (Smith *et al.,* 1975b).

Acute copper toxicosis has been studied in sheep by Isael *et al.* (1969) and by Wiener and Macleod (1970). They reported that 50 mg of copper subcutaneously administered produced death within 24 to 72 hours in sheep of various ages. They also found young sheep to be much more susceptible to acute copper toxicosis than older sheep. The intravenous administration of 50 mg copper as copper EDTA caused death among sheep within 3 to 7 days (Macleod and Watt, 1970), and a single oral dose of 0.7 to 1.5 g copper carbonate (estimated to supply 400 to 800 mg of copper) also caused death in sheep within 3 to 7 days (Sasu *et al.,* 1970).

The signs of acute oral copper toxicosis include nausea, vomition (in species capable of vomiting), salivation, violent abdominal pain, convulsions, paralysis, collapse, and death. Necropsy reveals marked gastroenteritis, necrotic hepatitis, splenic and renal congestion, and evidence of antemortem intravascular coagulation. Sheep dying acutely of subcutaneous injections of copper EDTA exhibit hydrothorax, hydroperitoneum, and hemorrhage into the alimentary tract. The toxic level of oral copper as $CuSO_4$ in sheep is believed to be between 9 and 20 mg/kg of body weight and approximately 200 mg/kg of body weight in cattle (Buck *et al.,* 1973).

Canadian geese ingesting pond water containing 100 ppm copper as $CuSO_4$ developed acute copper toxicosis with necrosis of the proventriculus and gizzard and a greenish discoloration of the lungs (Henderson and Winterfield, 1975).

Acute copper toxicosis in horses was studied by Bauer (1975), who found 125 mg $CuSO_4$ per kilogram of body weight in a single oral dose caused hypercupremia, hepatic and renal damage, and death within 2 weeks.

Eden and Green (1939) have conducted acute copper toxicity studies in rabbits. The level between no effect and $LD_{50}$ in rabbits for a single intravenous dose of copper is between 2.0 and 2.5 mg/kg of body weight. Five milligrams of copper per kilogram of body weight administered intravenously to rabbits was fatal within a few minutes, and 50 mg/kg of body weight administered to rabbits by a single oral drench was fatal in 6 hours.

Acute canine copper toxicosis has been produced (Gubler *et al.,*

1953) with 165 mg copper per kilogram of body weight in a single oral $CuSO_4$ dose. The toxicosis was characterized by vomition and death within 4 hours.

The $LD_{50}$ for $CuSO_4$ in rats is approximately 300 mg per kilogram of body weight (Stecher, 1968). In acute cases of copper toxicosis, copper analysis of feces is considered a more satisfactory diagnostic aid than hepatic or blood copper levels.

The acute toxic effects of intravenously administered copper have been studied in pregnant hamsters (Ferm and Hanlon, 1974). Both cupric sulfate and copper citrate caused resorption of fetuses and copper citrate at 250 $\mu$g per 100 g of body weight, when given on day 8 of gestation, caused fetal malformation.

### FACTORS INFLUENCING TOXICITY

The apparent differences in copper tolerance between ruminants and nonruminants is influenced significantly by concurrent dietary levels of iron, zinc, molybdenum, selenium, and inorganic sulfur. How each of these factors influences the toxic effects of dietary or parenterally administered copper is still being researched. The apparent differences between ruminants and nonruminants in their susceptibility to copper toxicity seems in large part determined by their differences in sulfur metabolism. Some variation in susceptibility to copper toxicity exists among various breeds within a species; for instance, Merino sheep are more tolerant of dietary copper than other breeds of sheep (Buck *et al.*, 1973). Development of an acquired tolerance to copper from previous exposure to copper, as occurs with cadmium, for instance, is believed not to occur.

## TISSUE LEVELS

Levels of copper in most tissues, except muscle and endocrine organs, are directly affected by copper intake, tend to decline with age, and are quite species-dependent (Underwood, 1977). Liver copper concentrations normally range, on a dry matter basis, from 15 to 30 ppm for a wide variety of monogastric mammals and domestic fowl (Beck, 1956), while liver copper levels of sheep, cattle, and ducks can range 10 times the above levels (Beck, 1961). Heart, hair, brain, and kidney tissue copper levels are intermediate in range (9 to 15 ppm dry basis) and also reflect copper intake rates (Underwood, 1977). Certain parts of the eye, especially the pigmented areas, contain copper concentrations exceeding that of the liver (Bowness *et al.,* 1952). In copper toxicosis hepatic concentrations of copper can be elevated to 2,000 to 3,000 ppm (Dick, 1954), especially in sheep and cattle, and are believed responsible for the hemolytic crises of copper toxicosis in these species.

Normal blood copper levels range between 50 to 150 $\mu$g/dl in many species (Beck, 1961). Increases in blood copper levels require rather exaggerated elevations in copper intake, while deficient copper intakes readily result in lowered plasma values (Underwood, 1977).

## MAXIMUM TOLERABLE LEVELS

The data reviewed suggest that the maximum tolerable levels of dietary copper during growth of various species approximate the following under normal levels of molybdenum, sulfate, zinc, and iron: sheep, 25 ppm; cattle, 100 ppm; swine, 250 ppm; horses, 800 ppm; chickens, 300 ppm; turkeys, 300 ppm; rabbits, 200 ppm; rats, 1,000 ppm; minnows, 100 ppb; trout, 100 ppb; lobsters, 15 ppb; crayfish, 15 ppb; and salmon, 20 ppb. In general, the maximum tolerable levels for copper in adults of the above species are expected to be greater than for the younger animals. The maximum tolerable level for copper in purified type diets is usually lower than in standard type diets. In the case of fish, the harder the tank water, the more tolerant the fish will be of copper.

## SUMMARY

Copper is an essential trace element primarily because of several copper-dependent enzymes involved with iron metabolism, elastin and collagen formation, melanin production, and integrity of the central nervous system. Copper toxicosis, for the most part, stems from the use of the metal as an antimicrobial and/or growth-promoting substance. Species vary widely in susceptibility to copper toxicity, in part due to differences in sulfur metabolism as well as differences in concurrent dietary levels of sulfur, molybdenum, zinc, iron, and selenium. Overt manifestations of copper toxicosis in ruminants are secondary to a hemolytic crisis triggered by severely elevated hepatic copper levels. The effects of copper toxicosis in other animals are less dramatic and include growth inhibition, anemia, muscular dystrophy, impaired reproduction, and decreased longevity. Copper levels below 1 ppm in waters inhabited by fish are toxic. Chronic dietary copper levels of 26 to 38 ppm for sheep can markedly elevate hepatic copper levels, whereas levels of 500 ppm copper in rat diets are well tolerated. Molybdenum is the most influential element affecting the level of copper tolerance in mammals and is used therapeutically in cases of copper toxicosis.

TABLE 16 Effects of Copper Administration in Animals

| Class and Number of Animals[a] | Age or Weight | Administration: Quantity of Element[b] | Source | Duration | Route | Effect(s) | Reference |
|---|---|---|---|---|---|---|---|
| Cattle—7 | 1 wk | 115 ppm | $CuSO_4$ | 13 wk | Diet | Hemolysis; icterus; hepatic necrosis | Shand and Lewis, 1957 |
| Cattle—5 | 46–56 kg | 300 ppm | $CuSO_4 \cdot 5H_2O$ | 129 d | Diet | Hemolytic crisis; thirst; apathy; icterus; hemoglobinuria; death | Weiss and Baur, 1968 |
| Cattle—32 | 6 wk | 0–900 ppm | $CuSO_4$ | Daily for 98 d | Diet | No observed effects | Felsman *et al.*, 1973 |
| Cattle—8 | 7 mo | 390 mg | $CuSO_4$ | 12–16 mo | Diet | No observed effects | Cunningham, 1946 |
| Cattle | Adult | 780 mg | $CuSO_4$ | Daily for 9 mo | Diet | Normal calving | |
| | | 1,950 mg | | | | Normal calving | |
| Cattle—6 | Adult | 468 mg | $CuSO_4$ | Daily for 5–18 wk | Diet | No observed effects | Ferguson, 1943 |
| | | 780 mg | | | | No observed effects | |
| Cattle—1 | 227 kg | 1.95 g | $CuSO_4$ | 122 d | Diet | No observed effects | Kidder, 1949 |
| Cattle—193 | 2–4 yr | 120 mg | Cu glycinate | Single dose | Subcutaneous | Subcutaneous abscess and skin necrosis | Smith *et al.*, 1975b |
| Sheep—18 | 20 kg | 7.3 ppm | $CuSO_4 \cdot 5H_2O$ | Daily to 36 kg live wt | Diet | No adverse effects | Hill and Williams, 1965 |
| Sheep | | 26.6 ppm | | | | Reduced gain | |
| | | 40.7 ppm | | | | Reduced gain; increased hepatic Cu; and mortality | |

| | | | | | | | |
|---|---|---|---|---|---|---|---|
| Sheep—170 | Growing | 38 ppm | $CuSO_4$ | 16 to 20 wk | Diet | Icterus; hepatitis; myocardial hemorrhage; hemolysis; liver Cu 1,000 ppm | Adamson *et al.*, 1969 |
| Sheep—6 | 6–12 wk | 80 ppm | $CuSO_4$ | Long term | Diet | Icterus; hepatic cirrhosis; renal tubular necrosis; death | Doherty *et al.*, 1969 |
| Sheep—2 | 6 mo | 50 mg | $CuSO_4 \cdot 5H_2O$ | Single dose | Intravenous | Hydrothorax; hydroperitoneum; gastrointestinal hemorrhage; death in 3–7 d | Macleod and Watt, 1970 |
| | 7 mo | 50 mg | CuEDTA | Single dose | Intravenous | Hydrothorax; hydroperitoneum; gastrointestinal hemorrhage; death in 3–7 d | |
| Sheep—1,629 | 15–44 kg | 50 mg | CuEDTA | Single dose | Subcutaneous | 1% mortality within 72 h | Wiener and Macleod, 1970 |
| Sheep—2,000 | Ewes | 50 mg | CuEDTA | Single dose | Subcutaneous | 3% mortality within 24–72 h; acute hepatic disease | Isael *et al.*, 1969 |
| Sheep—15 | | 114–228 mg | $CuSO_4$ | Daily 17 to 28 wk | Drench | Hemolytic crisis | Todd *et al.*, 1962 |
| Sheep—5 | 6 mo | 250 mg | $CuSO_4 \cdot 5H_2O$ | Daily for 10 wk | Drench | Hemolytic crisis within 2 wks; intramyelinic vacuoles; 80% deaths | Morgan, 1973 |
| Sheep—4 | Aged ewes | 384 mg/d followed by 576 mg/d | $CuSO_4$ | 95 d<br>9 d | Drench | Hemolytic crisis in 100 days; death | McCosker, 1968 |

TABLE 16 *Continued*

| Class and Number of Animals[a] | Age or Weight | Administration: Quantity of Element[b] | Source | Duration | Route | Effect(s) | Reference |
|---|---|---|---|---|---|---|---|
| Sheep—16 | 6 mo | 20 mg/kg | $CuSO_4 \cdot 5H_2O$ | Daily for 10 wk | Drench | Hemolytic crisis; soft feces; anorexia; icterus; hemoglobinuria | Gopinath *et al.*, 1974 |
| Sheep—8 | | 30 mg/kg | $CuSO_4 \cdot 5H_2O$ | Daily up to 83 d | Drench | Repeated hemolytic attacks and deaths; elevated serum enzymes and BUN | |
| Sheep | | 0.7–1.5 g/kg | $CuCO_3$ | Single dose | | Neutrophilia; pulmonary edema; cardiac hemorrhage; death in 3–7 d | Sasu *et al.*, 1970 |
| Swine—10 | 7 wk | 125 ppm | $CuSO_4$ | 12–15 wk | Diet | No observed effects | Ritchie *et al.*, 1963 |
| | 7 wk | 250 ppm | | | | Reduced hemoglobin and hepatic iron; elevated hepatic Cu and zinc | |
| Swine—6 | Growing | 250 ppm | $CuSO_4$ | To slaughter weight | Diet | Growth response | DeGoey *et al.*, 1971 |
| | 18 kg | 500 ppm | | | | Reduced growth rate; death | |
| Swine—24 | Young | 250 ppm | $CuSO_4$ | 11 wk | Diet | Reduced hemoglobin and growth | Gipp *et al.*, 1973a |
| Swine—34 | 3 wk | 250 ppm | $CuSO_4 \cdot 5H_2O$ | 9 wk | Diet | Reduced hemoglobin and growth; re- | Gipp *et al.*, 1973b |

| | | | | | | | |
|---|---|---|---|---|---|---|---|
| | | | | | | duced liver iron and increased liver zinc | |
| Swine—32 | 3 wk | 250 ppm | $CuSO_4 \cdot 5H_2O$ | 5 wk | Diet | All manifestations of iron deficiency | Gipp *et al.*, 1974 |
| Swine | 3 wk | 425 ppm | $CuSO_4$ | 6 wk | Diet | Reduced growth rate; skin ulcers | Suttle and Mills, 1966 |
| Swine—12 | 3 wk | 750 ppm | $CuCO_3$ | 6 wk | Diet | Reduced growth rate and hemoglobin levels | |
| Swine—20 | 26 kg | 250 ppm | $CuSO_4 \cdot 5H_2O$ | 57 d | Diet | No observed effects | Combs *et al.*, 1966 |
| | 500 ppm | | | | Diet | Reduced hemoglobin and hematocrit | |
| Swine—19 | 8–10 wk | 1,000 ppm | $CuSO_4$ | Daily | Diet | Reduced growth; unthrifty; mortality | Allcroft *et al.*, 1961 |
| Chicken—20 | Growing | 500 ppm | $CuO_3$ | 10 wk | Diet | Minimal toxic level | Mehring *et al.*, 1960 |
| | | 1,176 ppm | | | | Growth retardation 51% of controls | |
| Chicken | Young | 324 ppm | $CuSO_4$ | 4 wk | Diet | Muscular dystrophy; growth retardation | Mayo *et al.*, 1956 |
| | | 1,270 ppm | | | | Death | |
| Chicken—30 | 1.8 kg | 800 ppm | Cu acetate | 6 wk | Diet | Loss of weight; anemia | Goldberg *et al.*, 1956 |
| | | 1,600 ppm | | | | 33% mortality | |
| Turkey | Young | 300 ppm | $CuSO_4 \cdot 5H_2O$ | Daily | Diet | No observed effect | Supplee, 1964 |
| | 1 wk | 800 ppm | $CuSO_4 \cdot 5H_2O$ | Daily for 3 wk | | Reduced growth | |
| Turkey—10 | Poults | 400 ppm | $CuSO_4$ | 21 | Diet | No observed effect | Vohra and Kratzer, 1968 |
| | Poults | 800 ppm | $CuSO_4$ | | | Reduced growth | |
| | | 910 ppm | | | | Reduced growth | |

TABLE 16 *Continued*

| Class and Number of Animals[a] | Age or Weight | Administration: Quantity of Element[b] | Source | Duration | Route | Effect(s) | Reference |
|---|---|---|---|---|---|---|---|
| | Poults | 3,240 ppm | $CuSO_4$ | | | Death | |
| Turkey—21 | Young | 50 ppm | $CuCO_3$ | 21 | Diet | No adverse effect | Waibel *et al.*, 1964 |
| | Young | 100 ppm | $CuSO_3$ | | | Reduced growth | |
| | Young | 100 ppm | $CuSO_4$ | | | Reduced growth | |
| Duck—72 | 8 d | 100 ppm | $CuSO_4 \cdot 5H_2O$ | 8 wk | Diet | Growth promotion; thinned cecal wall | King, 1975 |
| Pony—5 | 137–419 kg | 19 mg | $CuCO_3$ | Daily for 6 mo | Diet | No effect | Smith *et al.*, 1975a |
| | | 618 mg | | | | No ill effect | |
| | | 1,080 mg | | | | No hemolytic crisis | |
| | | 1,866 mg | | | | Liver Cu levels, 3,445–4,294 ppm | |
| Horse—4 | 103–185 d | 8 ppm | $CuSO_4$ | 225 | Diet | No effect | Cupps and Howell, 1949 |
| Horse | | 109 ppm | | | | No effect | |
| Horse—32 | Mature | 60 mg/kg | $CuSO_4$ | Once | Gavage | Hypercupremia; hepatic and renal disease; death in 2 wk | Bauer, 1975 |
| Horse | | 65–99 mg/kg | | Divided in 5–9 doses | Gavage | Gastroenteritis; icterus; uremia; hemolytic crisis | Bauer, 1975 |
| | | 157–222 mg/kg | | Repeated doses | Gavage | Gastroenteritis; icterus; uremia; hemolytic crisis | |

| | | | | | | | |
|---|---|---|---|---|---|---|---|
| Rabbit—3 | 2 kg | 2 mg/kg | $CuSO_4$ | Single dose | Intravenous | No effect | Eden and Green, 1939 |
| Rabbit—2 | 2 kg | 2.5 mg/kg | | | | $MLD_{50}$ | |
| Rabbit | 2 kg | 5 mg/kg | | | | Death within minutes | |
| | 2 kg | 50 mg/kg | | | Drench | 50% fatal in 6 h | |
| Rabbit | Growing | 200 ppm | $CuSO_4 \cdot 5H_2O$ | Daily | Diet | Growth promotion, thinned cecal wall | Cited by King, 1975 |
| Minnow | | 66–118 ppb | Sewage-contaminated $H_2O$ | Continuous | Tank water | Maximal acceptable toxicant concentration (MATC) | Brungs *et al.*, 1976 |
| | | 600–980 ppb | | 96 h | | $TL_{50}$ | |
| | | 1.6–2.1 ppm | | 96 h | | Median tolerance limit (MTL) | |
| Minnow | | 147 ppb | $CuSO_4$ | 11 min | Flowing $H_2O$ | No effect level | Mount, 1968 |
| Minnow | | 30 ppb | $CuSO_4$ | 96 h | Static hard $H_2O$ | Acceptable toxicant concentration | Mount and Stephan 1969 |
| | | 70 ppb | | 96 h | | Acceptable toxicant concentration | |
| | | 130 ppb | $CuSO_4$ | 96 h | | Acceptable toxicant concentration | |
| | | 220 ppb | | 96 h | | Acceptable toxicant concentration | |
| Trout | 2 mo | 0.25 ppm | $CuSO_4$ | 48 h | Tank water | Progressive reduction in numbers of mucous cells with prolonged exposure | P'equignot and Moga, 1975 |
| | | 0.1 ppm | | 96 h | | Progressive reduction in numbers of mucous cells with | |

TABLE 16 *Continued*

| Class and Number of Animals[a] | Age or Weight | Administration: Quantity of Element[b] | Source | Duration | Route | Effect(s) | Reference |
|---|---|---|---|---|---|---|---|
| | | | | | | prolonged exposure | |
| | | 0.1 ppm | | 240 h | | Progressive reduction in numbers of mucous cells with prolonged exposure | |
| Salmon | Eggs (eyed) | 80 ppb | $CuSO_4 \cdot 5H_2O$ | Continuous | Flowing $H_2O$ | No effect on hatchability, decreased survival | Hazel and Meith, 1970 |
| | Fry | 20 ppb | | | | Growth inhibition and increased mortality | |
| | Fry | 40 ppb | | | | Acute toxicity and death | |
| Flounder | 2 yr | 80 ppb | $CuSO_4$ | 2 wk | Tank water | Reduced number of mucous cells in gill lamellae | Baker, 1969 |
| | | 500 ppb | | | | Reduced size of gill epithelial cells and fatty livers | |
| | | 1,000 ppb | | | | Enlarged chloride cells and impaired hematopoiesis | |
| | | 3,200 ppb | | | | Mortality | |

| | | | | | | | |
|---|---|---|---|---|---|---|---|
| Lobster—5 | 450 g | 56 ppb | $CuSO_4$ | Continuous up to 500 h 20–30% of saline | Flowing $H_2O$ | Lethal threshold ($LT_{50}$) | McLeese, 1974 |
| Crayfish—100 | New hatch | 60 ppb | $CuSO_4$ | Continuous | Flowing $H_2O$ | Acute toxicity level | Hubschman, 1967 |
| | | 125 ppb | | | | Acute toxicity level | |
| | Young | 15 ppb | | | | Retarded growth | |
| | Adult | 1 ppm | | | | Death in 16 days | |
| | Adult | 2.5 ppm | | 24 h | | Death in 16 days | |
| | | 3 ppm | | 96 h | | $MLD_{50}$ | |
| Dog—1 | Mature | 165 mg/kg | $CuSO_4$ | Single dose | Diet | Vomition and death in 4 h | Gubler *et al.*, 1953 |
| Rat—8 | 3 wk | 500 ppm | $CuSO_4$ | Daily for 7–70 d | Diet | No effect | Boyden *et al.*, 1938 |
| Rat—3 | | 1,000 ppm | | | | No effect | |
| Rat—8 | | 2,000 ppm | | | | Marked weight loss | |
| Rat—3 | | 4,000 ppm | | 7 d | | Severe anorexia and starvation | |
| Hamster—24 | Pregnant | 8 mg/kg | $CuSO_4$ | Once on 8th d gestation | Intra-venous | Increased fetal re-sorption | Ferm and Hanlon, 1974 |
| Hamster | Pregnant | 2.5 mg/kg | Cu citrate | | | Fetal malformation; increased resorption | |

[a] Number of animals per treatment.
[b] Quantity expressed in parts per million or as milligrams per kilogram of body weight, or parts per billion.

## REFERENCES

Adamson, A. H., D. A. Valks, M. A. Appleton, and W. B. Shaw. 1969. Copper toxicity in housed lambs. Vet. Rec. 85:368.

Allcroft, R., K. N. Burns, and G. Lewis. 1961.The effects of high levels of copper in rations for pigs. Vet. Rec. 73:714.

Baker, J. T. P. 1969. Histological and electron microscopical observations on copper poisoning in the winter flounder. J. Fish. Res. Bd. Can. 26:2785.

Bauer, M. 1975. Copper sulfate poisoning in horses. Vet. Arch. 45:257.

Beck, A. B. 1956. The copper content of the liver and blood of some vertebrates. Aust. J. Zool. 4:1.

Beck, A. B. 1961. Observations on the copper metabolism of the domestic fowl and duck. Aust. J. Agric. Res. 12:743.

Bowness, J. M., R. A. Morton, M. H. Shaker, and A. L. Stubbs. 1952. Distribution of copper and zinc in mammalian eyes. Occurrence of metals in melanin fractions of eye tissues. Biochem. J. 51:521.

Boyden, R., V. R. Potter, and C. A. Elvehjem. 1938. Effect of feeding high levels of copper to albino rats. J. Nutr. 15:397.

Brungs, W. A., J. R. Geckler, and M. Gast. 1976. Acute and chronic toxicity of copper to the fathead minnow in a surface water of variable quality. Water Res. 10:37.

Buck, W. B., G. D. Osweiler, and G. A. VanGelder. 1973. Clinical and Diagnostic Veterinary Toxicology. Kendall/Hunt Publishing Co., Dubuque, Iowa.

Bull, L. B., H. E. Albiston, G. Edgar, and A. T. Dick. 1956. Toxaemic jaundice of sheep: Phytogenous chronic copper poisoning, heliotrope poisoning and hepatogenous chronic copper poisoning. Aust. Vet. J. 32:229.

Cho, S. H. 1973. Morphological effect of excess copper sulfate on the adrenal gland and anterior hypophysis of the rat. Yonsei J. Med. Sci. 6:82.

Clarke, E. G. C., and M. L. Clarke. 1975. Veterinary Toxicology, p. 86. Williams & Wilkins, Baltimore, Md.

Comar, C. L. 1950. The use of radioisotopes of copper and molybdenum in nutritional studies. *In* W. D. McElroy and B. Glass, eds. Symposium on Copper Metabolism. Johns Hopkins Press, Baltimore, Md.

Combs, G. E., C. B. Ammerman, R. L. Shirley, and H. D. Wallace. 1966. Effects of source and level of dietary protein on pigs fed high-copper rations. J. Anim. Sci. 25:618.

Cunningham, I. J. 1946. The toxicity of copper to bovines. N.Z. J. Sci. Technol. 27A:372.

Cunningham, I. J., K. G. Hogan, and B. M. Lawson. 1959. The effect of sulfate and molybdenum on copper metabolism in cattle. N.Z. J. Agric. Res. 2:145.

Cupps, P. T., and C. E. Howell. 1949. The effects of feeding supplemental copper to growing foals. J. Anim. Sci. 8:286.

Davis, G. K. 1974. High-level copper feeding of swine and poultry and the ecology. Fed. Proc. 33:1194.

DeGoey, L. W., R. C. Wahlstrom, and R. J. Emerick. 1971. Studies of high levels of copper supplementation to rations for growing swine. J. Anim. Sci. 33:52.

Dick, A. T. 1954. Studies on the assimilation and storage of copper in crossbred sheep. Aust. J. Agric. Res. 5:511.

Doherty, P. C., R. M. Barlow, and K. W. Angus. 1969. Spongy changes in the brains of sheep poisoned by excess dietary copper. Res. Vet. Sci. 10:303.

Eden, A., and H. H. Green. 1939. The fate of copper in the blood stream. J. Comp. Pathol. Ther. 52:301.

Felsman, R. J., M. B. Wise, R. W. Harvey, and E. R. Barrick. 1973. Effect of added dietary levels of copper sulfate and an antibiotic on performance and certain blood constituents of calves. J. Anim. Sci. 36:157.

Ferguson, W. S. 1943. The teart pastures of Somerset. IV. The effect of continuous administration of copper sulfate to dairy cows. J. Agric. Sci. 33:116.

Ferm, V. H., and D. P. Hanlon. 1974. Toxicity of copper salts in hamster embryonic development. Biol. Reprod. 11:97.

Fontenot, J. P. 1972. Va. Polytec. Inst. State Univ. Res. Dir. Rep. 145:33.

Gipp, W. F., W. G. Pond, J. Tasker, D. VanCampen, L. Krook, and W. J. Visek. 1973a. Influence of level of dietary copper on weight gain, hematology and liver copper and iron storage in young pigs. J. Nutr. 103:713.

Gipp, W. F., W. G. Pond, and E. F. Walker. 1973b. Influence of diet, composition and mode of copper administration on the response of growing-finishing swine to supplemental copper. J. Anim. Sci. 36:91.

Gipp, W. F., W. G. Pond, F. A. Kallfelz, J. B. Tasker, D. R. VanCampen, L. Krook, and W. J. Visek. 1974. Effect of dietary copper, iron and ascorbic acid levels on hematology, blood and tissue copper, iron and zinc concentrations and $^{64}$Cu and $^{59}$Fe metabolism in young pigs. J. Nutr. 104:532.

Goldberg, A., C. B. Williams, R. S. Jones, M. Yanagita, G. E. Cartwright, and M. M. Wintrobe. 1956. Studies on copper metabolism. XXII. Hemolytic anemia in chickens induced by the administration of copper. J. Lab. Clin. Med. 48:442.

Gopinath, C., and J. M. Howell. 1975. Experimental chronic copper toxicity in sheep. Changes that follow cessation of dosing at onset of hemolysis. Res. Vet. Sci. 19:35.

Gopinath, C., G. A. Hall, and J. M. Howell. 1974. The effect of copper poisoning on the kidneys of sheep. Res. Vet. Sci. 16:47.

Gubler, C. J., M. E. Lahey, G. E. Cartwright, and M. M. Wintrobe. 1953. Studies on copper metabolism. IX. The transportation of copper in blood. J. Clin. Invest. 32:405.

Hart, E. B., H. Steenbock, J. Waddell, and C. A. Elvehjem. 1928. Iron in nutrition. VII. Copper as a supplement to iron for hemoglobin building in the rat. J. Biol. Chem. 77:797.

Hazel, C. R., and S. J. Meith. 1970. Bioassay of king salmon eggs and sac fry in copper solutions. Calif. Fish Game 56:121.

Henderson, B. M., and R. W. Winterfield. 1975. Acute copper toxicosis in the Canadian goose. Avian Dis. 19:385.

Hill, R., and H. L. Williams. 1965. The effects on intensively reared lambs of diets containing excess copper. Vet. Rec. 77:1043.

Hubschman, J. H. 1967. Effects of copper on the crayfish *Orconectes rusticus* (Girard). I. Acute toxicity. Crustaceana 12:33.

Isael, J., J. M. Howell, and P. J. Treeby. 1969. Deaths in ewes following the administration of copper calcium edetate for prevention of swayback. Vet. Rec. 85:205.

Kidder, R. W. 1949. Symptoms of induced copper toxicity in a steer. J. Anim. Sci. 8:623.

King, J. O. L. 1975. The feeding of copper sulfate to ducklings. Br. Poult. Sci. 16:409.

Luke, V. F., and B. Marquering. 1972. Untersuchungen uber den Mineralstaff gehalt in der Schafleber. I. Futterungsbedingte und genetische Einflusse auf den Cu-gehalt. Suchtungskunde 44:45.

Macleod, N. S., and J. A. Watt. 1970. Experimental copper poisoning in sheep. Vet. Rec. 86:375.

Mayo, R. J., S. M. Hauge, H. E. Parker, F. N. Andrews, and C. W. Carrick. 1956. Copper tolerance of young chickens. Poult. Sci. 35:1156.

McCosker, P. J. 1968. Observations on blood copper in the sheep. II. Chronic copper poisoning. Res. Vet. Sci. 9:103.

McHargue, J. S. 1925. The association of copper with substances containing the fat-soluble A vitamin. Am. J. Physiol. 72:583.

McLeese, D. W. 1974. Toxicity of copper at two temperatures and three salinities to the American lobster (*Homarus americanus*). J. Fish. Res. Bd. Can. 31:1949.

Mehring, A. L., J. H. Brumbaugh, A. J. Sutherland, and H. W. Titus. 1960. The tolerance of growing chickens for dietary copper. Poult. Sci. 39:713.

Morgan, K. T. 1973. Chronic copper toxicity of sheep: An ultrastructure study of spongiform leucoencephalopathy. Res. Vet. Sci. 15:88.

Mount, D. I., and C. E. Stephan. 1969. Chronic toxicity of copper to the fathead minnow (*Pimephales promelas*) in soft water. J. Fish. Res. Bd. Can. 26:2449.

Mount, P. L. 1968. Chronic toxicity of copper to fathead minnows (*Pimephales promelas,* rafinesque). Water Res. 2:215.

National Research Council. 1977. Medical and Biological Effects of Environmental Pollutants. Copper. National Academy of Sciences, Washington, D.C.

O'Dell, B. L. 1976. Biochemistry of copper. *In* R. E. Burch and J. F. Sullivan, eds. Symposium on Trace Elements. Med. Clin. North Am. 60:697.

P'equignot, J., and A. Moga. 1975. Effects of different toxic compounds (Pb, Cu, formal, $NH_4$) on the carp: Histologic changes in excretory and hematopoietic organs. Eur. J. Toxic. Environ. Hyg. 8:361.

P'equignot, J., R. Labat, A. Chatelet, and A. Moga. 1975. Action of copper sulfate on mucous epithelium cells in rainbow trout (*Salmo irideus*). Eur. J. Toxic. Environ. Hyg. 8:52.

Ritchie, H. D., R. W. Luecke, B. V. Baltzer, E. R. Miller, D. E. Ullrey, and J. A. Hoefer. 1963. Copper and zinc interrelationships in the pig. J. Nutr. 79:117.

Sasu, V., N. Hagiu, S. Tasca, O. Popescu, and E. Sasu. 1970. Clinical, hematologic and anatomohistopathological modifications in acute experimental intoxication with basic copper carbonate in the sheep. Inst. Agron. "Ion Ionescu De La Brad," Iasi 50:251.

Scheinberg, I. H., and I. Sternlieb. 1960. Copper metabolism. Pharm. Rev. 12:355.

Schroeder, H. A., A. P. Nason, I. H. Tipton, and J. J. Balassa. 1966. Essential trace metals in man: Copper. J. Chron. Dis. 19:1007.

Shand, A., and G. Lewis. 1957. Chronic copper poisoning in young calves. Vet. Rec. 69:618.

Smith, B., D. A. Woodhouse, and A. J. Frazer. 1975b. The effects of copper supplementation on stock health and production. I. Field investigations. N.Z. Vet. J. 23:73.

Smith, J. P., R. M. Jordan, and M. L. Nelson. 1975a. Tolerance of ponies to high levels of dietary copper. J. Anim. Sci. 41:1645.

Stecher, P. G. (ed.). 1968. The Merck Index. Merck & Co., Rahway, N.J.

Supplee, W. C. 1964. Observations on the effect of copper additions to purified turkey diets. Poult. Sci. 43:1599.

Suttle, N. F., and C. F. Mills. 1966. Studies of the toxicity of copper to pigs. I. Effects of oral supplements of zinc and iron salts on the development of copper toxicosis. Br. J. Nutr. 20:135.

Thompson, R. H., and J. R. Todd. 1974. Muscle damage in chronic copper poisoning of sheep. Res. Vet. Sci. 16:97.

Todd, J. R., J. F. Gracey, and R. H. Thompson. 1962. Studies on chronic copper poisoning. I. Toxicity of copper sulfate and copper acetate in sheep. Br. Vet. J. 118:482.

Underwood, E. J. 1977. Trace Elements in Human and Animal Nutrition, 4th ed. Academic Press, New York.

Vohra, P., and F. H. Kratzer. 1968. Zinc, copper and manganese toxicities in turkey poults and their alleviation by EDTA. Poult. Sci. 47:699.

Waibel, P. E., D. C. Snetsinger, R. A. Ball, and J. H. Sautter. 1964. Variation in tolerance of turkeys to dietary copper. Poult. Sci. 43:504.

Weiss, E., and P. Baur. 1968. Experimental studies on chronic copper poisoning in the calf. Zentralbl. Veterinaermed. 15:156.

Wiener, G., and N. S. Macleod. 1970. Breed, body weight and age as factors in the mortality rate of sheep following copper injection. Vet. Rec. 86:740.

# Fluorine

Fluorine (F), chemically bound as fluoride, is found in both igneous and sedimentary rock and constitutes about 0.06–0.09 percent of the upper layers of the earth's crust. Fluorine rarely occurs free in nature but combines chemically to form fluorides that are widely, but variably, distributed in the environment. An association between high-fluoride intakes and dental defects was first demonstrated in rats in 1925 (McCollum *et al.*, 1925). By 1931, chronic endemic fluorosis in man and livestock was identified in several parts of the world (Churchill, 1931; Smith *et al.*, 1931; Velu, 1931). Fluoride-bearing fumes and dusts from industrial plants processing fluoride-containing raw materials, such as bauxite or phosphate rock, were found to constitute a health hazard to man and animals living nearby (Roholm, 1937; Agate *et al.*, 1949). The use of unprocessed rock phosphates as mineral supplements subjected livestock to a further fluoride hazard. However, in the late 1930's it was discovered that fluoride had significant anticarieogenic properties, and subsequent research has explored both the toxic and essential character of this element. Excellent reviews are available (National Research Council, 1971, 1974).

## ESSENTIALITY

Whether fluorine is considered essential depends upon the criteria used. No one has yet produced an environment so low in this element that animal survival has been vitally threatened. However, fluorine was identified as a constant constituent of bones and teeth as early as 1805, and trace quantities are beneficial in development of caries-resistance and may be beneficial in inhibiting excessive demineralization of bone in the aged.

McClendon and Gershon-Cohen (1953) fed weanling rats for 66 days upon materials grown hydroponically in water said to be "fluorine-free." The rats weighed 51 g and had 10 carious molars per animal compared to fluoride-supplemented rats that weighed 128 g and had 0.5 carious molars per animal. No data on fluorine concentration in diets or tissues were presented. Maurer and Day (1957) purified dietary ingredients and produced a diet that contained about 0.007 ppm fluoride on which four generations of rats were raised without evidence of impaired general health, dental health, or weight gain as compared to rats raised on the same diet plus 2 ppm of fluoride in their drinking water. Doberanz *et al.* (1963) fed a diet containing less than 0.005 ppm fluoride (prepared from hydroponically grown soybeans and sorghum grain) and found no difference in general health or growth rate between rats fed this diet and rats fed the same diet plus 2 ppm of fluoride in their drinking water. A similar study (Weber, 1966) failed to find that fluoride was essential for mice raised through three generations. However, Messer *et al.* (1972a, 1973) have reported that fertility is impaired in female mice on a diet containing 0.1–0.3 ppm fluoride, and anemia in infant mice produced by low-fluoride females is more severe than when supplemental fluoride is provided (Messer *et al.*, 1972b). Tao and Suttie (1976) used the same low-fluoride diet fed to mice by Messer *et al.* (1973), found no impairment of reproduction, and suggested that the apparent essentiality of fluoride proposed by Messer and associates was due to a pharmacological effect of fluoride in improving iron utilization in mice fed a diet marginally sufficient in iron. Schwarz and Milne (1972), working in a filtered-air environment, reported a favorable growth response when small increments (1–2 ppm) of fluoride were added to a low-fluoride diet for rats.

## METABOLISM

Absorption of fluoride is presumed to be largely a passive process (National Research Council, 1974), although some researchers (Stookey *et al.*, 1964; Parkins *et al.*, 1966; Parkins, 1971) have suggested active transport on the basis of *in vitro* studies with inverted rat intestine. Sites of absorption include the stomach in man (Carlson *et al.*, 1960a) and rats (Stookey *et al.*, 1962; Wagner, 1962; Yeh *et al.*, 1970) and probably the rumen in sheep and cattle (Perkinson *et al.*, 1955). Soluble fluorides (e.g., NaF) and small amounts of poorly soluble fluorides (e.g., $CaF_2$) in solution are rapidly and rather efficiently absorbed (>90 percent). When $CaF_2$ was added to the diet in solid form, percentage absorption was 60–70 percent (Machle and Largent, 1942). Fluoride absorption from dietary bone ranged from 37 to 54 percent (Machle and Largent, 1942; Largent and Heyroth, 1949), and Underwood (1971) suggesting that about 50 percent of the fluoride in rock phosphate is absorbed. Calcium, aluminum, sodium chloride, and high fat levels are dietary factors that depress fluoride absorption (National Research Council, 1974).

About 75 percent of the fluoride in the blood is in the plasma (Carlson *et al.*, 1960a) with 15–70 percent (0.01–0.04 ppm) in ionic form (Singer and Armstrong, 1964). Nearly 5 percent of plasma fluoride is bound to protein, but most of the bound fluoride is associated with compounds having molecular weights less than albumin.

Homeostatic regulation of plasma fluoride concentration involves the kidneys and skeleton (Smith *et al.*, 1950; Singer and Armstrong, 1960, 1964). The dog kidney can concentrate plasma fluoride in urine by a factor of 10–20 times (Carlson *et al.*, 1960b) through glomerular filtration and variable tubular resorption. Bone has a great affinity for fluoride and incorporates it into hydroxyapatite to form fluorapatite. This results in a larger, less soluble, more stable apatite crystal (Zipkin *et al.*, 1964). The fluoride cannot be removed without resorption of this mineral unit.

Even low levels of fluoride intake will result in appreciable (but harmless) accumulations of fluoride in skeleton and teeth. These accumulations can increase, within limits, over a period of time without morphological evidence of pathology. However, in some cases of high-fluoride intake, structural bone changes develop (Shupe *et al.*, 1963b).

Most soft tissues do not accumulate much fluoride, even during high intakes, although tendon (Armstrong and Singer, 1970), aorta (Ericsson and Ullberg, 1958), and placenta have higher fluoride concentrations than other soft tissue, possibly associated with their relatively high levels of calcium and magnesium. Kidney will usually exhibit a high-fluoride concentration during high-fluoride ingestion due to urine retained in the tubules and collecting ducts.

Milk fluoride concentrations are affected only minimally by dietary fluoride, and Greenwood *et al.* (1964) found that when Holstein cows were fed 10, 29, 55, or 109 ppm fluoride from about 3 months of age to 7½ years, milk fluoride concentrations were 0.06, 0.10, 0.14, or 0.20

ppm, respectively. Fluoride crosses the placental barrier of cows, and fluoride levels in the bones of the offspring are correlated with the fluoride concentration of maternal blood (National Research Council, 1974). However, bone fluoride concentrations of calves born to cows consuming as much as 108 ppm fluoride (from sodium fluoride) were low (Hobbs and Merriman, 1962), and it appeared that neither placental fluoride transfer nor milk fluoride concentrations were sufficient to adversely affect the health of these calves.

## SOURCES

The primary fluorine-containing minerals are fluorspar ($CaF_2$), cryolite ($Na_3AlF_6$), and fluorapatite [$Ca_{10}F_2(PO_4)_6$]. Natural deposits of cryolite are currently of little economic importance, and cryolite for industrial purposes is synthesized in chemical plants. Fluorspar and fluorapatite deposits are widespread, and a 500-square-mile area at Bartow, Florida, contains deposits of phosphate rock that is chiefly fluorapatite. This is a major center for the production of phosphate fertilizers and calcium phosphates for animal feeds, although a number of other important phosphate deposits are found in the United States (U.S. Department of the Interior, 1970).

Soils may contain fluoride in several different minerals. The fluoride content tends to increase with depth, and the usual range in the United States is 20–500 ppm (average, 190 ppm) from 0–8 cm deep and 20–1,620 ppm (average, 292 ppm) from 0–30 cm deep (Robinson and Edgington, 1946). Some soils, unusually high in fluoride, have been found in Idaho (3,870 ppm) and in Tennessee (8,300 ppm).

Surface water in lakes and rivers generally contains less fluoride than water from springs or wells, unless that surface water is contaminated by dust from mining and processing of high-fluorine phosphate rock. The fluoride content of well water varies regionally, dependent on its origin. Much of the northeastern United States has water with natural fluoride concentrations ranging from 0.02 to 0.1 ppm. Farther west and south, concentrations tend to be above 0.2 ppm, but seldom over 1 ppm. In endemic fluorosis areas, deep well water may percolate through fluorapatite and frequently contains 3–5 ppm fluoride, and sometimes 10–15 ppm (Harvey, 1952; Cholak, 1959).

Active volcanoes and fumaroles, and certain industrial processes, may contribute significantly to local concentrations of fluoride. The latter furnish fluoride in one of three principal forms: hydrofluoric acid, silicon tetrafluoride, or fluoride-containing particulate matter. Direct inhalation of fluoride does not contribute significantly to fluoride accumulation in animals. However, these emissions may contaminate plants, soil, and water. Gaseous fluoride may be absorbed and incorporated into plant tissues. Particulate fluorides may accumulate on plant surfaces and be ingested as the plants are eaten. Rain may wash off some of the particles, and the particles are usually quite inert, with toxicity related largely to solubility.

Forages and grains are seldom a major factor in chronic fluorosis in animals or man, unless contaminated by fluoride-bearing dusts, fumes, or water. Most plants have a limited capacity to absorb fluoride from the soil. The tea plant and camellia are exceptions, and fluoride concentrations of 100 ppm or more have been reported (Underwood, 1977). Pasture plants have been shown to range from 2 to 16 ppm on a dry basis. Cereals and cereal by-products usually contain 1–3 ppm.

Animal by-products containing bone may contribute significant quantities of fluoride to animal diets, depending upon the amount of by-product used (and bone contained) and the dietary history of the animals from which the by-products were derived. Bone ash normally contains less than 1,500 ppm of fluoride and would contribute only minor amounts. However, cattle grazing fluoride-contaminated pastures can have bone ash containing over 10,000 ppm fluoride, or 5.5 parts of fluoride for each 100 parts of phosphorus.

Normally, the primary sources of dietary fluorides are the phosphorus supplements. These vary greatly in fluoride content, depending on origin and manufacturing processes. The majority of U.S. feed phosphates originate from rock phosphate deposits with fluoride levels of 2–5 percent (average, 3.5 percent) (VanWazer, 1961). When processed sufficiently to qualify as defluorinated, feed-grade phosphates must contain no more than 1 part of fluorine to 100 parts phosphorus (AAFCO, 1977). Processed low-fluoride, feed phosphates include mono-, di-, and tricalcium phosphates, mono- and diammonium phosphates, mono- and disodium phosphates, ammonium and sodium polyphosphates, feed-grade phosphoric acid, and defluorinated phosphate. Unprocessed feed phosphates, supplying substantial amounts of fluorine, include soft rock phosphate, ground rock phosphate, and ground low-fluoride rock phosphate. More dangerous sources of fluoride, when incorporated in animal diets, are undefluorinated, fertilizer-grade phosphates. Analyses of the above phosphates are presented in Table 17.

## TOXICOSIS

### LOW LEVELS

The precise dietary concentration at which fluoride ingestion becomes harmful is difficult to define. No single value is appropriate because low-level toxicosis depends upon duration of ingestion, solubility of the fluoride source, general nutritional status, species of animal, age when ingested, and toxicity-modifying components of the diet.

Diagnosis of fluoride toxicosis is also difficult, because there may be an extended interval of time between ingestion of elevated levels and the appearance of toxic signs (Shupe, 1970). Dietary history, clinical evidence, radiography, chemical analyses, necropsy findings, and histopathology are all important. The degree of dental fluorosis and osteofluorosis, evidence of intermittent lameness, and the concentration of fluoride in diet, urine, and bone are of particular diagnostic importance.

If excessive fluoride is ingested during tooth development, fluorotic lesions may be expected (Roholm, 1937). The period during which developing teeth in cattle are sensitive to excess fluoride is from approximately 6 months to 3 years of age. Teeth that have erupted are not influenced adversely by subsequent fluoride ingestion (Garlick, 1955), and cattle that are more than 3 years old will not develop typical dental lesions. Dental fluorosis is usually diagnosed by examining the incisors.

The degree of dental fluorosis that develops under experimental conditions has been correlated with the amount and duration of fluoride ingestion and the animal's age. Gross fluorotic lesions of the incisor enamel begin with slight mottling (white, chalky patches or striations) and progress to definite mottling, hypoplasia, and hypocalcification. The following scoring system for classification of dental fluorosis has been proposed (National Research Council, 1974):

| Score | Description |
|---|---|
| 0 | *Normal.* Smooth, translucent, glossy white enamel; tooth has normal shape. |
| 1 | *Questionable Effect.* Slight deviation from normal but cause not determinable; may have enamel flecks but is not mottled. |
| 2 | *Slight Effect.* Slight mottling of enamel, best observed as horizontal striations with transmitted light; may be slightly stained but no increase in normal rate of wear. |

| Score | Description |
|---|---|
| 3 | *Moderate Effect.* Definite mottling; large areas of chalky enamel or generalized mottling of entire tooth; tooth may have slightly increased rate of wear and may be stained. |
| 4 | *Marked Effect.* Definite mottling, hypoplasia, and hypocalcification; may have pitting of enamel; with use, tooth will have increased rate of wear and may be stained. |
| 5 | *Severe Effect.* Definite mottling, hypoplasia, and hypocalcification; with use, tooth will have excessive rate of wear, and may have eroded or pitted enamel. Tooth may be stained or discolored. |

The amount of fluoride stored in bone may increase over time with no apparent change in bone structure or function. However, if excess fluoride ingestion is sufficiently high, and over a sufficiently long period of time, morphological abnormalities will develop. In livestock, clinically palpable (bilateral) lesions usually develop first on the medial surface of the proximal third of the metatarsals. Subsequent lesions are seen on the mandible, metacarpals, and ribs. The osteofluorotic lesions tend to be more severe in those bones, and parts of bones, that are subject to the greatest physical stress. Radiographic evidence of osteoporosis, osteosclerosis, osteomalacia, hyperostosis, and osteophytosis, or any combination of these lesions, has been described (Johnson, 1965; Shupe and Alther, 1966; Shupe, 1969). Grossly, severely affected bones appear chalky white, are larger in diameter and heavier than normal, and have a roughened, irregular periosteal surface. In cattle poisoned by industrial fluoride emissions, Krook and Maylin (1979) contended that the primary target of fluoride was the resorbing osteocyte. Morphological signs of osteolysis were absent, and the failure of resorption caused osteopetrosis with retention of lamellar bone in the cortices.

Animal movement may be impaired by intermittent periods of stiffness and lameness, associated in advanced cases with calcification of periarticular structures and tendon insertions. In animals with marked periosteal hyperostosis, spurring and bridging of the joints may lead to rigidity of the spine and limbs.

Anorexia, unthriftiness, dry hair, and thick, nonpliable skin have been noted in fluorotic animals (Roholm, 1937; Shupe *et al.,* 1963a). Primary adverse effects on reproduction and lactogenesis have not been demonstrated, although milk production may decrease on high-fluoride intakes secondary to dental and skeletal damage and consequent reductions in feed and water intake (Stoddard *et al.,* 1963). Suttie *et al.* (1957b) have demonstrated that cows first exposed to fluoride at 4 months of age can consume 40–50 ppm of fluoride in their diet for two or three lactations without measurable effect on milk production. Milk production was reduced in the fourth and subsequent lactations. Higher dietary fluoride levels (93 ppm) affected milk production in the second lactation slightly and definitely reduced milk yield in subsequent lactations (Stoddard *et al.*, 1963). Irrespective of level or duration of fluoride intake, clinical signs of toxicosis will normally precede impaired milk production. No characteristic, unequivocal histologic or functional changes in blood or soft tissues have been correlated with fluoride intakes sufficient to induce chronic fluorosis of bones and teeth.

### HIGH LEVELS

Acute fluoride toxicosis is relatively rare and has usually resulted from accidental ingestion of compounds such as sodium fluosilicate, used as a rodenticide, or sodium fluoride, used as an ascaricide in swine. The rapidity with which toxic signs appear depends on the amount of fluoride ingested (Cass, 1961). Toxic signs include high-fluoride content of blood and urine, restlessness, stiffness, anorexia, reduced milk production, excessive salivation, nausea, vomiting, urinary and fecal incontinence, clonic convulsions, necrosis of gastrointestinal mucosa, weakness, severe depression, and cardiac failure. Death sometimes occurs within 12–14 hours (Krug, 1927).

### FACTORS INFLUENCING TOXICITY

The severity of fluoride toxicosis is influenced by the form in which the fluoride occurs, the nutritional status of animals consuming the fluoride source, variations in fluoride intake, and the presence of other dietary components.

In general, the toxicity of fluoride compounds that are most water-soluble is greater than that of compounds with lesser water solubility. Based on skeletal storage of fluoride by rats, Hobbs *et al.* (1954) concluded that the toxicity of fluoride compounds could be ranked in order from high to low as follows: potassium and sodium fluosilicate, potassium and sodium fluoride, rock phosphate, natural and synthetic cryolite, calcium and magnesium fluosilicates, and calcium fluoride. Hobbs and Merriman (1962) found that fluoride in rock phosphate was considerably less toxic to beef heifers than that in sodium fluoride. Ammerman *et al.* (1964) observed that fluoride storage in the bones of steers was least from calcium fluoride, intermediate from soft phosphate, and highest from sodium fluoride. Plumlee *et al.* (1958) found over 3 times as much fluoride in the femurs of swine fed soft phosphate as in swine fed calcium fluoride. Fluoride-contaminated forage growing near steel-processing or aluminum-reduction plants was found to be equal to (Shupe *et al.*, 1962) or somewhat less (Hobbs *et al.*, 1954) toxic than sodium fluoride to cattle. Naturally fluoridated water produced toxicity comparable to equal amounts of fluoride from sodium fluoride added to water (Wagner and Muhler, 1957) or to a dry diet (Harvey, 1952; Wuthier and Phillips, 1959).

Intakes of total digestible nutrients (TDN) that were 60 percent of recommended allowances resulted in greater incisor damage over a 4½-year period in Holstein heifers from amounts of fluoride less than or equal to (in milligrams per kilogram of body weight) those consumed by heifers consuming 100 percent of recommended TDN allowances (Suttie and Faltin, 1973).

Variations in fluoride intake, with alternating periods of high and low exposure, were more damaging to young cattle than were constant intakes providing the same annual amounts (Suttie *et al.*, 1972).

Other dietary components that have been shown to reduce fluoride toxicosis include calcium (Lawrenz and Mitchell, 1941; Ranganathan, 1941; Peters, 1948; Danowski, 1949; Weddle and Muhler, 1954, 1957; Boddie, 1957, 1960; Suttie *et al.*, 1957a) and sodium chloride (Ericsson, 1968). Boddie (1960) conducted a 3-year study with cattle on fluoride-contaminated pastures. All cattle received 900 g of concentrate daily, and half received in addition 28 g of calcium carbonate and 28 g of aluminum oxide. Cattle receiving the calcium and aluminum supplements had a less severe incisor fluorosis, but fluoride concentrations in the incisors, mandibles, and metacarpals were similar for both groups.

Suttie *et al.* (1957a) studied young dairy cows through five lactations that were fed a complete diet containing either 50 ppm fluoride (from sodium fluoride) alone or plus 200 g $CaCO_3$ per head daily. When $CaCO_3$ was fed, tooth fluorosis was less severe and similar to that in cows receiving 40 ppm fluoride. Fecal fluoride was increased by $CaCO_3$, but rib fluoride concentrations were unaffected.

Starting with 18-month-old beef heifers, the effect of feeding sodium fluoride at 8, 28, 38, 48, or 58 ppm dietary fluoride with or without 0.5 percent aluminum sulfate in the diet was studied for 8 years (Hobbs *et al.*, 1954; Hobbs and Merriman, 1959). The aluminum sulfate reduced bone fluoride deposition by 30–40 percent and reduced dental fluorosis. The two higher levels of fluoride produced hypertrophy of the mandibles and metatarsals in the absence of aluminum sulfate, but this pathology was prevented when aluminum sulfate was present. Similar findings have been reported by Greenwood *et al.* (1964) and Allcroft *et al.* (1965). Free choice access to aluminum sulfate in a mineral mixture offered to cattle grazing a fluoride-contaminated pasture was not effective in reducing bone fluoride deposition, perhaps because continuous intake of aluminum sulfate was inadequate (Merriman and Hobbs, 1962). It has been shown that aluminum compounds may adversely affect dietary phosphorus retention (Street, 1942; Hobbs *et al.*, 1954; Alsmeyer *et al.*, 1963; Storer and Nelson, 1968), and if they are used to alleviate fluoride toxicosis, increased levels of phosphorus may have to be fed. Aluminum chloride and aluminum acetate also appear to be effective in reducing fluorosis, but aluminum oxide produces only slight alleviation (Sharpless, 1936; Hobbs *et al.*, 1954). It should be noted, however, that the effectiveness of soluble aluminum salts may be dependent on feeding these compounds simultaneously with fluoride ingestion. Dietary aluminum compounds were ineffective in promoting depletion of fluoride previously deposited in the skeleton of the rat.

## TISSUE LEVELS

Plasma fluoride concentrations are maintained within narrow limits by regulatory mechanisms involving skeletal and renal tissues. Elevated intakes of fluoride will result in increased concentrations of fluoride in both urine and bone.

Urine fluoride levels are roughly correlated with dietary intake, although the duration of fluoride ingestion, sampling time, and total urinary output will introduce variation. Expression of urinary fluoride concentration on a common specific gravity basis will somewhat reduce the effect of variation in total urinary output. Shupe *et al.* (1963a) have suggested that relating fluoride to creatinine levels in the urine may even be more helpful. These workers found that by determining the concentration of fluoride in the urine and by combining this information with knowledge of the length of time fluoride had been ingested, the concentration of fluoride in ingested dry matter could be estimated. However, urinary fluoride concentration alone was an inadequate criterion for a definitive diagnosis of fluorosis in cattle.

In several long-term experiments with beef and dairy cattle, the skeletal retention of fluoride was approximately proportional to the concentration of fluoride (from sodium fluoride) in the diet. In these studies, relatively constant dietary fluoride concentrations were fed throughout the entire experimental period. When dietary fluoride concentrations vary widely over a study period, skeletal fluoride concentration may relate well to total fluoride intake, but correlations with dietary fluoride concentrations at a single time may be poor (Suttie *et al.,* 1972). In either case, there will be a decreasing rate of skeletal fluoride uptake with time (Shupe *et al.,* 1963b). Skeletal fluoride concentrations may be determined in the living animal by obtaining biopsies of ribs or coccygeal vertebrae (Burns and Allcroft, 1962; Purvance and Transtrum, 1967). Cancellous bones such as the frontal, ribs, vertebrae, and ilium have a higher fluoride concentration than the more compact metacarpals and metatarsals (Suttie and Phillips, 1959; Shupe *et al.,* 1963b), although correlations may be established between the fluoride levels in these different types of bone (Suttie, 1967). The diaphyseal portion of the metacarpals and metatarsals has a lower fluoride concentration than the metaphyseal portion (Shupe *et al.,* 1963a; Ammerman *et al.,* 1964).

## MAXIMUM TOLERABLE LEVELS

The following recommended maximum tolerable levels take into consideration the adverse biological and economic effects of excessive intakes of fluoride, plus the practical reality that many useful phosphorus supplements for livestock contain significant concentrations of fluoride. While small intakes of fluoride may be beneficial, or even essential, prolonged intakes of dry diet fluoride concentrations above these maximum tolerable levels may result in reduced performance. These levels are based on tolerances to sodium fluoride or other fluorides of similar toxicity (fluoride in certain phosphorus sources appears to be less toxic) and assume that the diet is essentially the sole source of fluoride. When water also contains appreciable fluoride (3 ppm or more), these dietary levels should be proportionately reduced.

Excessive exposure during tooth development in cattle may result in exaggerated tooth wear, impaired mastication, and sensitivity to cold drinking water. Thus, maximum levels for young cattle are set at 40 ppm. Minor morphological lesions may be seen in cattle teeth when dietary fluoride during tooth development exceeds 20 ppm, but a relationship between these lesions and animal performance has not been established. Mature dairy cattle tend to consume more feed in relation to body weight than mature beef cattle, so maximum dietary fluoride levels are set at 40 ppm for the former and at 50 ppm for the latter. Lifetime fluoride exposure for finishing cattle is less than for breeding cattle, so the maximum tolerable level for this productive class is set at 100 ppm. Maximum tolerable levels for other species are based on relatively limited published data in some cases, and the level of 40 ppm for horses and rabbits represents an extrapolation. Lifetime exposure of sheep, swine, and poultry is less than for cattle and horses, and poultry are relatively less sensitive to fluoride. Thus, maximum tolerable levels for breeding sheep and finishing sheep have been set at 60 and 150 ppm, respectively. Maximum tolerable levels are 150, 150, and 200 ppm for swine, turkey, and chicken, respectively.

## SUMMARY

Fluorine is chemically bound as fluoride in igneous and sedimentary rock. Most plants have a limited capacity to absorb fluoride from the soil, but significant quantities may be consumed by animals from contaminated surface water, deep well water percolating through fluorapatite, forages contaminated by fluoride-bearing dusts, fumes or water, animal by-products containing bone high in fluoride, and a variety of inorganic phosphate supplements. Animals normally ingest low levels without harm, and small amounts of fluoride may be beneficial and perhaps even essential. Ingestion of excessive fluoride induces characteristic lesions of the skeleton and teeth, resulting in intermittent lameness, excessive tooth wear, reduced feed and water intake, and decreased weight gain and milk production. Developing teeth and bone are particularly sensitive, and excessive exposure in early postnatal life is especially damaging. In general, those fluoride compounds that are most soluble are most toxic.

TABLE 17 Typical Elemental Concentrations in Fluoride-Containing Phosphate Compounds[a]

| Compound | Phosphorus (%) | Calcium (%) | Sodium (%) | Nitrogen (%) | Fluoride (%) |
|---|---|---|---|---|---|
| Defluorinated phosphates manufactured from defluorinated phosphoric acid[b] | | | | | |
| Monocalcium phosphate | 21.0 | 16.0 | — | — | 0.16 |
| Dicalcium phosphate | 18.5 | 21.0 | — | — | 0.14 |
| Defluorinated phosphate | 18.0 | 32.0 | — | 5.0 | 0.16 |
| Monoammonium phosphate | 24.0 | 0.5 | — | 11.0 | 0.18 |
| Diammonium phosphate | 20.0 | 0.5 | — | 18.0 | 0.16 |
| Ammonium polyphosphate solution | 14.5 | 0.1 | — | 10.0 | 0.12 |
| Defluorinated wet-process phosphoric acid | 23.7 | 0.2 | — | — | 0.18 |
| Defluorinated phosphates manufactured from furnace phosphoric acid[c] | | | | | |
| Monocalcium phosphate | 23.0 | 22.0 | — | — | 0.01 |
| Dicalcium phosphate | 18.5 | 26.0 | — | — | 0.01 |

| | | | | | |
|---|---|---|---|---|---|
| Tricalcium phosphate | 19.5 | 38.0 | — | — | 0.01 |
| Monosodium phosphate, anhydrous | 25.5 | — | 19.0 | — | 0.01 |
| Disodium phosphate | 21.5 | — | 32.0 | — | 0.01 |
| Sodium tripolyphosphate | 25.0 | — | 30.0 | — | 0.01 |
| Ammonium polyphosphate solution | 16.0 | — | — | 11.0 | 0.01 |
| Feed-grade phosphoric acid | 23.7 | — | — | — | 0.01 |
| High-fluoride phosphates | | | | | |
| Soft rock phosphate | 9.0 | 17.0 | — | — | 1.2 |
| Ground rock phosphate | 13.0 | 35.0 | — | — | 3.7 |
| Ground low-fluorine rock phosphate | 14.0 | 36.0 | — | — | 0.45 |
| Triple superphosphate | 21.0 | 16.0 | — | — | 2.0 |
| Diammonium phosphate (fertilizer grade) | 20.0 | 0.5 | — | 18.0 | 2.0 |
| Wet-process phosphoric acid (undefluorinated) | 23.7 | 0.2 | — | — | 2.5 |

[a] Adapted from the National Research Council (1974). Unpublished data supplied by International Minerals and Chemical Corp., Northbrook, Ill.

[b] Phosphoric acid produced by the action of sulfuric acid on phosphate rock followed by defluorination.

[c] Phosphoric acid produced by heating phosphate rock with coke in an electric arc furnace, capture of gaseous phosphorus, and oxidation with air. Fluoride concentrations of products are maximum values. Many are typically 1–10 ppm F (0.0001–0.001%).

TABLE 18 Effects of Fluorine Administration in Animals

| Class and Number of Animals[a] | Age or Weight | Administration: Quantity of Element[b] | Source | Duration | Route | Effect(s) | Reference |
|---|---|---|---|---|---|---|---|
| Cattle—2, heifers | 9–11 wk | 3–5 ppm | Forage, concentrate | 6 yr | Diet | No adverse effect | Suttie *et al.*, 1972 |
| Cattle—4, heifers | 2 yr | 3–5 ppm | Forage, concentrate | 5½ yr | Diet | No adverse effect | Suttie *et al.*, 1957a, b |
| | | 23–25 ppm | Forage, conc., NaF (20 ppm) | | | Slight to medium dental mottling, staining | |
| | | 33–35 ppm | Forage, conc., NaF (30 ppm) | | | Slight to heavy dental mottling, staining | |
| | | 43–45 ppm | Forage, conc., NaF (40 ppm) | | | Slight to heavy dental mottling, staining; slight to medium hypoplasia; slight metatarsal exostosis | Ramberg, et al., |
| | | 53–55 ppm | Forage, conc., NaF (50 ppm) | | | Slight to heavy dental mottling, staining; slight to medium hypoplasia; moderate to severe exostosis; intermittent lameness after 5 yr; refusal of F-containing feeds | Suttie *et al.*, 1957a, b, 1958 |
| Cattle—4, cows | 4–6 yr | 3–5 ppm | Forage, concentrate | 3 yr | Diet | No adverse effect | Suttie and Phillips, 1959 |
| | 33–35 ppm | | Forage, conc., NaF (30 ppm) | | | Slight to moderate metatarsal exostosis | |

| | | | | | | | |
|---|---|---|---|---|---|---|---|
| | 53–55 ppm | | Forage, conc., NaF (50 ppm) | | | Slight to extensive metatarsal exostosis | |
| Cattle—3, heifers | Yearlings | 8 ppm | Hay, pasture, concentrate | 8 yr | Diet | No adverse effect | Hobbs and Merriman, 1962 |
| | | | | 10 yr | | No adverse effect | |
| Cattle—6, heifers | Yearlings | 10 ppm | Pasture, hay, concentrate | 8 yr | | No adverse effect | |
| Cattle—10, heifers | | | | 10 yr | | No adverse effect | |
| Cattle—3, heifers | | 18 ppm | NaF (10 ppm) | | | Slight to medium dental mottling, staining | |
| | | 27 ppm | Hay, pasture, concentrate | | | Slight to heavy dental mottling, staining | |
| | | 28 ppm | NaF (20 ppm) | 8 yr | | Slight to medium dental mottling, staining | |
| | | 10 yr | | | | Slight to medium dental mottling, staining | |
| Cattle—12, heifers | | 31 ppm | Pasture, hay, concentrate | | | Slight to heavy dental mottling, staining | |
| | 16–20 mo | 6–22 ppm | Pasture | 7 yr | | No adverse effect | |
| | | 2–10 ppm | Hay | | | | |
| Cattle—3, heifers | Yearlings | 38 ppm | NaF (30 ppm) | 8 yr | | Slight to medium dental mottling, staining | |
| | | | | 10 yr | | Slight to heavy dental mottling, staining, negligible to medium wear | |
| | | 44 ppm | Hay, pasture, concentrate | | | Slight to heavy dental mottling, staining; trace to slight bone hypertrophy | |

TABLE 18 *Continued*

| Class and Number of Animals[a] | Age or Weight | Administration: Quantity of Element[b] | Source | Duration | Route | Effect(s) | Reference |
|---|---|---|---|---|---|---|---|
| Cattle—12, heifers | | | | | | Slight to heavy dental mottling, staining, negligible to medium wear; trace to slight bone hypertrophy | |
| Cattle—9, heifers | | 45 ppm | | 8 yr | | Slight to heavy dental mottling, staining, caries | |
| Cattle—3, heifers | | 48 ppm | NaF (40 ppm) | | | Slight to heavy dental mottling, staining | |
| | | | | 10 yr | | Slight to heavy dental mottling, staining, negligible to medium wear; trace to medium bone hypertrophy; decreased feed consumption | |
| | | 58 ppm | NaF (50 ppm) | 8 yr | | Slight to heavy dental mottling, staining, caries; decreased feed consumption after 3 yr | |
| | | | | 10 yr | | Dental mottling, staining, hypoplasia, slight to excessive wear; trace to medium bone | |

| | | | | | | | |
|---|---|---|---|---|---|---|---|
| | | | | | | hypertrophy; slight exostosis; decreased feed consumption | |
| Cattle—3, heifers | Yearlings | 78 ppm | NaF (70 ppm) | 10 yr | Diet | Dental mottling, staining, hypoplasia, slight to excessive wear; general bone hypertrophy, slight exostosis; decreased feed consumption | Hobbs and Merriman, 1962 |
| | | 108 ppm | NaF (100 ppm) | | | Dental mottling, staining, hypoplasia, slight to excessive wear; general bone hypertrophy; decreased feed consumption | |
| | | 110 ppm | | | | Dental mottling, staining, hypoplasia, slight to excessive wear; general bone hypertrophy, exostosis; decreased feed consumption | |
| Cattle—12, heifers | 16–20 mo | 162–474 ppm | Pasture | 409 d (then on 2–22 ppm F to 7 yr) | | Slight to heavy dental mottling, staining; slight bone hypertrophy in two | |
| | | 17 ppm | Hay | 639 d (then on 2–22 ppm F to 7 yr) | | Slight to heavy dental mottling, staining; slight bone hypertrophy | |

## TABLE 18 *Continued*

| Class and Number of Animals[a] | Age or Weight | Administration: Quantity of Element[b] | Source | Duration | Route | Effect(s) | Reference |
|---|---|---|---|---|---|---|---|
| | | | | 837 d (then on 2–22 ppm F to 7 yr) | | Slight to heavy dental mottling, staining; slight bone hypertrophy one | |
| Cattle—6, heifers | | 265–474 ppm<br>17 ppm | Pasture<br>Hay | 209 d (then on 2–22 ppm F to 7 yr) | | Slight to heavy dental mottling, staining; slight to medium bone hypertrophy | |
| | | 410–474 ppm<br>17 ppm | Pasture<br>Hay | 102 d (then on 2–22 ppm F to 7 yr) | | Slight to medium dental mottling, staining; slight bone hypertrophy one | |
| Cattle—4, heifers | 7–8 mo | 10 ppm | Hay | 588 d | Diet | No adverse effect | Shupe *et al.*, 1962 |
| | | 62 ppm | Hay (near steel plant) | | | Slight to definite dental mottling, staining, hypoplasia, wear; metatarsal hyperostosis | |
| | | 68 ppm | Hay, NaF (60 ppm) | | | Slight to definite dental mottling, staining, hypoplasia, wear; metatarsal hypero- | |

| | | | | | | | |
|---|---|---|---|---|---|---|---|
| | | | | | | stosis | |
| | | 69 ppm | Hay, $CaF_2$ (60 ppm) | | | Slight to definite dental mottling, staining; slight metatarsal hyperostosis | |
| Cattle—8, heifers | 3–4 mo, 105 kg | 12 ppm (dry basis) | Hay, concentrate | 7 yr | Diet | No adverse effect | Greenwood *et al.*, 1964; Harris *et al.*, 1964; Hoogstratten *et al.*, 1965 |
| | | 27 ppm (dry basis) | Hay, conc., NaF (15 ppm) | | | Slight dental mottling; slight histological bone changes | |
| | | 49 ppm (dry basis) | Hay, conc., NaF (37 ppm) | | | Moderate to marked dental mottling, staining, hypoplasia, slight to definite wear; intermittent lameness after 4½ yr; moderate bone changes; slightly decreased feed intake after 4 yr | |
| | | 93 ppm (dry basis) | Hay, conc., NaF (81 ppm) | | | Marked dental mottling, staining, hypoplasia, definite wear; marked bone changes; intermittent lameness after 2½ yr; decreased feed consumption after 4 yr; decline in milk production (not statistically signifi- | |

## TABLE 18 *Continued*

| Class and Number of Animals[a] | Age or Weight | Administration: Quantity of Element[b] | Source | Duration | Route | Effect(s) | Reference |
|---|---|---|---|---|---|---|---|
| | | | | | | cant); slight elevation in % eosinophils | |
| Cattle—10, heifers | 14 wk | 40 ppm | NaF | 32 mo | Diet | Mild hyperostosis of mandibles, metacarpals, and metatarsals; low TDN delayed incisor eruption and increased fluorosis of incisor 4 | Suttie and Faltin, 1973 |
| Cattle—5, heifers | | 43 ppm | Forage, conc., NaF (40 ppm) | 58 mo | | General dental mottling; mild hyperostosis of mandibles, metacarpals, and metatarsals | |
| Cattle—3 | 6 wk | 100 ppm | Pelleted forage and conc., NaF (100 ppm) | 11 mo | Water | Decreased feed intake, weight gain, calcium absorption and calcium balance | Ramberg *et al.*, 1970 |
| Cattle—20, heifers | Yearlings | 11–122 ppm<br>12–220 ppm<br>Trace | Pasture<br>Hay<br>Water | 5½ yr | Diet | No adverse effect | Merriman and Hobbs, 1962 |
| | | 11–122 ppm<br>12–220 ppm<br><br>0.7–2.6 ppm | Pasture<br>Hay<br><br>Water | | Diet and water | No adverse effect | |
| | | 27–735 ppm | Pasture | | Diet | Slight to medium dental | |

| | | | | | | | |
|---|---|---|---|---|---|---|---|
| | | 50–1,812 ppm | Hay | | | mottling, staining | |
| | | Trace | Water | | | | |
| | | 27–735 ppm | Pasture | | Diet and water | Slight to medium dental mottling, staining | |
| | | 50–1,812 ppm | Hay | | | | |
| | | 1–1.5 ppm | Water | | | | |
| Cattle—4, heifers | 2 yr | 0.15–0.3 mg/kg | Forage, concentrate | 6 yr | Diet | No adverse effect | Newell and Schmidt, 1958 |
| | | 1.0 mg/kg | NaF | | | No to slight dental mottling | |
| | | 1.5 mg/kg | | | | No to marked dental mottling, staining, wear | |
| | | 2.0 mg/kg | | | | No to marked dental mottling, staining, wear; occasional unthriftiness | |
| | | 2.5 mg/kg | | | | No to excessive dental mottling, staining; metatarsal exostosis; unthriftiness | |
| Cattle—3, heifers | 6–25 wk | 1.0 mg/kg | NaF | 2 lactations | Diet | Slight dental mottling | Suttie *et al.*, 1961 |
| Cattle—6, heifers | 6–24 wk | 1.2 mg/kg | | | | Slight dental mottling | |
| | 8–22 wk | | | | | Extensive dental mottling, mild enamel hypoplasia; intermittent lameness in second lactation | |
| Cattle—4, heifers | 6–21 wk | 1.6 mg/kg | | | | Extensive dental mottling, hypoplastic | |

TABLE 18 *Continued*

| Class and Number of Animals[a] | Age or Weight | Administration: Quantity of Element[b] | Source | Duration | Route | Effect(s) | Reference |
|---|---|---|---|---|---|---|---|
| | | | | | | enamel, excessive wear; intermittent lameness in second lactation | |
| Cattle—3, heifers | 6–27 wk | 2.0 mg/kg | | | | Extensive dental mottling, hypoplastic enamel, excessive wear; intermittent lameness in first lactation | |
| Cattle—2, heifers | 9–11 wk | 1.5 mg/kg | NaF | 6 yr | Diet | Moderate to general dental mottling, enamel hypoplasia; mild to severe metacarpal and metatarsal hyperostosis and exostosis | Suttie *et al.*, 1972 |
| Cattle—3, heifers | | 1.5 mg/kg (6 mo) No suppl (6 mo) | | | | Slight to general dental mottling, enamel hypoplasia; no to moderate metacarpal and metatarsal hyperostosis | |
| Cattle—4, heifers | 19–22 wk | 1.5 mg/kg | | | | Moderate to general dental mottling, enamel hypoplasia; | |

| | | | | | | | |
|---|---|---|---|---|---|---|---|
| | | | | | | mild to severe metacarpal and metatarsal hyperostosis and exostosis | |
| Cattle—3, heifers | | 3.0 mg/kg (4 mo)<br>0.75 mg/kg (8 mo) | | | | Moderate to general dental mottling, enamel hypoplasia, severe molar wear; mild to severe metacarpal and metatarsal hyperostosis and exostosis; intermittent lameness | |
| Cattle—40, cows | 4 yr or older | 0.5–1.7 mg/kg | NaF | 6 yr raised on diets with similar F | Diet | Dental mottling, staining, wear | Neeley and Harbaugh, 1954 |
| Cattle—3, heifers | 13 mo | 2.5 mg/kg (3 mo)<br>0 (32 mo) | NaF | 35 mo | Diet | Rapid increase in plasma F with rapid decrease postexposure; fluorotic damage to incisors 1 and 2 | Suttie and Faltin, 1971 |
| | 16 mo | 2.5 mg/kg (3 mo)<br>0 (29 mo) | | 32 mo | | Rapid increase in plasma F with rapid decrease postexposure; fluorotic damage to incisors 2 and 3 | |

TABLE 18 *Continued*

| Class and Number of Animals[a] | Age or Weight | Administration: Quantity of Element[b] | Source | Duration | Route | Effect(s) | Reference |
|---|---|---|---|---|---|---|---|
| Sheep—18, | Weanlings, wethers | 14 ppm 32 kg | NaF (50%) F on hay (50%) | 14 wk | Diet | No adverse effect | Harris *et al.*, 1958 |
| | | 18 ppm | | | | No adverse effect | |
| | | 32 ppm | | | | No adverse effect | |
| | | 58 ppm | | | | No adverse effect | |
| | | 112 ppm | | | | No adverse effect | |
| Sheep—18, wethers | Weanlings | 17 ppm | Natural diet | 84 d | Diet | No adverse effect | Harris *et al.*, 1963 |
| | | 50 ppm | NaF | | | No adverse effect | |
| | | 100 ppm | | | | No adverse effect | |
| | | 200 ppm | | | | Decreased feed intake, gain, and carcass grade | |
| Sheep—6 | 10–11 mo | 2.5 ppm, 5–7 mg/d | NaF | 3.5 yr | Water | No adverse effect | Peirce, 1952 |
| | | 5 ppm, 9–13 mg/d | | | | Slight dental mottling | |
| | | 10 ppm, 18–27 mg/d | | | | Dental mottling | |
| | | 20 ppm, 37–53 mg/d | | | | Marked dental lesions | |
| Sheep—16 | 2.5–3.5 yr | 10 ppm, 18 mg/d, 0.24 mg/kg | NaF | 26 mo | Water | No adverse effect | Peirce, 1954 |
| | | 20 ppm, | | | | No adverse effect | |

| | | | | | | | |
|---|---|---|---|---|---|---|---|
| | | 33 mg/d, 0.44 mg/kg | | | | | |
| Sheep—17 | Conception | 10 ppm | NaF | 7 yr | Water | Dental lesions and excessive wear; decreased wool production | Peirce, 1959 |
| Sheep—20 | | 20 ppm | | | | Dental lesions and excessive wear; decreased wool production; no effect on reproduction; slight increase in mortality | |
| Swine—8 | Weanlings, 17 kg | 100 ppm | NaF | 160 d | Diet | None on gain; increased femur density | Kick *et al.*, 1935 |
| | Weanlings, 18 kg | 100 ppm | Rock phosphate | 140 d | | No adverse effect | |
| | Weanlings, 18 kg | 150 ppm | Superphosphate | 140 d | | Slight decrease in feed intake and gain | |
| | Weanlings, 18 kg | 160 ppm | Rock phosphate | 160 d | | No adverse effect | |
| | Weanlings, 19 kg | 240 ppm | Phosphatic limestone | 148 d | | No adverse effect | |
| | Weanlings, 19 kg | 250 ppm | Rock phosphate | 148 d | | No adverse effect | |
| | Weanlings, 24 kg | 290 ppm | NaF | 144 d | | None on gain; increased femur breaking strength | |
| | Weanlings, 18 kg | 290 ppm | NaF | 160 d | | Decreased feed intake, gain, and femur breaking strength | |
| Swine—2 | Gilts | 290 ppm | NaF | 2 yr | | Femur hyperplasia; decreased feed intake | |

TABLE 18 *Continued*

| Class and Number of Animals[a] | Age or Weight | Administration: Quantity of Element[b] | Source | Duration | Route | Effect(s) | Reference |
|---|---|---|---|---|---|---|---|
| | | | | | | and poor lactation | |
| Swine—8 | Weanlings, 18 kg | 320 ppm | Rock phosphate | 160 d | | Decreased feed intake, gain, and femur breaking strength | |
| | Weanlings, 24 kg | 330 ppm | Rock phosphate | 144 d | | None on gain; increased femur breaking strength | |
| Swine—2 | Gilts | 330 ppm | | 2 yr | | None on gestation; decreased feed intake and poor lactation | |
| Swine—8 | Weanlings, 19 kg | 360 ppm | | 148 d | | No adverse effect | |
| | Weanlings, 18 kg | 580 ppm | NaF | 160 d | | Decreased feed intake, gain, and femur breaking strength | |
| | Weanlings, 24 kg | 580 ppm | | 144 d | | Decreased feed intake and gain | |
| Swine—2 | Gilts | 580 ppm | | 2 yr | | Femur hyperplasia; decreased feed intake and poor lactation | |
| Swine—8 | Weanlings, 24 kg | 650 ppm | Rock phosphate | 144 d | | Decreased feed intake, gain, and femur breaking strength | |
| Swine—5 | Gilts | 650 ppm | | 2 yr | | Femur hyperplasia; decreased feed consumption and poor | |

| | | | | | | | |
|---|---|---|---|---|---|---|---|
| | | | | | | lactation | |
| Swine—8 | Weanlings, 23 kg | 700 ppm | | 144 d | | Decreased feed intake, gain, and femur breaking strength | |
| | Weanlings, 18 kg | 710 ppm | | 140 d | | Decreased feed intake, gain, and femur breaking strength | |
| | Weanlings, 19 kg | 710 ppm | | 148 d | | Decreased feed intake and gain; femoral exostosis | |
| Swine—8 | Weanlings, 18 kg | 910 ppm | $CaF_2$ | 140 d | Diet | No adverse effect | Kick *et al.*, 1935 |
| | Weanlings, 24 kg | 970 ppm | NaF | 144 d | | Decreased feed intake, gain, and femur breaking strength | |
| Swine—2, barrows | 120–150 d, 27–36 kg | 0.4 g/d | NaF | 40 d | Diet | Reduced gain, gain/feed, and bone growth | Comar *et al.*, 1953 |
| | | 0.4 g/d, 200 ppm | | 75 d | | No adverse effect | |
| | | 1.4 g/d, 1,000 ppm | | | | Reduced gain, gain/feed, and bone growth | |
| | | 1.6 g/d | | 40 d | | Reduced gain, gain/feed, and bone growth | |
| | | 1.6 g/d | | 75 d | | Reduced gain, gain/feed, and bone growth | |
| Swine—6, females | 28 kg | 1 mg/kg | NaF | 3 lactations | Diet | Slight mottling and staining of incisors | Spencer *et al.*, 1971 |
| Swine | Market pigs | 11–14 mg/kg | Rock phosphate | 3 generations, | Diet | Dental fluorosis; slight crease in feed intake | Fargo *et al.*, 1938 |
| | | 5 yr | | | | | |
| | 19–26 mg/kg | | | | | Dental fluorosis; reluctance to masticate whole corn or drink | |

TABLE 18 *Continued*

| Class and Number of Animals[a] | Age or Weight | Administration: Quantity of Element[b] | Source | Duration | Route | Effect(s) | Reference |
|---|---|---|---|---|---|---|---|
| | | | | | | cold water; decrease in feed intake; no effect on reproduction | |
| Chicken—50, females | 1 d | 423 ppm | Natural diet including superphosphate | 8 wk | Diet | Slightly lower weight at 6 wk; no effect by 8 wk | Gerry *et al.*, 1947 |
| | | 462 ppm | Natural diet including colloidal phosphate | | | Slightly lower weight at 6 and 8 wk | |
| Chicken—100, females | 1 d | 530–546 ppm | Natural diet including raw rock phosphate | 38 wk | | Lower weight at 6 and 8 wk; no weight difference at 12, 16, or 38 wk; egg production, hatchability and growth of chicks to 8 wk unaffected | |
| | | 546 ppm | | 8 wk | | Lower weight at 6 and 8 wk | |
| Chicken—50, females | 1 d | 681 ppm | | | | Lower weight at 6 and 8 wk | |
| | | 721 ppm | | | | Lower weight at 6 and 8 wk | |
| Chicken—100, females | 1 d<br>16 wk | 40 ppm<br>1,017 ppm | Natural diet (plus raw | 24 wk | | No adverse effect | |

| | | | | | | | |
|---|---|---|---|---|---|---|---|
| | | | rock phosphate) | | | | |
| Chicken—100, females | 1 d | 1,017 ppm | Natural diet including raw rock phospate | | | Lower weight at 6, 8, 16, and 24 wk | |
| | | 1,019 ppm | | 8 wk | | Lower weight at 6 and 8 wk | |
| Chicken—135, females | 20 wk | 27 ppm | Natural diet including mix of defluor. and nondefluor. dical. phos. | 51 wk | Diet | No adverse effect | Smith *et al.*, 1970 |
| | | 53 ppm | Natural diet including mix of defluor. and nondefluor. dical. phos. | 51 wk | | No adverse effect | |
| | | 87 ppm | Natural diet including mix of defluor. and dical. phos. | | | No effect on egg production; reduced tibia bone ash phosphorus concentration | |
| Chicken—50, females | 1 d | 37 ppm | Natural diet | 8 wk | Diet | No adverse effect | Gerry *et al.*, 1947 |
| | | 40 ppm | Natural diet | 24 wk | | No adverse effect | |
| | | 43 ppm | Natural diet | 8 wk | | No adverse effect | |
| Chicken—100, females | 1 d | 43 ppm | Natural diet | 38 wk | | No adverse effect | |
| Chicken—50, females | 1 d | 99 ppm | Natural diet including melted rock | 8 wk | | No adverse effect | |

TABLE 18 *Continued*

| Class and Number of Animals[a] | Age or Weight | Administration: Quantity of Element[b] | Source | Duration | Route | Effect(s) | Reference |
|---|---|---|---|---|---|---|---|
| | | | phosphate | | | | |
| | | 318 ppm | Natural diet including partially defluorinated rock phosphate | 8 wk | | No adverse effect | |
| Chicken—20 | 1 d | 250 ppm | Rock phosphate | 8 wk | Diet | No adverse effect | Kick *et al.*, 1933 |
| Chicken—40 | | 360 ppm | Rock phosphate | | | No adverse effect | |
| | | | NaF | | | Decreased blood clotting time | |
| Chicken—20 | | 360 pm | Rock phos- | | | No adverse effect | |
| | | | NaF | | | No adverse effect | |
| | | 450 ppm | NaF | | | No adverse effect | |
| Chicken—60 | | 710 ppm | Rock phosphate | | | Reduced feed consumption, gain, and blood clotting time | |
| | | 720 ppm | NaF | | | Reduced feed consumption, gain, and blood clotting time | |
| Chicken—20 | | 890 ppm | | | | Reduced feed consumption and gain | |
| | | 950 ppm | Rock phosphate | | | Reduced feed consumption and gain | |

| | | | | | | | |
|---|---|---|---|---|---|---|---|
| Chicken—80 | | 1,070 ppm | Rock phosphate | | | Reduced feed consumption, gain, and blood clotting time | |
| Chicken—60 | | 1,080 ppm | NaF | | | Reduced feed consumption, gain, and blood clotting time | |
| Chicken—20 | | 1,080 ppm | $CaF_2$ | | | No adverse effect | |
| | | 1,580 ppm | Rock phosphate | | | Reduced feed consumption and gain | |
| Chicken—20 | 1 d | 1,780 ppm | Rock phosphate | 8 wk | Diet | Reduced feed consumption and gain | Kick *et al.*, 1933 |
| | | 1,810 ppm | | | | Reduced feed consumption and gain | |
| | | 1,820 ppm | $CaF_2$ | | | No adverse effect | |
| | | 2,210 ppm | NaF | 5 wk | | Reduced feed consumption and gain; death by 5 wk | |
| | | 2,710 ppm | $CaF_2$ | 8 wk | | No adverse effect | |
| | | 4,420 ppm | NaF | 3 wk | | Reduced feed consumption and gain; death by 3 wk | |
| Chicken—20 | 1 d | 250 ppm | Rock phosphate | 8 wk | Diet | Decreased gain (?) | Kick *et al.*, 1935 |
| Chicken—60 | | 360 ppm | Rock phosphate | | | No adverse effect | |
| | | 360 ppm | NaF | | | No adverse effect | |
| Chicken—20 | | 450 ppm | NaF | | | No adverse effect | |
| Chicken—60 | | 710 ppm | Rock phosphate | | | Decreased feed intake and gain | |
| | | 720 ppm | NaF | | | Decreased feed intake and gain | |
| Chicken—20 | | 890 ppm | NaF | | | Decreased feed intake and gain | |

TABLE 18 *Continued*

| Class and Number of Animals[a] | Age or Weight | Administration: Quantity of Element[b] | Source | Duration | Route | Effect(s) | Reference |
|---|---|---|---|---|---|---|---|
| | | 950 ppm | Rock phosphate | | | Decreased feed intake and gain | |
| | | 1,070 ppm | Rock phosphate | | | Decreased feed intake and gain | Kick *et al.*, 1935 |
| Chicken—60 | | 1,080 ppm | NaF | | | Decreased feed intake and gain | |
| Chicken—20 | | 1,080 ppm | $CaF_2$ | | | No adverse effect | |
| | | 1,080 ppm | Rock phosphate | | | Decreased gain | |
| | | 1,580 ppm | | | | Decreased feed intake | |
| | | 1,780 ppm | | | | Decreased feed intake | |
| | | 1,810 ppm | | | | Decreased feed intake | |
| | | 1,820 ppm | $CaF_2$ | | | No adverse effect | |
| | | 2,210 ppm | NaF | 5 wk | | Decreased feed intake and gain; death in 5 wk | |
| | | 2,710 ppm | $CaF_2$ | 8 wk | | No adverse effect | |
| | | 4,420 ppm | NaF | 3 wk | | Decreased feed intake and gain; death in 3 wk | |
| Chicken—18, pullets | 0.66 kg | 444 ppm | Rock phosphate | 2 yr | Diet | No adverse effect | Snook, 1958 |
| Chicken—48, males and females | 1 d | 500 ppm | NaF | 4 wk | Diet | Reduction in body weight, 9% | Weber *et al.*, 1969 |
| | | 1,000 ppm | | | | Reduction in body wt., 21%; increased | |

| | | | | | | | |
|---|---|---|---|---|---|---|---|
| | | | | | | plasma alk. phos-phosphatase activity | |
| Chicken—44, males | 1 d | 800 ppm | NaF | 6 wk | Diet | Decreased body wt.; increased proven-triculus wt. | Gardiner *et al.*, 1959 |
| Chicken—25, males | | 1,000 ppm | | | | Decreased body wt.; increased proven-triculus wt. | |
| Chicken—30, males | 5 wk | 800 ppm | NaF | 4 wk | Diet | Decreased gain, feed consumption, and gain/feed | Gardiner *et al.*, 1968 |
| Chicken—10 | 1 d | 900 ppm | NaF | 49 d | Diet | Reduced feed intake and gain | Phillips *et al.*, 1935 |
| Chicken—6 | | 1,000 ppm | | 30–35 d | | Reduced feed intake and gain | |
| Chicken—10 | | 34–40 mg/kg | | 30–35 d | IP inj. | Depressed gain | |
| Turkey—28, males and females | 10–12 wk | 43 ppm | Natural diet | 16 wk | Diet | No adverse effect | Anderson *et al.*, 1955 |
| | | 100 ppm | Natural diet + NaF | | | No adverse effect | |
| | | 200 ppm | | | | Decreased gain in males | |
| | | 400 ppm | | | | Decreased gain in males | |
| | | 800 ppm | | | | Decreased gain in males and decreased gain/feed | |
| | | 1,600 ppm | | | | Decreased gain in both sexes, feed consump-tion and gain/feed | |
| Dog—4 | Weanlings | 200 ppm | NaF | 7 wk | Diet (30 ppm | Slight reduction in feed intake and gain; pre- | Bunce *et al.*, 1962 |

TABLE 18 *Continued*

| Class and Number of Animals[a] | Age or Weight | Administration: Quantity of Element[b] | Source | Duration | Route | Effect(s) | Reference |
|---|---|---|---|---|---|---|---|
| | | | | | Mg) | vented aortic calcification of Mg deficiency | |
| Dog—5 | | | | | Diet (180 ppm Mg) | Slight reduction in feed intake and gain | |
| Dog—4 to 6 | | 250 ppm | NaF | 3+ wk | Diet (low Mg) | Weight gain reduced 50–75% at 3 wk as compared to low Mg basal; no gain after 3 wk; muscular weakness, convulsions, low serum Mg (due to low Mg?) but no aortic lesions | |
| Dog—36 | Puppies | 5 mg/kg | Bonemeal | 1 yr | Diet | No adverse effect | Greenwood *et al.*, 1946 |
| Dog—22 | | | Defluorinated rock phosphate | | | No adverse effect | |
| Dog—20 | | | NaF | | | Dental fluorosis | |
| Rat—5 | 24 d, 55 g | 44 ppm | NaF | 19 wk | Diet | No adverse effect | Kick *et al.*, 1935 |
| | 24 d, 56 g | 46 ppm | $CaF_2$ | | | No adverse effect | |
| | 24 d, | 88 ppm | NaF | | | None on gain; excess | |

| | | | | | | | |
|---|---|---|---|---|---|---|---|
| | 55 g | | | | | wear of mandibular incisors | |
| | 24 d, 54 g | 92 ppm | $CaF_2$ | | | No adverse effect | |
| | 24 d | 100 ppm | Rock phosphate | Through 3 litters and 4 generations | | No adverse effect | |
| | 24 d, 51 g | 220 ppm | NaF | 19 wk | | Decreased feed intake and gain; excess wear of mandibular incisors | |
| | 24 d, 61 g | 230 ppm | $CaF_2$ | | | None on gain; excess wear of mandibular incisors | |
| | 24 d | 240 ppm | Phosphatic limestone | Through 3 litters and 4 generations | | Decreased gain; dental fluorosis | |
| | | 240 ppm | $CaF_2$ | | | No adverse effect | |
| | | 250 ppm | Rock phosphate | | | No adverse effect | |
| Rat—5 | 24 d | 480 ppm | Phosphatic limestone | Through 3 litters and 4 generations | Diet | No adverse effect | Kick *et al.*, 1935 |
| | | 480 ppm | $CaF_2$ | | | No adverse effect | |
| | | 510 ppm | Rock phos- | | | None on gain; failure to reproduce in second generation | |
| | | 710 ppm | | | | No adverse effect | |
| | | 720 ppm | | | | Decreased gain; reproductive failure in | |

TABLE 18 *Continued*

| Class and Number of Animals[a] | Age or Weight | Administration: Quantity of Element[b] | Source | Duration | Route | Effect(s) | Reference |
|---|---|---|---|---|---|---|---|
| | | | | | | third generation | |
| | | 750 ppm | $CaF_2$ | | | No adverse effect | |
| Rat—8, males | Weanlings | 200 ppm | NaF | 6 wk | Diet (30 ppm Mg) | No decreased gain as compared to low Mg basal; calcinosis of low Mg not prevented | Bunce *et al.*, 1962 |
| | | | | | Diet (180 ppm Mg) | No adverse effect | |
| Rat—females | Weanlings | 450 ppm | NaF | 30–35 d | Diet | Reduced gain; altered pattern of food consumption (more "nibbling") and reduced glycogen metabolism | Zebrowski and Suttie, 1966 |

[a] Number of animals per treatment.
[b] Quantity expressed in parts per million or as milligrams per kilogram of body weight.

# REFERENCES

Agate, J. N., G. H. Bell, G. F. Boddie, R. G. Bowler, M. Buckell, E. A. Cheeseman, T. H. Douglas, H. A. Druett, J. Garrad, J. Hunter, K. Perry, J. D. Richardson, and J. B. Weir. 1949. Med. Res. Counc. (Gr. Brit.) Memo. No. 22.

Allcroft, R., K. N. Burns, and C. N. Hebert. 1965. Fluorosis in cattle. II. Development and alleviation: Experimental studies. *In* Ministry of Agriculture, Fisheries and Food. Animal Disease Surveys Report No. 2. Her Majesty's Stationery Office, London. 58 pp.

Alsmeyer, W. L., B. G. Harmon, D. E. Becker, A. H. Jensen, and H. W. Norton. 1963. Effects of dietary Al and Fe in phosphorus utilization. J. Anim. Sci. 22:1116.

Ammerman, C. B., L. R. Arrington, R. L. Shirley, and G. K. Davis. 1964. Comparative effects of fluorine from soft phosphate, calcium fluoride and sodium fluoride on steers. J. Anim. Sci. 23:409.

Anderson, J. O., J. S. Hurst, D. C. Strong, H. M. Nielsen, D. A. Greenwood, W. Robinson, J. L. Shupe, W. Binns, R. A. Bagley, and C. I. Draper. 1955. Effect of feeding various levels of sodium fluoride to growing turkeys. Poult. Sci. 34:1147.

Armstrong, W. D., and L. Singer. 1970. Distribution in body fluids and soft tissues, pp. 94–104. *In* Fluorides and Human Health. World Health Organization, Geneva.

Association of American Feed Control Officials (AAFCO). 1977. Official Publication of the Association of American Feed Control Officials, Inc. Baton Rouge, La.

Boddie, G. F. 1957. Fluorine alleviators. II. Trials involving rats. Vet. Rec. 69:483.

Boddie, G. F. 1960. Fluorine alleviators. III. Field trials involving cattle. Vet. Rec. 72:441.

Bunce, G. E., Y. Chiemchaisri, and P. H. Phillips. 1962. The mineral requirements of the dog. IV. Effect of certain dietary and physiologic factors upon the magnesium deficiency syndrome. J. Nutr. 76:23.

Burns, K. N., and R. Allcroft. 1962. The use of tail bone biopsy for studying skeletal deposition of fluorine in cattle. Res. Vet. Sci. 3:215.

Carlson, C. H., W. D. Armstrong, and L. Singer. 1960a. Distribution and excretion of radiofluoride in the human. Proc. Soc. Exp. Biol. Med. 104:205.

Carlson, C. H., W. D. Armstrong, L. Singer, and L. B. Hinshaw. 1960b. Renal excretion of radiofluoride in the dog. Am. J. Physiol. 198:829.

Cass, J. S. 1961. Fluorides: A critical review. IV. Response of livestock and poultry to absorption of inorganic fluorides. J. Occup. Med. 3:471, 527.

Cholak, J. 1959. Fluorides: A critical review. I. The occurrence of fluoride in air, food, and water. J. Occup. Med. 1:501.

Churchill, H. N. 1931. Occurrence of fluorides in some waters of the U.S. Ind. Eng. Chem. 23:996.

Comar, C. L., W. J. Visek, W. E. Lotz, and J. H. Rust. 1953. Effects of fluorine on calcium metabolism and bone growth in pigs. Am. J. Anat. 92:361.

Danowski, T. S. 1949. Cancellation of fluoride inhibition of blood glucose metabolism. Yale J. Biol. Med. 22:31.

Doberanz, A. R., A. A. Kurnick, E. B. Kurtz, A. R. Kemmerer, and B. L. Reid. 1963. Minimal fluoride diet and effect on rats. Fed. Proc. 22:554. (Abstr.)

Ericsson, Y. 1968. Influence of sodium chloride and certain other food components on fluoride absorption in the rat. J. Nutr. 96:60.

Ericsson, Y., and S. Ullberg. 1958. Autoradiographic investigations of the distribution of $F^{18}$ in mice and rats. Acta Odontol. Scand. 16:363.

Fargo, J. M., G. Bohstedt, P. H. Phillips, and E. B. Hart. 1938. The effect of fluorine in rock phosphate on growth and reproduction in swine. Proc. Am. Soc. Anim. Prod. 31:122.

Gardiner, E. E., F. N. Andrews, R. L. Adams, J. C. Rogler, and C. W. Carrick. 1959. The effect of fluorine on the chicken proventriculus. Poult. Sci. 38:1423.

Gardiner, E. E., K. S. Winchell, and R. Hironaka. 1968. The influence of dietary sodium fluoride on the utilization and metabolizable energy value of a poultry diet. Poult. Sci. 47:1241.

Garlick, N. L. 1955. The teeth of the ox in clinical diagnosis. IV. Dental fluorosis. Am. J. Vet. Res. 16:38.

Gerry, R. W., C. W. Carrick, R. E. Roberts, and S. M. Hauge. 1947. Phosphate supplements of different fluorine contents as sources of phosphorus for chickens. Poult. Sci. 26:323.

Greenwood, D. A., J. R. Blayney, O. K. Skinsnes, and P. C. Hodges. 1946. Comparative studies of the feeding of fluorides as they occur in purified bone meal powder, defluorinated phosphate and sodium fluoride in dogs. J. Dent. Res. 25:311.

Greenwood, D. A., J. L. Shupe, G. E. Stoddard, L. E. Harris, H. M. Nielsen, and L. E. Olson. 1964. Fluorosis in Cattle. Utah Agric. Exp. Stn. Spec. Rep. 17. Logan, Utah. 36 pp.

Harris, L. E., M. A. Madsen, D. A. Greenwood, J. L. Shupe, and R. J. Raleigh. 1958. Effect of various levels and sources of fluorine in the fattening ration of Columbia, Rambouillet, and Targhee lambs. J. Agric. Food Chem. 6:365.

Harris, L. E., R. J. Raleigh, G. E. Stoddard, D. A. Greenwood, J. L. Shupe, and H. M. Nielsen. 1964. Effects of fluorine on dairy cattle. III. Digestion and metabolism trials. J. Anim. Sci. 23:537.

Harris, L. E., R. J. Raleigh, M. A. Madsen, J. L. Shupe, J. E. Butcher, and D. A. Greenwood. 1963. Effect of various levels of fluorine, stilbestrol, and oxytetracycline in the fattening ration of lambs. J. Anim. Sci. 22:51.

Harvey, J. M. 1952. Chronic endemic fluorosis of Merino sheep in Queensland. Queensl. J. Agric. Sci. 9:47.

Hobbs, C. S., and G. M. Merriman. 1959. The effects of eight years continuous feeding of different levels of fluorine and alleviators on feed consumption, teeth, bones and production of cows. J. Anim. Sci. 18:1526. (Abstr.)

Hobbs, C. S., and G. M. Merriman. 1962. Fluorosis in Beef Cattle. Tenn. Agric. Exp. Stn. Bull. 351. Knoxville, Tenn. 183 pp.

Hobbs, C. S., R. P. Moorman, Jr., J. M. Griffith, J. L. West, G. M. Merriman, S. L. Hansard, and C. C. Chamberlain, with the collaboration of W. H. MacIntire, L. J. Jones, and L. S. Jones. 1954. Fluorosis in Cattle and Sheep. Tenn. Agric. Exp. Stn. Bull. 235. Knoxville, Tenn. 163 pp.

Hoogstratten, B., N. C. Leone, J. L. Shupe, D. A. Greenwood, and J. Lieberman. 1965. Effect of fluorides on the hematopoietic system, liver and thyroid gland in cattle. J. Am. Med. Assoc. 192:26.

Johnson, L. C. 1965. Histogenesis and mechanisms in the development of osteofluorosis, pp. 424–441. *In* J. H. Simons, ed. Fluorine Chemistry, vol. 4. Academic Press, New York.

Kick, C. H., R. M. Bethke, and P. R. Record. 1933. Effect of fluorine in the nutrition of the chick. Poult. Sci. 12:382.

Kick, C. H., R. M. Bethke, B. H. Edgington, O. H. M. Wilder, P. R. Record, W. Wilder, T. J. Hill, and S. W. Chase. 1935. Fluorine in Animal Nutrition. Ohio Agric. Exp. Stn. Bull. 558. 77 pp.

Krook, L., and G. A. Maylin. 1979. Industrial fluoride pollution. Chronic fluoride poisoning in Cornwall Island cattle. Cornell Vet. 69 (Suppl. 8).

Krug, O. 1927. Eine Vergiftung von Milchohen durch Kilselfluornatrium. Z. Fleish. Milchhyg. 37:38.

Largent, E. J., and F. F. Heyroth. 1949. Absorption and excretion of fluorides: Further observations on the metabolism of fluorides. J. Ind. Hyg. Toxic. 31:134.

Lawrenz, M., and H. H. Mitchell. 1941. The effect of dietary calcium and phosphorus on the assimilation of dietary fluorine. J. Nutr. 22:91.

Machle, W., and E. J. Largent. 1942. Absorption and excretion of fluorides. Part 1. The normal fluoride balance. J. Ind. Hyg. Toxic. 24:7.

Maurer, R. L., and H. G. Day. 1957. The non-essentiality of fluorine in nutrition. J. Nutr. 62:561.

McClendon, J. F., and J. Gershon-Cohen. 1953. Water-culture crops designed to study deficiencies in animals. J. Agric. Food Chem. 1:464.

McCollum, E. V., N. Simmonds, J. E. Becker, and R. W. Bunting. 1925. The effect of additions of fluorine to the diet of the rat on the quality of the teeth. J. Biol. Chem. 63:553.

Merriman, G. M., and C. S. Hobbs. 1962. Bovine Fluorosis from Soil and Water Sources. Tenn. Agric. Exp. Stn. Bull. 347. Knoxville, Tenn. 46 pp.

Messer, H. H., W. D. Armstrong, and L. Singer. 1972a. Fertility impairment in mice on a low fluoride intake. Science 177:893.

Messer, H. H., K. Wong, M. Wegner, L. Singer, and W. D. Armstrong. 1972b. Effect of reduced fluoride intake by mice on hematocrit values. Nature New Biol. 240:218.

Messer, H. H., W. D. Armstrong, and L. Singer. 1973. Influence of fluoride intake on reproduction in mice. J. Nutr. 103:1319.

National Research Council (NRC). 1971. Fluorides. Biologic Effects of Atmospheric Pollutants. National Academy of Sciences, Washington, D.C. 295 pp.

National Research Council (NRC). 1974. Effects of Fluorides in Animals. National Academy of Sciences, Washington, D.C. 70 pp.

Neeley, K. L., and F. G. Harbaugh. 1954. Effects of fluoride ingestion on a herd of dairy cattle in the Lubbock, Texas area. J. Am. Vet. Med. Assoc. 124:344.

Newell, G. W., and H. J. Schmidt. 1958. The effects of feeding fluorine as sodium fluoride to dairy cattle—A six-year study. Am. J. Vet. Res. 19:363.

Parkins, F. M. 1971. Active F transport: Species and age effects with rodent intestine, *in vitro*. Biochem. Biophys. Acta 241:507.

Parkins, F. M., J. W. Hollifield, A. J. McCaslin, S. L. Wu, and R. G. Faust. 1966. Active transport of fluoride by the rat intestine, *in vitro*. Biochem. Biophys. Acta 126:513.

Peirce, A. W. 1952. Studies on fluorosis of sheep. I. The toxicity of water-borne fluoride for sheep maintained in pens. Aust. J. Agric. Res. 3:326.

Peirce, A. W. 1954. Studies on fluorosis of sheep. II. The toxicity of water-borne fluoride for mature grazing sheep. Aust. J. Agric. Res. 5:545.

Peirce, A. W. 1959. Studies on fluorosis of sheep. III. The toxicity of water-borne fluoride for the grazing sheep throughout its life. Aust. J. Agric. Res. 10:186.

Perkinson, J. D., Jr., I. B. Whitney, R. A. Monroe, W. E. Lotz, and C. L. Comar. 1955. Metabolism of fluorine-18 in domestic animals. Am. J. Physiol. 182:383.

Peters, J. H. 1948. Therapy of acute fluoride poisoning. Am. J. Med. Sci. 216:278.

Phillips, P. H., H. E. English, and E. B. Hart. 1935. The augmentation of the toxicity of fluorosis in the chick by feeding desiccated thyroid. J. Nutr. 10:399.

Plumlee, M. P., C. E. Jordan, M. H. Kennington, and W. M. Beeson. 1958. Availability of the phosphorus from various phosphate materials for swine. J. Anim. Sci. 17:73.

Purvance, G. T., and L. G. Transtrum. 1967. Vertebral biopsy in cattle. J. Am. Vet. Med. Assoc. 151:716.

Ramberg, C. F., Jr., J. M. Phang, G. P. Mayer, A. I. Norberg, and D. S. Kronfeld. 1970.

Inhibition of calcium absorption and elevation of calcium removal rate from bone in fluoride treated calves. J. Nutr. 100:981.

Ranganathan, S. 1941. Calcium intake and fluorine poisoning in rats. Ind. J. Med. Res. 29:693.

Robinson, W. O., and G. Edgington. 1946. Fluorine in soils. Soil Sci. 61:341.

Roholm, K. 1937. Fluorine Intoxication: A Clinical–Hygienic Study with a Review of the Literature and Some Experimental Investigations. H. K. Lewis & Co., Ltd., London. 364 pp.

Schwarz, K., and D. B. Milne. 1972. Fluorine requirement for growth in the rat. Bioinorg. Chem. 1:331.

Sharpless, G. R. 1936. Limitation of fluorine toxicosis in the rat with aluminum chloride. Proc. Soc. Exp. Biol. Med. 34:562.

Shupe, J. L. 1969. Levels of toxicity to animals provide sound basis for fluoride standards. *In* A symposium: The technical significance of air quality standards. Environ. Sci. Technol. 3:721.

Shupe, J. L. 1970. Fluorosis, pp. 288–301. *In* W. J. Gibbons, E. J. Catcott, and J. F. Smithcors (eds.). Bovine Medicine and Surgery. Am. Vet. Publ., Wheaton, Ill.

Shupe, J. L., and E. W. Alther. 1966. The effects of fluorides on livestock, with particular reference to cattle, pp. 307–354. *In* O. Eichler, A. Faran, H. Herken, A. D. Welch, and F. A. Smith (eds.). Handbook of Experimental Pharmacology, vol. 20, pt. 1. Springer-Verlag, New York.

Shupe, J. L., M. L. Miner, L. E. Harris, and D. A. Greenwood. 1962. Relative effects of feeding hay atmospherically contaminated by fluoride residue, normal hay plus calcium fluoride, and normal hay plus sodium fluoride to dairy heifers. Am. J. Vet. Res. 23:777.

Shupe, J. L., L. E. Harris, D. A. Greenwood, J. E. Butcher, and H. M. Nielsen. 1963a. The effect of fluorine on dairy cattle. V. Fluorine in the urine as an estimator of fluorine intake. Am. J. Vet. Res. 24:300.

Shupe, J. L., M. L. Miner, D. A. Greenwood, L. E. Harris, and G. E. Stoddard. 1963b. The effect of fluorine on dairy cattle. II. Clinical and pathological effects. Am. J. Vet. Res. 24:964.

Singer, L., and W. D. Armstrong. 1960. Regulation of human plasma fluoride concentration. J. Appl. Physiol. 15:508.

Singer, L., and W. D. Armstrong. 1964. Regulation of plasma fluoride in rats. Proc. Soc. Exp. Biol. Med. 117:686.

Smith, F. A., D. E. Gardner, and H. C. Hodge. 1950. Investigation on the metabolism of fluoride. II. Fluoride content of blood and urine as a function of fluorine in drinking water. J. Dent. Res. 29:596.

Smith, M. C., E. M. Lantz, and H. V. Smith. 1931. The Cause of Mottled Enamel, a Defect of Human Teeth. Ariz. Agric. Exp. Stn. Tech. Bull. 32. Tucson, Ariz. 253 pp.

Smith, S. B., N. S. Cowen, J. W. Dodge, L. S. Mix, G. L. Rumsey, A. A. Warner, and D. F. Woodward. 1970. Effect of added levels of fluorine on selected characteristics of egg shells and bones from caged layers. Poult. Sci. 49:1438. (Abstr.)

Snook, L. C. 1958. The use of rock phosphate from Christmas Island in poultry rations. J. Dep. Agric. West Aust. 7:545.

Spencer, G. R., F. I. el-Sayed, G. H. Kroening, K. L. Pell, N. Shoup, D. F. Adams, M. Franke, and J. E. Alexander. 1971. Effects of fluoride, calcium, and phosphorus on porcine bone. Am. J. Vet. Res. 32:1751.

Stoddard, G. E., G. Q. Bateman, L. E. Harris, J. L. Shupe, and D. A. Greenwood. 1963. Effects of fluorine on dairy cattle. IV. Milk production. J. Dairy Sci. 46:720.

Stookey, G. K., D. B. Crane, and J. C. Muhler. 1962. Effect of molybdenum on fluoride absorption. Proc. Soc. Exp. Biol. Med. 109:580.

Stookey, G. K., E. L. Dellinger, and J. C. Muhler. 1964. *In vitro* studies concerning fluoride absorption. Proc. Soc. Exp. Biol. Med. 115:298.

Storer, N. L., and T. S. Nelson. 1968. The effect of various aluminum compounds on chick performance. Poult. Sci. 47:244.

Street, H. R. 1942. The influence of aluminum sulfate and aluminum hydroxide upon the absorption of dietary phosphorus by the rat. J. Nutr. 24:111.

Suttie, J. W. 1967. Vertebral biopsies in the diagnosis of bovine fluoride toxicosis. Am. J. Vet. Res. 28:709.

Suttie, J. W., and E. C. Faltin. 1971. The effect of a short period of fluoride ingestion on dental fluorosis in cattle. Am. J. Vet. Res. 32:217.

Suttie, J. W., and E. C. Faltin. 1973. Effects of sodium fluoride on dairy cattle: Influence of nutritional state. Am. J. Vet. Res. 34:479.

Suttie, J. W., and P. H. Phillips. 1959. Studies on the effects of dietary sodium fluoride on dairy cows. V. A three-year study on mature animals. J. Dairy Sci. 42:1063.

Suttie, J. W., R. F. Miller, and P. H. Phillips. 1957a. Effects of dietary sodium fluoride on dairy cows. I. The physiological effects and the development symptoms of fluorosis. J. Nutr. 63:211.

Suttie, J. W., R. F. Miller, and P. H. Phillips. 1957b. Studies of the effects of dietary sodium fluoride on dairy cows. II. Effects on milk production. J. Dairy Sci. 40:1485.

Suttie, J. W., R. Gesteland, and P. H. Phillips. 1961. Effects of dietary sodium fluoride on dairy cows. VI. In young heifers. J. Dairy Sci. 44:2250.

Suttie, J. W., J. R. Carlson, and E. C. Faltin. 1972. Effects of alternating periods of high- and low-fluoride ingestion on dairy cattle. J. Dairy Sci. 55:790.

Tao, S., and J. W. Suttie. 1976. Evidence for a lack of an effect of dietary fluoride level on reproduction in mice. J. Nutr. 106:1115.

Underwood, E. J. 1971. Trace Elements in Human and Animal Nutrition, 3rd ed. Academic Press, New York. 543 pp.

Underwood, E. J. 1977. Trace Elements in Human and Animal Nutrition, 4th ed. Academic Press, New York. 545 pp.

U.S. Department of the Interior (USDI). 1970. Industrial and chemical minerals, p. 184. *In* The National Atlas of the United States of America. Washington, D.C.

VanWazer, J. R. 1961. Technology, biological functions, and applications, pp. 1025–1075. *In* Phosphorus and Its Compounds, vol. 2. Interscience, New York.

Velu, H. 1931. Troubles des aux phosphates naturels et cachexie fluorique de au fluorure de calcium. C. R. Soc. Biol. 108:750.

Wagner, M. J. 1962. Absorption of fluoride by the gastric mucosa in the rat. J. Dent. Res. 41:667.

Wagner, M. J., and J. C. Muhler. 1957. The metabolism of natural and artificial fluoridated waters. J. Dent. Res. 36:552.

Weber, C. W. 1966. Fluoride in the nutrition and metabolism of experimental animals. Ph.D. thesis. University of Arizona, Tucson.

Weber, C. W., A. R. Doberanz, and B. L. Reid. 1969. Fluoride toxicity in the chick. Poult. Sci. 48:230.

Weddle, D. A., and J. C. Muhler. 1954. The effects of inorganic salts on fluorine storage in the rat. J. Nutr. 54:437.

Weddle, D. A., and J. C. Muhler. 1957. The metabolism of different fluorides in the rat. I. Comparisons between sodium fluoride, sodium silicofluoride, and stannous fluoride. J. Dent. Res. 36:386.

Wuthier, R. E., and P. H. Phillips. 1959. The effects of longtime administration of small amounts of fluoride in food or water on caries-susceptible rats. J. Nutr. 67:581.

Yeh, M. C., L. Singer, and W. D. Armstrong. 1970. Roles of kidney and skeleton in regulation of body fluid fluoride concentrations. Proc. Soc. Exp. Biol. Med. 135:421.

Zebrowski, E. J., and J. W. Suttie. 1966. Glucose oxidation and glycogen metabolism in fluoride-fed rats. J. Nutr. 88:267.

Zipkin, I., E. D. Eanes, and J. L. Shupe. 1964. Effect of prolonged exposure to fluoride on the ash, fluoride, citrate, and crystallinity of bovine bone. Am. J. Vet. Res. 25:1591.

# Iodine

It is believed that the Chinese, many centuries B.C., had learned by trial and error that substances in certain marine products exerted beneficial effects upon the thyroid. Burnt sponges and seaweed were added to the diet during the time of Hippocrates (460–370 B.C.) to relieve enlarged thyroids. It was not known until discovered by Davy in 1815 that the efficacy of burnt sponges and other marine products were due to the presence of iodine (I), an element discovered by Courtois in 1811. Probably the first to recommend the use of iodine in salt as a means of preventing goiter was Koestl, who in 1895 began its use in Austria. The view that iodine is an essential component of a protein molecule synthesized by the thyroid began to take form shortly before 1900. By 1914 Kendall had isolated crystalline thyroxine from thyroid tissue. The empirical formula for thyroxine was established in 1926 by Harrington, who estimated that 40 percent of the total iodine present in the thyroid is contained in thyroxine.

The use of iodine in livestock production is not limited to its role as a nutrient in feed. Iodine in the form of ethylenediaminedihydriodide (EDDI) is used at relatively high levels to prevent or treat foot rot and soft tissue lumpy jaw in cattle (Miller and Tillapaugh, 1966). Iodine-containing products such as iodophors are widely used in the dairy industry as teat dips and udder washes. Iodophor solutions are also used as sanitizing agents for cleansing equipment.

## ESSENTIALITY

Iodine is an essential element for animals and man. Although nearly every cell in the body contains iodine, the thyroid gland is the main location of iodine reserve. The thyroid hormones, which contain iodine, are known to have a role in thermoregulation, intermediary metabolism, reproduction, growth and development, hematopoiesis and circulation, and neuromuscular functioning. The role of iodine in thyroid function and the manifestation of iodine deficiency in various species have been described by Evvard (1928), Riggs (1952), and Berson (1956).

## METABOLISM

Iodine occurs in foods largely as inorganic iodide and is absorbed in this form from all levels of the gastrointestinal tract (Underwood, 1977). In the ruminant the rumen is the major site of absorption of iodine and the abomasum the major site of endogenous secretion (Barua *et al.,* 1964). After absorption the iodide is rapidly distributed throughout the body. The major sites of iodine concentration are the thyroid and the kidney. In addition, iodine is concentrated by the salivary glands, stomach, skin and hair, mammary gland, placenta, and ovary (Gross, 1962). The iodide trapped by the thyroid is rapidly oxidized and converted to organic iodine by combination with tyrosine. This process also occurs in the lactating mammary gland and to a very small extent in the ovum within the ovary. In the other sites the element remains in the form of iodide. The iodide pool is replenished continuously, exogenously from the diet and endogenously from the saliva, the gastric juice, and the breakdown of hormones produced by the thyroid. Iodine is lost from the iodide pool by the activities of the thyroid, kidneys, salivary glands, and gastric glands, which compete for the available iodine. Iodine is lost from the body mainly in urine and milk, with smaller amounts appearing in the feces and sweat.

## SOURCES

Iodine is widely distributed in nature, but it is present in both organic and inorganic substances in very small amounts. Only in a few substances, such as the saltpeter deposits of Chile and some marine products, do concentrations of up to 1,000 to 2,000 ppm occur. Iodine is present in soil, air, and water and becomes a constituent of plants and animals used for food. The iodine content of water reflects the iodine content of the rocks and soils of the region. Plants vary widely in iodine content, depending on the species of plant and the iodine content of the soil. Hemken *et al.* (1972), in a study of milk iodine and dairy cattle performance, collected feed samples from Maryland and Illinois farms. Samples of hay from Maryland farms contained 1.31 to 2.54 ppm iodine, while those from Illinois contained 0.62 to 1.02 ppm iodine. The Chilean Iodine Educational Bureau (1952) reported that oilseed meals (soybean, cottonseed, linseed, and peanut) contained 0.11 to 0.2 ppm iodine. Products of animal origin, other than fish meal, do not contain significant levels of iodine, unless animals from which they were obtained ingested large amounts of the element. Iodine sources permitted as feed additives include calcium iodate, calcium iodobehenate, cuprous iodide, 3,5-diiodosalicylic acid, ethylenediaminedihydriodide (EDDI) pentacalcium orthoperiodate, potassium iodate, potassium iodide, sodium iodate, sodium iodide, and thymol iodide. A review of the biological availability of some iodine compounds is presented by Ammerman and Miller (1972).

## TOXICOSIS

More than a century ago, in practically all of the goiter areas of Europe there occurred a wave of enthusiasm for the use of some form of inorganic iodine in the treatment and prevention of goiter. It appears that indiscriminate use of various iodine preparations was practiced with many cases of poisoning. In 1860 Rillet presented to the French Academy of Medicine a classical description of the toxic symptoms that follow overdosage of iodine. Iodine toxicity has been studied in many laboratory animals, dogs, poultry, swine, and cattle. Significant species differences exist in tolerance to high levels of iodine. Prolonged administration of large doses of iodine markedly reduces iodine uptake by the thyroid, thus causing antithyroidal or goitrogenic effects. All species appear to have a wide margin of safety for this element.

### LOW LEVELS

In a series of trials, Newton *et al.* (1974) fed graded levels of calcium iodate to give iodine levels ranging from 10 to 200 ppm iodine to calves having an initial weight of about 100 kg. Elevated levels of dietary iodine depressed growth rate and feed intake, with the depression being significant for diets containing 50, 100, or 200 ppm added iodine. The feeding of either 100 or 200 ppm iodine, and in some cases lower levels, produced toxic signs that included coughing and nasal discharge. All levels of added iodine increased serum iodine, and calves fed 200 ppm had significantly lower blood hemoglobin and serum calcium. Calves fed diets with added iodine tended to have heavier adrenal glands, but there was no consistent iodine effect on the weight of the thyroid glands. Based on trends in growth rate and adrenal weights, Newton *et al.* (1974) concluded that 25 ppm iodine was undesirable, and 50 ppm appeared to be the minimum toxic level for calves.

Fish and Swanson (1977) found that calves weighing about 100 kg tolerated 20 and 40 ppm iodine (from EDDI) with no untoward effects, but daily gains were slightly depressed at 86 and 174 ppm. Iodine levels of 71, 140, and 283 ppm had no effect, but a level of 435 ppm depressed daily gains in yearling (320 kg) heifers (Fish and Swanson, 1977). These authors fed lactating dairy cows levels of iodine as high as 314 ppm for 12 weeks and found no adverse effect on milk production. Convey *et al.* (1978) showed that lactating cows receiving about 200 ppm iodine (from EDDI) for 49 weeks exhibited no aberrations in thyroid or pituitary function. When EDDI supplied iodine at levels of 2.5 mg/kg body weight and below to pregnant cows, there was no significant effect on cows or their calves. Levels of EDDI that supplied 5.0 and 7.5 mg iodine per kilogram of body weight increased the incidence of premature calving, weak or abnormal calves at birth, and stillborn calves (E. W. Swanson, University of Tennessee, personal communication).

Calves having an initial weight of 120 kg were given doses of 0, 50, 250, and 1,250 mg iodine (as EDDI) per head per day for 6 months (Haggard, 1978). Determinations included titers to brucellosis, leptospirosis, and infectious bovine rhinotracheitis (IBR) vaccinations. This author found that the brucellosis and leptospirosis titers of calves in the control and two lower levels of iodine were significantly higher than those of calves given 1,250 mg iodine per day. The levels of iodine had no effect on IBR titers. Haggard (1978) also showed that the white blood cells with plasma and without plasma from calves of the control group demonstrated greater *in vitro* phagocytic activity than white blood cells from calves on all iodine levels. The white blood cell counts of calves dosed with either 250 or 1,250 mg iodine per day were less than control calves and calves dosed with 50 mg iodine per day. Rosiles *et al.* (1975) found that calves (192 kg) fed 500 mg EDDI per day coughed more, had greater nasal discharge, and exhibited greater lacrimation than those fed a daily dose of 50 mg. Neither the 50- nor 500-mg level of EDDI had any effect on growth rate of the calves.

McCauley *et al.* (1973) administered by capsule iodine, either in the form of potassium iodide or EDDI, to lambs weighing about 30 kg. Iodine was given daily for 22 days at levels of 150, 300, 450, and 600 mg per lamb per day. Coughing was observed in lambs given large doses of iodide, and these animals had higher mean rectal temperatures. Body weight gains were depressed by daily intakes of 393 mg potassium iodide (300 mg iodine) or 562 mg EDDI (450 mg iodine) per lamb per day.

Pigs are more tolerant of excess iodine than cattle. Newton and Clawson (1974) fed levels of iodine ranging from 10 to 1,600 ppm to growing–finishing pigs and found that the minimum toxic level was between 400 and 800 ppm. Growth rate, feed intake, and hemoglobin levels were depressed at 800 and 1,600 ppm iodine, and liver iron levels were significantly depressed at 400 ppm. Arrington *et al.* (1965) fed pregnant sows either 1,500 or 2,500 ppm iodine for 30 days before farrowing. These levels of iodine did not adversely affect reproductive performance.

Using potassium iodide as the iodine source, Wilgus *et al.* (1953) found no adverse effect on the performance of chicks fed 500 ppm iodine up to 6 weeks of age followed by 180 ppm from 6 weeks through maturity and the laying period. These workers found that 50 ppm in the breeder ration caused a reduction and delay in hatchability. Excessive levels of dietary iodine were shown to have a profound effect on egg production and hatchability (Perdomo *et al.,* 1966; Arrington *et al.,* 1967; Marcilese *et al.,* 1968). When laying hens were fed 625 to 5,000 ppm iodine, egg production varied inversely with level of iodine and ceased with intakes of 5,000 ppm (Arrington *et al.,* 1967). The fertility of the eggs was not affected, but early embryonic death, reduced hatchability, and delayed hatching resulted. Egg production commenced within 1 week after cessation of iodine feeding. Roland *et al.* (1977) reported that serum calcium was significantly increased in laying hens that received diets containing 5,000 ppm iodine. This level of iodine caused a marked reduction in egg production and in the size of ovaries and oviducts.

A high incidence (3 to 50 percent) of goiter was reported in thoroughbred foals born on two farms in Maryland and on one farm in central Ontario, Canada (Baker and Lindsey, 1968). The dietary intake of iodine by mares bearing goitrous foals ranged from 48 to 432 mg/day. Plasma iodine levels were elevated in the goitrous foals as well as in the mares fed rations containing high levels of iodine. Enlarged thyroids and leg weakness were reported in four foals born to mares fed 83 mg iodine daily (Drew *et al.,* 1975).

Marked differences exist between rabbits, hamsters, and rats in their tolerance to high intakes of iodine. Mortality was high in the offspring of rabbits fed 250 ppm iodine in late gestation. On the other hand, the feeding of diets containing 2,500 ppm iodine to hamsters during gestation did not affect death loss in the offspring (Arrington *et al.,* 1965). The survival of the offspring of rats was not affected by feeding gestating female rats 500 ppm of iodine, but high mortality of the young was found when the gestation diets contained 1,000 ppm iodine (Ammerman *et al.,* 1964). Webster *et al.* (1959) found no gross lesions or abnormalities in mice or guinea pigs that received 5,000 ppm of potassium iodate in their drinking water for several weeks. Microscopic examination, however, showed hemosiderin deposits in the renal convoluted tubules of nearly all the mice.

### HIGH LEVELS

Webster *et al.* (1966) determined the minimum lethal dose and the maximum allowable dose of potassium iodate for dogs. The iodine was administered in gelatin capsules in single doses supplying either 100, 200, or 250 mg potassium iodate per kilogram of body weight. The 100 mg/kg level caused brief anorexia and occasional vomiting but all dogs lived. The effects of feeding the 200 and 250 mg/kg doses were very pronounced, and death preceded by anorexia, prostration, and coma occurred at these levels. Fatty changes in the viscera and necrotic lesions in the liver, kidney, and mucosa of the gastrointestinal tract were sometimes present. Retinal changes were noted in one dog given the intermediate level of iodine. Highman *et al.* (1955) reported severe retinal degenerative changes in rabbits and guinea pigs injected intraperitoneally with potassium iodate, but no such retinal changes were observed in guinea pigs given potassium iodate in the drinking water.

### FACTORS INFLUENCING TOXICITY

Many plants and plant products used for animal feeds or for forage are known to contain substances that can induce goiter in animals. More than 300 natural or synthetic chemicals possess goitrogenic properties that may have an effect on iodine bioavailability (Talbot *et al.,* 1976). Among the naturally occurring goitrogens, the best characterized are the glucosinolate derivatives isolated from the *Brassica* species. The occurrence of a potent goitrogen in soybean products is well documented. Thiocyanates, perchlorates, and rubidium salts are known to interfere with iodine uptake by the thyroid, and high levels of arsenic can induce goiter in rats (Underwood, 1977). Bromide, fluoride, cobalt, manganese, and nitrate may also inhibit normal iodine uptake (Talbot *et al.*, 1976).

Few studies have been conducted comparing the relative toxicity of the various iodine compounds. Arrington *et al.* (1965) reported no difference in toxicity to rats between sodium and potassium iodide. Using white Swiss mice, Webster *et al.* (1957) compared the toxicity of single doses of sodium iodide, sodium iodate, potassium iodide, and potassium iodate given either orally, intraperitoneally, or intravenously. In these studies the iodate salts were more toxic than the iodide salts. Miller and Swanson (1973) found that ethylenediaminedihydriodide was absorbed at least as well as sodium or potassium iodide by dairy cows. Also, the iodine from EDDI was retained in most organs and tissues longer than iodine from sodium iodide.

Webster *et al.* (1959) showed that the daily consumption of potassium iodate in the drinking water by mice and guinea pigs at times exceeded the estimated oral $LD_{50}$ values for single doses of iodate given by stomach tube. This suggests a marked increase in tolerance to iodine when it is given in divided, small doses. These same workers reported that the presence of food in the stomach greatly decreased the acute toxic effects of orally administered iodine.

## TISSUE LEVELS

The level of iodine in milk is influenced by iodine intake, season, level of milk production, and the use of iodine-containing disinfectants. Hemken *et al.* (1972) reported that daily supplementation of the diet of lactating dairy cows with either 0, 6.8, or 68.0 mg potassium iodide resulted in milk containing 0.008, 0.081, and 0.694 ppm iodine, respectively. In the same study, Hemken *et al.* (1972) reported that the iodine content of milk from 13 Illinois farms averaged 0.425 ppm and that from 8 Maryland farms averaged 0.457 ppm. The range in iodine content within each location was wide and could be explained largely by the level of supplemental iodine used in the diets of the lactating cows. Miller and Swanson (1973) fed dairy cows daily doses of either 106 mg potassium iodide or 100 mg EDDI and obtained iodine levels in the milk of 0.379 and 0.895 ppm, respectively. Feeding a level of 500 mg EDDI per day caused the iodine content of the milk to reach 2.036 ppm. Feeding 16 and 164 mg iodine per head per day to cows resulted in milk iodine levels of 0.370 and 2.2 ppm, respectively (Convey *et al.*, 1977). The average iodine content of milk from 111 herds was 0.646 ppm, with a range of 0.04 to 4.84 ppm (Hemken, 1978). The high iodine levels in milk were the result of either feeding high levels of dietary iodine or the use of iodine as a sanitizing agent. Hemken (1978) reported that iodine as udder washes caused the iodine content of milk to increase by 0.035 ppm.

Fisher and Carr (1974) reported that the iodine content of beef, pork, and mutton was low (0.027 to 0.045 ppm). The studies by Miller *et al.* (1975) showed that of the nonthyroid tissues skeletal muscle is the poorest concentrator of radioiodine. Eggs from hens fed 0.022 ppm dietary iodine had 0.001 ppm of iodine in the liquid egg, whereas those from hens fed 5 ppm dietary iodine had 5 ppm iodine in the liquid egg (Wilgus *et al.,* 1953). Marcilese *et al.* (1968) fed high concentrations of iodine (100 mg/day) to laying hens and found that the iodine content of the egg increased linearly for 10 days and reached a plateau of approximately 3 mg/egg at that time. The iodine concentration in the eggs from hens fed 500 mg/day increased rapidly to an average of 7 mg/egg by 8 days, at which time egg production in most hens ceased.

## MAXIMUM TOLERABLE LEVELS

Newton *et al.* (1974) showed that 50 ppm iodine significantly reduced growth rate and feed intake of calves weighing about 100 kg. Fish and Swanson (1977) found that calves weighing about 100 kg tolerated 20 and 40 ppm iodine (from EDDI) with no untoward effects. Yearling heifers weighing about 320 kg were not affected by iodine levels of 71, 140, and 283 ppm (Fish and Swanson, 1977). McCauley *et al.* (1973) showed that daily iodine intakes of 300 mg (from EDDI) or 150 mg (from potassium iodide) depressed growth rate of lambs. Based on available information, the maximum tolerable level of iodine for cattle and sheep is 50 ppm. Although cattle can tolerate 50 ppm iodine, it should be understood that this level in the diet may result in undesirably high levels of iodine in the milk. Fish and Swanson (1977) showed that dairy cows receiving 47 ppm iodine produced milk containing 2.4 ppm iodine. The Food and Nutrition Board of the National Research Council (1970) has stated that iodine intakes between 50 and 1,000 $\mu$g/day are estimated as safe, but intakes between 100 and 300 $\mu$g/day are desirable.

Newton and Clawson (1974) reported that 400 ppm iodine had no adverse effect on the performance of pigs weighing about 17 kg. Because 625 ppm iodine reduced egg production and hatchability, the maximum tolerable level for iodine in poultry diets appears to be about 300 ppm.

Horses are less tolerant of excess iodine than cattle, sheep, swine, and poultry. A high incidence of goiter was found in the offspring of mares consuming 48 to 432 mg iodine per day. Assuming that mares consume 10 kg dry matter daily, the maximum tolerable level for iodine in horse diets is 5 ppm.

## SUMMARY

Iodine is an essential element for all animals. Its only known function in the body is in the synthesis of the thyroid hormones. Species differ widely in their susceptibility to iodine toxicity, but all animals can tolerate iodine levels far in excess of their requirements for this element. Feeding excessive levels of iodine has resulted in decreased egg production in hens, inhibition of lactation in rats, decreased hemoglobin levels in pigs, necrotic lesions in the liver of dogs, and goiter and reduced thyroid hormone synthesis in several species. Increasing the iodine intake of lactating cows and laying hens increases the levels of iodine in milk and eggs.

TABLE 19 Effects of Iodine Administration in Animals

| Class and Number of Animals[a] | Age or Weight | Administration: Quantity of Element[b] | Source | Duration | Route | Effect(s) | Reference |
|---|---|---|---|---|---|---|---|
| Cattle—8 | 83 kg | 10 ppm | $Ca(IO_3)_2$ | 144 d | Diet | No adverse effect | Newton *et al.*, 1974 |
| | | 100 ppm | | | | Reduced gains, feed intake, and hemoglobin | |
| | | 200 ppm | | | | Reduced gains and feed intake; coughing and nasal discharge | |
| | 112 kg | 25 ppm | | 112 d | Diet | Reduced gains and feed intake; enlarged adrenals | |
| | | 50 ppm | | | | Coughing and nasal discharge | |
| | | 100 ppm | | | | Coughing and nasal discharge | |
| Cattle—6 | 100 kg | 20 ppm | EDDI | 12 wk | Diet | No adverse effect | Fish and Swanson, 1977 |
| | | 42 ppm | | | | No adverse effect | |
| | | 86 ppm | | | | No adverse effect on growth; reduced feed intake | |
| | | 174 ppm | | | | Slightly reduced growth and feed intake | |
| Cattle—2 (lactating) | Mature | 40 mg/d | EDDI | 7 wk | Diet | No adverse effect | Miller and Swanson, 1973 |
| | | 80 mg/d | KI | | | No adverse effect; milk contained 0.38 ppm I | |
| | | 80 mg/d | EDDI | | | No adverse effect; milk contained 0.36 ppm I | |
| | | 160 mg/d | | | | No adverse effect; milk contained 1.6 ppm I | |
| | | 400 mg/d | | | | No adverse effect; milk | |

| | | | | | | | |
|---|---|---|---|---|---|---|---|
| | | | | | | contained 2 ppm I | |
| | | 800 mg/d | | | | No toxic signs; milk contained 2.4 ppm I | |
| Cattle—5 (lactating) | 600 kg | 47 ppm | EDDI | 12 wk | Diet | No adverse effect; milk contained 2.4 ppm I | Fish and Swanson, 1977 |
| | | 93 ppm | | | | No adverse effect; milk contained 2.8 ppm I | |
| | | 198 ppm | | | | No adverse effect; milk contained 5.0 ppm I | |
| | | 314 ppm | | | | No adverse effect; milk contained 6.4 ppm I | |
| Sheep—4 | 30 kg | 75 mg/d | EDDI | 22 d | Capsule | No adverse effect | McCauley *et al.*, 1973 |
| | | 150 mg/d | | | | No adverse effect | |
| | | 300 mg/d | | | | Slightly reduced daily gains | |
| | | 450 mg/d | | | | Anorexia and reduced gains | |
| | | 150 mg/d | KI | | | Slightly reduced gains | |
| | | 300 mg/d | | | | Reduced gains and feed intake | |
| | | 450 mg/d | | | | Reduced gains and feed intake | |
| Swine—8 | 17 kg | 400 ppm | $Ca(IO_3)_2$ | 97 d | Diet | No adverse effect on performance; increased thyroid weight and reduced liver iron | Newton and Clawson, 1974 |
| | | 800 ppm | | | | Decreased gains and feed conversion | |
| Swine—2 | Mature | 1,500 ppm | KI | 30 d | Diet | No adverse effect on reproduction | Arrington *et al.*, 1965 |
| Swine—3 | | 2,500 ppm | | 30 d | | No adverse effect on reproduction | |
| Chicken—10 | Young | 500 ppm | Iodinated casein | 6 wk | Diet | No adverse effect on growth | Wilgus *et al.*, 1953 |

TABLE 19 *Continued*

| Class and Number of Animals[a] | Age or Weight | Administration: Quantity of Element[b] | Source | Duration | Route | Effect(s) | Reference |
|---|---|---|---|---|---|---|---|
| | | | or potassium iodide | | | | |
| Chicken—5 | 27 wk | 625 ppm | KI | 6 wk | Diet | Reduced egg production and hatchability | Arrington *et al.*, 1967 |
| | | 2,500 ppm | | | | Increased thyroid weights in chicks | |
| | Mature | 625 to 2,500 ppm | | | | Similar to pullets but less pronounced | |
| Chicken—15 | 27 wk | 5,000 ppm | KI | 10 d | Diet | Reduced egg production and serum calcium | Roland *et al.*, 1977 |
| Horse—165 | Mature | 48 to 55 mg I/mare | Iodized salt | | Diet | 3% incidence of goiter in foals | Baker and Lindsey, 1968 |
| Horse—60 | | 56 to 69 mg I/mare | Kelp | Several months | | 10% incidence of goiter in foals | |
| Horse—6 | | 288 to 432 mg I/mare | Kelp | Several months | | 50% incidence of goiter in foals | |
| Rabbit—9 | Mature | 250 ppm | NaI or KI | 2 d before parturition | Diet | Only 30% survival in young to 3 d | Arrington *et al.*, 1965 |
| Rabbit—19 | Mature | 500 ppm | NaI or KI | 5 d before parturition | | Only 3% survival in young to 3 d | |
| Dog—4 | 8–16 kg | 36 mg/kg | $KIO_3$ | Several months | Oral (milk) | Some dogs vomited | Webster *et al.*, 1966 |

| | | | | | | | |
|---|---|---|---|---|---|---|---|
| | 10–15 kg | 59 mg/kg | $KIO_3$ | Single dose | Capsule | No effect | |
| Dog—3 | | 118 mg/kg | | | | 1 of 3 died within 1 wk | |
| | | 148 mg/kg | | | | All died within 1 wk | |
| Rat—9 | Mature | 500 ppm | KI | 12 wk before mating | Diet | Slight increase in mortality of young | Ammerman *et al.*, 1964 |
| | | 1,000 ppm | | | | 66% litter survival to 5 d | |
| | | 1,500 ppm | | | | 36% litter survival to 5 d | |
| | | 2,000 ppm | | | | 16% litter survival to 5 d | |
| Mouse—80 to 140 | 15–20 g | 698 mg/kg | $KIO_3$ | | 6% $KIO_3$ soln. (not fasted) | | Webster *et al.*, 1957 |
| | | 483 mg/kg | $KIO_3$ | | 6% $KIO_3$ soln. (fasted) | $LD_{50}$ | |
| | | 1,580 mg/kg | KI | | 6% KI soln. (not fasted) | $LD_{50}$ | |
| Mouse—130 | 15–20 g | 1,550 mg/kg | KI | | 6% KI soln. (fasted) | $LD_{50}$ | |
| Guinea pig—12 | 250 g | | $KIO_3$ | 28 d | 0.05% $KIO_3$ soln. | No adverse effect | Webster *et al.*, 1959 |
| | | | | 28 d | 0.5% $KIO_3$ soln. | No adverse effect | |
| Hamster—31 | Mature | 250 ppm | KI | 12 d | Diet | Slight reduction in feed intake and decreased weaning weight of offspring | Arrington *et al.*, 1965 |

[a] Number of animals per treatment.

[b] Quantity expressed as parts per million or as milligrams per kilogram of body weight.

## REFERENCES

Ammerman, C. B., and S. M. Miller. 1972. Biological availability of minor mineral ions: A review. J. Anim. Sci. 35:681.

Ammerman, C. B., L. R. Arrington, A. C. Warnick, J. L. Edwards, R. L. Shirley, and G. K. Davis. 1964. Reproduction and lactation in rats fed excessive iodine. J. Nutr. 84:108.

Arrington, L. R., R. N. Taylor, Jr., C. B. Ammerman, and R. L. Shirley. 1965. Effects of excess dietary iodine upon rabbits, hamsters, rats and swine. J. Nutr. 87:394.

Arrington, L. R., R. A. Santa Cruz, R. H. Harms, and H. R. Wilson. 1967. Effects of excess dietary iodine upon pullets and laying hens. J. Nutr. 92:325.

Baker, H. J., and J. R. Lindsey. 1968. Equine goiter due to excess dietary iodide. J. Am. Vet. Med. Assoc. 153:1618.

Barua, J., R. G. Gragle, and J. K. Miller. 1964. Sites of gastrointestinal-blood passage of iodide and thyroxine in young cattle. J. Dairy Sci. 47:539.

Berson, S. A. 1956. Pathways of iodine metabolism. Am. J. Med. 20:653.

Chilean Iodine Educational Bureau. 1952. Iodine Content of Foods: Annotated Bibliography 1825–1951, with Review Tables. Chilean Iodine Educational Bureau, London.

Convey, E. M., L. Chapin, J. S. Kesner, D. Hillman, and A. R. Curtis. 1977. Serum thyrotropin and thyroxine after thyrotropin releasing hormone in dairy cows fed varying amounts of iodine. J. Dairy Sci. 60:975.

Convey, E. M., L. T. Chapin, J. W. Thomas, K. Leung, and E. W. Swanson. 1978. Serum thyrotropin, thyroxine and tri-iodothyronine in dairy cows fed varying amounts of iodine. J. Dairy Sci. 61:771.

Drew, B., W. P. Barber, and D. G. Williams. 1975. The effect of excess iodine on pregnant mares and foals. Vet. Rec. 97:93.

Evvard, J. M. 1928. Iodine deficiency symptoms and their significance in animal nutrition and pathology. Endocrinology 12:529.

Fish, R. E., and E. W. Swanson. 1977. Iodine tolerance of calves, yearlings, dry cows, and lactating cows. J. Dairy Sci. 60 (Suppl. 1):151.

Fisher, K. D., and C. J. Carr. 1974. Iodine in Foods: Chemical Methodology and Sources of Iodine in the Human Diet. Life Sciences Research Office, Federation of American Societies for Experimental Biology, Bethesda, Md. PB-233 599. National Technical Information Service, Springfield, Va.

Gross, J. 1962. Iodine and bromine. *In* C. L. Comar and Felix Bronner, eds. Mineral Metabolism. Academic Press, New York.

Haggard, D. L. 1978. Immunologic effects of experimental iodine toxicity in cattle. M. S. thesis. Michigan State University, East Lansing.

Hemken, R. W. 1978. Factors that influence the iodine content of milk and meat: A review. J. Anim. Sci. 48:981.

Hemken, R. W., J. H. Vandersall, M. A. Oskarsson, and L. R. Fryman. 1972. Iodine intake related to milk iodine and performance of dairy cattle. J. Dairy Sci. 55: 931.

Highman, B., S. H. Webster, and M. E. Rice. 1955. Degeneration of retina and gastric parietal cells and other pathologic changes following administration of iodates. Fed. Proc. 14:407. (Abstr.)

Marcilese, N. A., R. H. Harms, R. M. Valsechhi, and L. R. Arrington. 1968. Iodine uptake by ova of hens given excess iodine and effect upon ova development. J. Nutr. 94:117.

McCauley, E. H., J. G. Linn, and R. D. Goodrich. 1973. Experimentally induced iodine toxicosis in lambs. Am. J. Vet. Res. 34:65.

Miller, J. I., and K. Tillapaugh. 1966. Iodide Medicated Salt for Beef Cattle. Cornell Feed Service No. 62. Cooperative Extension Service, Cornell University, Ithaca, N.Y.

Miller, J. K., and E. W. Swanson. 1973. Metabolism of ethylenediaminedihydriodide and sodium or potassium iodide by dairy cows. J. Dairy Sci. 56:378.

Miller, J. K., E. W. Swanson, and G. E. Spalding. 1975. Iodine absorption, excretion, recycling and tissue distribution in the dairy cow. J. Dairy Sci. 58:1578.

National Research Council (NRC), Food and Nutrition Board (FNB). 1970. Iodine Nutrition in the United States. National Academy of Sciences, Washington, D.C.

Newton, G. L., and A. J. Clawson. 1974. Iodine toxicity: Physiological effects of elevated dietary iodine on pigs. J. Anim. Sci. 39:879.

Newton, G. L., E. R. Barrick, R. W. Harvey, and M. B. Wise. 1974. Iodine toxicity. Physiological effects of elevated dietary iodine on calves. J. Anim. Sci. 38:449.

Perdomo, J. T., R. H. Harms, and L. R. Arrington. 1966. Effect of dietary iodine upon egg production, fertility and hatchability. Proc. Soc. Exp. Biol. Med. 122:758.

Riggs, D. S. 1952. Quantitative aspects of iodine metabolism in man. Pharmacol. Rev. 4:284.

Roland, D. A., S. T. McGready, R. H. Stonerock, and R. H. Harms. 1977. Hypercalcemic effect of potassium iodide on serum calcium in domestic fowl. Poult. Sci. 56:1310.

Rosiles, R., W. B. Buck, and L. N. Brown. 1975. Clinical infectious bovine rhinotracheitis in cattle fed organic iodine and urea. Am. J. Vet. Res. 36:1447.

Talbot, J. M., K. D. Fisher, and C. J. Carr. 1976. A Review of the Effects of Dietary Iodine on Certain Thyroid Disorders. Life Sciences Research Office, Federation of American Societies for Experimental Biology, Bethesda, Md.

Underwood, E. J. 1977. Trace Elements in Human and Animal Nutrition, 4th ed. Academic Press, New York.

Webster, S. H., M. E. Rice, B. Highman, and W. F. Von Oettingen. 1957. The toxicology of potassium and sodium iodate: Acute toxicity in mice. J. Pharmacol. Exp. Ther. 120:171.

Webster, S. H., M. E. Rice, B. Highman, and E. F. Stohlman. 1959. The toxicology of potassium and sodium iodate. II. Subacute toxicity of potassium iodate in mice and guinea pigs. Toxicol. Appl. Pharmacol. 1:87.

Webster, R. H., E. F. Stohlman, and B. Highman, 1966. The toxicology of potassium and sodium iodate. III. Acute and subacute oral toxicology of potassium iodate in dogs. Toxicol. Appl. Pharmacol. 8:185.

Wilgus, H. F., F. X. Gassner, A. P. Patton, and G. S. Harshfield. 1953. The Iodine Requirements of Chickens. Colo. Agric. Exp. Stn. Tech. Bull. 49. Fort Collins, Colo.

# Iron

Iron (Fe) and iron salts have been used as medicinal agents for centuries. The ancient Greeks, Egyptians, and Hindus prescribed iron as a treatment for general weakness, diarrhea, and constipation. The role of iron in blood formation became apparent in the seventeenth century when it was shown that iron salts were of value in the treatment of chlorosis, now known as iron-deficiency anemia, in young women. Lemery and Geoffy demonstrated in 1713 the presence of iron in blood, and in 1746 Menghini reported that blood iron levels could be increased by the feeding of iron-rich foods. Various iron compounds continue to be used in the prevention and treatment of iron-deficiency anemia in man and animals.

## ESSENTIALITY

Iron is essential to every form of life from plants to man. Hemoglobin, myoglobin, the cytochromes, and many other enzyme systems contain iron. The conjugated proteins help maintain the vital cellular activities of respiration and oxygen transport. The frequency of iron-deficiency anemia in many human populations, including those in affluent countries, emphasizes the importance of dietary iron as a marginally adequate nutrient. Iron-deficiency anemia in young pigs has been recognized for almost a century, and the nature and pathology of the anemia were described more than 50 years ago (McGowan and Crichton, 1923; Doyle *et al.,* 1928). Other young animals that receive milk as the sole diet can suffer from iron-deficiency anemia (Blaxter *et al.,* 1957).

## METABOLISM

Iron absorption takes place almost exclusively in the small intestine. Ferrous iron entering the blood plasma is quickly oxidized to the ferric state. The ferric form then immediately complexes with a specific $B_1$-globulin (transferrin), in which form it is transported to various parts of the body as required for use or storage. Iron is stored intracellularly in the liver, spleen, bone marrow, and other tissues as ferritin and hemosiderin. Iron is a component of myoglobin in skeletal muscles. The body has limited capacity to excrete iron. It is excreted mainly in the bile and desquamated mucosal epithelial cells sloughed from the duodenal villi (Dubach *et al.,* 1955; Braude *et al.,* 1962). Small amounts of iron are lost in urine and sweat. Various aspects of iron metabolism are covered in the articles and reviews by Moore (1961), Bothwell and Finch (1962), Christopher *et al.* (1974), Forth (1974), Jacobs (1976), and Underwood (1977).

## SOURCES

Iron is more abundant in the earth's crust than are the macronutrients calcium, magnesium, potassium, phosphorus, sulfur, and nitrogen. All plant materials commonly used in the feeding of animals contain variable amounts of iron. The concentration of iron in plants is a reflection of the species and soil upon which the plants grow. The iron content of cultivated grasses and legumes ranges from 100 to 700 ppm, although values in excess of 1,000 ppm have been reported (Beeson, 1941). Most cereal grains contain between 30 and 60 ppm iron (Miller, 1958). Feeds of animal origin, other than milk and milk products, are rich sources of iron. Many of the minerals used to supply the calcium and phosphorus needs of animals contain iron.

Most of the iron in plant products is in the ferric form in organic combinations from which it must be released in the gastrointestinal tract to permit absorption. The bioavailability of iron in certain grasses was studied by Thompson and Raven (1959) and Raven and Thompson (1959). Inorganic iron as ferric chloride was significantly more available than iron present in either the grasses or legumes. Assuming that ferric chloride resulted in an improvement in total hemoglobin equal to 100 percent, the grasses yielded improvements of 48 to 63 percent in one study and the legumes 47 to 57 percent in another study. Numerous iron compounds are used as dietary sources of iron. Some iron oxide is used in animal feeds as a coloring agent. The bioavailability of the iron in these different compounds varies greatly (Nesbit and Elmslie, 1960; Ammerman *et al.,* 1967; Harmon *et al.,* 1969; Fritz *et al.,* 1970). Ammerman *et al.* (1974) showed the bioavailability of iron in several ferrous carbonates was correlated with *in vitro* solubility.

## TOXICOSIS

### LOW LEVELS

Characteristic signs of chronic iron toxicosis for most species are reduced feed intake, growth rate, and efficiency of feed conversion. Standish and Ammerman (1971) reported that feeding 1,600 ppm iron to lambs reduced plasma copper. Iron levels in the range of 4,000 to 5,000 ppm in the diet produced signs of phosphorus deficiency in swine and poultry (Deobald and Elvehjem, 1935; O'Donovan *et al.,* 1963; Furugouri, 1972).

Standish *et al.* (1969) added ferrous sulfate at levels to supply either 400 or 1,600 ppm added iron in the diets of steers weighing about 235 kg. The level of 1,600 ppm added iron caused significant reductions in daily gains and feed intake. Koong *et al.* (1970), using iron citrate as the source of added iron, conducted two experiments in which six levels (100 to 4,000 ppm) of dietary iron were fed to calves weighing about 125 kg. The animals fed 4,000 ppm performed poorly (poor gains and diarrhea) and were changed to a level of 2,000 ppm after being on test for 6 weeks. A level of 2,500 ppm iron significantly reduced feed intake and daily gains. The body weight gains and feed consumption data for calves receiving 1,000 ppm iron were not significantly different from those receiving 100 ppm, but there was a trend towards poorer performance at dietary iron levels of 500 ppm or more. Lawlor *et al.* (1965) obtained no effects on weight gains, hemoglobin, packed cell volume, or red blood cell values from supplementing lambs' diets with 280 ppm iron. Following the fourth week of the experiment, diarrhea occurred among the lambs receiving 210 and 280 ppm iron. Postmortem examination of two lambs did not attribute death to any particular causes other than severe generalized edema. The authors point out that it is unlikely that dietary levels of 210 and 280 ppm should approach toxicity levels for lambs. Standish and Ammerman (1971) found that lambs did not consume adequate feed to maintain body weight when fed rations containing 1,600 ppm iron, which reduced plasma copper.

Furugouri (1972) reported no significant effect on body weight gains of young pigs fed either 1,102 or 3,103 ppm iron. When the dietary iron levels were increased to 5,102 and 7,102 ppm, body weight gains and feed intake were decreased. Signs of phosphorus deficiency were noted in pigs receiving either 5,102 or 7,102 ppm iron, even though the diets contained 0.92 percent phosphorus. O'Donovan *et al.* (1963) reported that 5,000 ppm iron caused a significant decrease in rate of gain, a slight decrease in serum inorganic phosphorus, but failed to reduce femur ash. McGhee *et al.* (1965) found that chicks could tolerate 1,600 ppm iron when adequate copper was included in the diet; when the diet contained only 5 ppm copper, decreased gains and increased mortality were reported with dietary iron levels of 200 ppm. Growth rate appeared to be depressed when diets contained 800 ppm iron and 80 ppm copper. Deobald and Elvehjem (1935) found that 4,500 ppm iron produced rickets in young chicks. Woerpel and Balloun (1964) showed no consistent adverse effect on the growth rate of turkey poults from the addition of 440 ppm iron.

Goldberg *et al.* (1957) studied the effects of administering a dose of 1,650 mg iron per kilogram of body weight intramuscularly to rats for an extended period. The only significant findings were those characteristic of vitamin E deficiency. Tollerz and Lannek (1964) reported that vitamin E gave protection against iron toxicosis in mice and young pigs.

## HIGH LEVELS

The clinical signs of acute toxicosis of iron include anorexia, oliguria, diarrhea, hypothermia, diphasic shock, metabolic acidosis, and death (Boyd and Shanas, 1963). Vascular congestion of the gastrointestinal tract, liver, kidneys, heart, lungs, brain, spleen, adrenals, and thymus are the dominant histopathologic findings. Elevated serum iron levels are found in iron toxicosis.

Cornelius and Harmon (1976) administered oral doses of 200 mg iron from either ferric ammonium citrate, ferrous sulfate, ferric oxide, or an iron dextran complex to piglets within 6 hours of birth. The ferric ammonium citrate proved highly toxic, with only 33.3 percent of the piglets surviving to 21 days. The rate of survival in piglets given the other three iron compounds ranged from 82 to 100 percent. Ferrous sulfate given in a large oral dose to dogs caused vomiting, and the median emetic dose was between 19 and 29 mg iron per kilogram of body weight (Weaver *et al.*, 1961). Gastric intolerance, as indicated by emesis, was much less for an iron polysaccharide complex than for several other salts. When ferrous sulfate was given to dogs at levels to supply 150 to 600 mg iron per kilogram of body weight, various disorders ranging from diarrhea and vomiting to irritation of the gastrointestinal tract occurred (Reissman and Coleman, 1955; D'Arcy and Howard, 1962). In rabbits, 750 mg ferrous sulfate per kilogram of body weight caused hepatic congestion within 24 to 48 hours (Luongo and Bjornson, 1954). These workers reported a dose of 2,000 mg ferrous sulfate per kilogram of body weight caused death in all rabbits within a few hours of administration.

## FACTORS INFLUENCING TOXICITY

Iron that is added to the plasma in excess of the physiological iron-binding capacity is bound more loosely than the $B_1$-globulin iron. The loosely bound iron is rapidly removed from the plasma, and it is this fraction that causes toxic reactions in the organism. Because of the limited capacity of the body to excrete iron, the toxicity of iron is governed largely by its absorption. Although iron is absorbed by the cells of the intestinal mucosa in the ferrous state, substances in the gastric and intestinal secretions can reduce the ferric ions to the ferrous state. The solubility of the iron appears to be as important as the valence, because some insoluble ferrous compounds are less available than the more soluble ferric compounds (Fritz *et al.*, 1970). In reviewing the toxicity of iron compounds, Herbert (1965) concluded that all iron compounds are probably equally toxic per unit of soluble iron.

Among the dietary factors that have been shown to influence iron toxicity are levels of copper (McGhee *et al.*, 1965), phosphorus (O'Donovan *et al.*, 1963), and vitamin E (Tollerz and Lannek, 1964). Enhanced iron absorption has been seen with certain amino acids, e.g., valine and histidine (El Hawary *et al.*, 1975). Ascorbic acid alone or in combination with vitamin E increased iron absorption (Greenberg *et al.*, 1957; Monsen and Page, 1978). A number of organic acids, including succinic, lactic, pyruvic, and citric are effective in increasing iron absorption (Van Campen, 1974). Some simple sugars such as fructose and sorbitol increase iron absorption, whereas the effects of more complex carbohydrates are somewhat variable (Herndon *et al.*, 1958; Jacobs and Miles, 1969). It has been postulated that some of the compounds mentioned above form complexes with iron that keep the iron in solution during transit through the upper part of the small intestine, where absorption most rapidly occurs.

Treatment for iron poisoning aims at precipitating the iron as insoluble hydroxide. Preparations such as milk of magnesia and milk of lime are recommended (Garner, 1961). Szabuniewicz *et al.* (1971) suggested desferrioxamine (deferoxamine) as a treatment for iron toxicosis.

## TISSUE LEVELS

Standish *et al.* (1969) reported that livers and spleens of steers fed 400 ppm iron contained significantly more iron than those from steers fed no additional iron. The change in the level of iron in liver, spleen, and heart in response to dietary iron was almost linear for steers fed 0, 400, and 1,600 ppm. On a dry matter basis, the liver, spleen, kidney, heart, and muscle from steers fed no supplemental iron contained 185, 1,219, 315, 291, and 91 ppm iron, respectively; the same organs and tissue from steers receiving 1,600 ppm iron contained 605, 8,941, 410, 329, and 98 ppm, respectively. The iron content of the muscle was not increased by feeding 1,600 ppm. Thoren-Tolling (1975) showed that the liver is the main storage site in young pigs receiving oral iron. The iron deposits in the liver were almost depleted 19 days after the oral iron treatment was given. Estimates of the distribution of iron in various species are given by Moore and Dubach (1962).

## MAXIMUM TOLERABLE LEVELS

Pigs are more tolerant of excess iron than cattle, sheep, or poultry. Based on available information, the maximum tolerable levels of dietary iron are 3,000 ppm for swine and 1,000 ppm for cattle and poultry. The more limited data available for sheep suggest a maximum tolerable level of 500 ppm dietary iron. The values listed above assume that the biological availability of the dietary iron is high. All species can probably tolerate much higher levels when the iron is supplied from sources with low bioavailability.

## SUMMARY

Iron is essential to every form of life from plants to animals. It is concerned with the vital cellular activities of respiration and oxygen transport. The frequency of iron-deficiency anemia in many human populations, and in some young animals that rely on milk as the sole diet, emphasizes the importance of dietary iron as a marginally adequate nutrient. Although a wide variation in the susceptibility of various species of livestock to iron toxicosis exists, most species have a high tolerance.

TABLE 20 Effects of Iron Administration in Animals

| Class and Number of Animals[a] | Age or Weight | Administration | | | | Effect(s) | Reference |
|---|---|---|---|---|---|---|---|
| | | Quantity of Element[b] | Source | Duration | Route | | |
| Cattle—6 | 235 kg | 477 ppm | $FeSO_4$ | 84 d | Diet | Slight decrease in gains and feed conversion | Standish *et al.*, 1969 |
| | | 1,677 ppm | | | | Significant reduction in growth and feed intake | |
| Cattle—12 | 198–234 kg | 1,000 ppm | $FeSO_4$ | 77 d | Diet | Significant reduction in growth, feed intake, and plasma P | Standish *et al.*, 1971 |
| Cattle—8 | 125 kg | 100 ppm | Iron citrate | 98 d | Diet | No adverse effects | Koong *et al.*, 1970 |
| | | 1,000 ppm | | | | No significant effect on growth | |
| | | 2,000 ppm | | | | No significant effect on growth; reduced serum P | |
| | | 2,500 ppm | | | | Reduced growth, feed intake, and serum P | |
| | | 4,000 ppm | | | | Reduced growth, feed intake, and serum P | |
| Sheep—6 | 23 kg | 140 ppm | $FeSO_4 \cdot 7H_2O$ | 49 d | Diet | No adverse effect | Lawlor *et al.*, 1965 |
| | | 210 ppm | | | | No effect on growth | |
| | | 280 ppm | | | | No effect on growth | |
| Sheep—4 | 34 kg | 1,600 ppm | Iron sulfate or citrate | 44 d | Diet | Reduced feed intake | Standish and Ammerman, 1971 |
| Swine—12 | Neonatal | 200 mg | Ferric ammonium citrate | Single dose | Gavage | 33.3% survival to 21 days | Cornelius and Harmon, 1976 |

TABLE 20 *Continued*

| Class and Number of Animals[a] | Age or Weight | Administration: Quantity of Element[b] | Source | Duration | Route | Effect(s) | Reference |
|---|---|---|---|---|---|---|---|
| | | 200 mg | Ferrous sulfate | | | 81.8% survival to 21 days | |
| | | 200 mg | Ferric oxide | | | 91.7% survival to 21 days | |
| | | 200 mg | Iron dextran complex | | | 100% survival to 21 days | |
| Swine—6 | 6 kg | 1,000 ppm | $FeSO_4 \cdot 2H_2O$ | 56 d | Diet | No adverse effects | O'Donovan *et al.*, 1963 |
| | | 2,000 ppm | | | | No adverse effects | |
| | | 3,000 ppm | | | | No adverse effects | |
| | | 4,000 ppm | | | | Reduced growth and serum P | |
| | | 5,000 ppm | | | | Reduced growth and rickets | |
| Swine—6 | 15 kg | 1,100 ppm | Ferrous sulfate | 30 d | Diet | No adverse effects | Furugouri, 1972 |
| | | 3,100 ppm | | | | No adverse effects | |
| | 20 kg | 5,100 ppm | | 60 d | | Reduced growth, feed intake, serum P, and bone ash | |
| | | 7,100 ppm | | | | Reduced growth, feed intake, serum P, and bone ash | |
| Chicken—20 | Day old | 400 ppm | $FeSO_4 \cdot 7H_2O$ | 28 d | Diet | No adverse effect with adequate copper; reduced growth with 5 ppm Cu in diet | McGhee *et al.*, 1965 |

| | | | | | | | |
|---|---|---|---|---|---|---|---|
| | | 800 ppm | | | | Reduced growth | |
| | | 1,600 ppm | | | | Reduced growth | |
| Chicken—11 | | 4,500 ppm | $Fe_2(SO_4)_3$ | | Diet | Rickets | Deobald and Elvehjem, 1935 |
| Turkey | 1 wk old | 440 ppm | | 12 wk | Diet | Slight reduction in bone ash; effects on growth were variable | Woerpel and Balloun, 1961 |
| Rabbit—5 | 1,800 g | 275 mg/kg | Ferrous sulfate | Single | Gavage | Hepatic congestion | Luongo and Bjornson, 1954 |
| | | 460 mg/kg | | Single | | Severe hemorrhagic necrosis of liver | |
| Dog—1 | 7–12 kg | 10 mg/kg | $FeSO_4 \cdot 7H_2O$ | 10 d | Capsule | Vomiting | Hoppe *et al.*, 1955a |
| | | 20.1 mg/kg | | | | Diarrhea | |
| | | 40.2 mg/kg | | | | Diarrhea | |
| Dog—1 | | 150 mg/kg | $FeSO_4$ | Single | Gavage | Metabolic acidosis | Reissman and Coleman, 1955 |
| Dog—1 | | 150 mg/kg | Ferrous sulfate | | | Survived | Franklin *et al.*, 1958 |
| Dog—10 | | 250 mg/kg | | | | All died | |
| Dog—1 | 6–14 kg | 300 mg/kg | Ferrous sulfate | | Tablets | Diarrhea, emesis, and gastrointestinal damage | D'Arcy and Howard, 1962 |
| Cat—2 | Adult | 5 mg/kg | $FeSO_4 \cdot 7H_2O$ | 10 d | Capsule | Occasional emesis | Hoppe *et al.*, 1955a |
| | | 10 mg/kg | | | | Occasional emesis and diarrhea | |
| | | 20 mg/kg | | | | Frequent emesis and diarrhea | |
| | | 40.2 mg/kg | | | | Frequent emesis and diarrhea | |
| Cat | Adult | 100 mg/kg | $FeSO_4 \cdot 7H_2O$ | Single | Oral | $LD_{50}$ | Hoppe *et al.*, 1955b |
| Rat—10 | Adult | 1,000 mg/kg | Ferrous sulfate | Single | Gavage | $LD_{50}$ | Shanas and Boyd, 1969 |
| | | 1,000 mg/kg | Ferrous chloride | Single | | $LD_{50}$ | |

TABLE 20 *Continued*

| Class and Number of Animals[a] | Age or Weight | Administration: Quantity of Element[b] | Source | Duration | Route | Effect(s) | Reference |
|---|---|---|---|---|---|---|---|
| Mouse | 22 g | 306 mg/kg | Ferrous sulfate | Single | Oral | $LD_{50}$ at 24 h | Hoppe *et al.*, 1955a |
| | | 429 mg/kg | Ferrous gluconate | | Oral | $LD_{50}$ at 24 h | |
| Guinea pig | | 300 mg/kg | $FeSO_4 \cdot 7H_2O$ | Single | | $LD_{50}$ | Hoppe *et al.*, 1955b |
| | | 300 mg/kg | Ferrous gluconate | | | $LD_{50}$ | |
| | | 200 mg/kg | Ferric chloride | | | $LD_{50}$ | |
| | | 350 mg/kg | Ferric ammonium citrate | | | $LD_{50}$ | |

[a] Number of animals per treatment.

[b] Quantity expressed in parts per million, as milligrams per kilogram of body weight, or milligrams per head per day.

## REFERENCES


Ammerman, C. B., J. M. Wing, B. G. Dunavant, W. K. Robertson, J. P. Feaster, and L. R. Arrington. 1967. Utilization of inorganic iron by ruminants as influenced by form of iron and iron status of the animal. J. Anim. Sci. 26:404.

Ammerman, C. B., J. F. Standish, C. E. Holt, R. H. Houser, S. M. Miller, and G. E. Combs. 1974. Ferrous carbonates as sources of iron for weanling pigs and rats. J. Anim. Sci. 38:52.

Beeson, K. C. 1941. The Mineral Composition of Crops with Particular Reference to the Soils in Which They Were Grown. USDA Misc. Publ. No. 369.

Blaxter, K. L., G. A. M. Sharman, and A. M. MacDonald. 1957. Iron-deficiency anemia in calves. Br. J. Nutr. 11:234.

Bothwell, T. H., and C. A. Finch. 1962. Iron Metabolism. Little, Brown and Co., Boston.

Boyd, E. M., and S. N. Shanas. 1963. The acute oral toxicity of reduced iron. Can. Med. Assoc. J. 89:171.

Braude, R., A. G. Chamberlein, M. Kotarbinska, and K. G. Mitchell. 1962. The metabolism of iron on piglets given labelled iron either orally or by injection. Br. J. Nutr. 16:427.

Christopher, J. P., J. C. Hegenauer, and P. D. Saltman. 1974. Iron metabolism as a function of chelation. *In* W. G. Hoekstra, J. W. Suttie, H. E. Ganther, and W. Mertz (eds.). Trace Element Metabolism in Animals—2. University Park Press, Baltimore, Md.

Cornelius, S. G., and B. G. Harmon. 1976. Sources of oral iron for neonatal piglets. J. Anim. Sci. 42:1350. (Abstr.)

D'Arcy, P. F., and E. M. Howard. 1962. The acute toxicity of ferrous salts administered to dogs by mouth. J. Pathol. Bacteriol. 83:65.

Deobald, H. J., and C. A. Elvehjem. 1935. The effect of feeding high amounts of soluble iron and aluminum salts. Am. J. Physiol. 111:118.

Doyle, L. P., F. P. Mathews, and R. A. Whiting. 1928. Anemia in young pigs. J. Am. Vet. Med. Assoc. 72:491.

Dubach, R., C. V. Moore, and S. Callender. 1955. Studies in iron transportation and metabolism. IX. The excretion of iron as measured by the isotope technique. J. Lab. Clin. Med. 45:599.

El-Hawary, M. F. S., F. A. El-Shobaki, T. Kholeif, R. Sakr, and M. El-Bassoussy. 1975. The absorption of iron, with or without supplements of single amino acids and of ascorbic acid in healthy and Fe-deficient children. Br. J. Nutr. 33:351.

Forth, W. 1974. Iron absorption, a medicated transport across the mucosal epithelium. *In* W. G. Hoekstra, J. W. Suttie, H. E. Ganther, and W. Mertz (eds.). Trace Element Metabolism in Animals—2. University Park Press, Baltimore, Md.

Franklin, M., W. G. Rohse, J. de la Huerga, and C. R. Kemp. 1958. Chelate iron therapy. J. Am. Med. Assoc. 166:1685.

Fritz, J. C., G. W. Pla, T. Roberts, J. W. Boehne, and E. L. Hove. 1970. Biological availability in animals of iron from common dietary sources. J. Agric. Food Chem. 18:647.

Furugouri, K. 1972. Effect of elevated dietary levels of iron on iron store in liver, some blood constituents and phosphorus deficiency in young swine. J. Anim. Sci. 34:573.

Garner, R. J. 1961. Veterinary Toxicology, 2d. ed. The Williams & Wilkins Co., Baltimore, Md.

Goldberg, L., J. P. Smith, and L. E. Martin. 1957. Effects of massive iron overload in the rat. Nature 179:734.

Greenberg, S. M., R. G. Tucker, A. E. Heming, and J. K. Mathues. 1957. Iron absorption and metabolism. J. Nutr. 63:19.

Harmon, B. G., D. E. Hoge, A. H. Jensen, and D. H. Baker. 1969. Efficacy of ferrous carbonate as a hematinic for swine. J. Anim. Sci. 29:706.

Herbert, V. 1965. Drugs effective in iron-deficiency and other hypochromic anemias. *In* L. S. Goodman and R. Gilman (eds.). Pharmacological Basis for Therapeutics. Macmillan Company, New York.

Herndon, J. G., T. G. Rice, R. G. Tucker, E. J. Van Loon, and S. M. Greenberg. 1958. Iron absorption and metabolism. III. The enhancement of iron absorption in rats by D-sorbitol. J. Nutr. 64:615.

Hoppe, J. O., G. M. A. Marcelli, and M. L. Tainter. 1955a. An experimental study of the toxicity of ferrous gluconate. Am. J. Med. Sci. 230:491.

Hoppe, J. O., G. M. A. Marcelli, and M. L. Tainter. 1955b. A review of the toxicity of iron compounds. Am. J. Med. Sci. 230:558.

Jacobs, A. 1976. Sex differences in iron absorption. Proc. Nutr. Soc. 35:159.

Jacobs, A., and P. M. Miles. 1969. Intraluminal transport of iron from stomach to small intestinal mucosa. Br. Med. J. 4:778.

Koong, L-J., M. B. Wise, and E. R. Barrick. 1970. Effect of elevated dietary levels of iron on the performance and blood constituents of calves. J. Anim. Sci. 31:422.

Lawlor, J. J., W. H. Smith, and W. M. Beeson. 1965. Iron requirement of the growing lamb. J. Anim. Sci. 24:742.

Luongo, M. A., and S. S. Bjornson. 1954. The liver in ferrous sulfate poisoning. A report of three fatal cases in children and an experimental study. N. Engl. J. Med. 251:995.

McGhee, F., C. R. Greger, and J. R. Couch. 1965. Copper and iron toxicity. Poult. Sci. 44:310.

McGowan, J. P., and J. Crichton. 1923. On the effect of deficiency of iron in the diets of pigs. Biochem. J. 17:240.

Miller, D. F. 1958. Composition of Cereal Grains and Forages. Natl. Acad. Sci.–Natl. Res. Counc. Publ. 585. National Academy of Sciences, Washington, D.C.

Monsen, E. R., and J. F. Page. 1978. Effects of EDTA and ascorbic acid on the absorption of iron from an isolated rat intestinal loop. J. Agric. Food Chem. 26:223.

Moore, C. V. 1961. Iron metabolism and nutrition. Harvey Lectures (1959–1960) 55:67.

Moore, C. V., and R. Dubach. 1962. Mineral Metabolism, vol. 2, part B. Academic Press, New York.

Nesbit, A. H., and W. P. Elmslie. 1960. Biological availability to the rat of iron and copper from various compounds. Trans. Ill. State Acad. Sci. 53:101.

O'Donovan, P. B., R. A. Pickett, M. P. Plumlee, and W. M. Beeson. 1963. Iron toxicity in the young pig. J. Anim. Sci. 22:1075.

Raven, A. M., and A. Thompson. 1959. The availability of iron in certain grasses, clover and herb species. I. Perennial ryegrass, cocksfoot and timothy. J. Agric. Sci. 52:177.

Reissman, K. R., and T. S. Coleman. 1955. Acute intestinal iron intoxication. II. Metabolic, respiratory and circulatory effects of absorbed iron salts. Blood 10:46.

Shanas, M. N., and E. M. Boyd. 1969. Powdered iron from 1681 to 1968. Clin. Toxicol. 2:37.

Standish, J. F., and C. B. Ammerman. 1971. Effect of excess dietary iron as ferrous sulfate and ferric citrate on tissue mineral composition of sheep. J. Anim. Sci. 33:481.

Standish, J. F., C. B. Ammerman, C. F. Simpson, F. C. Neal, and A. Z. Palmer. 1969. Influence of graded levels of dietary iron, as ferrous sulfate, on performance and tissue mineral composition of steers. J. Anim. Sci. 29:496.

Standish, J. F., C. B. Ammerman, A. Z. Palmer, and C. F. Simpson. 1971. Influence of dietary iron and phosphorus on the performance, tissue mineral composition and mineral absorption in steers. J. Anim. Sci. 33:171.

Szabuniewicz, M., E. M. Bailey, and D. O. Wiersig. 1971. Treatment of some common poisonings in animals. VM/SAC 66:1197.

Thompson, A., and A. M. Raven. 1959. The availability of iron in certain grass, clover and herb species. II. Alsike, broad red clover, Kent wild white clover, trefoil and lucerne. J. Agric. Sci. 53:224.

Thoren-Tolling, K. 1975. Studies on the absorption of iron after oral administration in piglets. Acta Vet. Scand. Suppl. 54.

Tollerz, G., and N. Lannek. 1964. Protection against iron toxicity in vitamin E-deficient piglets and mice by vitamin E and synthetic antioxidants. Nature 201:846.

Underwood, E. J. 1977. Trace Elements in Human and Animal Nutrition, 4th ed. Academic Press, New York.

Van Campen, D. 1974. Regulation of iron absorption. Fed. Proc. 33:100.

Weaver, L. C., R. W. Gardier, V. B. Robinson, and C. A. Bunde. 1961. Comparative toxicology of iron compounds. Am. J. Med. Sci. 241:296.

Woerpel, H. R., and S. L. Balloun. 1964. Effects of iron and magnesium on manganese metabolism. Poult. Sci. 43:1135.

# Lead

The symbol for lead, Pb, is derived from the Latin *plumbum*. This heavy, pliable metal is a bright bluish color, although easily tarnished to dull gray with an oxide film. Lead rarely occurs in the native form, but is usually found in the sulfide form in its chief ore, galena. The other common inorganic salts, lead carbonate (cerussite), lead sulfate (anglesite), and lead chlorophosphate (pyromorphite), are highly insoluble. The industrial use of lead in the United States has doubled in the last 30 years to a stable annual consumption around 1,300,000 tons (National Research Council, 1972) with the storage battery industry as the leading consumer. Substantial amounts are also used in gasoline additives, pigments, ceramics, pesticides, and plumbing (Paone, 1970).

Lead is considered to be one of the major environmental pollutants and has been incriminated as a cause of accidental poisoning in domestic animals more than any other substance (National Research Council, 1972). One of the primary sources of lead contamination in the air, soil, and water is combustion of fuel containing lead additives. Underwood (1977) and the National Research Council (1972) have excellent reviews.

## ESSENTIALITY

Lead is generally not considered to be an essential mineral for animals. However, in a recent study by Schwarz (1974), the addition of 1 ppm lead as lead subacetate increased the growth of rats by 16 percent (1.79 vs. 2.08 g/day) over controls receiving no supplemental lead. Increasing the lead level from 1.0 to 2.5 ppm decreased this growth response by 33 percent from 0.29 to 0.19 g/day over controls. Lead oxide and lead nitrate produced similar responses.

## METABOLISM

Many reviews are available concerning the metabolism of lead (National Research Council, 1972; Vallee and Ullmer, 1972; Hammond, 1973; Neathery and Miller, 1975). The bioavailability of lead may be altered by diet, growth rate, and physiological stresses, such as malnutrition, pregnancy, and lactation (White *et al.,* 1943; Allcroft, 1950; Blaxter, 1950a,b; Jones, 1965).

Lead tends to accumulate in the bones, consequently, the majority of body lead (about 90 percent) can be accounted for in the skeleton (Schroeder and Tipton, 1968) and appears to be relatively immobile. Nonruminant animals absorb approximately 10 percent of dietary lead, and ruminants absorb less than 3 percent (National Research Council, 1972). Balance studies on humans ingesting environmental quantities of lead have shown that only 5 to 10 percent of ingested lead is absorbed (Kehoe, 1964; Thompson, 1971). During chronic exposure a steady state appears to be reached in which metabolic excretion, by way of urinary and fecal excretion, approximately equals absorption. This occurs after an initial tissue saturation level is reached; therefore, levels of lead in many tissues and body fluids have been shown to increase with increasing exposure to lead. Lead absorption in the human infant and rat pup occurs at considerably higher rates than in the adult. In the rat pup, high lead absorption was associated with lactation period (Kostial *et al.,* 1971, 1974; Forbes and Reina, 1972). A decrease in absorption of oral $^{212}Pb$ in rats from nearly 90 percent to 15 percent occurred within 20 to 30 days of age (Forbes and Reina, 1972).

Oral administration of 25 IU daily of cholecalciferol to weanling rats increased absorption of lead acetate 33 percent (Smith *et al.,* 1978). Vitamin $D_3$ injected intraperitoneally to young rats (20,000 IU) 48 hours before oral administration of radioactive lead increased lead absorption from the intestine and its deposition into both kidney and bone (Hart and Smith, 1979). Experiments with various vitamin D metabolites (Mahaffey, 1979) showed that 1,25-dihydroxyvitamin $D_3$ caused the greatest increase in gastrointestinal lead absorption in rats.

In dogs studies with $^{203}Pb$, following acute administration of lead, a significant fraction of plasma lead was ultrafilterable, and a large fraction of the filtered lead underwent tubular reabsorption in the kidney. The results provide no direct evidence for kidney tubular secretion of lead (Vander *et al.,* 1977). When 30 mg lead as lead acetate was injected intravenously in rabbits, the total quantities recovered from tissues were bone marrow—14.1, liver—10.1, bone—6.2, and muscle—1.1 mg after 4 days (Blaxter, 1950b).

The absorption of metallic lead in rats was inversely related to particle size, which ranged from 6 to 200 $\mu$ (Barltrop and Meek, 1979).

## SOURCES

Environmental lead is largely airborne but returns to soil, water, and plants as dust and can become a hazard, especially to grazing livestock. Recent studies (Bolter *et al.,* 1975) indicated that lead deposits from smelters, as PbS, $PbSO_4$, $PbO \cdot PbSO_4$, or elemental lead, were from 2 to 7 times more soluble in organic acids of decaying foliage than in water. Lead from automobile exhaust was primarily lead bromochloride (PbBrCl) (Olson and Skogerboe, 1975), but the lead halides were converted to other compounds, primarily lead sulfate ($PbSO_4$), and deposited in soil.

Alkylated lead compounds are extremely unstable upon exposure to air and light. Acute toxicity of the alkyl lead compounds is greater and clinically different from that of inorganic lead compounds, but quantitative toxic dosages are similar for the different compounds with long-term subacute exposure (Hammond, 1973).

Most of the investigations on effects of lead in animals have been conducted with inorganic salts or with the more soluble compounds such as lead acetate. This limits the direct comparison of these results to actual environmental situations where lead occurs in other forms. Lead sulfate ($PbSO_4$), the inorganic lead compound that appears to be the major form contributing to the environmental burden and that may be more soluble in an organic medium than an aqueous solution, has not been studied in animals.

The main source of excess lead intake for cattle was that in paint (Garner and Papworth, 1967) until the restrictions on the use of lead-base pigments in paints. Other sources of ingestible lead include storage battery plates, putty, linoleum, asphalt roofing, engine oil, insecticide baits, and contaminated feeds (Garner and Papworth, 1967; Blood and Henderson, 1968; Buck, 1970; Christian and Tryphonas, 1971; Aronson, 1972). A major source of poisoning in wild water fowl is spent lead shot (Rac and Crisp, 1954; Trainer and Hunt, 1965; Cook and Trainer, 1966; Grandy *et al.*, 1968).

## TOXICOSIS

Clinical toxicosis in animals exposed chronically to lead is indirect and probably results through interference in normal metal-dependent enzyme functions at specific cellular sites characterized by apparent clinical abnormalities in hematological, neural, renal, or skeletal systems. Alleviation and diagnosis of toxicosis will depend on clarification of the mechanisms involved. Lead poisoning in livestock is well documented and reviews have been published by Lillie (1970), the National Research Council (1972), Ammerman *et al.* (1973), Clarke (1973), Bremmer (1974), MacLeavey (1977), and Forbes and Sanderson (1978).

Lead toxicosis is characterized by one or more of several clinical signs and underlying pathophysiological effects (Ammerman *et al.*, 1977). The main clinical signs in various species are:

1. microcytic hypochromic anemia;
2. anorexia, fatigue, depression;
3. intestinal colic (constipation, diarrhea, abdominal pain);
4. vomiting, increased salivation, esophageal paralysis in dogs;
5. nephropathy;
6. irritability, peripheral neuropathy, encephalopathy, blindness in cattle, laryngeal paralysis in horses;
7. weight loss;
8. abortion; and
9. maniacal excitement in young calves.

The main pathological effects are:

1. derangement of porphyrin and heme synthesis;
2. interference in protein and globin synthesis;
3. increased mechanical fragility of cell membranes resulting in shortened life of RBC's;
4. enzyme changes where small concentrations of lead ($10^{-6}M$) may inhibit or enhance activities;
5. renal tubular intranuclear inclusion bodies, containing protein-bound lead, calcium, and phosphorus;
6. basophilic stippling of erythrocytes and inhibitors of hemoglobin synthesis; and
7. altered endocrine function.

### LOW LEVELS

Relatively large amounts of absorbed lead can be sequestered preferentially in the skeleton with subsequent gradual release to the blood for excretion during long-term, low-level consumption. Chronic lead toxicosis is rarely seen in ruminants, but is more common in the nonruminant. It is usually recognized, however, only when distinct signs of poisoning are apparent.

Lead poisoning was produced in cattle within 6 to 8 weeks when fed lead acetate at 6 to 7 mg lead per kilogram of body weight daily (Buck *et al.,* 1961; Hammond and Aronson, 1964). No adverse effects were observed (Allcroft, 1950) when cattle were fed 1 to 2 g lead daily as lead acetate, carbonate, or sulfide over a 2-year period. Dinius *et al.* (1973) fed calves a concentrate diet containing 0, 10, and 100 ppm added lead as lead chromate for 100 days and saw no effect on feed consumption or weight gain. There was increased accumulation of lead in the liver and kidney with 100 ppm lead. Kelliher *et al.* (1973) observed reduced growth and feed utilization when calves were fed 15 mg lead (lead acetate) per kilogram of body weight for 283 days. There was no adverse effect on performance when lead acetate $[Pb(C_2H_3O_2)_2 \cdot 3H_2O]$ was fed to lambs at added levels of 10, 100, 500, or 1,000 ppm lead (Fick *et al.,* 1976).

Coburn *et al.* (1951) fed lead nitrate at 6 mg lead per kilogram of body weight daily for 137 days to ducks and did not observe any adverse effects, but when the dose was increased to 8 to 12 mg/kg, the survival periods averaged 28 and 25 days, respectively. Damron *et al.* (1969) studied the effects of feeding 0, 10, 100, 1,000, and 2,000 ppm added lead as lead acetate on feed intake and weight gain of broilers during a 4-week period. Decreased weight gain, feed efficiency, and feed intake were noted at 1,000 and 2,000 ppm.

There is evidence indicating that horses may be more susceptible to chronic lead toxicosis than cattle. Horses were poisoned on pastures adjacent to a smelter and succumbed to the toxicity following a lead intake during the winter period of 2.4 mg lead per kilogram of body weight daily (Hammond and Aronson, 1964). Horses exposed to a daily intake as low as 1.7 mg/kg body weight (approximately 80 ppm Pb in forage dry matter) were poisoned (Aronson, 1972).

Due to the interference of lead in the biosynthesis of heme (Hammond, 1973), assays on urine for δ-aminolevulinic acid (ALA) or its dehydrase (ALAD) in red blood cells (McSherry *et al.,* 1971; McIntire *et al.,* 1973; Lauwerys *et al.,* 1974), and assays for erythrocyte zinc protoporphyrin (APP) (Lamola and Yamane, 1974; Lamola *et al.,* 1975), which may be present in amounts other than normal, have reflected subclinical effects of lead ingestion. Blood lead level, ALA in urine and plasma, and urine porphyrin concentrations were indicative of chronic accidental lead exposure to paint by cattle that were being monitored prior to lead administration (Hilliard *et al.,* 1973). The level of red blood cell ALAD is also a sensitive test for lead exposure in Japanese quail (Stone *et al.,* 1977). For most mammalian species, blood lead levels in excess of 40 to 50 μg/dl are associated with recognizable effects of toxicosis of lead (Hsu *et al.,* 1975). Elevated blood lead may persist for long periods of time after withdrawal of the lead source, as the body burden is slowly depleted through lead excretion.

Lead-induced anemia in rats, a microcytic hypochromic type, has been shown to result from an interference with copper and iron metabolism (Klauder and Petering, 1977). Copper may be the target upon which ingested lead has its antagonistic effect on hematopoiesis.

### HIGH LEVELS

Acute lead toxicosis represents the greatest incidence of accidental poisoning in domestic animals. Horses appear to be more susceptible to lead poisoning than cattle, although cattle and dogs are the animals most frequently diagnosed with lead intoxication. The problem has been observed rarely in swine and sheep and it is uncommon in cats, goats, and zoo animals (Priester and Hayes, 1974; Staples, 1975). Chickens are very resistant to lead poisoning (Damron *et al.,* 1969; Vengris and Maré, 1974). In the foal (Willoughby *et al.,* 1972a), calf (Aronson, 1972; Buck, 1975), canine pup (Zook, 1972), and child under 3 years (King, 1971; Green *et al.,* 1973; Kolbye *et al.,* 1974; Bryce-Smith and Waldron, 1974), "pica," or consumption of nonfood items, has been implicated in the majority of accidental lead poisonings. Lead pica can be habitual in some cases and requires removal of the source.

Allcroft (1951) stated that 200 to 400 mg of lead as acetate, basic carbonate, or oxide per kilogram of body weight ingested in 1 day were sufficient to cause death in calves up to 4 months old. Single oral doses of 600 to 800 mg/kg may be a lethal dose to older cattle (Buck, 1970). Blood and Henderson (1968) reported that 30 g of lead acetate as a single dose are lethal to sheep, which agrees with results of Blaxter (1950a). Death was reported in sheep (Bennett and Schwartz, 1971) with an accumulative dose of 417 mg/kg lead (lead arsenate, $PbHAsO_4$) after 7 months.

Lead intoxications in swine are usually accidental and acute (Cristea, 1967). Link and Pensinger (1966) reported that pigs were relatively resistant to the toxic action of lead acetate administered orally. Neither 11 nor 66 mg of elemental lead per kilogram of body weight produced acute toxicosis in 7-week-old pigs. Three pigs weighing 20 kg were dosed weekly with 12 g of lead acetate for 3 weeks. No clinical signs other than slight hypersensitivity were observed; however, liver levels of 10 to 20 ppm lead (dry basis) were found (Nelson, 1971). Damron *et al.* (1969) also found chickens to be relatively resistant to lead poisoning at levels up to 2,000 ppm.

Blood and Henderson (1968) reported that a single oral dose of 500 g of lead acetate was lethal to horses. Most common sources of lead for sheep and cattle are usually not a problem for horses, because they are less likely to lick old paint cans, storage batteries, peeling paint, or motor oil (Aronson, 1972). Knight and Burau (1973) also observed lead poisoning in horses grazing pastures near a smelter that contained 325 ppm lead (dry basis).

## FACTORS INFLUENCING TOXICITY

Dietary calcium has been used for decades to decrease lead toxicity. Voluntary ingestion of lead by weanling rats was increased during periods of calcium deficiency, suggesting that this deficiency contributes to lead pica (Snowdon and Sanderson, 1974). Adult rats consuming 3 or 200 ppm lead (as acetate) in drinking water for 10 weeks showed less toxicosis when 0.7 percent than when 0.1 percent calcium was in the diet (Mahaffey *et al.,* 1973). Excessive dietary calcium and phosphorus decreased lead absorption in rats or lambs, and dietary calcium decreased retention of lead in bone and tissues. (Morrison *et al.,* 1974; Quarterman and Morrison, 1975; Quarterman *et al.,* 1978).

Foals consuming 0.25 or 0.6 percent calcium and 0.3, 0.4, or 0.6 percent phosphorus, and challenged with 30 ppm dietary lead for 14 weeks, showed increased liver lead only with the lower calcium and phosphorus levels (Willoughby *et al.,* 1972b). Increased dietary calcium (1.1 versus 0.7 percent) for 13 weeks also protected weanling swine against toxicity from 1,000 ppm dietary lead as acetate (Hsu *et al.,* 1975). Tissue and blood lead levels were decreased, and bone ash and specific gravity were increased by higher calcium.

Lead toxicosis in animals may be complicated by simultaneous exposure to excessive mercury, cadmium, zinc, molybdenum, copper, or other microelements. Addition of dietary copper at 1, 5, and 20 ppm increased accumulation of lead in liver and kidney tissue of rats when lead was ingested at 200 ppm (Cerklewski and Forbes, 1977). The beneficial effect of high zinc on lead toxicity has been described in horses (Schmitt *et al.*, 1971; Willoughby *et al.*, 1972b), in rats (Cerklewski and Forbes, 1976a; Cerklewski, 1979), and in swine (Hsu *et al.*, 1975). Iron supplementation also decreased lead deposition in tissues of rats (Six and Goyer, 1972). Rats deficient in vitamin E and challenged with high levels of lead had toxic signs more severe than vitamin E-supplemented rats. These signs included decreased hematocrit, increased reticulocyte count, and splenic enlargement (Levander *et al.*, 1975, 1977b). Further studies (Levander *et al.*, 1977a) revealed that spherocytes develop more rapidly in vitamin E-deficient, lead-poisoned rats than in vitamin E-supplemented, nonpoisoned rats and may help explain the splenomegaly, increased erythrocyte mechanical fragility, and decreased red cell filterability observed.

Dietary selenium at 1 ppm did not reduce toxic effects in Japanese quail fed 500 or 1,000 ppm lead (Stone and Soares, 1976). In rats 1 ppm selenium increased the toxic effects of 200 ppm lead (Cerklewski and Forbes, 1976b).

When 300 ppm fluorine as NaF and 200 ppm lead as $Pb(C_2H_3O_2)_2$ were fed in combination to rats, there was severe weight loss and a 30 percent death rate, which were not observed when either element was fed alone (Mahaffey and Stone, 1976).

There are conflicting reports concerning the effect of protein on lead toxicity. Early studies by Baernstein and Grand (1942) indicated that low dietary protein enhanced susceptibility to lead toxicity in rats. Conversely, Milev *et al.* (1970) reported that increasing dietary protein from 20 to 60 percent increased the retention of a single oral dose of $^{212}Pb$ from 7 to 49 percent. The protein may influence the retention of lead by decreasing absorption. Isocaloric protein-free diets also enhanced the retention of lead in comparison to the 20 percent protein diet, but only by a factor of 2. Gontzea *et al.* (1964) observed that pair-fed rats on a 9 percent protein diet had higher lead concentrations in blood, liver, and kidney than rats fed 18 percent protein. Gontzea *et al.* (1964) suggested that the aminoaciduria caused by lead may increase protein deficiency in low-protein diets, but an adequate level of dietary protein might enhance elimination of lead by the kidneys.

Acute lead poisoning in animals is usually fatal if the animals are not treated promptly. Attempts should be made to remove the lead from the gastrointestinal tract, and sedatives can be used to relieve convulsions (Garner and Papworth, 1967; Blood and Henderson, 1968). Repeated infusions of calcium–EDTA have been used for diagnosis and treatment of lead toxicosis in cattle (Holm *et al.*, 1953a,b; Lewis and Meikle, 1956; Hammond and Sorensen, 1957; Aronson *et al.*, 1968). Renal intranuclear inclusion bodies have been proposed as detoxification sequestrations of lead. They contain protein-bound lead, calcium, and phosphorus. In rat studies (Goyer and Wilson, 1975), EDTA administration dislodged the bodies increasing urinary lead excretion. These "inclusion bodies" have been observed in kidney, liver, brain, and osteoblasts of bone marrow in lead poisoned animals and in nuclei of plant leaf cells grown on soil high in lead.

## TISSUE LEVELS

No significant changes in the tissue level of lead were found in liver, kidney, heart, spleen, brain, bone, or muscle of sheep when dietary lead as lead acetate was 100 ppm (Fick *et al.*, 1976). Tissue levels increased when dietary levels were 500 or 1,000 ppm (Figures 2 and 3). Similar results were obtained for calves fed 100 ppm lead for 100 days (Dinius *et al.*, 1973). Liver and kidney contained 2.3 and 4.7 ppm lead (wet weight), and none was detected in muscle. From slaughter animals in Canada, 256 samples of beef and pork liver and kidney and poultry

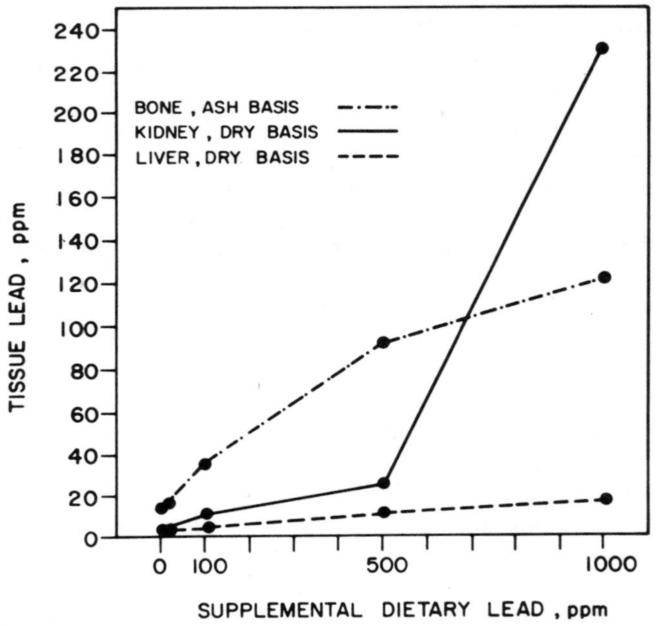


FIGURE 2 Influence of dietary lead on lead deposition in bone, kidney, and liver in sheep after 84 days.

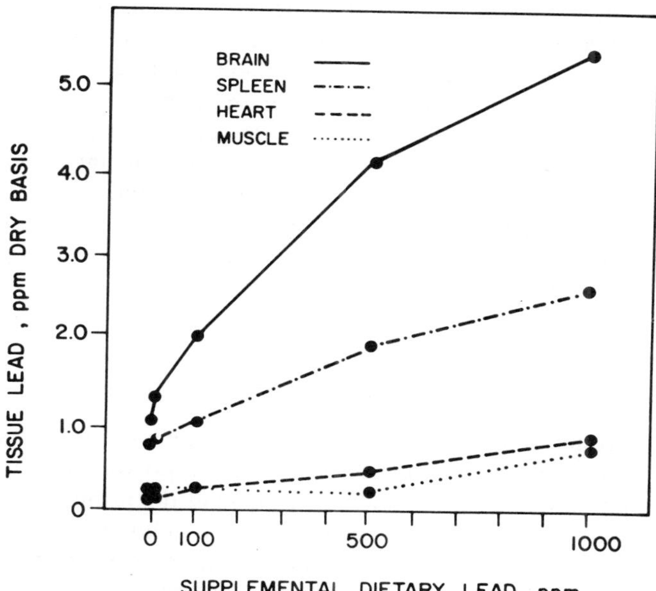


FIGURE 3 Influence of dietary lead on lead deposition in brain, spleen, heart, and muscle in sheep after 84 days.

liver ranged from 0.46–1.77 ppm lead (wet weight) (Prior, 1976). Fenstermacher *et al.* (1946) concluded that 10 ppm (dry weight) or more lead in liver should be considered suspicious of lead poisoning and 0–3 ppm normal. Kidney cortex lead levels above 25 ppm (dry weight) are considered as being of diagnostic significance (Todd, 1962; Garner and Papworth, 1967. Lead levels in milk and urine are variable and usually low.

## MAXIMUM TOLERABLE LEVELS

Cattle, sheep, and chickens have been fed 10 ppm supplemental lead in a soluble form for extended periods without adverse effects. Significant increases in tissue lead levels occurred when 100 ppm lead was fed to the same species. Dietary lead at 1,000 ppm has been tolerated by ruminants and poultry for several months with no visible signs of toxicosis. Approximately 300 ppm dietary lead resulted in observable signs of toxicosis in horses of various ages. Young growing pigs fed 11 mg lead per kilogram of body weight suffered from diarrhea, and 33 mg resulted in decreased growth and muscle tremors. Death occurred with a dietary intake of 66 mg lead per kilogram of body weight. With regard to acute toxicosis, the ingestion of 200 to 400 mg lead (as acetate) per kilogram of body weight caused acute death in calves and lambs up to 4 months old. In older cattle and sheep, the lethal single oral dose was 600 to 800 mg/kg of body weight. A single oral dose of 500 g lead acetate (700 mg lead per kilogram of body weight) was lethal to horses.

The maximum tolerable dietary level for lead is considered to be 30 ppm for most species, although detectable increases in lead concentration may occur in certain tissues.

## SUMMARY

Lead is considered to be one of the major environmental pollutants and has been incriminated as a cause of accidental poisoning in domestic animals more than any other substance. One of the primary sources of lead contamination in air, soil, and water is combustion of fuels containing lead additives. Young animals are more susceptible to lead toxicosis because they are more prone to lead pica and have a higher rate (90 percent) of absorption from the intestinal tract. Adult nonruminants, however, absorb only 10 percent of ingested lead and ruminants may absorb less than 3 percent. Clinical toxicosis appears to be exerted through interference in normal metal-dependent enzyme functions and is characterized by abnormalities in hematological, neural, renal, or skeletal systems.

TABLE 21 Effects of Lead Administration in Animals

| Class and Number of Animals[a] | Age or Weight | Administration: Quantity of Element[b] | Administration: Source | Administration: Duration | Administration: Route | Effect(s) | Reference |
|---|---|---|---|---|---|---|---|
| Cattle—16 | 16 wk | 10 ppm | Lead chromate | 100 d | Diet | No adverse effect | Dinius *et al.*, 1973 |
| | 120 kg | 100 ppm | | | | Increased lead in liver and kidney | |
| Cattle—5 | 43 kg | 1.5 mg/kg | $PbCO_3$ | 3x/wk | Capsule | Increased blood lead at 49 days | Lynch *et al.*, 1976a |
| | | 3.0 mg/kg | | | | Increased blood lead; decreased hemoglobin (Hb) | |
| | | 6.0 mg/kg | | | | Increased blood lead; decreased Hb | |
| Cattle—19 | 3 d<br>10 mo | 5–6 mg/kg | Lead acetate, carbonate, or sulfide | 2 yr | Diet | No adverse effect | Allcroft, 1950 |
| Cattle | | 6–7 mg/kg | Lead acetate | 42–54 d | Diet | Toxic | Buck *et al.*, 1961 |
| Cattle—2 | 61 kg | 9 mg/kg | $PbCO_3$ | 3x/wk | Capsule | Incoordination; death in 84 days; decreased gain | Lynch *et al.*, 1976b |
| | | 18 mg/kg | | | | Decreased gain and feed intake at 84 days | |
| Cattle | 155 kg | 15 mg/kg | Lead acetate | 282 d | Diet | Decreased growth and feed utilization | Kelliher *et al.*, 1973 |
| Sheep—4 | 36 kg | 10 ppm | Lead acetate $[Pb(C_2H_3O_2)_2 \cdot 3H_2O]$ | 84 d | Diet | No adverse effect | Fick *et al.*, 1976 |
| | | 100 ppm | | | | Increased lead in liver and kidney | |

TABLE 21 *Continued*

| Class and Number of Animals[a] | Age or Weight | Administration: Quantity of Element[b] | Source | Duration | Route | Effect(s) | Reference |
|---|---|---|---|---|---|---|---|
| | | 500 ppm | | | | Increased lead in liver, spleen, brain, and bone | |
| | | 1,000 ppm | | | | Increased lead in liver, kidney, and muscle | |
| Sheep—5 | 25 kg | 22 mg/kg | Lead arsenate ($PbHAsO_4$) | Once/mo | Diet | Increased lead in liver (283 d) | Bennett and Schwartz, 1971 |
| | | 44 mg/kg | | | | Increased lead in liver (283 d) | |
| | | 88 mg/kg | | | | Death (210–283 d) | |
| Sheep—10 | Yearling | 550 ppm | Metallic lead | 189 d | Diet | No clinical signs; increased lead in blood from 0.064 to 0.171 ppm | Carson *et al.*, 1973 |
| | | 1,000 ppm | | | | No clinical signs; increased blood lead from 0.064 to 0.320 ppm | |
| Sheep | 40 kg | 750–1,000 mg/kg | Lead acetate | 1 d | Diet | Death | Blaxter, 1950a |
| Goat—1, pregnant | 46.4 kg | 50 mg/kg | Lead acetate | 24 d | Drench | Aborted after 6 days; death after 41 days | Dollahite *et al.*, 1975 |
| Goat—1 | 41.6 kg | 100 mg/kg | | 24 d | | No adverse effect | |
| | 25 kg | 400 mg/kg | | 1 d | | Death after 23 days | |

| | | | | | | | |
|---|---|---|---|---|---|---|---|
| Goat—1, pregnant | 34.5 kg | 400 mg/kg | | 1 d | | Aborted after 8 days; death after 14 days | |
| | 35 kg | 800 mg/kg | | 1 d | | Aborted after 8 days; death | |
| Goat—1 | 18 kg | 1,600 mg/kg | | 1 d | | No adverse effect | |
| Goat—1, pregnant | 19 kg | 6,400 mg/kg | | 1 d | | Aborted after 7 days; death after 8 days | |
| Swine—4 | 7 wk | 11 mg/kg | Lead acetate | 14 d | Diet | Mild diarrhea | Link and Pensinger, 1966 |
| | | 22 mg/kg | | 14 d | | Mild diarrhea | |
| | 9 wk | 33 mg/kg | | 90 d | | Decreased growth and feed intake; muscle tremors | |
| | 9 wk | 66 mg/kg | | 64–90 d | | Death | |
| Swine—24 | 7.5 kg | 1,000 ppm | Lead acetate | 91 d | Diet | Anorexia; decreased growth | Hsu *et al.*, 1975 |
| Swine—3 | 20 kg | 12 g/wk | Lead acetate | 21 d | Diet | Hypersensitivity | Nelson, 1971 |
| Chicken—4 | 22 wk | 1 ppm | Lead oxide | 56 d | Diet | No adverse effect | Hermayer *et al.*, 1977 |
| | | 10 ppm | | | | No adverse effect | |
| | | 100 ppm | | | | No adverse effect | |
| | | 1,000 ppm | | | | Decreased egg production | |
| Chicken—16 | 4 wk | 10 ppm | Lead acetate | 28 d | Diet | No adverse effect | Damron *et al.*, 1969 |
| | | 100 ppm | | | | No adverse effect | |
| | | 1,000 ppm | | | | Decreased gain and feed utilization | |
| | | 2,000 ppm | | | | Decreased growth and feed utilization | |
| Chicken—10 | 4 wk | 5,000 ppm | Lead acetate | 21 d | Diet | Decreased growth and feed utilization; 10% death rate | Simpson *et al.*, 1970 |
| | | 10,000 ppm | | | | Decreased growth | |

TABLE 21 *Continued*

| Class and Number of Animals[a] | Age or Weight | Administration: Quantity of Element[b] | Source | Duration | Route | Effect(s) | Reference |
|---|---|---|---|---|---|---|---|
| | | | | | | and feed utilization; 30% death rate; necrosis of of renal tubules | |
| Chicken—12 | 6 wk | 20 mg/kg | Lead acetate | 35 d | Diet | Increased lead in bone | Vengris and Maré, 1974 |
| | | 40 mg/kg | | | | Increased lead in bone | |
| | | 80 mg/kg | | | | Increased lead in bone | |
| | | 160 mg/kg | | | | Increased lead in blood and bone | |
| | | 320 mg/kg | | 11–30 d | | Lethargy; weakness; 50% death rate | |
| | | 640 mg/kg | | 6–34 d | | Death | |
| Duck—6 | Mature | 3 mg/kg | Lead nitrate | 137 d | Diet | No adverse effect | Coburn *et al.*, 1951 |
| | | 6 mg/kg | | 137 d | | No adverse effect | |
| | | 8 mg/kg | | 24–41 d | | Death | |
| | | 12 mg/kg | | 19–27 d | | Death | |
| Japanese quail | Adult | 500 ppm | | 32 d | Diet | Decreased gain and egg production | Stone and Soares, 1974 |
| | | 1,000 ppm | | | | Decreased gain and egg production; increased number of soft-shell eggs | |
| Japanese quail—20 | Day old | 1 ppm | Lead acetate | 42 d | Diet | No adverse effect | Morgan *et al.*, 1975 |

|  |  |  |  |  |  |  |  |
|---|---|---|---|---|---|---|---|
|  |  | 10 ppm |  |  |  | No adverse effect |  |
|  |  | 100 ppm |  |  |  | No adverse effect |  |
|  |  | 1,000 ppm |  |  |  | Decreased growth; anemia |  |
| Japanese quail—20 | Day old | 10 ppm | Lead acetate | 35 d | Diet | No adverse effect | Morgan *et al.*, 1975 |
|  |  | 100 ppm |  |  |  | No adverse effect |  |
|  |  | 500 ppm |  |  |  | Decreased growth; anemia |  |
|  |  | 1,000 ppm |  |  |  | Decreased growth; anemia |  |
| Horse |  | 1.7 mg/kg | Natural from smelter |  | Diet | Toxic | Aronson, 1972 |
| Horse—17 | 4 wk | 0.5–30 ppm | Lead carbonate | 105 d | Diet | No adverse effect | Willoughby *et al.*, 1972b |
| Horse—9 | 2–3 wk | 800 ppm | Lead carbonate | 182–196 d | Diet | Pharyngeal and laryngeal paralysis; increased lead in soft tissue | Willoughby *et al.*, 1972a |
| Horse—20 | Various | 325 ppm | Lead contaminated pasture | 300–540 d | Diet | Anorexia; weight loss; weakness; laryngeal hemiplegia; mortality | Knight and Burau, 1973 |
| Pony—1 | 170 kg | 100 mg/kg | Lead acetate | 28 d | Drench | Death | Dollahite *et al.*, 1975 |
|  | 177 kg | 200 mg/kg |  | 9 d |  | Death |  |
|  | 160 kg | 1,000 mg/kg |  | 1 d |  | Anorexia, 81.7%; decrease weight |  |
|  | 177 kg | 1,000 mg/kg |  | 1 d |  | Anorexia; 23.7%; decrease weight |  |

[a] Number of animals per treatment group.
[b] Quantity expressed in parts per million or as milligrams per kilogram of body weight.

## REFERENCES


Allcroft, R. 1950. Lead as a nutritional hazard to farm livestock. IV. Distribution of lead in the tissues of bovines after ingestion of various lead compounds. J. Comp. Pathol. 60:190.

Allcroft, R. 1951. Lead poisoning in cattle and sheep. Vet. Rec. 63:583.

Ammerman, C. B., K. R. Fick, S. L. Hansard II, and S. M. Miller. 1973. Toxicity of Certain Minerals to Domestic Animals. A Review. Fla. Agric. Exp. Stn. Res. Bull. AL73-6. University of Florida, Gainesville.

Ammerman, C. B., S. M. Miller, K. R. Fick, and S. L. Hansard II. 1977. Contaminating elements in mineral supplements and their potential toxicity: A review. J. Anim. Sci. 44:485.

Aronson, A. L. 1972. Lead poisoning in cattle and horses following long-term exposure to lead. Am. J. Vet. Res. 33:627.

Aronson, A. L., P. B. Hammond, and A. C. Strafuss. 1968. Studies with calcium ethylenediaminetetraacetate in calves: Toxicity and use in bovine lead poisoning. Toxicol. Appl. Pharmacol. 12:337.

Baernstein, H. D., and J. A. Grand. 1942. The relation of protein intake to lead poisoning in rats. J. Pharmacol. Exp. Ther. 74:18.

Barltrop, D., and F. Meek. 1979. Effect of particle size on lead absorption from the gut. Arch. Environ. Health 34:280.

Bennett, D. G., Jr., and T. E. Schwartz. 1971. Cumulative toxicity of lead arsenate in phenothiazine given to sheep. Am. J. Vet. Res. 32:727.

Blaxter, K. L. 1950a. Lead as a nutritional hazard to farm livestock. II. The absorption and excretion of lead by sheep and rabbits. J. Comp. Pathol. 60:140.

Blaxter, K. L. 1950b. Lead as a nutritional hazard to farm livestock. III. Factors influencing the distribution of lead in the tissues. J. Comp. Pathol. 60:177.

Blood, D. C., and J. A. Henderson. 1968. Veterinary Medicine, 3rd ed. Williams & Wilkins Co., Baltimore, Md.

Bolter, E., T. R. Butz, and J. F. Arseneau. 1975. Heavy metal mobilization by natural organic acids. International Conference on Heavy Metals in the Environment, Toronto, Canada. p. C-81.

Bremmer, I. 1974. Heavy metal toxicities. Q. Rev. Biophys. 7:75.

Bryce-Smith, D., and H. A. Waldron. 1974. Lead in food—Are today's regulations sufficient? Chem. Br. 10:202.

Buck, W. B. 1970. Lead and organic pesticide poisonings in cattle. J. Am. Vet. Med. Assoc. 156:1468.

Buck, W. B. 1975. Toxic minerals and neurological disease in cattle. J. Am. Vet. Med. Assoc. 166:222.

Buck, W. B., L. F. James, and W. Binns. 1961. Changes in serum transaminase activities associated with plant and mineral toxicity in sheep and cattle. Cornell Vet. 51:568.

Carson, T. L., G. A. Van Gelder, W. B. Buck, L. J. Hoffman, D. L. Mick, and K. R. Long. 1973. Effects of low level lead ingestion in sheep. Clin. Toxicol. 6:389.

Cerklewski, F. L. 1979. Influence of dietary zinc on lead toxicity during gestation and lactation in the female rat. Fed. Proc. 38:606. (Abstr.)

Cerklewski, F. L., and R. M. Forbes. 1976a. Influence of dietary zinc on lead toxicity in the rat. J. Nutr. 106:689.

Cerklewski, F. L., and R. M. Forbes. 1976b. Influence of dietary selenium on lead toxicity in the rat. J. Nutr. 106:778.

Cerklewski, F. L., and R. M. Forbes. 1977. Influence of dietary copper on lead toxicity in the young male rat. J. Nutr. 107:143.

Christian, R. G., and L. Tryphonas. 1971. Lead poisoning in cattle: Brain lesions and hematologic changes. Am. J. Vet. Res. 32:203.

Clarke, E. G. C. 1973. Lead poisoning in small animals. J. Small Anim. Pract. 14:183.

Coburn, D. R., D. W. Metzler, and R. Treichler. 1951. A study of absorption and retention of lead in wild water fowl in relation to clinical evidence of lead poisoning. J. Wildl. Manage. 15:186.

Cook, R. S., and D. O. Trainer. 1966. Experimental lead poisoning of Canada geese. J. Wildl. Manage. 30:1.

Cristea, J. 1967. Acute lead poisoning in swine. Rec. Med. Vet. 143:749.

Damron, B. L., C. F. Simpson, and R. H. Harms. 1969. The effect of feeding various levels of lead on the performance of broilers. Poult. Sci. 48:1507.

Dinius, D. A., T. H. Brinsfield, and E. E. Williams. 1973. Effect of subclinical lead intake on calves. J. Anim. Sci. 37:169.

Dollahite, J. W., L. D. Rowe, and J. C. Reagor. 1975. Experimental lead poisoning in horses and Spanish goats. Southwest Vet. 28:40.

Fenstermacher, R., B. S. Pomeroy, M. H. Roepke, and W. L. Boyd. 1946. Lead poisoning of cattle. J. Am. Vet. Med. Assoc. 108:1.

Fick, K. R., C. B. Ammerman, S. M. Miller, C. F. Simpson, and P. E. Loggins. 1976. Effect of dietary lead on performance, tissue mineral composition and lead adsorption in sheep. J. Anim. Sci. 42:515.

Forbes, G. B., and J. C. Reina. 1972. Effect of age on gastrointestinal absorption (Fe, Sr, Pb) in the rat. J. Nutr. 102:647.

Forbes, R. M., and G. C. Sanderson. 1978. Lead toxicity in domestic animals and wildlife. Top. Environ. Health, p. 225.

Garner, R. J., and D. S. Papworth. 1967. Garner's Veterinary Toxicology, 3rd ed. Williams & Wilkins Co., Baltimore, Md.

Gontzea, I., P. Sutzesco, D. Cocora, and D. Lungu. 1964. Importance of protein intake for resistance to lead poisoning. Arch. Sci. Physiol. 18:211; Nutr. Abstr. Rev. 35(746):126.

Goyer, R. A., and M. H. Wilson. 1975. Lead-induced inclusion bodies. Results of ethylenediaminetetraacetic acid treatment. Lab. Invest. 32:149.

Grandy, J. W. IV, L. N. Locke, and G. E. Bagley. 1968. Relative toxicity of lead and five proposed substitute shot types to pen-reared mallards. J. Wildl. Manage. 32:483.

Green, V. A., G. W. Wise, and N. W. Smull. 1973. Lead survey of selected children in Kansas City and some unusual cases. Clin. Toxicol. 6:29.

Hammond, P. B. 1973. Metabolism and metabolic action of lead and other heavy metals. Clin. Toxicol. 6:353.

Hammond, P. B., and A. L. Aronson. 1964. Lead poisoning in cattle and horses in the vicinity of a smelter. Ann. N.Y. Acad. Sci. 111:595.

Hammond, P. B., and D. K. Sorensen. 1957. Recent observations on the course and treatment of bovine lead poisoning. J. Am. Vet. Med. Assoc. 130:23.

Hart, M. H., and J. L. Smith. 1979. Effect of vitamin D on lead absorption and retention. Fed. Proc. 38:384. (Abstr.)

Hermayer, K. L., P. E. Stake, and R. L. Shippe. 1977. Evaluation of dietary zinc, cadmium, tin, lead, bismuth and arsenic toxicity in hens. Poult. Sci. 56:1721.

Hilliard, E. P., D. B. R. Poole, and J. D. Collins. 1973. Accidental lead intoxication of cattle; further evidence of an interference in heme biosynthesis. Br. Vet. J. 129:83.

Holm, L. W., E. A. Rhode, J. D. Wheat, and G. Firch. 1953a. Treatment of acute lead poisoning in calves with calcium disodium ethylenediaminetetraacetate. J. Am. Vet. Med. Assoc. 123:528.

Holm, L. W., J. D. Wheat, E. A. Rhode, and G. Firch. 1953b. The treatment of chronic lead poisoning in horses with calcium disodium ethylenediaminetetraacetate. J. Am. Vet. Med. Assoc. 123:383.

Hsu, F. S., L. Krook, W. G. Pond, and J. R. Duncan. 1975. Interactions of dietary calcium with toxic levels of lead and zinc in pigs. J. Nutr. 105:112.

Jones, L. M. 1965. Veterinary Pharmacology and Therapeutics, 3rd ed. Iowa State University Press, Ames.

Kehoe, R. A. 1964. Normal metabolism of lead. Arch. Environ. Health 8:44.

Kelliher, D. J., E. P. Hilliard, D. B. R. Poole, and J. D. Collins. 1973. Chronic lead intoxication in cattle: Preliminary observations on its effect on the erythrocyte and on porphyrin metabolism. Irish J. Agric. Res. 12:61.

King, B. C. 1971. Maximum daily intakes of lead without excessive body lead burden in children. Am. J. Dis. Child. 122:337.

Klauder, D. S., and H. G. Petering. 1977. Anemia of lead intoxication: A role for copper. J. Nutr. 107:1779.

Knight, H. D., and R. G. Burau. 1973. Chronic lead poisoning in horses. J. Am. Vet. Med. Assoc. 162:781.

Kolbye, A. C., Jr., K. R. Mahaffey, J. A. Fiorino, P. C. Corneliussen, and C. F. Jelinek. 1974. Food exposures to lead. Environ. Health Perspect. 8:65.

Kostial, K., I. Simonovic, and M. Pisonic. 1971. Lead absorption from the intestine in newborn rats. Nature 233:564.

Kostial, K., T. Maljkovic, and S. Jogo. 1974. Lead acetate toxicity in rats in relation to age and sex. Arch. Toxicol. 31:265.

Lamola, A. A., and T. Yamane. 1974. Zinc protoporphyrin in the erythrocytes of patients with lead intoxication and iron deficiency anemia. Science 186:936.

Lamola, A. A., M. Joselow, and T. Yamane. 1975. Zinc protoporphyrin (ZPP). A simple sensitive fluorometric screening test for lead poisoning. Clin. Chem. 12:93.

Lauwerys, R., J. P. Buchet, H. A. Roels, and D. Materne. 1974. Relationship between urinary δ-aminolevulinic acid excretion and the inhibition of red cell δ-aminolevulinate dehydrase by lead. Clin. Toxicol. 7:383.

Levander, O. A., V. C. Morris, D. J. Higgs, and R. J. Ferretti. 1975. Lead poisoning in vitamin E-deficient rats. J. Nutr. 105:1481.

Levander, O. A., M. Fisher, V. C. Morris, and R. J. Ferretti. 1977a. Morphology of erythrocytes from vitamin E-deficient lead-poisoned rats. J. Nutr. 107:1828.

Levander, O. A., V. C. Morris, and R. J. Ferretti. 1977b. Comparative effects of selenium and vitamin E in lead-poisoned rats. J. Nutr. 107:378.

Lewis, E. F., and J. C. Meikle. 1956. The treatment of acute lead poisoning in cattle with calcium versenate. Vet. Rec. 68:98.

Lillie, R. J. 1970. Air Pollutants Affecting the Performance of Domestic Animals. A Literature Review. Agricultural Handbook No. 380. U.S. Department of Agriculture, Agricultural Research Service, Washington, D. C.

Link, L. P., and R. R. Pensinger. 1966. Lead toxicosis in swine. Am. J. Vet. Res. 27:759.

Lynch, G. P., E. D. Jackson, C. A. Kiddy, and D. F. Smith. 1976a. Responses of young calves to low doses of lead. J. Dairy Sci. 59:1490.

Lynch, G. P., D. F. Smith, M. Fisher, T. L. Pike, and B. T. Weinland. 1976b. Physiological responses of calves to cadmium and lead. J. Anim. Sci. 42:410.

MacLeavey, B. J. 1977. Lead poisoning in dogs. N.Z. Vet. J. 25:395.

Mahaffey, K. R. 1979. Stimulation of gastrointestinal lead absorption by 1,25-dihydroxyvitamin $D_3$. Fed. Proc. 38:384. (Abstr.)

Mahaffey, K. R., and C. L. Stone. 1976. Effect of high fluorine (F) intake on tissue lead (Pb) concentrations. Fed. Proc. 35:256. (Abstr.)

Mahaffey, K. R., R. Goyer, and J. K. Haseman. 1973. Dose–response to lead ingestion in rats fed low dietary calcium. J. Lab. Clin. Med. 82:92.

McIntire, M. S., G. L. Wolf, and C. R. Angle. 1973. Red cell lead and δ-aminolevulinic acid dehydrase. Clin. Toxicol. 6:183.

McSherry, G. J., R. A. Willoughby, and R. G. Thomson. 1971. Urinary delta amino levulinic acid (ALA) in the cow, dog and cat. Can. J. Comp. Med. 35:136.

Milev, N., E. L. Sattler, and E. Menden. 1970. Uptake and deposition of Pb in the body in different nutritional states. 1. Effect of different protein intake on uptake of $^{212}Pb$ in rats. Med. Ernahrung. 11:29.

Morgan, G. W., F. W. Edens, P. Thaxton, and C. R. Parkhurst. 1975. Toxicity of dietary lead in Japanese quail. Poult. Sci. 54:1636.

Morrison, J. N., J. Quarterman, and W. R. Humphries. 1974. Lead metabolism in lambs and the effect of phosphate supplements. Proc. Nutr. Soc. 33:88A. (Abstr.)

National Research Council. 1972. Lead: Airborne Lead in Perspective. National Academy of Sciences, Washington, D.C.

Neathery, M. W., and W. J. Miller. 1975. Metabolism and toxicity of cadmium, mercury and lead in animals: A review. J. Dairy Sci. 58:1967.

Nelson, H. A. 1971. Lead poisoning. J. Am. Vet. Med. Assoc. 158:258.

Olson, K. W., and R. K. Skogerboe. 1975. Identification of soil lead compounds from automotive sources. Environ. Sci. Technol. 9:227.

Paone, J. 1970. Mineral Facts and Problems. Bureau of Mines Bull. 650. U.S. Department of the Interior, Washington, D. C.

Priester, W. A., and H. M. Hayes. 1974. Lead poisoning in cattle, horses, cats and dogs as reported by 11 colleges of veterinary medicine in the United States and Canada from July, 1968, through June, 1972. Am. J. Vet. Res. 35:567.

Prior, M. G. 1976. Lead and mercury residues in kidney and liver of Canadian slaughter animals. Can. J. Comp. Med. 40:9.

Quarterman, J., and J. N. Morrison. 1975. The effects of dietary calcium and phosphorus on the retention and excretion of lead in rats. Br. J. Nutr. 34:351.

Quarterman, J., J. N. Morrison, and W. R. Humphries. 1978. The influence of high dietary calcium and phosphate on lead uptake and release. Environ. Res. 17:60.

Rac, R., and C. S. Crisp. 1954. Lead poisoning in domestic ducks. Aust. Vet. J., 30:145.

Schmitt, N., G. Brown, E. L. Devlin, A. A. Larsen, E. D. McCausland, and J. M. Saville. 1971. Lead poisoning in horses. Arch. Environ. Health 23:185.

Schroeder, H. A., and I. H. Tipton. 1968. The human body burden of lead. Arch. Environ. Health 17:965.

Schwarz, K. 1974. New essential trace elements (Sn, V, F, Si): Progress report and outlook. *In* W. G. Hoekstra, J. W. Suttie, H. E. Ganther, and W. Mertz (eds.). Trace Element Metabolism in Animals—2. University Park Press, Baltimore, Md.

Simpson, C. F., B. L. Damron, and R. H. Harms. 1970. Abnormalities of erythrocytes and renal tubules of chicks poisoned with lead. Am. J. Vet. Res. 31:515.

Six, K. M., and R. A. Goyer. 1972. The influence of iron deficiency on tissue content and toxicity of ingested lead in the rat. J. Lab. Clin. Med. 79:128.

Smith, C. M., H. F. DeLuca, Y. Tanaka, and K. R. Mahaffey. 1978. Stimulation of lead absorption by vitamin D administration. J. Nutr. 108:843.

Snowdon, C. T., and B. A. Sanderson. 1974. Lead pica produced in rats. Science 183:92.

Staples, L. J. 1975. Lead poisoning still kills. N.Z. J. Agric. 130:21.

Stone, C., and J. H. Soares, Jr. 1974. Studies on the metabolism of lead in Japanese quail. Poult. Sci. 53:1982. (Abstr.)

Stone, C. L., and J. H. Soares, Jr. 1976. The effect of dietary selenium level on lead toxicity in Japanese quail. Poult. Sci. 55:341.

Stone, C. L., M. R. S. Fox, A. L. Jones, and K. R. Mahaffey. 1977. $\mu$-Aminolevulinic acid dehydratase—A sensitive indicator of lead exposure in Japanese quail. Poult. Sci. 56:174.

Thompson, J. A. 1971. Balance between intake and output of lead in normal individuals. Br. J. Ind. Med. 28:189.

Todd, J. R. 1962. A knackery survey of lead poisoning incidence in cattle in northern Ireland. Vet. Rec. 74:116.

Trainer, D. O., and R. A. Hunt. 1965. Lead poisoning of whistling swans in Wisconsin. Avian Dis. 9:252.

Underwood, E. J. 1977. Trace Elements in Human and Animal Nutrition, 4th ed. Academic Press, New York.

Vallee, B. L., and D. D. Ullmer. 1972. Biochemical effect of mercury, cadmium and lead. Annu. Rev. Biochem. 41:91.

Vander, A. J., D. L. Taylor, K. Kalitis, D. R. Mouw, and W. Victery. 1977. Renal handling of lead in dogs: Clearance studies. Am. J. Physiol. 233:532.

Vengris, V. E., and C. J. Maré. 1974. Lead poisoning in chickens and the effect of lead on interferon and antibody production. Can. J. Comp. Med. 38:328.

White, W. B., P. A. Clifford, and H. O. Calvey. 1943. Lethal dose of lead for the cow: The elimination of ingested lead through milk. J. Am. Vet. Med. Assoc. 102:292.

Willoughby, R. A., E. MacDonald, B. J. McSherry, and G. Brown. 1972a. Lead and zinc poisoning and the interaction between Pb and Zn poisoning in the foal. Can. J. Comp. Med. 36:348.

Willoughby, R. A., T. Thirapatsakum, and B. J. McSherry. 1972b. Influence of rations low in calcium and phosphorus on blood and tissue lead concentrations in the horse. Am. J. Vet. Res. 33:1165.

Zook, B. C. 1972. The pathologic anatomy of lead poisoning in dogs. Vet. Pathol. 9:310.

# Magnesium

---

Magnesium (Mg) is an alkaline earth metal belonging to Group IIA of the periodic table. It ranks eighth in abundance in the earth's crust. Industrially, it is important as a structural material such as for fire bricks in production of steel.

Magnesium is one of the major minerals recognized as essential for animals. Approximately 60 percent of total body magnesium is located in bone, where the function of the element is not known (Pike and Brown, 1975). About one-third of that in the bone is combined with phosphate, and the remainder is adsorbed loosely on the surface of the mineral structure. Magnesium occurs intra- and extracellularly in soft tissues. The small amount present in extracellular fluid is exchanged easily with that adsorbed on the bone surface. The serum level of magnesium varies, usually between 1 and 3 mg/dl. Within the cells of soft tissues, magnesium is found in larger concentrations than any other element except potassium.

Supplemental magnesium is used at fairly high levels for dairy cows, beef cows, and ewes. Although recommended levels are not toxic, magnesium may be toxic when excessively high levels are accidentally used.

## ESSENTIALITY

Magnesium is essential for cellular respiration. It is necessary for all phosphate transfer reactions, and in certain tissues it is complexed with adenosine triphosphate (ATP), adenosine diphosphate (ADP), and adenosine monophosphate (AMP). It is an activator for all thiamine pyrophosphate (TPP) requiring reactions. Also, for certain reactions magnesium is involved in the metabolism of fat and protein. Magnesium is an essential nutrient for all animals. The amount required varies among species and between classes of animals within species.

Signs of deficiency in all animals include loss of appetite, lower rate of body weight gain, and hyperexcitability. Among farm animals the disturbance most frequently associated with lack of magnesium is hypomagnesemic tetany in ruminants, also known as grass tetany, grass staggers, winter tetany, and wheat pasture poisoning. In the United States this disturbance usually occurs in beef cows in the early stages of lactation, but also occurs in ewes and dairy cows (Fontenot *et al.*, 1973). The disturbance is more prevalent in older animals, which may be due to less labile magnesium in older than in young animals (Thomas, 1965). The signs of the condition are of a neuromuscular nature. Usually, the disease is fatal if the animals are not treated. Hypomagnesemic tetany appears to be caused by a physiological deficiency of magnesium, which may result from a simple dietary deficiency or lowered efficiency of utilization of the element. However, there is some evidence that a shift of magnesium ion inside the body is responsible, at least in part, for hypomagnesemia (Larvor, 1976). Perhaps the main reason that ruminants are more susceptible to this disturbance than nonruminants is the generally lower magnesium levels in roughages than concentrates and lower bioavailability of the magnesium, especially from certain "tetany-prone" roughages. Calves on all-milk diets are quite susceptible to magnesium deficiency (Duncan *et al.*, 1935). This appears to be due to a low magnesium level in milk. Magnesium deficiency was produced in baby pigs fed diets containing 125 ppm magnesium or less (Miller *et al.*, 1965).

## METABOLISM

Magnesium is absorbed from the small intestine in simple-stomached animals and from the first three compartments of the ruminant stomach (Grace *et al.*, 1974). There is considerable excretion into the lower digestive tract (Cragle, 1973). The magnesium in bone can be mobilized to a limited extent, especially in younger animals, but apparently this is not under hormonal control (Rayssiguier *et al.*, 1977). In older animals mobilization is very limited (Thomas, 1965). Approximately one-third of magnesium in bone is in combination with phosphorus, but its function is not known.

Urinary excretion is usually a reflection of quantity of magnesium absorbed. It has been suggested that in ruminants magnesium absorbed in excess of requirement is excreted via the urine (Rook and Storry, 1962). Chicco *et al.* (1972) reported a high correlation (r = 0.95) between magnesium absorption and urinary excretion.

## SOURCES

Magnesium is present in variable amounts in common feedstuffs (National Research Council, 1979). Generally, concentrates contain higher levels than roughages. There is a large degree of variability among forages, presumably due to soil availability (Reid *et al.,* 1970). Legumes are generally higher in magnesium than grasses.

There are a number of supplemental sources available, with the most commonly used being magnesium oxide. The bioavailability of magnesium from this product is very good, and there is variation in the availability of salts of magnesium. Another important consideration is the level of magnesium in the different supplemental sources. For example, the magnesium content is more than 50 percent in feed-grade magnesium oxide, but only about 12 percent for magnesium carbonate.

Bioavailability in cattle and sheep of magnesium from crude products such as dolomitic limestone and magnesite has been shown to be very low (Gerken and Fontenot, 1967; Ammerman *et al.,* 1972). Furthermore, supplementing with dolomitic limestone results in a large depression in apparent digestibility of energy, resulting mainly from depressions in digestibility of the carbohydrate components (Gerken and Fontenot, 1967).

## TOXICOSIS

Toxicosis due to ingestion of natural feedstuffs has not been reported and does not appear likely. Thus, toxicosis would occur from using excess levels of supplementary magnesium.

### LOW LEVELS

Ingestion of excess levels of magnesium has generally resulted in decreased growth rate in chicks (Nugara and Edwards, 1963; Chicco *et al.*, 1967), guinea pigs (Morris and O'Dell, 1963), and sheep (Kerk, 1973). The decrease in performance appears to be caused at least partly by decreased feed intake. Scouring is also a problem with high dietary magnesium levels (Peirce, 1959; Care, 1960). Generally, the high intake of magnesium increased blood serum magnesium.

Supplemental magnesium oxide was administered orally by capsule to supply up to 5.3 percent magnesium, dry basis, to yearling wethers (personal communication, J. P. Fontenot, Virginia Polytechnic Institute and State University, Blacksburg). High levels of supplemental magnesium resulted in depressed feed intake, elevated serum magnesium, and diarrhea. The time required to produce diarrhea was inversely related to the dietary level. Diarrhea was observed 24 hours, 48 hours, and 6 days after initiation of treatment in sheep receiving 5.3, 2.0, and 0.8 percent magnesium, respectively. Feeding 2.3 or 4.3 percent magnesium to Holstein bull calves resulted in severe diarrhea and decreased feed intake and rate of gain (Gentry *et al.*, 1978). Mucus was voided in feces of calves fed the high magnesium levels. In studies designed to establish the magnesium requirement of beef cows, levels as high as 20 g per day (0.29 percent) were fed to cows during gestation (O'Kelley and Fontenot, 1973) and 42 g (0.29 percent) to lactating beef cows (O'Kelley and Fontenot, 1969) with no deleterious effects. Increasing the level of magnesium in the diet from 0.16 to 0.22 percent lowered rate and efficiency of gain in growing or finishing swine when they weighed 20 to 45 kg, but had no effect thereafter (Krider *et al.*, 1975).

### HIGH LEVELS

The presence of high magnesium levels in water (about 1 percent) was reported to cause a weakening effect on men and livestock in an area including parts of Minnesota, the Dakotas, and Montana (Allison, 1930). He reported that cattle and hogs could not be fattened for market while drinking this water. Cattle developed a "run-down-ragged appearance," and many died prematurely. A degeneration of the bones occurred. Calves were stunted and many never matured. The cows developed depressed appetites.

Hypertonic magnesium sulfate enemas produced adverse effects in young lambs (Andrews *et al.*, 1965). Administration of 10 ml of 50 percent magnesium sulfate to five newborn lambs resulted in death in 23 to 46 minutes. Signs included lack of reflexes, anesthesia, and cardiorespiratory depression. Plasma magnesium was 3.70 to 5.72 mg/dl at death. Administration of 10 ml of a 25 percent solution resulted in loss of deep tendon reflexes, deep sleep, and cyanosis in two of five lambs, but only one lamb died. When 10 ml of 50 percent solution were administered to five 2- to 3-months old lambs, three had absent reflexes and slept, and only one died. Postmortem examinations showed congestion of the lungs, heart, liver, spleen, and kidney.

Intravenous infusion of high levels of magnesium ions resulted in disruption of motor function in horses, cattle, and dogs (Bowen *et al.*, 1970). The levels required to produce the effect were 0.13–0.14 g magnesium sulfate per kilogram (0.026–0.028 g magnesium per kilogram) of body weight. Calcium gluconate and ethylenediaminetetraacetic acid (EDTA) ameliorated the effects of the magnesium ions. Infusion of 0.22 g magnesium sulfate per kilogram (0.044 g magnesium per kilogram) of body weight was lethal in one horse due to respiratory paralysis and cardiac arrest.

Feeding 1.2 percent magnesium resulted in high mortality and depressed growth in guinea pigs fed diets containing 0.9 percent calcium and 1.7 percent phosphorus (Morris and O'Dell, 1963). The animals were lethargic, suffered from diarrhea, and exhibited a poor general appearance. When the calcium level was increased to 2.5 percent and the phosphorus was held at 1.7 percent, the toxic effects were alleviated. The level of 1.2 percent magnesium had no effect on guinea pigs when the diet contained 0.9 percent calcium and 0.6 percent phosphorus.

Accidental feeding of high-magnesium pellets to sheep instead of concentrate pellets resulted in acute metabolic disorders, including loss of appetite and severe diarrhea (Kerk, 1973). In some ewes a syndrome similar to milk fever occurred. Growth of lambs was poor, due to lower milk production by the ewes. Following withdrawal of the pellets, disorders of the gastrointestinal tract disappeared. The adverse effects were ascribed to high levels of magnesium oxide in the pellets.

Ingestion of water with 0.2–0.3 percent magnesium chloride (1.05–0.69 percent sodium chloride, respectively) was harmful to sheep (Peirce, 1959). There was occasional diarrhea, which was more frequent with animals receiving higher concentrations of magnesium chloride. Drenching cattle with 170–342 g magnesium oxide (102–205 g magnesium) per day resulted in severe scouring in 24–48 hours (Care, 1960). However, administration of up to 114 g (68 g magnesium) per day did not affect condition or produce scouring.

Supplementing 0.64 percent or 1.28 percent magnesium to chicks increased mortality and depressed growth rate (Nugara and Edwards, 1963).

### FACTORS INFLUENCING TOXICITY

Increased calcium and phosphorus levels in the diet have been shown to increase the magnesium requirement of chicks (Nugara and Edwards, 1963) and guinea pigs (Morris and O'Dell, 1963). Increasing the phosphorus level from 0.6 to 1.7 percent in a diet containing 0.9 percent calcium and 1.2 percent magnesium produced high mortality and poor growth in guinea pigs (Morris and O'Dell, 1963). Increasing the calcium level to 2.5 percent alleviated the deleterious effect. Addition of 0.2 or 0.4 percent magnesium tended to overcome the adverse effects of deficiencies of both calcium and phosphorus in chicks (Chicco *et al.,* 1967). However, when 0.6 percent magnesium was supplemented, growth and bone mineralization were adversely affected regardless of the calcium and phosphorus levels. High levels of calcium and phosphorus have been shown to depress magnesium absorption in sheep (Chicco *et al.,* 1973; Pless *et al.,* 1973). Metastatic calcification in hearts and kidneys of rats administered high levels of vitamin D was aggravated by high dietary levels of magnesium (Whittier and Freemen, 1971).

The interrelationships between magnesium and calcium and phosphorus suggest that hormones and enzymes involved with bone metabolism may be related to magnesium metabolism. Supplementation of 1.68 g of magnesium as magnesium sulfate per kilogram of diet to rats increased the serum alkaline phosphatase activity (Moinuddin and Lee, 1960). Administration of magnesium chloride subcutaneously to nephrectomized rats resulted in a decrease in ionic calcium in plasma, but no such effect was observed in parathyroidectomized rats (Gitelman *et al.,* 1968). Based on these results, it was suggested that hypermagnesemia may inhibit parathyroid gland activity. Hypermagnesemia in dogs resulted in decreased filtered phosphorus excretion and serum calcium (Massry *et al.,* 1970). These effects were reversed by administration of parathyroid extracts, suggesting that hypermagnesemia suppresses parathyroid gland activity. The magnitude of involvement of dietary magnesium level on parathyroid function is not clear, however, since magnesium deficiency in calves has not affected plasma parathyroid hormone (PTH) levels in calves (Rayssiguier *et al.,* 1977).

High dietary potassium depresses magnesium absorption in ruminants (Newton *et al.,* 1972).

## TISSUE LEVELS

There is little information concerning the tissue levels of magnesium. A three- to fivefold increase in magnesium in kidney was found from feeding high levels of magnesium to sheep (personal communication, J. P. Fontenot, Virginia Polytechnic Institute and State University, Blacksburg). The level in the unsupplemented controls was 0.18–0.21 percent. There was also an increase of magnesium in bone from 4.14 percent in controls to 4.95 percent in those administered the highest magnesium level. There was no increase in magnesium levels in muscle, liver, and heart.

## MAXIMUM TOLERABLE LEVELS

In cattle, up to 0.39 percent magnesium was fed without problems (O'Kelley and Fontenot, 1969). Oral administering of 0.5 percent magnesium to yearling wethers did not produce toxicity, whereas, administering 0.8 percent or higher resulted in signs of toxicosis (Fontenot *et al.*, unpublished). Cattle and sheep should be able to tolerate 0.5 percent magnesium. Feeding 0.6 percent magnesium to chicks decreased growth rate and bone calcification (Nugara and Edwards, 1963; Chicco *et al.*, 1967). Feeding diets with 0.32 or 0.4 percent magnesium was without effect. The maximum tolerable level for poultry and swine appears to be 0.3 percent.

## SUMMARY

Magnesium is a required element for maintaining normal health and well-being of animals and is present in soft tissue and bone. It functions as a component or activator of all enzymes involved in cell respiration, but its function in bone is not clear. Magnesium occurs in most natural feeds. Supplemental sources are magnesium oxide or salts of magnesium.

Magnesium is toxic when administered at high levels. The signs are lethargy, disturbance in locomotion, diarrhea, lowered feed intake and performance, and death. Toxicosis is not likely except by accident in mixing feeds or feeding animals. Certain levels of calcium and phosphorus in the diet protect the animals from toxicosis.

TABLE 22 Effects of Magnesium Administration in Animals

| Class and Number of of Animals[a] | Age or Weight | Administration | | | | Effect(s) | Reference |
|---|---|---|---|---|---|---|---|
| | | Quantity of Element[b] | Source | Duration | Route | | |
| Cattle—6, gestating beef cows | 13 yr | 0.05% | Natural feeds | 10 d | Diet | Decreased serum Mg | O'Kelley and Fontenot, 1973 |
| | | 0.11% | 55% from magnesium oxide | | | Decreased serum Mg | |
| | | 0.17% | 71% from magnesium oxide | | | Decreased serum Mg | |
| | | 0.29% | 83% from magnesium oxide | | | No adverse effect | |
| Cattle—6, lactating beef cows | 12 yr | 0.09% | Natural feeds | 10 d+ | Diet | Decreased serum Mg | O'Kelley and Fontenot, 1969 |
| | | 0.19% | 53% from magnesium oxide | | | No adverse effect | |
| | | 0.29% | 69% from magnesium oxide | | | No adverse effect | |
| | | 0.39% | 77% from magnesium oxide | | | No adverse effect | |
| Cattle—4, steers | 19 mo | 0.76% | Calcined magnesite | 8 d | Diet | No adverse effect | Care, 1960 |
| | | 1.15% | | 48 h | Diet and drench | Diarrhea | |
| | | 1.52% | | | | Diarrhea | |
| | | 2.28% | | | | Diarrhea | |

| | | | | | | | |
|---|---|---|---|---|---|---|---|
| Cattle—4 | 96 d, 81 kg | 1.3% | MgO | 14 d and 6 wk | Diet | Diarrhea; mucus in feces; lower feed intake and rate of gain | Gentry *et al.*, 1978 |
| | 160 d, 151 kg | 2.3% | | 14 d | | Diarrhea; mucus in feces; lower feed intake and rate of gain | |
| | 92 d, 86 kg | 4.3% | MgO | 14 d | | Diarrhea; mucus in feces; lower feed intake and rate of gain | |
| Cattle—3, steers | | 0.026 g/kg[c] | $MgSO_4 \cdot 7H_2O$ | | I.V. | Recumbent | Bowen *et al.*, 1970 |
| Cattle—2, steers | | 0.052 g/kg[c] (in two doses) | | | | Respiratory failure; cardiac arrest; death | |
| Sheep—12, wethers | 3–6 yr, 45–65 kg | 0.004 g/kg[c] | MgC1 | 16 mo | Water | Lower feed intake; diarrhea | Peirce, 1959 |
| | | 0.008 g/kg[c] | | 16 mo | | Lower feed intake; diarrhea | |
| Sheep—5 | 1 d | 0.54 g/kg[c] | $MgSO_4 \cdot 7H_2O$ | <10 min | Enema | Loss of deep tendon reflexes; deep sleep; cyanosis; death (1 of 5) | Andrews *et al.*, 1965 |
| | 1 d | 1.08 g/kg[c] | | | | Lack of reflexes; anaesthesia; cardio-respiratory depression; death (2 of 5); increased plasma Mg | |
| | 2–3 mo | 1.08 g/kg[c] | | | | Loss of reflexes; | |

TABLE 22 *Continued*

| Class and Number of Animals[a] | Age or Weight | Administration: Quantity of Element[b] | Source | Duration | Route | Effect(s) | Reference |
|---|---|---|---|---|---|---|---|
| | | | | | | sleeping; death (1 of 5); increased plasma Mg | |
| Sheep—1 | 1 yr+ | 0.18% | Natural feeds | 15 d | Diet | No adverse effect | Fontenot *et al.*, unpublished |
| | | 0.54% | 67% from magnesium oxide | | | Decreased rumen bacteria | |
| | | 0.90% | 80% from magnesium oxide | | | Decreased rumen bacteria and decreased feed intake | |
| | | 1.6% | 89% from magnesium oxide | | | Decreased rumen bacteria; decreased dry matter digestibility; dilation of duodenal glands | |
| Sheep—2 | 1 yr+ | 0.2% | Natural feeds | 18 d | Diet | No adverse effect | |
| | | 0.5% | 60% from magnesium oxide | | Diet and capsule | No adverse effect | |
| | | 1.9% | 89% from magnesium oxide | | | Decreased feed intake; diarrhea; higher serum Mg | |
| | | 5.1% | 96% from | | | Decreased feed | |

| | | | | | | | |
|---|---|---|---|---|---|---|---|
| | | | magnesium oxide | | | intake; higher serum Mg; diarrhea | |
| Chickens—180 | 1 d | 0.2% (added) | $MgCO_3$ | 4 wk | Diet | No adverse effect | Chicco *et al.*, 1967 |
| | | 0.4% (added) | | | | No adverse effect | |
| | | 0.6% (added) | | | | Decreased growth rate; lower bone calcification; higher bone Mg | |
| Chicken—24 | | 0.16% | Magnesium carbonate | 3 wk | Diet | No increased gain; lower mortality | Nugara and Edwards, 1963 |
| | | 0.32% | | | | Similar to 0.16% | |
| | | 0.64% | | | | Decreased gain and increased mortality, compared to 0.16% | |
| | | 1.28% | | | | Decreased gain and increased mortality, compared to 0.16% | |
| Horse—1 | | 0.028 g/kg[c] | $MgSO_4 \cdot 7H_2O$ | | I.V. | Recumbent posture | Bowen *et al.*, 1970 |
| Dog—8 | | 0.028 g/kg[c] | $MgSO_4 \cdot 7H_2O$ | | I.V. | Recumbent posture | Bowen *et al.*, 1970 |

[a] Number of animals per treatment.
[b] Quantity expressed as percent or as milligrams per kilograms of body weight.
[c] Values refer to milligrams per kilogram of body weight.

## REFERENCES

Allison, I. S. 1930. The problem of saline drinking waters. Science 71:559.

Ammerman, C. B., C. F. Chicco, P. E. Loggins, and L. R. Arrington. 1972. Availability of different salts of magnesium to sheep. J. Anim. Sci. 34:122.

Andrews, B. F., D. R. Campbell, and P. Thomas. 1965. Effects of hypertonic magnesium–sulphate enemas on newborn and young lambs. Lancet 2:64.

Bowen, J. M., D. M. Blackman, and J. E. Heavener. 1970. Effect of magnesium ions on neuromuscular transmission in the horse, steer and dog. J. Am. Vet. Med. Assoc. 157:164.

Care, A. D. 1960. The effect on cattle of high level magnesium supplementation of their diet. Vet. Rec. 72:517.

Chicco, C. F., C. B. Ammerman, P. A. van Walleghem, P. W. Waldroup, and R. H. Harms. 1967. Effects of varying dietary ratios of magnesium, calcium and phosphorus in growing chicks. Poult. Sci. 46:368.

Chicco, C. F., C. B. Ammerman, W. G. Hillis, and L. R. Arrington. 1972. Utilization of dietary magnesium by sheep. Am. J. Physiol. 222:1469.

Chicco, C. F., C. B. Ammerman, J. P. Feaster, and B. G. Dunavant. 1973. Nutritional interrelationships of dietary calcium, phosphorus and magnesium in sheep. J. Anim. Sci. 36:986.

Cragle, R. G. 1973. Dynamics of mineral elements in the digestive tract of ruminants. Fed. Proc. 32:1910.

Duncan, C. W., C. F. Huffman, and C. S. Robinson. 1935. Magnesium studies in calves. I. Tetany produced by a ration of milk or milk with various supplements. J. Biol. Chem. 108:35.

Fontenot, J. P., M. B. Wise, and K. E. Webb, Jr. 1973. Interrelationships of potassium, nitrogen and magnesium in ruminants. Fed. Proc. 32:1925.

Gentry, R. P., W. J. Miller, D. G. Pugh, M. W. Neathery, and J. B. Bynoum. 1978. Effects of feeding high magnesium to young dairy calves. J. Dairy Sci. 61:1750.

Gerken, H. J., Jr., and J. P. Fontenot. 1967. Availability and utilization of magnesium from dolomitic limestone and magnesium oxide in steers. J. Anim. Sci. 32:789.

Gitelman, H. J., S. Kukolj, and L. G. Welt. 1968. Inhibition of parathyroid gland activity by hypermagnesemia. Am. J. Physiol. 215:483.

Grace, N. D., M. J. Ulyatt, and J. C. Macrae. 1974. Quantitative digestion of fresh herbage by sheep. III. The movement of Mg, Ca, P, K and Na in the digestive tract. J. Agric. Sci. 82:321.

Kerk, P. V. D. 1973. Metabolic disorders in sheep and cattle caused by magnesium oxide in the concentrate feed. Tijdschr. Diergeneesk. 98:1166 (via Nutr. Abstr. Rev. 44:799).

Krider, J. L., J. L. Albright, M. P. Plumlee, J. H. Conrad, C. L. Sinclair, L. Underwood, R. G. Jones, and R. B. Harrington. 1975. Magnesium supplementation, space and docking effects on swine performance and behavior. J. Anim. Sci. 40:1027.

Larvor, P. 1976. $^{28}$Mg kinetics in ewes fed normal or tetany prone grass. Cornell Vet. 66:413

Massry, S. G., J. W. Coburn, and C. R. Kleeman. 1970. Evidence for suppression of parathyroid gland activity by hypermagnesemia. J. Clin. Invest. 49:1619.

Miller, E. R., D. E. Ullrey, C. L. Zutout, B. V. Baltzer, D. A. Schmidt, J. A. Hoefer, and R. W. Luecke. 1965. Magnesium requirement of the baby pig. J. Nutr. 85:13.

Moinuddin, J. F., and H. W. Lee. 1960. Alimentary, blood and other changes due to feeding $MnSo_4$, $MgSO_4$ and $Na_2SO_4$. Am. J. Physiol. 199:77.

Morris, E. R., and B. L. O'Dell. 1963. Relationship of excess calcium and phosphorus to magnesium requirement and toxicity in guinea pigs. J. Nutr. 81:175.

National Research Council. 1979. Nutrient Requirements of Domestic Animals. No. 2. Nutrient Requirements of Swine. National Academy of Sciences, Washington, D.C.

Newton, G. L., J. P. Fontenot, R. E. Tucker, and C. E. Polan. 1972. Effects of high dietary potassium intake on the metabolism of magnesium by sheep. J. Anim. Sci. 35:440.

Nugara, D., and H. M. Edwards, Jr. 1963. Influence of dietary Ca and P levels on the Mg requirement of the chick. J. Nutr. 80:181.

O'Kelley, R. E., and J. P. Fontenot. 1969. Effects of feeding different magnesium levels to drylot-fed lactating beef cows. J. Anim. Sci. 29:959.

O'Kelley, R. E., and J. P. Fontenot. 1973. Effects of feeding different magnesium levels to drylot-fed gestating beef cows. J. Anim. Sci. 36:994.

Peirce, A. W. 1959. Studies on salt tolerance of sheep. II. The tolerance of sheep for mixtures of sodium chloride and magnesium chloride in the drinking water. Aust. J. Agric. Res. 10:725.

Pike, R. L., and M. L. Brown. 1975. Nutrition: An integrated approach, 2nd ed. John Wiley & Sons, New York.

Pless, C. D., J. P. Fontenot, and K. E. Webb, Jr. 1973. Effect of dietary calcium and phosphorus levels on magnesium utilization in sheep. Va. Polytech. Inst. State Univ. Res. Div. Rep. 153:104.

Rayssiguier, Y., J. M. Garel, M. J. Prat, and J. P. Barlet. 1977. Plasma parathyroid hormone and calcitonin levels in hypocalcaemic magnesium deficient calves. Ann. Rech. Vet. 8:267.

Reid, R. L., A. J. Post, and G. A. Jung. 1970. Mineral composition of forages. W. Va. Univ. Agric. Exp. Stn. Bull. 589T.

Rook, J. A. F., and J. E. Storry. 1962. Magnesium in the nutrition of farm animals. Nutr. Abstr. Rev. 32:1055.

Thomas, J. W. 1965. Mechanisms responsible for grass tetany, p. 14. *In* Proc. Ga. Nutr. Conf. Feed Manuf.

Whittier, P. C., and R. M. Freeman. 1971. Potentiation of metastatic calcification in vitamin D-treated rats by magnesium. Am. J. Physiol. 220:209.

# Manganese

Manganese (Mn) is a steel gray lustrous metal that is hard and brittle. In chemical properties it is similar to iron; its two most important valence states in biological systems are II and III. Manganese constitutes 0.10 percent of the earth's crust and is the twelfth most abundant element. The most important manganese ore is pyrolusite; however, manganese occurs in hausmannite, manganite, manganosite, and braunite. Manganese is an important component of metallic nodules that are found on the ocean floors. Manganese is used in the manufacture of steel, cast iron, alloys of copper and aluminum, pigments for glass and ceramics, dry cell batteries, and a wide range of chemicals. In 1970, the tons of manganese ore (at least 35 percent manganese) used in the United States were as follows: manganese alloys and metals, 2,099,426; pig iron and steel, 107,733; and dry cells, chemicals, and miscellaneous, 156,778 (National Research Council, 1973). Since animals and plants require manganese, it is incorporated into poultry and livestock feeds and frequently into fertilizers.

Reviews cover various aspects of manganese function, metabolism, toxicity, and uses (National Research Council, 1973; Leach, 1974, 1976; Hurley, 1976; Matrone *et al.,* 1977; Underwood, 1977; and Leach and Lilburn, 1978).

## ESSENTIALITY

Manganese was first recognized as an essential nutrient for animals when it was shown to be required for growth and reproduction in rats and mice (Kemmerer *et al.*, 1931; Orent and McCollum, 1931). The principal signs of manganese deficiency in several species include reduced growth rate, skeletal abnormalities, abnormal reproductive function in males and females, and ataxia in the newborn. A susceptibility to convulsions has been observed in manganese deficient guinea pigs. Righting ability was related to abnormal otoliths, resulting from defective bone formation.

The manganese requirements vary considerably between species. In terms of dietary concentration (ppm), the requirements of young animals have been estimated as follows: dog, 4.5; rabbit, 8.5; pig, 4; calf, 40; sheep, 30; rat, 50; chick, 55; and turkey, 55.

## METABOLISM

Manganese is absorbed throughout the intestinal tract. The homeostatic mechanism for regulating tissue levels involves the excretion of manganese via bile and the intestine. The details of these processes are not understood. Manganese absorption is decreased by feeding isolated soy protein or excess levels of calcium, phosphorus, and iron. The movement of manganese within the body is highly specific for the element. Biogenic amines can influence the metabolism of manganese, apparently by effects of cyclic AMP.

The bone defects associated with manganese deficiency appear to be related to chondrogenesis rather than osteogenesis. Manganese is the preferred metal cofactor for a group of glucosyl transferases involved in mucopolysaccharide synthesis. There is evidence that manganese functions in carbohydrate and lipid metabolism and in the metabolism of the brain. The mitochondria generally contain high concentrations of manganese. Pyruvate carboxylase and superoxide dismutase are two important metalloenzymes that contain manganese.

## SOURCES

Underwood (1977) reviewed information on manganese content of forage and other plant sources. The manganese concentration can vary widely in relation to species, variety, type of soil, and manganese concentration in the soil. Typical values for various grasses, clover, etc., ranged from 60 to more than 800 ppm on a dry basis. As with many other elements, whole seeds contain significant concentrations of manganese (Schroeder *et al.*, 1966); however, refining processes remove much of the element. Typical values for whole cereal grains are 30–50 ppm and 30–40 ppm for soybean meal. Protein supplements obtained from animal sources are generally low in manganese, approximately 5–15 ppm.

The manganese content of surface waters collected at 140 U.S. sampling stations averaged 29.4 ppb between 1957 and 1969 (National Research Council, 1974). Minimum and maximum values were 0.20 and 3,230 ppb, and the mean was 29.4 ppb. Consumption of surface water by domestic animals would not contribute significantly on the average to the requirement; however, consumption of water containing the maximum manganese content could supply 300–600 percent of the requirement of cattle. The estimated maximum amount for swine was 20–40 percent of requirement, and for poultry the amount was negligible.

Ammerman and Miller (1972) reviewed information on bioavailability of manganese from various concentrated forms. Data with chicks showed that feeding reagent grade chemicals resulted in equal bioavailability of manganese from the sulfate, chloride, carbonate, and dioxide. Manganous oxide and manganous sulfate are the two most commonly used forms in animal feeds. The sulfate is used when greater ease of solubility is important.

## TOXICOSIS

The manganese requirements of the common domestic animals and fowl range from 20 to 55 ppm in the diet. The effects in five species of animals continuously fed excess levels of dietary manganese, ranging from 35 to 7,586 ppm, are summarized in Table 23. Data for some of the same and three additional species appear at the end of the table. For these, the experimental conditions differed from the bulk of available data. The basal diets fed to the animals were generally adequate but varied greatly in concentration of essential nutrients, other components, and even in manganese content. Measurements of varying sensitivity have been used to detect adverse effects; however, the most common has been growth of young animals.

### LOW LEVELS

The data in Table 23 show that feeding excess manganese at levels as high as 1,000 ppm produced a serious health problem in only 1 of 21 experimental groups, including several species. In one study, pigs fed

500 ppm manganese exhibited retarded growth, limb stiffness, and a stilted gait (Grummer *et al.,* 1950). It is possible that these adverse responses may have been related to other components of the manganese source, which contained 65 percent manganous sulfate, or to the composition of the diet. Similarly severe effects were not observed by other workers in pigs fed higher manganese levels.

Metabolic deviations from control animals have been shown with fairly low levels of excess manganese. These include slightly decreased copper absorption by the calf fed 50 ppm manganese above 12 ppm in the basal diet (Ivan and Grieve, 1976) and negative calcium balance during early lactation in cows fed 70 ppm manganese (Reid *et al.,* 1947). Increased fecal phosphorus was observed without other changes.

The observations at these levels of manganese, 1,000 ppm or less, were made with the rat, calf, cow, poultry, and pig. A few experiments were long-term, including egg production of chickens and reproduction in mammals. Similar effects were observed in sheep fed sequentially increasing amounts of manganese up to 2,500 ppm manganese (Hartman *et al.,* 1955).

### HIGH LEVELS

Beginning around 2,000 ppm excess manganese, significant adverse health effects, such as growth depression, were observed in some experiments. These and higher levels of manganese caused some mortality and decreased levels of hemoglobin. It is remarkable that in a 240-day study with rats growth and reproduction were normal with 4,990 ppm manganese, and only growth was adversely affected at 9,980 ppm manganese. Otherwise, 4,080 ppm manganese fed to poults is the highest level that had no effect on growth (Vohra and Kratzer, 1968). Effects of dietary manganese between 1,000 and 2,000 ppm have not been studied.

Cunningham *et al.* (1966) investigated the effect of feeding 5,000 ppm manganese to a rumen-fistulated cow. The manganese produced a marked change in rumen bacterial species. The *in vitro* production of propionic and total volatile fatty acids was depressed in flasks innoculated with flora from the manganese-fed cow as compared with a cow fed the basal diet. The suppressive effects of manganese added *in vitro* were also greater with the innoculum from the manganese-fed cow. These data suggest that at least part of the adverse effects of excess manganese fed to ruminants is due to effects on the rumen microflora.

Intubation of guinea pigs with a high daily dose of manganese, 4.37 mg/kg of body weight, produced some mortality and lesions of the gastrointestinal tract of survivors (Chandra and Imam, 1973). The lowest published lethal dose of manganese, as the sulfate, was 182 mg/kg for hamsters (Fairchild *et al.*, 1977). Young rabbits that were given large daily doses of manganese in drinking water lost weight and developed a transient paralysis and prolonged anesthesia of the extremities (Umarji *et al.*, 1969). It is surprising that neurological damage, which occurs in humans exposed to airborne manganese dusts such as ores, has not been observed more frequently in animals.

## FACTORS INFLUENCING TOXICITY

As noted above, excess manganese affected the metabolism of several elements. Generally a mineral antagonism is characterized by reciprocal effects: i.e., a deficiency of the second or antagonized element enhances the toxicity of the first element, and, conversely, an excess of the second element protects against toxicity of the first.

The primary antagonism of importance in manganese toxicity is the effect on iron. Low hemoglobin levels were found by several workers in animals fed excess manganese (Table 23). This anemia was accompanied by low levels of tissue iron and elevated levels of liver copper (Hartman *et al.*, 1955). Matrone *et al.* (1959) showed that excess manganese interfered with hemoglobin regeneration in rabbits and baby pigs. Data from two experiments suggested that the minimal level of excess manganese to depress hemoglobin regeneration in baby pigs was 125 ppm manganese or less. A supplement of 400 ppm iron in the diet completely counteracted the effect of 2,000 ppm manganese in depressing hemoglobin regeneration in baby pigs. Chandra and Tandon (1973) found that iron deficiency in rats increased manganese levels and pathology in the liver and kidneys when excess manganese was given orally.

With 14,000 ppm excess manganese in the diet of rats, Diez-Ewald *et al.* (1968) observed decreased liver iron stores and increased absorption of iron; however, they also found blood loss into the gastrointestinal tract. Under more physiological conditions, increased manganese absorption has been reported in iron deficiency (Borg and Cotzias, 1958; Pollack *et al.*, 1965; Diez-Ewald *et al.*, 1968). Thomson and Valberg (1972) showed that iron and manganese each interfered with the absorption of the other from perfusate in open-ended duodenal loops of iron-deficient rats. There is evidence that manganese can be incorporated *in vivo* into the porphyrin of red blood cells under conditions of iron deficiency (Borg and Cotzias, 1958).

These diverse studies demonstrate the importance of iron status in modifying manganese toxicity. Much remains to be learned about the nature of the interactions and the practical relation of iron status, ranging from deficiency to excess, in defining resistance to the spectrum of excess manganese intakes.

The effect of supplemental form on toxicity has not been systematically investigated; however, it appears that inorganic salts have similar effects. Abnormalities of reproduction in dairy cattle have been associated with 200 ppm or higher concentrations of manganese in forages of Costa Rica (Fonseca and Davis, 1969). Tentatively, 100 ppm or more of manganese in forage was designated as high. The toxicity of manganese in forage merits detailed study.

## TISSUE LEVELS

Underwood (1971, 1977) reviewed data on concentrations of manganese in animal tissues. The liver and pituitary contain the highest concentrations, each with approximately 2.5 ppm. The manganese in hair, wool, and feathers reflects dietary levels from deficiency to excess in a more sensitive manner than organs or internal tissues. Feeding manganese in the diet of laying hens at 13 and 1,000 ppm resulted in total egg yolk values of 4 and 33 $\mu$g, respectively. In general, tissue manganese remains relatively constant over a wide range of intakes (Cotzias, 1958). The efficient homeostatic mechanisms for eliminating excess manganese from the body would seem to preclude significant accumulation in tissues of domestic animals (Watson *et al.,* 1973). Doyle and Spaulding (1978) have summarized the data on manganese content in liver, kidney, heart, and muscle in normal cattle, sheep, swine, and chickens.

## MAXIMUM TOLERABLE LEVELS

Levels of 1,000 ppm manganese produced some metabolic deviations from normal but almost no effects on growth or other indications of toxicosis in most experiments. Whether the metabolic changes would become threats to health on a long-term basis would probably depend on the diet composition, age, or physiological status of the animal and on the mechanism of the adverse effect.

The data with iron deficiency show clearly that small amounts of excess manganese, as low as 125 ppm, are undesirable. Most studies of manganese toxicity were carried out with stock diets. Whereas these have practical significance, they make it impossible to assess the exact nutrient composition of each diet, which may explain apparent discrepancies between adverse effects and dose level, particularly at 2,000 ppm manganese and above. Most manganese toxicity studies were carried out many years ago. Due to changes in diet formulation and genetic characteristics of domestic animals, the old results may not be entirely applicable now.

With a well-balanced, adequate diet, it appears that 1,000 ppm dietary manganese is the maximum tolerable level, at least under short-term conditions, for cattle and sheep and 2,000 ppm for poultry. Some data indicated a greater sensitivity of swine, so the maximum tolerable level was set at 400 ppm.

## SUMMARY

Manganese is an essential element for animals and plants. It functions in mucopolysaccharide synthesis and carbohydrate and lipid metabolism. A variety of bone disorders, retarded growth, and reproductive failure have been observed in manganese-deficient animals. Water-soluble salts of manganese are readily available to meet the animal's needs.

In general, adverse health effects have not occurred in most species with dietary concentrations of 1,000 ppm manganese or less, although some metabolic alterations have occurred. These do not appear serious and probably would not occur in animals receiving a well-balanced adequate diet. Swine appear to be more sensitive to manganese than cattle, sheep, or poultry. At 2,000 ppm and above, growth retardation, anemia, gastrointestinal lesions, and sometimes neurological signs have been observed. Many studies have been reported in which no adverse effects were observed at high levels of manganese intake. Manganese and iron are mutually antagonistic. With low iron intake, animals are much more sensitive to manganese toxicity; conversely, excess iron is protective. The tissues, apart from skin, hair, and feathers, do not accumulate large amounts of the element. Homeostatic mechanisms maintain most tissue manganese concentrations within fairly narrow limits, primarily by excretion of excess manganese via bile or the small intestine.

## TABLE 23 Effects of Manganese Administration in Animals

| Class and Number of Animals[a] | Age or Weight | Administration: Quantity of Element[b] | Source | Duration | Route | Effect(s) | Reference |
|---|---|---|---|---|---|---|---|
| Cattle—8 | Young | 50 ppm | Manganous sulfate | 11 wk | Diet | Decreased net absorption of Cu, primarily in the large intestine | Ivan and Grieve, 1976 |
| Cattle—8 | Young | 250 ppm | Manganese sulfate | | Diet | Increased fecal P; positive P and Ca balances | Gallup *et al.*, 1952 |
| | | 500 ppm | | | | Increased fecal P; positive Ca and P balances | |
| | | 1,000 ppm | | | | Increased fecal Ca | |
| | | 2,000 ppm | | | | Increased fecal Ca | |
| Cattle—7 | Young | 820 ppm | Manganese sulfate | 84 d | Diet | No adverse effect | Cunningham *et al.*, 1966 |
| | | 2,460 ppm | | | | Decreased growth and feed intake | |
| | | 4,920 ppm | | | | Decreased growth and feed intake | |
| Cattle—6 | Young | 1,000 ppm | Manganese sulfate | 100 d | Diet | No adverse effect | |
| | | 2,000 ppm | | | | No adverse effect | |
| | | 3,000 ppm | | | | Decreased hemoglobin | |
| Cattle—4 | Adult | 70 ppm | $MnSO_4 \cdot H_2O$ | Three 1-wk balances | Diet | Negative Ca balance; decreased Ca utilization | Reid *et al.*, 1947 |
| Cattle—5 | Adult | 75 ppm[c] | Manganous sulfate | 4.5 mo | Diet | No adverse effect | Fain *et al.*, 1952 |
| | | 100 ppm | | | | Transient decrease in | |

TABLE 23 *Continued*

| Class and Number of Animals[a] | Age or Weight | Administration: Quantity of Element[b] | Source | Duration | Route | Effect(s) | Reference |
|---|---|---|---|---|---|---|---|
| | | | | | | serum Mg | |
| | | 150 ppm | | | | No adverse effect | |
| | | 200 ppm | | | | No adverse effect | |
| Sheep—4 | Young | 15 ppm[d] | $MnSO_4 \cdot H_2O$ | 12 wk | Diet | No adverse effect | Hartman *et al.*, 1955 |
| | | 30 ppm | | 8 wk | | No adverse effect | |
| | | 450 ppm | | 4 wk | | No adverse effect | |
| | | 2,500 ppm | | 11 wk | | Decreased Fe in liver, kidney, and spleen | |
| | | 45 ppm[d] | | 12 wk | | Decreased hemoglobin and serum Fe by 6 wk | |
| | | 90 ppm | | 12 wk | | Decreased hemoglobin and serum Fe | |
| | | 900 ppm | | 4 wk | | Decreased hemoglobin | |
| | | 5,000 ppm | | 11 wk | | Decreased hemoglobin and Fe in serum, liver, kidney, and spleen; increased Cu in liver | |
| Sheep—4 | 34 kg | 4,000 ppm | Manganese carbonate | 11 wk | Diet | Increased Mn in seven tissues, Cu in liver and bone; decreased Fe and Zn in liver and retention of $^{54}$Mn administered after feeding high Mn | Watson *et al.*, 1973 |
| Swine—8 | Young | 500 ppm | Manganous sulfate, 65% | | Diet | Decreased food intake and growth; limb stiffness; | Grummer *et al.*, 1950 |

| | | | | | | | |
|---|---|---|---|---|---|---|---|
| | | | | | | stilted gait | |
| Swine—6 | Young | 75 ppm | Manganese sulfate | 2–8 wk | Diet | No adverse effect | Leibholz *et al.*, 1962 |
| | | 225 ppm | | | | No adverse effect | |
| | | 675 ppm | | | | Small increase bone Mn | |
| | | 2,025 ppm | | | | Decreased growth and hemoglobin; small increase bone Mn | |
| Swine—15 | Young | 400 ppm | | 2–12 wk | Diet | Ca. 30% increase hair Mn | |
| | | 4,000 ppm | | | | Decreased growth; eightfold increase hair Mn | |
| Chicken—25 | Young | 4,779 ppm | $MnCO_3$ | | Diet | Decreased growth, 52% mortaility | Heller and Penquite, 1937 |
| Chicken—25 | Young | 50 ppm | $MnCO_3$ | 20 wk | Diet | No adverse effect | Gallup and Norris, 1939 |
| | | 500 ppm | | | | No adverse effect | |
| | | 1,000 ppm | | | | No adverse effect | |
| Turkey—2 | Young | 35 ppm | $MnCO_3$ | Diet | | No adverse effect | Mussehl and Ackerson, 1939 |
| | | 350 ppm | | | | No adverse effect | |
| | | 35 ppm | Manganese sulfate | | | No adverse effect | |
| | | 350 ppm | | | | No adverse effect | |
| Turkey—10 | Young | 510 ppm | $MnSO_4 \cdot H_2O$ | 21 d | Diet | No adverse effect | Vohra and Kratzer, 1968 |
| | | 1,020 ppm | | | | No adverse effect | |
| | | 2,040 ppm | | | | No adverse effect | |
| | | 3,000 ppm | | | | No adverse effect | |
| | | 3,060 ppm | | | | No adverse effect | |
| | | 3,620 ppm | | | | No adverse effect | |

## TABLE 23 *Continued*

| Class and Number of Animals[a] | Age or Weight | Administration: Quantity of Element[b] | Source | Duration | Route | Effect(s) | Reference |
|---|---|---|---|---|---|---|---|
| | | 4,080 ppm | | | | No adverse effect | |
| | | 4,800 ppm | | | | Decreased growth | |
| Rabbit—2 | Young | 2.3 mg/d[e] | Manganese sulfate | 500 d | Water[f] | Small decrease in growth; four-fold increase in hair Mn by 100 d | Umarji *et al.*, 1969 |
| | | 24.4 mg/d | | | | Weight loss by 180 d with transient paralysis and continuing anesthesia in extremeties; increased Mn in hair (10-fold) by 100 d, subsequently declined with weight loss | |
| Rat | Young | 499 ppm | $MnCl_2 \cdot 4H_2O$ | 240 d | Diet | No adverse effect | Becker and McCollum, 1938 |
| | | 998 ppm | | | | No adverse effect | |
| | | 2,495 ppm | | | | No adverse effect | |
| | | 4,990 ppm | | | | No adverse effect | |
| | Young | 9,980 ppm | $MnCl_2 \cdot 4H_2O$ | 240 d | Diet | Depressed growth; no effect on reproduction | |
| Rat—6 | Young | 475 ppm[g] | $MnSO_4 \cdot H_2O$ | 9 d | Diet | No adverse effect | Moinuddin and Lee, 1960 |
| | | 950 ppm | | 7 d | | No adverse effect | |
| | | 1,900 ppm | | 8 d | | No adverse effect | |
| | | 3,800 ppm | | 4 d | | No adverse effect | |
| | | 7,586 ppm | $MnSO_4 \cdot H_2O$ | 4 wk | Diet | Decreased feed intake, feed | |

| | | | | | | | |
|---|---|---|---|---|---|---|---|
| | | | | | | efficiency, body weight, hemoglobin, serum P; loss of incisor pigment; increased water consumption, urine volume, and RBC count | |
| Guinea pig—30 | 350 g | 4.37 mg/kg | $MnCl_2$ | 30 d | Gavage | Six deaths; gastric mucosa: patchy necrosis; decreased mucin and pepsinogen granules; activities for adenosine triphosphatase and glucose-6-phosphatase; intestinal mucosa: patchy necrosis; decreased activities for adenosine triphosphatase and glucose-6-phosphatase; increased acid phosphatase | Chandra and Imam, 1973 |
| Hamster | | 182 mg/kg | $MnSO_4$ | 1 dose | Gavage | $LDL_0$ | Fairchild *et al.*, 1977 |

[a] Number of animals per treatment.
[b] Quantity expressed as parts per million (concentration in diet) or as milligrams per kilogram of body weight. All amounts are for the total added to the basal diet.
[c] Levels were fed sequentially over 4.5 months to the same cows.
[d] Each series of Mn levels was fed sequentially to the same lambs.
[e] Assumed $MnSO_4 \cdot H_2O$.
[f] Drinking water.
[g] Levels 475 through 3,800 ppm were fed sequentially to the same rats.

## REFERENCES


Ammerman, C. B., and S. M. Miller. 1972. Biological availability of minor mineral ions: A review. J. Anim. Sci. 35:681.

Becker, J. E., and E. V. McCollum. 1938. Toxicity of $MnCl_2 \cdot 4H_2O$ when fed to rats. Proc. Soc. Exp. Biol. Med. 38:740.

Borg, D. C., and G. C. Cotzias. 1958. Incorporation of manganese into erythrocytes as evidence for a manganese porphyrin in man. Nature 182:1677.

Chandra, S. V., and Z. Imam. 1973. Manganese induced histochemical and histological alterations in gastrointestinal mucosa of guinea pigs. Acta Pharmacol. Toxicol. 33:449.

Chandra, S. V., and S. K. Tandon. 1973. Enhanced manganese toxicity in iron deficient rats. Environ. Physiol. Biochem. 3:230.

Cotzias, G. C. 1958. Manganese in health and disease. Physiol. Rev. 38:503.

Cunningham, G. N., M. B. Wise, and E. R. Barrick. 1966. Effect of high dietary levels of manganese on the performance and blood constituents of calves. J. Anim. Sci. 25:532.

Diez-Ewald, M., L. R. Weintraub, and W. H. Crosby. 1968. Interrelationship of iron and manganese metabolism. Proc. Soc. Exp. Biol. Med. 129:448.

Doyle, J. J., and J. E. Spaulding. 1978. Toxic and essential trace elements in meats—A review. J. Anim. Sci. 47:398.

Fain, P., J. Dennis, and F. G. Harbaugh. 1952. The effect of added manganese in feed on various mineral components of cattle blood. Am. J. Vet. Res. 13:348.

Fairchild, E. J., R. J. Lewis, and R. L. Tatkin (eds.). 1977. Registry of Toxic Effects of Chemical Substances, vol. 2, p. 524. DHEW Publ. No. (NIOSH) 78-104-B.

Fonseca, H. A., and G. K. Davis. 1969. Manganese content of some forage crops in Costa Rica and its relation to cattle fertility, p. 371. *In* Proc. 2nd World Conf. Anim. Prod.

Gallup, W. D., and L. C. Norris. 1939. The amount of manganese required to prevent perosis in the chick. Poult. Sci. 18:76.

Gallup, W. D., J. A. Nance, A. B. Nelson, and A. E. Darlow. 1952. Forage manganese as a possible factor affecting calcium and phosphorus metabolism of range beef cattle. J. Anim. Sci. 11:783.

Grummer, R. H., O. G. Bentley, P. H. Phillips, and G. Bohstedt. 1950. The role of manganese in growth, reproduction and lactation in swine. J. Anim. Sci. 9:170.

Hartman, R. H., G. Matrone, and G. H. Wise. 1955. Effects of high dietary manganese on hemoglobin formation. J. Nutr. 57:429.

Heller, V. G., and R. Penquite. 1937. Factors producing perosis in chickens. Poult. Sci. 16:243.

Hurley, L. S. 1976. Manganese and other essential elements, p. 345. *In* Present Knowledge in Nutrition, 4th ed. The Nutrition Foundation, Inc., New York.

Ivan, M., and C. M. Grieve. 1976. Effects of zinc, copper and manganese supplementation of high-concentrate ration on gastrointestinal absorption of copper and manganese in Holstein calves. J. Dairy Sci. 59:1764.

Kemmerer, A. R., C. A. Elvehjem, and E. B. Hart. 1931. Studies on the relation of manganese to the nutrition of the mouse. J. Biol. Chem. 92:623.

Leach, R. M., Jr. 1974. Biochemical role of manganese, p. 51. *In* W. G. Hoekstra, J. W. Suttie, H. E. Ganther, and W. Mertz, eds. Trace Element Metabolism in Animals—2. University Park Press, Baltimore, Md.

Leach, R. M., Jr. 1976. Metabolism and function of manganese, p. 235. *In* A. S. Prasad (ed.). Trace Elements in Human Health and Disease, vol. II. Academic Press, New York.

Leach R. M., and M. S. Lilburn. 1978. Manganese metabolism and its function. World Rev. Nutr. Dietet. 32:123.

Leibholz, J. M., V. C. Speer, and V. W. Hays. 1962. Effects of dietary manganese on baby pig performance and tissue manganese levels. J. Anim. Sci. 21:772.

Matrone, G., R. H. Hartman, and A. J. Clawson. 1959. Studies of a manganese–iron antagonism in the nutrition of rabbits and baby pigs. J. Nutr. 67:309.

Matrone, G., E. A. Jenne, J. Kubota, I. Mena, and P. M. Newberne. 1977. Manganese, p. 29. *In* Geochemistry and the Environment, vol. II. The Relation of Other Selected Trace Elements to Health and Disease. National Academy of Sciences, Washington, D.C.

Moinuddin, J. F., and H. W. T. Lee. 1960. Alimentary, blood and other changes due to feeding $MnSO_4$, $MgSO_4$, and $Na_2SO_4$. Am. J. Physiol. 199:77.

Mussehl, F. E., and C. W. Ackerson. 1939. The effect of adding manganese to a specific ration for growing poults. Poult. Sci. 18:408. (Abstr.)

National Research Council. 1973. Medical and Biological Effects of Environmental Pollutants. Manganese. National Academy of Sciences, Washington, D.C.

National Research Council. 1974. Nutrients and Toxic Substances in Water for Livestock and Poultry. National Academy of Sciences, Washington, D.C.

Orent, E. R., and E. V. McCollum. 1931. Effects of deprivation of manganese in the rat. J. Biol. Chem. 92:651.

Pollack, S., J. N. George, R. C. Reva, R. M. Kaufman, and W. H. Crosby. 1965. The absorption of nonferrous metals in iron deficiency. J. Clin. Invest. 44:1470.

Reid, J. T., K. O. Pfau, R. L. Salisbury, C. B. Bender, and G. M. Ward. 1947. Mineral metabolism studies in dairy cattle. I. The effect of manganese and other trace elements on the metabolism of calcium and phosphorus during early lactation. J. Nutr. 34:661.

Schroeder, H. A., J. J. Balassa, and I. H. Tipton. 1966. Essential trace metals in man: Manganese. A study in homeostasis. J. Chron. Dis. 19:145.

Thomson, A. B. R., and L. S. Valberg. 1972. Intestinal uptake of iron, cobalt and manganese in the iron-deficient rat. Am. J. Physiol. 223:1327.

Umarji, G. M., K. G. Anantanarayan, and R. A. Bellare. 1969. Content of manganese in rabbit hair in the course of oral chronic administration of manganese sulfate. C. R. Soc. Biol. 162:1725.

Underwood, E. J. 1971. Trace Elements in Human and Animal Nutrition, 3rd ed. Academic Press, New York.

Underwood, E. J. 1977. Trace Elements in Human and Animal Nutrition, 4th ed. Academic Press, New York.

Vohra, P., and F. H. Kratzer. 1968. Zinc, copper and manganese toxicities in turkey poults and their alleviation by EDTA. Poult. Sci. 47:699.

Watson, L. T., C. B. Ammerman, J. P. Feaster, and C. E. Roessler. 1973. Influence of manganese intake on metabolism of manganese and other minerals in sheep. J. Anim. Sci. 36:131.

# Mercury

Mercury (Hg), a heavy silver-white element also called quicksilver, was named for Mercury, the Roman god of commerce and gain. It is the only metallic element found in the liquid phase at normal room temperature. Cinnabar and calomel, the principal ores of mercury, have been mined for 2,300 years (D'Itri, 1971). The most important industrial uses for the metal involve electrical apparatus, chloralkali production, water-base paint, and agricultural fungicides.

Mercury toxicosis has received considerable attention recently because of poisonings that have occurred in the human population (Curley *et al.,* 1971; Won, 1973; Clarkson *et al.,* 1976). Animals can be exposed to mercury contamination from air, soil, and water, as well as from that which may be ingested with feed. The concentration of mercury in the environment is, in part, the result of waste products from manufacturing processes that utilize mercury or of the disposal of products containing mercury. Fossil fuel combustion (Billings and Matson, 1972), smelting of commercial ores, and agricultural fungicides also contribute mercury to the environmental burden (D'Itri, 1971). Recent reviews on mercury include Underwood (1977), D'Itri (1971), and Nelson *et al.* (1971).

## ESSENTIALITY

Based on present evidence, mercury is not considered an essential element for living organisms (Underwood, 1977).

## METABOLISM

Metabolism of mercury has been reviewed by MacGregor and Clarkson (1974). Living organisms can concentrate mercury when an excess is available but the level of concentration depends on the type of organism and the form of mercury contamination (D'Itri, 1971). Inorganic mercury can be biomethylated before or after ingestion (Gage, 1975); consequently, it is necessary to discuss metabolism of the element in both forms. The radical, in which mercury is attached directly to a carbon atom, is the organic form (MacGregor and Clarkson, 1974). The mercuric ion ($Hg^{2+}$) is a potential *in vivo* metabolite of any of the major mercurials (Hammond, 1973).

The short-chain alkylmercurials are more readily absorbed and thus more toxic than other mercurial compounds when ingested. Degree of toxicity decreases with increased length of the carbon chain. Alkylmercurials are more toxic to living organisms because they are more stable biologically than other forms of mercury (70-day biological half-time) and resist degradation to inorganic mercury, which can be eliminated from the body. Also, the alkylmercurials can readily cross the blood–brain barrier and attack the central nervous system (D'Itri, 1971). Inorganic mercury follows metabolic pathways similar to those of zinc and cadmium, but interrelationships between the mercuric ion and other trace elements have not been investigated to any extent. Elemental mercury vapor is more likely to cross the blood–brain barrier than inorganic mercury salts (Magos, 1968), but only traces (30 ppb) of either appear in the fetus, milk, or eggs (Clarkson *et al.,* 1973).

Since organic mercurials are more highly lipid-soluble, the organic mercury salts are absorbed more completely than inorganic mercury salts. Methylmercury is lipid-soluble, and absorption is 60 to 100 percent of intake for all species studied. Clarkson (1970) reported that only 2 percent of ingested inorganic mercuric compounds was absorbed, while, in contrast, Fitzhugh *et al.* (1950) found that rats can absorb 50 percent of oral mercuric acetate. The mercuric ion was absorbed at 15 percent or less of intake with highest tissue accumulation occurring in the liver and kidney (Ellis and Fang, 1967). Methylmercury was absorbed more readily (15 to 35 times) than inorganic mercury from ligated segments of rat gut. The relative order of methylmercury absorption was as follows: duodenum > stomach = ileum > jejunum. There was no difference in absorption of inorganic mercury among sections of the intestinal tract (Sasser *et al.,* 1978). Studies by Rubenstein and Soares (1979) based on intestinal wall mercury concentration indicated that the upper small intestine appeared to be the site of preferential absorption of $CH_3Hg^+$, while $Hg^{2+}$ was absorbed in the ileum and lower intestine in broiler chicks.

Following ingestion, methylmercury was distributed widely in all tissues, including muscle, fetus, milk, eggs, hair, and feathers, but the concentrations were all below those of the target organs, brain, liver, and kidney. More than 97 percent of the total $^{203}Hg$ in egg white from chickens fed 20 ppm mercury as $CH_3HgCl$ was associated with ovalbumin (Magat and Sell, 1979). Phenyl- and ethylmercuric salts tend to dissociate more rapidly than methylmercuric salts *in vivo* (Clarkson *et al.*, 1973). When methylmercuric chloride was double-labeled and fed to rats, the $^{203}Hg$–$^{14}C$ bond was cleaved *in vivo*, 8 percent in kidney and 6 percent in liver and brain (Garcia *et al.*, 1974b).

Ingested labeled $^{203}Hg^{2+}$ accumulated in the rumen wall of calves (Ansari *et al.*, 1973), suggesting a rapid attachment of $Hg^{2+}$ to tissue protein of the gastrointestinal tract. Labeled $CH_3Hg^+$ did not exhibit this property (Neathery and Miller, 1975). Apparent absorption of either form appears to be lower in ruminants than in nonruminants, but the biological half-time retentions for the ruminant (78 to 88 and 22 days, respectively, for $HgCl_2$ and $CH_3HgCl$) were similar to those estimated for some other species (Norseth and Clarkson, 1970; Skerfving, 1972, 1974; Khan, 1974; Hollins *et al.*, 1975; Sell and Davidson, 1975). Effects of mercury on ruminal microflora or epithelium appear not to have been investigated.

Mercuric ions initially enter the serum fraction of blood and organic mercury enters the erythrocytes (Alberg *et al.*, 1969), but with time the distribution becomes similar. The major mercury excretion route is fecal, regardless of mercurial form. With repeated exposure even to low mercury levels, the rate of mercury accretion may exceed the rate of excretion, resulting in tissue accumulation of the metal and eventual impairment of physiological function. Methylmercury is not continuously accumulated if given over a long period of time. A steady state is reached where excretion equals intake. Time required to reach the steady state depends on the half-time and species involved. In rats, all tissues except hair showed saturation kinetics with repeated intake of methylmercury (Salvaterra *et al.*, 1975). Thiol groups in tissue proteins have a high affinity for mercurials, and a major excretory mechanism appears to be via extrusion of intestinal epithelial cells (Norseth and Clarkson, 1971).

Calves were given a single intravenous dose of mercury as $CH_3{}^{203}HgCl$ or $^{203}HgCl_2$ (Stake *et al.*, 1975). Mercury excretion in total feces and urine was higher for $HgCl_2$ (28.3 versus 8.1 percent of the dose) than for $CH_3{}^{203}HgCl$. Retention of $^{203}Hg$ from $HgCl_2$ was 2–8 times greater in kidney, liver, spleen, lung , bone, serum, and intestine but lower in brain, muscle, heart, and red blood cells than that from $CH_3HgCl$. Biliary excretion of mercury as $^{203}HgCl_2$ or $CH_3{}^{203}HgCl$ following intravenous injection is a minor route. Over a 2-hour period, less than 0.5 percent of the amount administered (0.03, 0.1, 0.3, 1.0, and 3.0 mg mercury per kilogram of body weight), regardless of dose or form of mercury, was excreted in the bile of rats. Pretreatment with pregnenolone −16 α-carbonitrile (PCN) doubled the amount of mercury excreted in bile (Klassen, 1975a). With time or increased dosage, this compound was reabsorbed completely within 1 hour in rats (Norseth and Clarkson, 1970, 1971; Tichy *et al.*, 1975) and resulted in redistribution of methylmercury in the animal. Only one-half the mercuric ion in higher-molecular-weight biliary proteins was reabsorbed in rats 1 hour after dosing. However, in studies with broiler chicks (Rubenstein and Soares, 1979), bile served as an important route of mercury excretion for $HgCl_2$ but a less important route for $CH_3HgCl_2$. These studies seem to contradict those of Norseth (1974), in which the organomercurials were more readily excreted into bile. The difference may be the mode of administration. Methylmercury glutathione and methylmercury cysteine are the major mercury components in bile following injection of $CH_3HgCl$ (Norseth and Clarkson, 1971), but, when this mercury source is given orally, $Hg^{2+}$ appears as the major form in bile (Berlin *et al.*, 1975).

Formation of metallothioneins has been suggested as a means of detoxification for inorganic cadmium and mercury (Shaikh *et al.*, 1973; MacGregor and Clarkson, 1974; Piotrowski *et al.*, 1974a,b) but has not appeared to affect detoxification of methylmercury (Chen *et al.*, 1973, 1975a). The metalloproteins appear to have a limit of elemental saturability (Cousins, 1974; Colucci *et al.*, 1975). The amount of mercury required to saturate the sites was 300–500 mg mercury and corresponded to levels required to elicit frank nephrotoxicity (MacGregor and Clarkson, 1974). Although inorganic mercury is absorbed at 5 percent or less of intake, it is sequestered preferentially in tissue proteins in the kidney, liver, and gastrointestinal tissue. With repeated low dosage, rate of absorption can exceed the excretory capacity, and, at tissue elemental saturation, extrusion of damaged cell proteins bearing the toxic element may occur. The potential capacity for biomethylation or demethylation of inorganic microelements *in vivo* appears to influence the degree and mode of toxicity.

Metallothioneins loaded with "abnormal" metals apparently are incapable of significant turnover in tissue cytoplasm, and this fact suggests that high concentrations of "abnormal" metals in association with metallothionein in target tissues are due to the apparent absence of turnover (Mills, 1974; Chen *et al.*, 1974, 1975b) rather than to increased *de novo* synthesis (Fowler and Nordberg, 1975; Nordberg and Nordberg, 1975; Shaikh and Smith, 1975; Webb, 1975).

## SOURCES

The more important sources of mercury under practical feeding conditions would be fish protein concentrates and contaminated seed grain, which may be used accidentally. Fish concentrate methylmercury by ingestion of contaminated food, as well as by direct uptake from the water. Johnels (1967) reported that the biological magnification of each organism appears to be a function of its metabolic rate and that pike have exhibited muscle mercury levels that were 3,000 times greater than the level of the water from which they were taken. Rucker and Amend (1969) found that rainbow trout exposed to water containing 60 ppb of methylmercury daily for 1 hour over 10 days had mercury levels of 4,000 and 17,300 ppb (dry basis) in muscle and kidney, respectively. Fish taken from Korean waters contained from 20 to 580 ppb (fresh basis) (Won, 1973), while walleye and pike from Ball Lake, Ontario, contained 3.24 and 5.55 ppm mercury on a wet tissue basis (Annett *et al.*, 1975).

There are several citations of poisoning in human populations from treated seed grains (Haq, 1963; Ordonez *et al.*, 1966; Curley *et al.*, 1971; Bakir *et al.*, 1973; Clarkson *et al.*, 1976).

The mercury content of cow's milk can range from 3 to 10 ppb (Mullen *et al.*, 1975; Roh *et al.*, 1975). At 24 days following an 8-day exposure, goat's milk had 1.22 and 0.22 percent of total oral dosages, respectively, of organic and inorganic mercury (Sell and Davidson, 1975). All mercury from either source was in milk proteins. Similar results were reported for $CH_3Hg^+$ in milk from rats (Garcia *et al.*, 1974a,b).

Mercury concentration in hair and feathers of animals has been highly correlated with tissue turnover of mercury (Nelson *et al.*, 1971; Herigstad *et al.*, 1972; Skerfving, 1972; Huckabee *et al.*, 1973), and these tissues are major excretion routes that should be included in estimates of mercury retention (Hollins *et al.*, 1975). Accumulation of mercury in hair and feathers could cause contamination of processed hair and feather meals used as protein supplements for livestock.

## TOXICOSIS

Mercury toxicosis has been reviewed by Mills (1974), Buck (1975), Neathery and Miller (1975), Ammerman *et al.* (1977), and Gruber *et al.* (1978). Accumulation of metalloproteins during mercury intoxication to intolerable tissue threshold levels could be responsible for renal cortical tubular epithelial damage and subsequent renal failure (Fowler, 1972). Epithelial damage has been reported also in intestinal walls following mercury dosage to rats and swine (Norseth and Clarkson, 1971; Piper *et al.*, 1971). Lysosomes were the major sites of renal tubular mercury deposition in rats fed $HgCl_2$ or $CH_3HgOH$ (Fowler *et al.*, 1975; Madsen and Christensen, 1975). Ingestion of 20 ppm mercury by rats appeared to result in exocytosis of the lysosomal material into the tubular lumen. These effects were thought to be involved in the appearance of urinary protein. Mercurials have a high affinity for sulfhydryl groups altering SH-containing molecules (MacGregor and Clarkson, 1974).

### LOW LEVELS

The onset of chronic mercury toxicosis is variable and slow. The manifestations include dysfunction of the central nervous, digestive, genitourinary, respiratory, and muscular systems, as well as skin and visual problems (D'Itri, 1971). Daily consumption of methylmercury at 0.1 mg mercury per kilogram of body weight was tolerated by 4-week-old calves for 90 days, but 0.2–0.4 mg/kg produced methylmercury toxicosis in 75 days (Herigstad *et al.*, 1972). Methylmercury dicyandiamide was toxic for cattle and sheep at 0.225 mg mercury per kilogram of body weight. Animals displayed signs of incoordination and unsteady gait within 40 to 60 days (Wright *et al.*, 1973).

Tryphonas and Nielsen (1970) fed pigs phenylmercuric chloride for 90 days and observed no problems with 0.19 mg mercury per kilogram of body weight; but increased tissue accumulation of mercury occurred with 0.38 and 0.76 mg/kg, while 2.28 and 4.56 mg mercury per kilogram resulted in weight loss and kidney and colon necrosis.

With 0.075 or 0.150 mg mercury as methylmercury dicyandiamide per kilogram of body weight, chickens showed increased mercury accumulation in tissues (Wright *et al.*, 1973). Miller *et al.* (1967) fed mercuric chloride and phenylmercuric acetate to day-old chicks at 2 or 20 ppm mercury. Feeding 2 ppm as either form produced mercury accumulation in the liver and kidney in 20 days; however, 20 ppm mercury produced the same result in 5 days. Laying hens given 10 ppm mercury as $CH_3HgCl$ for 70 days accumulated 55 percent of the mercury in the eggs with 80 percent of that amount associated with the albumin (Sell *et al.,* 1974). Methylmercury dicyandiamide (33 ppm mercury) produced a death rate in 30 days of 90 percent in pheasants, 85 percent in ducks, and 7.5 percent in chickens (Gardiner, 1972). Daily consumption of 1 ppm mercury as mercuric or methylmercuric chloride by mice for life did not affect health and longevity, but 5 ppm was toxic (Schroeder and Mitchener, 1975). Daily consumption of greater than 1 mg mercury per kilogram of body weight as alkylmercurials was toxic for rats and young swine (Tryphonas and Nielsen, 1970, 1973; Khera and Tabacova, 1973).

Adult mink were not affected by 0.1 ppm mercury as $CH_3HgCl$ for 93 days, but 1.1 ppm increased tissue mercury and levels from 1.8 to 15 ppm proved lethal (Wobeser *et al.,* 1976b). Sperm from steelhead trout exposed to 1 ppm or greater mercury as $CH_3HgCl$ showed decreased ability to fertilize eggs (McIntyre, 1973). Mercury in the axial muscle of large benthopelagic fish taken at 2,500 m deep ranged from 0.03 to 0.76 ppm on a wet weight basis (Barber *et al.,* 1972). Tuna and swordfish have been found to have tissue levels of mercury in excess of 0.5 ppm (Ganther *et al.,* 1972).

### HIGH LEVELS

Comparisons among species for tolerance to cumulative toxic elements indicate that tissue saturation kinetics are related to body mass and duration of exposure. With high body burdens of methylmercury, pathology of many tissues may be simultaneous, but nerve tissue is particularly vulnerable and critical. Ultrastructural pathology usually occurs well in advance of the clinical signs for peripheral neuropathy.

The acute signs that result from ingestion of mercury include nausea, vomiting of blood-stained mucus, severe gastrointestinal irritation and abdominal pain, shock, and cardiac arrhythmias. From 1 day to 2 weeks following exposure, reactions include excessive salivation, foul breath, loose teeth, soft spongy gums, and a blue-black gum line caused by a mercury–sulfhydryl complex. Death is usually caused by uremia (D'Itri, 1971). The primary clinical lesion in mercury toxicosis is acute renal failure due to injury in renal epithelial tubular cortical tissue (Bulger and Siegel, 1975; Preuss *et al.*, 1975). Buck (1975) described clinical signs that were similar for cattle acutely poisoned with organic or inorganic mercury exposure. From the onset of clinical signs, the average time to death was 20 days but ranged from 1 to 43 days.

Palmer *et al.* (1973) produced mercury toxicosis in cattle, sheep, and turkeys with an alkylmercury fungicide administered daily in capsules. Cattle and sheep receiving 0.48 mg mercury per kilogram of body weight died within 7 to 27 and 13 to 31 days, respectively. Turkeys developed weakness and incoordination in 13 to 14 days with 0.16 mg mercury per kilogram of body weight.

Pigs tolerated a single oral dose of 2.5 mg mercury per kilogram of body weight as methylmercury dicyandiamide. With 5 and 10 mg/kg, anorexia, reduced gain, central nervous system depression, vomiting, muscular tremors, and increased tissue mercury concentration were observed. Death occurred in 7.5 to 29 days, 7 days, 24 hours, and 12 hours for 20, 40, 80, and 160 mg/kg of body weight, respectively (Piper *et al.*, 1971).

Differences in tolerance to organic mercury among sex and strain of chicks, swine, and rats have been reported (Miller *et al.*, 1970; Piper *et al.*, 1971; Parizek *et al.*, 1974). When hens were fed 10 ppm mercury as $CH_3HgCl$ for 10 days, eggs contained 55 percent of the total hen dose after 70 days (Sell *et al.*, 1974). Egg mercury levels increased sharply to 12 days and declined during the next 58 days. At 10 ppm, mercury methylmercury produced 50 percent mortality in 16 weeks with Japanese quail (El-Begearmi *et al.*, 1974). Dietary methylmercury was acutely toxic at 20 ppm for Japanese quail (Stoewsand *et al.*, 1974). In Japanese quail and chicks, however, 25 ppm mercury as mercuric ion had no effect on growth, fertility, or egg hatchability but did increase mortality (Thaxton and Parkhurst, 1973a; Thaxton *et al.*, 1974).

## FACTORS INFLUENCING TOXICITY

Studies with one broiler strain and three White Leghorn strains indicate genetic differences in the degree of tissue concentration of mercury from dietary fish meals (March *et al.*, 1974). Mercuric chloride up to 500 ppm expresses a more toxic effect in Japanese quail when incorporated in the diet as a dry salt rather than as a solution, regardless of the solvent system (ethanol, methanol, or water) (El-Begearmi *et al.*, 1979). Selenite or selenate administered to rats orally or parenterally (Parizek *et al.*, 1974) or Japanese quail (El-Begearmi *et al.*, 1977a) reduced acute or chronic toxicosis of mercuric or methylmercuric ions by redistribution of tissue mercury as opposed to increased excretion of the element. Dietary selenium (5 ppm selenium) protected rats against toxicity of otherwise acute lethal doses of methylmercury and mercuric mercury (Potter and Matrone, 1974). Selenite increased the percentage of mercury retained in liver and spleen but decreased that in kidney compared with animals untreated with selenium. In selenium-deficient rats injected simultaneously with $^{75}SeO_3^{2-}$ and $^{203}Hg^{2+}$, a protein of sulfhydrylselenium–mercuric components was identified in plasma 20 hours postinjection, suggesting that the protein was formed after isotopes were metabolized (Burk *et al.*, 1974). Reduction of kidney mercury with redistribution in other tissues appears to reduce toxicosis (Ganther and Sunde, 1974; Stillings *et al.*, 1974; Klassen, 1975b). Similar results were reported with rats receiving selenium and organic or inorganic mercury (Chen *et al.*, 1974; Moffitt and Clary, 1974; Chen *et al.*, 1975a; Ohi *et al.*, 1975). It has been postulated that cystine or thiols provide protein–sulfur binding sites for mercury, that selenium catalyzes mercury to change to a less damaging form, or that selenium reacts directly with mercury.

The addition of 5 ppm selenium as $Na_2SeO_3$ to diets containing 20 ppm mercury as $CH_3HgCl$ reduced the death rate in Japanese quail by 78 percent (Stoewsand *et al.*, 1974). Other authors have indicated that simultaneous equimolar ratios of selenium and mercury are necessary to prevent toxicity of either (Ganther and Sunde, 1974; Moffitt and Clary, 1974). Selenium as sodium selenite at 8 ppm alleviated reduction in egg production induced by feeding 20 ppm mercury as $CH_3HgCl$ in Japanese quail, but not in chickens (Sell, 1977). Either 4 or 8 ppm selenium partially prevented decreased egg production and hatchability in chickens produced by 10 ppm mercury as $CH_3HgCl$ (Emerick *et al.*, 1976). Selenium as $Na_2SeO_3$ at 0.5 ppm increased weight gain in rats receiving 1, 5, 10, and 25 ppm mercury as $CH_3HgOH$ in drinking water (Ganther *et al.*, 1972).

Blackstone *et al.* (1974) fed maintenance levels of ascorbic acid to guinea pigs and provided mercuric ion (8 mg mercury per kilogram of body weight as $HgCl_2$) in drinking water. Ascorbic acid levels were depressed in brain, adrenals, and spleen. Mercury deposition in kidney and liver increased with ascorbic acid level.

Vitamin E has been shown to protect against the toxic effects of methylmercury in Japanese quail (Welsh and Soares, 1975) and rats (Welsh, 1976, 1979) and organic mercury in Japanese quail (El-Begearmi *et al.*, 1977b).

The specific antidote for mercury poisoning is dimercaprol, which can be used in conjunction with proteins such as milk and eggs to bind mercury still in the gastrointestinal tract. Gastric lavage with sodium formaldehyde sulfoxalate will reduce divalent mercury to the less toxic monovalent form (Siegmund and Fraser, 1973).

## TISSUE LEVELS

Samples of kidney and liver obtained from slaughter animals in Canada (pork, poultry, and beef) ranged from undetectable (<0.01 ppm) to

0.097 ppm mercury in wet tissue for 265 samples (Prior, 1976). Muscle accounted for 72 percent and liver 7 percent of a tracer dose of methylmercury in ruminants 1 week following ingestion (Neathery *et al.*, 1974). At 24 days following an 8-day exposure, milk of goats accounted for 1.22 and 0.22 percent of total oral organic and inorganic mercury, respectively (Sell and Davidson, 1975). Similar results were reported for $CH_3{}^{203}Hg$ in milk of rats (Garcia *et al.*, 1974a,b). Mercury content of cow's milk may range from 3 to 10 ppb (Mullen *et al.*, 1975; Roh *et al.*, 1975). Methylmercury ($CH_3Hg^+$) comprises 75 to 90 percent of mercury in fish and is transferred through the food chain with the C–Hg bond intact. The methylmercury was not removed by boiling fish (Westoo, 1966). When breast of ducks that had received single oral doses of methylmercury was cooked by dry or moist heat, mercury levels of meat and drippings were not different on a dry matter basis from uncooked meat (Hough and Zabik, 1973).

## MAXIMUM TOLERABLE LEVELS

Dairy calves tolerated mercury as methylmercury at a level of 0.1 mg per kilogram of body weight (about 3 ppm in their diet) for 90 days without visible adverse effects, while a level of 0.2 mg/kg resulted in toxicosis. Yearling sheep receiving 0.22 mg mercury per kilogram of body weight orally as methylmercury dicyandiamide exhibited incoordination and unsteady gait after 40–50 days of exposure. Swine have received 0.38 mg mercury per kilogram of body weight daily by capsule in the form of either phenylmercuric chloride or methylmercuric dicyandiamide for 60–90 days without visible adverse effects. Increased mercury in tissue and signs of mercury toxicosis occurred with a level of 0.76 mg/kg body weight. Chickens, turkeys, ducks, and pheasants tolerated 3.3 ppm supplemental dietary mercury without evidence of adverse effects, although increased tissue mercury has been shown at levels lower than this. Elemental mercury was tolerated at considerably higher levels than this by Japanese quail. Daily consumption of drinking water containing 1 ppm mercury as $CH_3HgCl$ by mice did not affect health or longevity, but 5 ppm of either $CH_3HgCl$ or $HgCl_2$ resulted in toxicosis and death.

The suggested maximum tolerable dietary level for domestic animals is 2 ppm mercury for both the organic and inorganic forms. Research with several species indicates that animals can tolerate higher dietary quantities of the inorganic form, but the maximum tolerable level for this form was not increased because of the possibility of elevated tissue levels of the element. Studies with rats and mice support the proposed tolerance level for domestic animals, but limited research with mink suggests that this species is much more sensitive to mercury.

## SUMMARY

Mercury toxicity has received considerable attention because of poisonings that have occurred in the human population. The metal is not essential in animal or human nutrition, and its level of concentration in living organisms depends on the type of organism and the form of mercury to which the organism is exposed. Biomethylation of inorganic mercury occurs in the environment or the animal and increases the potential for toxicity. Animals can be exposed to mercury contamination from air, soil, or water, while the major feed sources are fish protein concentrates or the accidental use of treated seed grain.

Acute toxic signs include nausea, vomiting, severe gastrointestinal irritation and pain, shock, and cardiac arrhythmias. Death usually results from uremia, caused by damage to renal epithelial tubular cortical tissue. Dietary selenium has been reported to decrease mercury toxicity.

TABLE 24 Effects of Mercury Administration in Animals

| Class and Number of Animals[a] | Age or Weight | Administration: Quantity of Element[b] | Source | Duration | Route | Effect(s) | Reference |
|---|---|---|---|---|---|---|---|
| Cattle—1 | 45.4 kg | 0.025 mg/kg | Methylmercury | 40 d | Diet | No adverse effect | Herigstad *et al.*, 1972 |
| | 52.2 kg | 0.05 mg/kg | | 96 d | | No adverse effect | |
| | 54.5 kg | 0.1 mg/kg | | 96 d | | No adverse effect | |
| | 56.8 kg | 0.2 mg/kg | | 91 d | | Ataxia; prostration | |
| | 49.9 kg | 0.4 mg/kg | | 33 d | | Ataxia; prostration | |
| Cattle—10 | Yearling | 0.225 mg/kg | Methylmercury dicyandiamide | 56 d | Oral | Incoordination; unsteady gait | Wright *et al.*, 1973 |
| Cattle—8 | 200 kg | 0.48 mg/kg | Alkylmercury | 7–27 d | Capsule | Incoordination; death | Palmer *et al.*, 1973 |
| Sheep—12 | Yearling | 0.225 mg/kg | Methylmercury dicyandiamide | 42–49 d | Oral | Incoordination; unsteady gait | Wright *et al.*, 1973 |
| Sheep—8 | 34 kg | 0.48 mg/kg | Alkylmercury fungicide | 13–31 d | Capsule | Death | Palmer *et al.*, 1973 |
| Swine—5 | 5 wk | 0.19 mg/kg | Phenyl-HgCl | 90 d | Capsule | No adverse effect | Tryphonas and Nielsen, 1970 |
| Swine—7 | 5 wk | 0.38 mg/kg | Phenyl-HgCl | 90 d | Capsule | No adverse effect | Tryphonas and Nielsen, 1970 |
| | | 2.28 mg/kg | | 14–63 d | | Decreased growth; kidney and colon necrosis; diarrhea | |
| | | 4.56 mg/kg | | 10–51 d | | Decreased growth; diarrhea | |
| Swine—4 | 5 wk | 0.19 mg/kg | Methylmercuric dicyandiamide | 60 d | Capsule | No adverse effect | Tryphonas and Nielsen, 1973 |
| Swine—3 | | 0.38 mg/kg | | 60 d | | No adverse effect | |

TABLE 24 *Continued*

| Class and Number of Animals[a] | Age or Weight | Administration: Quantity of Element[b] | Source | Duration | Route | Effect(s) | Reference |
|---|---|---|---|---|---|---|---|
| Swine—4 | | 0.76 mg/kg | | 41–46 d | | Increased mercury in tissues; necrosis of nerve tissue | |
| Swine—2 | 8 wk | 1.7 mg/kg | Methylmercuric dicyandiamide | 32 d | Capsule | Increased tissue mercury | Piper *et al.*, 1971 |
| | | 3.4 mg/kg | | 33 d | | Increased tissue mercury; anorexia; vomiting; CNS depression | |
| Swine—4 | | 6.7 mg/kg | | 29–34 d | | Increased tissue mercury; diarrhea; CNS depression | |
| Swine—2 | | 10.1 mg/kg | | 34 d | | Increased tissue mercury; vomiting; CNS depression | |
| Swine—4 | | 13.4 mg/kg | | 7–35 d | | Death | |
| Swine—2 | | 26.9 mg/kg | | 7 d | | Anorexia; CNS depression; incoordination; diarrhea | |
| | | 53.8 mg/kg | | Single dose | | Death in 24 h | |
| | | 107.4 mg/kg | | Single dose | | Death in 12 h | |
| Swine—4 | | 0.5 ppm | $HgCl_2$ | 27 d | Diet | No adverse effect | Chang *et al.*, 1977 |
| | | 5 ppm | | | | Increased tissue Hg | |
| | | 50 ppm | | | | "Fatty" livers; enlarged lymph nodes | |
| | | 0.5 ppm | $CH_3HgCl$ | | | "Fatty" livers; en- | |

| | | | | | | | |
|---|---|---|---|---|---|---|---|
| | | | | | | larged lymph nodes | |
| | | 5 ppm | | | | "Fatty" livers; enlarged lymph nodes, periglandular hemorrhage of lymph nodes | |
| | | 50 ppm | | | | Anorexia; blindness; ataxia; death in 24–27 d | |
| Chicken—24 | Day old | 0.5 ppm | $CH_3HgCl$ | 49 d | Diet | No adverse effect | Soares *et al.*, 1973 |
| | | 1.0 ppm | | 49 d | | No adverse effect | |
| | | 2.2 ppm | | 49 d | | No adverse effect | |
| | | 5.0 ppm | | 33 d | | Death—50% | |
| | | 7.8 ppm | | 22 d | | Death—100% | |
| | | 16.9 ppm | | 17 d | | Death—100% | |
| Chicken—12 right *et al.*, 1973 | 6 wk | 0.075 mg/kg | dicyandiamide | Methylmercuric | | mercury | 84 d Oral Increased tissue W |
| | | 0.15 mg/kg | | | Oral | Increased tissue mercury | |
| Chicken—5 | 68-wk laying hen | 8 ppm | $CH_3HgCl$ | 70 d | Diet | 55% mercury in eggs of which 80% in albumin | Sell *et al.*, 1974 |
| Chicken—20 | 1 d | 2 ppm | $HgCl_2$ or phenyl-Hg acetate | 20 d | Diet | Increased tissue mercury | Miller *et al.*, 1967 |
| | | 20 ppm | | 5 d | | Increased tissue mercury | |
| Chicken—40 | 5 d | 0.33 ppm | Methylmercuric dicyandiamide | 35 d | Diet | No adverse effect | Gardiner, 1972 |
| | | 3.3 ppm | | | | No adverse effect | |
| | | 33 ppm | | | | 7.5% death rate; reduced growth | |

TABLE 24 *Continued*

| Class and Number of Animals[a] | Age or Weight | Administration: Quantity of Element[b] | Source | Duration | Route | Effect(s) | Reference |
|---|---|---|---|---|---|---|---|
| Chicken—60 | 1 yr | 25 ppm | Ethyl-HgCl | 88 d | Diet | Increased Hg in eggs, liver, and kidney | Al-Fayadh *et al.*, 1976 |
| | | 50 ppm | | | | Increased Hg in eggs, liver, and kidney | |
| | | 100 ppm | | | | Increased Hg in eggs, liver, and kidney | |
| Chicken—120 | Day old | 5 ppm | $HgCl_2$ | 42 d | Water | No adverse effect | Thaxton *et al.*, 1973 |
| | | 25 ppm | | | | No adverse effect | |
| | | 125 ppm | | | | No adverse effect | |
| | | 250 ppm | | | | No adverse effect | |
| Chicken—40 | Young | 5 ppm | $HgCl_2$ | 98 d | Water | Death—7.5% | Parkhurst and Thaxton, 1973 |
| | | 25 ppm | | | | Death—5.0% | |
| | | 125 ppm | | | | Death—7.5% | |
| | | 250 ppm | | | | Decreased growth and feed intake; increased heart weight; death—48% | |
| | | 500 ppm | | | | Decreased growth and feed intake; increased heart weight; death—100% | |
| Chicken—80 | Young | 5 ppm | $HgCl_2$ | 42 d | Water | No adverse effect | Thaxton *et al.*, 1975 |
| | | 25 ppm | | | | No adverse effect | |

| | | | | | | | |
|---|---|---|---|---|---|---|---|
| | | 125 ppm | | | | Decreased growth | |
| | | 250 ppm | | | | Decreased growth and adrenal weight | |
| | | 500 ppm | | | | Decreased growth and adrenal weight | |
| Chicken—20 | Mature | 100 ppm | $HgSO_4$ | | | No adverse effect | Scott *et al.*, 1975 |
| | | 200 ppm | | | | Decreased hatchability | |
| | | 10 ppm | $CH_3HgCl$ | | | Decreased gain, egg weight, production, and fertility | |
| | | 20 ppm | | | | Decreased gain, egg weight, production, and fertility | |
| Turkey—9 | 7 kg | 0.16 mg/kg | Alkylmercury fungicide | 8–42 d | Capsule | Weakness; ataxia; incoordination | Palmer *et al.*, 1973 |
| Duck—40 | 5 d | 0.33 ppm | Methylmercuric dicyandiamide | 35 d | Diet | No adverse effect | Gardiner, 1972 |
| | | 3.3 ppm | | | | No adverse effect | |
| | | 33 ppm | | | | Reduced growth; 85% death rate | |
| Pheasant—40 | 5 d | 0.33 ppm | Methylmercuric dicyandiamide | 35 d | Diet | No adverse effect | Gardiner, 1972 |
| | | 3.3 ppm | | | | No adverse effect | |
| | | 33 ppm | | | | Reduced growth; 90% death rate | |
| Japanese quail | 40 d | 2 ppm | $HgCl_2$ | 1 yr | Diet | No adverse effect | Hill and Shaffner, 1976 |
| | | 4 ppm | | | | No adverse effect | |
| | | 8 ppm | | | | Decreased egg fertilization | |
| | | 16 ppm | | | | Decreased egg fertilization | |
| | | 32 ppm | | | | Egg production increased 13.5% | |

TABLE 24 *Continued*

| Class and Number of Animals[a] | Age or Weight | Administration: Quantity of Element[b] | Source | Duration | Route | Effect(s) | Reference |
|---|---|---|---|---|---|---|---|
| Japanese quail | | 10 ppm | $CH_3HgCl$ | 112 d | Diet | 50% death rate | El-Begearmi *et al.*, 1974 |
| Japanese quail | Immature | 25 ppm | $Hg^{2+}$ | 42 d | Diet | No adverse effect | Thaxton and Parkhurst, 1973a |
| | | 125 ppm | | | | Reduced growth; death | |
| Rat—20 | Weanling | 0.1 ppm | Phenylmercuric acetate | 2 yr | Diet | No adverse effect | Fitzhugh *et al.*, 1950 |
| | | 0.5 ppm | | | | No adverse effect | |
| | | 2.5 ppm | | | | Increased tissue mercury | |
| Rat—24 | | 10 ppm | | | | Decreased growth; increased tissue mercury | |
| | | 40 ppm | | | | Decreased growth; increased tissue mercury | |
| | | 160 ppm | | 1½ yr | | Decreased growth; increased tissue mercury; death | |

| | | | | | | | |
|---|---|---|---|---|---|---|---|
| Rat—20 | Weanling | 0.5 ppm | Mercuric acetate | 2 yr | Diet | No adverse effect | Fitzhugh *et al.*, 1950 |
| | | 2.5 ppm | | | | No adverse effect | |
| Rat—24 | | 10 ppm | | | | Increased tissue mercury | |
| | | 40 ppm | | | | Increased tissue mercury | |
| | | 160 ppm | | | | Reduced growth; increased tissue mercury | |
| Mouse—72 | Immature | 1 ppm | $CH_3HgCl$ | Life-term | Water | No adverse effect | Schroeder and Mitchener, 1975 |
| Mouse—108 | | 5 ppm | $HgCl_2$ | Life-term | | No adverse effect | |
| | | 5 ppm | $CH_3HgCl$ | 90 d | | Death | |
| Mink—107 | Immature–adult | 0.22 ppm | Fish | 145 d | Diet | Increased Hg in liver and kidney | Wobeser *et al.*, 1976a |
| Mink—5 | Adult | 1.1 ppm | $CH_3HgCl$ | 93 d | Diet | Increased tissue mercury | Wobeser *et al.*, 1976b |
| | | 1.8 ppm | | | | Anorexia; death | |
| | | 4.8 ppm | | | | Ataxia; death | |
| | | 8.3 ppm | | | | Convulsion; blindness; death | |
| | | 15 ppm | | | | Convulsion; death | |

[a] Number of animals per treatment group.

[b] Quantity expressed in parts per million or as milligrams per kilogram of body weight.

## REFERENCES

Alberg, G., L. Ekman, R. Falk, U. Greitz, G. Persson, and J. Snibs. 1969. Metabolism of methyl mercury ($^{203}$Hg) compounds in man. Arch. Environ. Health 19:478.

Al-Fayadh, H., A. W. R. Mehdi, K. Al-Soudi, A. K. Al-Khazraji, N. A. Al-Jiboori, and S. Al-Muraib. 1976. Effects of feeding ethyl mercury chloride to chickens. Poult. Sci. 55:772.

Ammerman, C. B., S. M. Miller, K. R. Fick, and S.L. Hansard II. 1977. Contaminating elements in mineral supplements and their potential toxicity: A review. J. Anim. Sci. 44:485.

Annett, C. S., F. M. D'Itri, J. R. Ford, and H. H. Prince. 1975. Mercury in fish and water fowl from Lake Ball, Ontario. J. Environ. Qual. 4:219.

Ansari, M. S., W. J. Miller, R. P. Gentry, M. W. Neathery, and P. E. Stake. 1973. Tissue $^{203}$Hg distribution in young Holstein calves after single tracer oral doses in organic and inorganic forms. J. Anim. Sci. 36:415.

Bakir, F., S. F. Damluji, L. Amin-Zaki, M. Murtadha, A. Khalidi, N. Y. Al-Rawi, S. Tikriti, H. I. Dhahir, T. W. Clarkson, J. C. Smith, and R. A. Doherty. 1973. Methylmercury poisoning in Iraq. Science 681:230.

Barber, R. T., A. Vijayakumar, and F. A. Cross. 1972. Mercury concentrations in recent and ninety-year-old benthopelagic fish. Science 178:636.

Berlin, M., J. Carlson, and T. Norseth. 1975. The dose-dependence of methylmercury metabolism. Arch. Environ. Health 30:307.

Billings, C. E., and W. R. Matson. 1972. Mercury emissions from coal combustion. Science 176:1232.

Blackstone, S., R. J. Hurley, and R. E. Hughes. 1974. Some inter-relationships between vitamin C (L-ascorbic acid) and mercury in the guinea pig. Food Cosmet. Toxicol. 12:511.

Buck, W. B. 1975. Toxic materials and neurological diseases in cattle. J. Am. Vet. Med. Assoc. 166:222.

Bulger, R. E., and F. L. Siegel. 1975. Alterations of the renal papilla during mercuric chloride-induced acute tubular necrosis. Lab. Invest. 33:712.

Burk, R. F., K. A. Foster, P. M. Greenfield, and K. W. Kiker. 1974. Binding of simultaneously administered inorganic selenium and mercury to a rat plasma protein. Proc. Soc. Exp. Biol. Med. 145:782.

Chang, C. W. J., R. M. Nakamura, and C. C. Brooks. 1977. Effect of varied dietary levels and forms of mercury on swine. J. Anim. Sci. 45:279.

Chen, R. W., H. E. Ganther, and W. G. Hoekstra. 1973. Studies on the binding of methyl mercury by thionein. Biochem. Biophys. Res. Commun. 51:383.

Chen, R. W., P. D. Whanger, and S. C. Fang. 1974. Diversion of mercury binding in rat tissues by selenium: A possible mechanism of protection. Pharmacol. Res. Commun. 6:571.

Chen, R. W., V. L. Lacy, and P. D. Whanger. 1975a. Effect of selenium on methylmercury binding to subcellular and soluble proteins in rat tissues. Res. Commun. Chem. Pathol. Pharmacol. 12:297.

Chen, R. W., P. D. Whanger, and P. H. Weswig. 1975b. Selenium-induced redistribution of cadmium binding to tissue proteins: A possible mechanism of protection against cadmium toxicity. Bioinorg. Chem. 4:125.

Clarkson, T. W. 1970. Epidemiological aspects of lead and mercury contamination of food. Presented at a Symposium on "Chemical Contaminants in Foods—Hazard or Not?" sponsored by the Food and Drug Directorate, Department of National Health and Welfare, Ottawa, Ontario, and held in Ottawa on June 18–19, 1970. Also published as "Epidemiological and experimental aspects of lead and mercury contamination of food." Food Cosmet. Toxicol. 9:229, April 1971.

Clarkson, T. W., L. Magos, and G. G. Berg. 1973. Mercury compounds. Science 176:1074.

Clarkson, T. W., L. Amin-Zaki, and S. K. Al-Tikriti. 1976. An outbreak of methylmercury poisoning due to consumption of contaminated grain. Fed. Proc. 35:2395.

Colucci, A. V., D. Winge, and J. Krasno. 1975. Cadmium accumulation in rat liver. Arch. Environ. Health 80:153.

Cousins, R. J. 1974. Influence of cadmium on the synthesis of liver and kidney cadmium-binding protein. *In* W. G. Hoekstra, J. W. Suttie, H. E. Ganther, and W. Mertz (eds.). Trace Element Metabolism in Animals—2. University Park Press, Baltimore, Md.

Curley, A., V. A. Sedlak, E. F. Girling, R. E. Hawk, W. F. Barthel, P. E. Pierce, and W. H. Likosky. 1971. Organic mercury identified as the cause of poisoning in humans and hogs. Science 172:65.

D'Itri, F. M. 1971. The Environmental Mercury Problem. Michigan Legislative Report HR-424. Lansing.

El-Begearmi, M., H. E. Ganther, and M. L. Sunde. 1974. Effect of some sulfur amino acids, selenium and arsenic on mercury toxicity using Japanese quail. Poult. Sci. 53:1921. (Abstr.)

El-Begearmi, M. M., M. L. Sunde, and H. E. Ganther. 1977a. A mutual protective effect of mercury and selenium in Japanese quail. Poult. Sci. 56:313.

El-Begearmi, M. M., H. E. Ganther, and M. L. Sunde. 1977b. Protective effect of vitamin E against inorganic mercury and methylmercury toxicity. Poult. Sci. 56:1711.

El-Begearmi, M. M., H. E. Ganther, and M. L. Sunde. 1979. Toxicity of mercuric chloride as affected by method of incorporation into the diet. Fed. Proc. 38:869. (Abstr.)

Ellis, R. W., and S. C. Fang. 1967. Elimination, tissue accumulation and cellular incorporation of mercury in rats receiving an oral dose of mercury-203-labeled phenylmercuric acetate and mercuric acetate. Toxicol. Appl. Pharmacol. 11:104.

Emerick, R. J., S. Palmer, C. W. Carlson, and R. A. Nelson. 1976. Mercury–selenium interrelationships in laying hens. Fed. Proc. 35:577. (Abstr.)

Fitzhugh, O. G., A. A. Nelson, E. P. Laug, and F. M. Kunze. 1950. Chronic oral toxicities of mercuri-phenyl and mercuric salts. AMA Arch. Ind. Hyg. Occup. Med. 2:433.

Fowler, B. A. 1972. Ultrastructural evidence for nephropathy induced by long-term exposure to small amounts of methyl mercury. Science 175:780.

Fowler, B. A., and G. F. Nordberg. 1975. The renal toxicity of cadmium metallothionein. Int. Conf. Heavy Metals Environ. Toronto, Canada (Abstr.).

Fowler, B. A., H. W. Brown, G. W. Lucier, and M. R. Krigman. 1975. The effects of chronic oral methyl mercury exposure on the lysosome system of rat kidney. Lab. Invest. 32:313.

Gage, J. C. 1975. Mechanisms for the biodegradation of organic mercury compounds: The actions of ascorbate and of soluble proteins. Toxicol. Appl. Pharmacol. 32:225.

Ganther, H. E., and M. L. Sunde. 1974. Effect of tuna fish and selenium on the toxicity of methylmercury: A progress report. J. Food Sci. 39:1.

Ganther, H. E., C. Goudie, M. L. Sunde, M. J. Kopecky, P. Wagner, S. H. Oh, and W. G. Hoekstra. 1972. Selenium: Relation to decreased toxicity of methylmercury added to diets containing tuna. Science 175:1122.

Garcia, J. D., M. G. Yang, J. H. C. Wang, and P. S. Belo. 1974a. Carbon–mercury bond breakage in milk, cerebrum, liver and kidney of rats fed methyl mercuric chloride. Proc. Soc. Exp. Biol. Med. 146:190.

Garcia, J. D., M. G. Yang, J. H. C. Wang, and P. S. Belo. 1974b. Translocation and fluxes of mercury in neonatal and maternal rats treated with methyl mercuric chloride during gestation. Proc. Sci. Exp. Biol. Med. 147:224.

Gardiner, E. E. 1972. Differences between ducks, pheasants and chickens in tissue mercury retention, depletion and tolerance to increasing levels of dietary mercury. Can. J. Anim. Sci. 52:419.

Gruber, T. A., P. Costigan, G. T. Wilkinson, and A. A. Seawright. 1978. Chronic methylmercurialism in the cat. Aust. Vet. J. 54:155.

Hammond, P. B. 1973. Metabolism and metabolic action of lead and other heavy metals. Clin. Toxicol. 6:353.

Haq, I. U. 1963. Agrosan poisoning in man. Br. J. Med. 5345:1579.

Herigstad, R. R., C. K. Whitehair, N. Beyer, O. Mickelsen, and M. J. Zabik. 1972. Chronic methylmercury toxicosis in calves. J. Am. Vet. Med. Assoc. 160:173.

Hill, E. F., and C. S. Shaffner. 1976. Sexual maturation and productivity of Japanese quail fed graded concentrations of mercuric chloride. Poult. Sci. 55:1449.

Hollins, J. G., R. F. Willes, F. R. Bryce, S. M. Charbonneau, and I. C. Munro. 1975. The whole body retention and tissue distribution of $^{203}$Hg methylmercury in adult cats. Toxicol. Appl. Pharmacol. 33:438.

Hough, E. J., and M. E. Zabik. 1973. Mercury residues in duck breast tissue after moist and dry heat cooking. J. Sci. Food Agric. 24:107.

Huckabee, J. W., F. O. Cartan, G. S. Kenhington, and F. J. Camenzind. 1973. Mercury concentration in the hair of coyotes and rodents in Jackson Hole, Wyoming. Bull. Environ. Contam. Toxicol. 9:37.

Johnels, A. G. 1967. Mercury content of *Exox lucius* as indicator of pollution. Oikos. 18:323.

Khan, J. M. 1974. Compartmental analysis for the evaluation of biological half-lives of cadmium and mercury in mouse organs. Environ. Res. 7:54.

Khera, K. S., and S. A. Tabacova. 1973. Effects of methylmercuric chloride on the progency of mice and rats treated before or during gestation. Food Cosmet. Toxicol. 11:245.

Klaasen, C. D. 1975a. Biliary excretion of mercury compounds. Toxicol. Appl. Pharmacol. 33:356.

Klaasen, C. D. 1975b. Effect of spisonolactone on the distribution of mercury. Toxicol. Appl. Pharmacol. 33:336.

MacGregor, J. T., and T. W. Clarkson. 1974. Distribution, tissue binding and toxicity of mercurials. Adv. Exp. Med. Biol. 48:463.

Madsen, K., and E. I. Christensen. 1975. Effects of mercury on kidney lysosome function. Int. Conf. Heavy Metals Environ. Toronto, Canada.

Magat, W., and J. L. Sell. 1979. Distribution of mercury and selenium in egg components and egg-white proteins. Proc. Soc. Exp. Biol. Med. 161:458.

Magos, L. 1968. The uptake of mercury by the brain. Br. J. Ind. Med. 25:315.

March, B. E., R. Soong, E. Bilinski, and R. E. E. Jonas. 1974. Effect on chickens of chronic exposure to mercury at low levels through dietary fish meal. Poult. Sci. 53:2175.

McIntyre, J. D. 1973. Toxicity of methyl mercury for steelhead trout sperm. Bull. Environ. Conf. Toxicol. 9:98.

Miller, V. L., D. V. Larkin, G. E. Bearse, and C. M. Hamilton. 1967. The effects of dosage and administration of two mercurials on mercury retention in two strains of chickens. Poult. Sci. 46:142.

Miller, V. L., G. E. Bearse, and E. Csonka. 1970. Mercury retention in several strains and strain crosses of chickens. Poult. Sci. 49:1101.

Mills, C. F. 1974. Trace element interactions: Effect of dietary composition on the development of imbalance and toxicity. *In* W. G. Hoekstra, J. W. Suttie, H. E. Ganther, and W. Mertz (eds.). Trace Element Metabolism in Animals—2. University Park Press, Baltimore, Md.

Moffitt, A. E., Jr., and J. J. Clary. 1974. Selenite-induced binding of inorganic mercury in blood and other tissues in the rat. Res. Commun. Chem. Pathol. Pharmacol. 7:593.

Mullen, A. L., R. E. Stanley, S. R. Lloyd, and A. A. Moghessi. 1975. Absorption, distribution and milk secretion of radionuclides by the dairy cow. IV. Inorganic radiomercury. Health Phys. 28:685.

Neathery, M. W., and W. J. Miller. 1975. Metabolism and toxicity of cadmium, mercury and lead in animals: A review. J. Dairy Sci. 58:1767.

Neathery, M. W., W. J. Miller, R. P. Gentry, P. E. Stake, and D. M. Blackmon. 1974. Cadmium-109 and methyl mercury-203 metabolism, tissue distribution and secretion into milk of cows. J. Dairy Sci. 57:1177.

Nelson, N., T. C. Byerly, A. C. Kolbye, Jr., L. T. Kurland, R. E. Shapiro, S. I. Shibko, W. H. Stickel, J. E. Thompson, L. A. Vanden Berg, and A. Weissler. 1971. Hazards of mercury. Environ. Res. 4:1.

Nordberg, G. F., and M. Nordberg. 1975. Metabolism and toxicity of metallothionein-bound cadmium. Int. Conf. Heavy Metals Environ. Toronto, Canada.

Norseth, T. 1974. The effect of diethyldithiocarbamate on biliary transport, excretion and organic distribution of mercury in the rat after exposure to methylmercuric chloride. Acta Pharmacol. Toxicol. 34:76.

Norseth, T., and T. W. Clarkson. 1970. Studies on the biotransformation of $^{203}$Hg-labelled methyl mercury chloride in rats. Arch. Environ. Health 21:717.

Norseth, T., and T. W. Clarkson. 1971. Intestinal transport of $^{203}$Hg-labeled methyl mercury chloride. Arch. Environ. Health 22:568.

Ohi, G., S. Nishigaki, H. Seki, Y. Tamura, T. Maki, H. Maeda, S. Ochiai, H. Yamada, Y. Shimamura, and H. Yagyu. 1975. Interaction of dietary methylmercury and selenium on accumulation and retention of these substances in rat organs. Toxicol. Appl. Pharmacol. 32:527.

Ordonez, J. V., J. A. Carrillo, and M. Miranda. 1966. Epidemiological study of a disease in the Guatemalan highlands believed to be encephalitis. Bol. Of. Sanit. Panam. 60:510.

Palmer, J. S., F. C. Wright, and M. Haufler. 1973. Toxicologic and residual aspects of an alkyl mercury fungicide to cattle, sheep and turkeys. Clin. Toxicol. 6:425.

Parizek, J., J. Kalouskova, A. Babicky, J. Benes, and L. Pavlik. 1974. Interaction of selenium with mercury, cadmium and other toxic metals. *In* W. G. Hoekstra, J. W. Suttie, H. E. Ganther, and W. Mertz (eds.). Trace Element Metabolism in Animals—2. University Park Press, Baltimore, Md.

Parkhurst, C. R., and P. Thaxton. 1973. Toxicity of mercury to young chickens. 1. Effect on growth and mortality. Poult. Sci. 52:273.

Piotrowski, J. K., B. Trojanowska, J. M. Wisniewska-Knypl, and W. Bolanowska. 1974a. Mercury binding in the kidney and liver of rats repeatedly exposed to mercuric chloride: Induction of metallothionein by mercury and cadmium. Toxicol. Appl. Pharmacol. 27:11.

Piotrowski, J. K., B. Trojanowska, and A. Sapota. 1974b. Binding of cadmium and mercury by metallothionein in the kidneys and liver of rats following repeated administration. Arch. Toxicol. 32:351.

Piper, R. C., V. L. Miller, and E. O. Dickenson. 1971. Toxicity and distribution of mercury in pigs with acute methylmercurialism. Am. J. Vet. Res. 32:263.

Potter, S., and G. Matrone. 1974. Effect of selenite on the toxicity of dietary methyl mercury and mercuric chloride in the rat. J. Nutr. 104:638.

Preuss, H. G., A. Tourkantonis, C. H. Hsu, P. S. Shim, P. Barzyk, F. Tio, and G. E. Schriiner. 1975. Early events in various forms of experimental acute tubular necrosis in rats. Lab. Invest. 32:286.

Prior, M. G. 1976. Lead and mercury residues in kidney and liver of Canadian slaughter animals. Can. J. Comp. Med. 40:9.

Roh, J. K., R. L. Bradley, Jr., T. Richardson, and K. G. Weckel. 1975. Distribution and removal of added mercury in milk. J. Dairy Sci. 58:1782.

Rubenstein, D. A., and J. H. Soares, Jr. 1979. The effect of selenium on the biliary excretion and tissue deposition of two forms of mercury in the broiler chick. Poult. Sci. 58:1289.

Rucker, R. R., and D. F. Amend. 1969. Absorption and retention of organic mercurials by rainbow trout and chinook and sockeye salmon. Prog. Fish Cult. 31:197.

Salvaterra, P., E. J. Massaro, J. B. Morganti, and B. A. Lown. 1975. Time-dependent tissue/organ uptake and distribution of $^{203}$Hg in mice exposed to multiple sublethal doses of methyl mercury. Toxicol. Appl. Pharmacol. 32:432.

Sasser, L. B., G. E. Jarboe, B. K. Walter, and B. J. Kelman. 1978. Absorption of mercury from ligated segments of the rat gastrointestinal tract. Proc. Soc. Exp. Biol. Med. 157:57.

Schroeder, H. A., and M. Mitchener. 1975. Life-term effects of mercury, methylmercury and nine other trace metals on mice. J. Nutr. 105:452.

Scott, M. L., J. R. Zimmerman, S. Marinsky, P. A. Mullenhoff, G. L. Rumsey, and R. W. Rice. 1975. Effects of PCB's, DDT, and mercury compounds upon egg production, hatchability and shell quality in chickens and Japanese quail. Poult. Sci. 54:350.

Sell, J. L. 1977. Comparative effects of selenium on metabolism of methylmercury by chickens and quail: Tissue distribution and transfer into eggs. Poult. Sci. 56:939.

Sell, J. L., and K. L. Davidson. 1975. Metabolism of mercury, administration as methylmercuric chloride or mercuric chloride by lactating ruminants. J. Agric. Food Chem. 23:803.

Sell, J. L., W. Guenter, and M. Sifri. 1974. Distribution of mercury among components of eggs following the administration of methylmercuric chloride to chickens. J. Agric. Food Chem. 22:248.

Shaikh, Z., and J. C. Smith. 1975. Mercury induced synthesis of renal metallothionein. Int. Conf. Heavy Metals Environ. Toronto, Canada.

Shaikh, Z. A., R. L. Coleman, and O. J. Lucis. 1973. Sequestration of mercury by cadmium-induced metallothionein. Trace Subst. Environ. Health 7:313.

Siegmund, O. H., and C. M. Fraser (eds.). 1973. The Merck Veterinary Manual, 4th ed. Merck and Co., Inc., Rahway, N.J.

Skerfving, S. 1972. Mercury in fish—Some toxicological considerations. Food Cosmet. Toxicol. 10:545.

Skerfving, S. 1974. Methylmercury exposure, mercury levels in blood and hair, and health status in Swedes consuming contaminated fish. Toxicology 2:3.

Soares, J. H., Jr., D. Miller, H. Lagally, B. R. Stillings, P. Bauersfeld, and S. Cuppett. 1973. The comparative effect of oral ingestion of methyl mercury on chicks and rats. Poult. Sci. 52:452.

Stake, P. E., M. W. Neathery, W. J. Miller, and R. P. Gentry. 1975. $^{203}$Hg excretion and tissue distribution in Holstein calves following single tracer intravenous doses of methyl mercury chloride or mercuric chloride. J. Anim. Sci. 40:720.

Stillings, B. R., H. Lagally, P. Bauersfeld, and J. Soares. 1974. Effect of cystine, selenium and fish protein on the toxicity and metabolism of methylmercury in rats. Toxicol. Appl. Pharmacol. 30:243.

Stoewsand, B. S., C. A. Bache, and D. J. Lisk. 1974. Dietary selenium protection of methylmercury intoxication of Japanese quail. Bull. Environ. Contam. Toxicol. 11:152.

Thaxton, P., and C. R. Parkhurst. 1973a. Abnormal mating behavior and reproductive dysfunction caused by mercury in Japanese quail. Proc. Soc. Exp. Biol. Med. 144:252.

Thaxton, P., and C. R. Parkhurst. 1973b. Toxicity of mercury to young chickens. 2. Gross changes in organs. Poult. Sci. 52:277.

Thaxton, P., L. A. Cogburn, and C. R. Parkhurst. 1973. Dietary mercury as related to the blood chemistry in young chickens. Poult. Sci. 52:1212.

Thaxton, P., P. S. Young, L. A. Cogburn, and C. R. Parkhurst. 1974. Hematology of mercury compounds in young chickens. Bull. Environ. Contam. Toxicol. 12:46.

Thaxton, P., C. R. Parkhurst, L. A. Cogburn, and P. S. Young. 1975. Adrenal function in chickens experiencing mercury toxicity. Poult. Sci. 54:578.

Tichy, M., J. Haurdova, and M. Cekrt. 1975. Comments on the mechanism of excretion of mercury compounds via bile in rats. Arch. Toxicol. 33:267.

Tryphonas, L., and N. O. Nielsen. 1970. The pathology of arylmercurial poisoning in swine. Can. J. Comp. Med. 34:181.

Tryphonas, L., and N. O. Nielsen. 1973. Pathology of chronic alkylmercurial poisoning in swine. Am. J. Vet. Res. 34:379.

Underwood, E. J. 1977. Trace Elements in Human and Animal Nutrition, 4th ed. Academic Press, New York.

Webb, M. 1975. Toxicity of cadmium–thionein. Int. Conf. Heavy Metals Environ. Toronto, Canada (Abstr.).

Welsh, S. O. 1976. Influence of vitamin E on mercury poisoning in rats. Fed. Proc. 35:761. (Abstr.)

Welsh, S. O. 1979. The protective effect of vitamin E and N, N'-diphenyl-p-phenylenediamine (DPPD) against methylmercury toxicity in the rat. J. Nutr. 109:1673.

Welsh, S. O., and J. H. Soares, Jr. 1975. The effects of selenium and vitamin E on methyl mercury toxicity in the Japanese quail. Fed. Proc. 34:913 (Abstr.).

Westoo, G. 1966. Determination of methylmercury compounds in food stuffs. I. Methylmercury compounds in fish, identification and determination. Acta Chem. Scand. 20:2131.

Wobeser, G., N. O. Nielsen, and B. Schiefer. 1976a. Mercury and mink. I. The use of mercury contaminated fish as a food for ranch mink. Can. J. Comp. Med. 40:30.

Wobeser, G., N. O. Nielsen, and B. Schiefer. 1976b. Mercury and mink. II. Experimental methyl mercury intoxication. Can. J. Comp. Med. 40:34.

Won, J. H. 1973. The concentration of mercury, cadmium, lead and copper in fish and shell fish of Korea. Bull. Korean Fish. Soc. 6:1.

Wright, F. C., J. S. Palmer, and J. C. Riner. 1973. Accumulation of mercury in tissues of cattle, sheep and chickens given the mercurial fungicide, Panogen 15 orally. J. Agric. Food Chem. 21:414.

# Molybdenum

Molybdenum (Mo), discovered about 1782, now is a recognized ubiquitous element in the earth's surface and living matter. Molybdenite, the principal molybdenum ore, is found in close association with tin ore and is used industrially in the manufacture of various alloys. Present knowledge regarding the biological importance of molybdenum developed, in large part, from studies of its metabolic interrelationship with copper. The true nature of this interrelationship is still being researched. Several comprehensive reviews on molybdenum exist, including those of Schroeder *et al.* (1970) and Underwood (1976). Aspects of molybdenosis have been described by Ammerman and Miller (1975), Buck *et al.* (1973), Clarke and Clarke (1975), Kubota (1976), Ward (1976), and Poitevint and Nelson (1978).

## ESSENTIALITY

A biological requirement for molybdenum was first demonstrated by Bortels (1930), who found molybdenum to be an essential media nutrient for the growth of *Azotobacter* sp. Subsequently, nitrogen-fixing bacteria, such as the symbiotic *Rhizobia* sp. in legume roots, were shown to require molybdenum (Steinberg, 1936). Molybdenum-deficient soils were subsequently recognized, and significant improvements in their yields of pasture legumes were made by appropriate molybdenum applications.

When a long recognized severe diarrhea of cattle (teart scours) in England was shown to be caused by molybdenum toxicosis, attention turned to the metabolic significance of molybdenum in animals. The copper–molybdenum interrelationships were revealed when Ferguson *et al.* (1938) found molybdenum toxicosis could be controlled by copper supplementation and when Dick and Bull (1945) showed chronic copper toxicosis of sheep in Australia could be alleviated by molybdenum supplementation.

The major biochemical role of molybdenum in animals is currently believed to be in the formation and activity of xanthine oxidase (xanthine dehydrogenase), a molybdenum-containing metalloprotein essential for the metabolic degradation of purines to uric acid (DeRenzo *et al.*, 1953; Richert and Westerfield, 1953). Xanthine oxidase is present in microorganisms, animal tissues, and milk, especially cow's milk. Aldehyde and sulfide oxidases are also molybdenum-dependent enzymes present in animal tissues.

## METABOLISM

Most dietary forms of molybdenum, except molybdenite ($MoS_2$), are absorbed from the gastrointestinal tract, but the rates of absorption and routes of excretion may differ with species (Underwood, 1977). In swine, peak blood levels of molybdenum occur within 4 hours after an oral dose of $^{99}$molybdenum, and the urinary tract is the main excretory route for absorbed molybdenum. In cattle, peak blood levels from an equivalent dose of $^{99}$molybdenum are not reached until 96 hours post-administration, and the main route of molybdenum excretion is via the feces (Bell *et al.*, 1964). Peak plasma molybdenum levels in cattle fed 100 ppm supplemental molybdenum in cottonseed meal for 12 months were not reached until 7 months after the start of the feeding trial (Lesperance and Bohman, 1963). Absorbed molybdenum is also excreted via the milk from cattle and sheep in proportion to the levels of orally or parenterally administered molybdenum.

The rates of absorption, retention, and excretion of molybdenum are inversely related to the level of dietary inorganic sulfate. In sheep, for instance, increasing the dietary sulfate from 0.1 to 0.3 percent in a diet supplemented with 10 mg molybdenum per day decreased the molybdenum retention from 37 to 4 percent. A working hypothesis for the effect of sulfate on molybdenum retention is that sulfate inhibits membrane transport of molybdenum, thus decreasing absorption of molybdenum in the intestine and decreasing reabsorption of molybdenum by the renal tubules (Dick, 1956b). Other factors influencing molybdenum metabolism, in addition to copper and sulfate, include dietary levels of manganese, zinc, iron, lead, tungstate, ascorbic acid, methionine cysteine, and protein.

## SOURCES

Naturally growing herbage usually reflects the molybdenum content of the soil. Concentrations of molybdenum in normal herbage often range from 0.1 to 3 ppm (Underwood, 1977) on a dry weight basis. The molybdenum in herbage is present as water-soluble sodium and ammonium molybdate and as insoluble molybdenum oxide ($MoO_3$), calcium molybdate ($CaMoO_4$) and molybdenum sulfide ($MoS_2$). Only $MoS_2$ appears to be very poorly absorbed. Plants growing on soils industrially contaminated with molybdenum or containing naturally high levels of molybdenum have contained up to 231 ppm molybdenum (Gardner and Hall-Patch, 1962).

## TOXICOSIS

### LOW LEVELS

Manifestations of molybdenum toxicosis in cattle include diarrhea, anorexia, achromotrichia, and posterior weakness. Natural feedstuffs containing up to 6.2 ppm molybdenum were found by Smith *et al.* (1975) to be associated with bone malformations in calves. Cunningham *et al.* (1953) have reported that natural forages containing 25.6 ppm molybdenum were responsible for diarrhea, emaciation, anemia, achromotrichia, and even death in several age-groups of cattle. Huber *et al.* (1971) and Vanderveen and Keener (1964) reported that molybdenum levels up to 100 ppm had no effect in cattle, yet Gardner and Hall-Patch (1962) found cattle grazing industrially contaminated forages containing 85 ppm molybdenum developed diarrhea and locomotor disturbances. The achromotrichia, diarrhea, and reduced weight gains have also been demonstrated in cattle consuming molybdenum at 100 ppm (Lesperance and Bohman, 1963), at 173–300 ppm (Huber *et al.*, 1971), at 400 ppm (Cunningham *et al.*, 1953), and at 2.34 g per day (Britton and Goss, 1946). Molybdenum toxicosis has been observed in young lactating cattle consuming as little as 40 ppm molybdenum when the diets contained 0.3 percent sulfate (Vanderveen and Keener, 1964). These authors also reported that 200 ppm molybdenum, in conjunction with 0.3 percent dietary sulfate, produced posterior paresis in the young lactating cattle. It appears that 100–200 ppm dietary molybdenum are required to significantly increase the molybdenum content of milk (Cunningham *et al.*, 1953). Cattle with molybdenosis have also been reported to have an increased incidence of parturient hemoglobinuria (Goold and Smith, 1975). Thomas and Moss (1951) have observed decreased libido and testicular degeneration in young bulls fed 1–2 g sodium molybdate dihydrate daily for a period of 120 days.

The effects of molybdenum on hepatic copper levels depend upon the levels of dietary molybdenum and copper (Lesperance and Bohman, 1963). Levels of molybdenum up to 40 mg per day tend to decrease hepatic copper levels, while dietary molybdenum levels beyond 40 mg per day may alter hepatic copper levels very little (Ammerman and Miller, 1975). Dick (1956a) has reported that dietary copper levels of 8–10 ppm protect cattle against dietary molybdenum levels of approximately 5–6 ppm.

Sheep appear more resistant to molybdenosis than cattle and tolerate plasma molybdenum levels of 0.1–0.2 mg/dl, or approximately 20–40 times the normal plasma molybdenum levels, without affecting ceruloplasmin levels. This is true providing the dietary sulfate intake is about 0.1 percent (Dick, 1953a; Suttle, 1975). The manifestations of molybdenum-induced, secondary hypocuprosis include reduced crimp and pigmentation of wool, anemia, alopecia, and reduced weight gains. Neonates born to hypocupremic dams exhibit enzootic ataxia (swayback), a demyelinating disease that may also be accompanied by blindness. Dick and Bull (1945) found young ewes that consumed 10–100 mg molybdenum per day as ammonium molybdate for a 6-month period had significantly lower than normal liver copper levels regardless of the dietary copper level. Thus, the daily intake of molybdenum, which will alter liver copper levels in sheep, approximates 10 mg. In cases of elevated liver copper levels, Ross (1970) found that 100 mg molybdenum per day for 12 weeks caused a significant reduction in hepatic copper levels in sheep. Sheep fed 120 mg molybdenum and 7.4 g sulfate daily for 20 months were found to have reduced hemoglobin levels, reduced levels of copper in the wool, and increased levels of albumin-bound copper in blood (Bingley, 1974). Normal plasma copper levels were noted at molybdenum intakes of 12 mg/day.

Goodrich and Tillman (1966) investigated the effect of 2 and 8 ppm molybdenum on lambs receiving either 10 or 40 ppm copper and either 0.1 or 0.4 percent sulfate. Eight parts per million molybdenum eliminated the detrimental effects of the high sulfate level upon rate of gain and feed efficiency and also reduced liver copper levels. The latter effect was reversed by the addition of 40 ppm copper. A direct effect of molybdenum upon rumen sulfate levels has been demonstrated by Gawthorne and Nader (1976) in rumen fistulated sheep. Marcilese *et al.* (1976) suggested a higher turnover rate of ceruloplasmin in cattle than in sheep accounted for the difference in molybdenum tolerance between these two species.

Other ruminants are even more tolerant of molybdenum than sheep or cattle. For instance Nagy *et al.* (1975) found that mule deer tolerated up to 1,000 mg molybdenum per day without clinical signs. Daily molybdenum intakes by the mule deer of 2,500 to 5,000 mg caused diarrhea, and 5,000 to 7,500 mg caused anorexia. Appetite returned as soon as dietary molybdenum supplementation ceased.

Gipp *et al.* (1967) and Kline *et al.* (1973) have demonstrated little to no effect of 26 to 50 ppm molybdenum upon swine growth in the presence of supplemental copper and sulfate, while Davis (1950) reports no apparent effect of 1,000 ppm molybdenum in growing swine. Standish *et al.* (1975) have fed 1,500 ppm molybdenum to growing swine for 69 days in the presence of 17.8 ppm copper. These levels of molybdenum and copper caused a marked reduction in the rate of gain of these swine; however, the effects were reversible by 0.4 percent sulfate. High molybdenum levels in swine diets tend to promote copper storage in liver and kidney in contrast to an opposite effect in ruminants.

Avian species appear comparable to rodents in their susceptibility to molybdenum. Kratzer (1952) demonstrated a slight growth inhibition in young chickens fed 200 ppm molybdenum and a 25 percent growth inhibition in poults fed 300 ppm molybdenum. Davies *et al.* (1960) fed molybdenum to young chicks at levels ranging from 500 to 8,000 ppm. The effects ranged from growth depression and anemia at the low levels to 61 percent mortality at the highest level. Arthur *et al.* (1958) also induced anemia in young birds fed 4,000 ppm molybdenum for 8 weeks. Lepore and Miller (1965) indicated 500 ppm molybdenum in laying hens caused decreased hatchability, and 1,000 ppm molybdenum caused decreased egg production. Davies *et al.* (1960) have also indicated that ammonium molybdate is more toxic for birds than sodium molybdate.

Horses seem resistant to molybdenosis, for they can graze, without apparent problems, the same pastures that are known to cause diarrhea in cattle. However, clinical cases of rickets in foals and yearlings have been thought to be due to molybdenosis from pasture or dam's milk (Walsh and O'Moore, 1953).

Arrington and Davis (1953) have fed up to 4,000 ppm molybdenum to rabbits consuming a basal diet containing approximately 16.4 ppm copper. Rabbits fed 1,000 or more ppm molybdenum experienced anorexia, loss of weight, alopecia, a slight dermatosis, anemia, splayed front legs, and premature deaths. Manifestations of molybdenum toxicosis became apparent within 4 weeks in young rabbits and after a longer period in older rabbits. The toxic manifestations of molybdenum were alleviated by 200 ppm copper for at least a 4-month period. Molybdenosis in rabbits may decrease phosphorus absorption, increase phosphorus excretion, and result in a ricketic syndrome.

The tolerance of rats for dietary molybdenum is, as with other species, dependent upon the dietary levels of copper and sulfate. Gray and Daniel (1964) found young copper-deficient rats experienced reductions in growth rate, liver copper, and blood hemoglobin levels when fed 10–1,000 ppm molybdenum. Dietary copper supplementation for these rats at 3 ppm caused their growth rate, liver copper, and blood hemoglobin levels to return to normal. Miller *et al.* (1972) reported reduced growth rate of young rats fed 100 ppm molybdenum could be prevented by $SO_4$ supplemention. The reports of Whanger and Weswig (1970), Gray and Ellis (1950), Gray and Daniel (1954), Halverson *et al.* (1960), Comperé *et al.* (1965), and Nielands *et al.* (1948) indicate that dietary molybdenum levels in excess of 500 ppm in rats impair growth, increase blood and liver copper levels, decrease ceruloplasmin, and increase tissue molybdenum levels. In general, all these effects are usually alleviated by copper and/or sulfate supplementation.

In guinea pigs fed 100 ppm molybdenum in the presence of adequate copper, Smith and Wright (1975) found a tricholoracetic acid-insoluble, copper–molybdenum complex in plasma, which they felt accounted for the absence of a significant elevation in liver molybdenum during molybdenosis in this species.

### HIGH LEVELS

Very few molybdenum toxicity studies in which sufficient molybdenum was used to cause death of the animals have been reported. Davies *et al.* (1960) found 6,000–8,000 ppm molybdenum as sodium or ammonium molybdate caused 30–60 percent mortality in a 4-week period in growing chickens, and the ammonium molybdate appeared to be the more toxic. Robitaille and Bilek (1976) found the $LD_{50}$ for molybdenum in tank water for fish to be 7,340 ppm for trout in 96 hours. The $LD_{50}$ for oral ammonium molybdate in guinea pigs is reported to be 2.2 g/kg of body weight, and the MLD of sodium molybdate intraperitoneally in rats is 290 mg/kg (Stecher *et al.,* 1968).

### FACTORS INFLUENCING TOXICITY

As has been implied, the effects of excess molybdenum are essentially those of copper deficiency. The integumental changes, including rough hair coat, achromotrichia, and loss of crimp in the wool, are related to a deficiency of the copper-dependent enzyme, tyrosinase. The anemia of molybdenum toxicosis is related to a deficiency of a second copper-dependent enzyme, ferroxidase. The skeletal and/or collagenous manifestations of molybdenum toxicosis are also related to a deficiency of a third copper-dependent enzyme, dopamine $\beta$ hydroxylase. It is also probable that the general growth retardation and anorexia associated with molybdenosis may relate to deficiencies of a fourth copper-dependent enzyme, cytochrome C oxidase.

Inorganic sulfate supplements appear to reverse all the manifestations of molybdenosis except the increased copper storage by the liver. There is also some indication that the effects of molybdenum on copper levels of milk, ceruloplasmin, albumin, and urine are not reversed by sulfate. The apparent effects of molybdenum are also influenced by manganese, zinc, iron, lead, tungstate, ascorbic acid, methionine, cysteine, protein, and alkalinity of soils. The bases for many of these interactions are yet unexplained.

Miller *et al.* (1972) present paradoxical data suggesting that ruminal processes decrease the biological availability of molybdenum. Gawthorne and Nader (1976) report that molybdenum and sulfate must be supplied in the diets simultaneously before copper concentration in ruminant liver is decreased. In rodents, endogenous sulfur from the metabolism of sulfur-containing amino acids appears not to function as dietary sulfate in altering the effects of molybdenum as occurs in ruminants (Mills *et al.,* 1958; Cook *et al.,* 1966).

## TISSUE LEVELS

Solid tissue, blood, and milk levels of molybdenum are readily altered by changes in dietary molybdenum levels, and the magnitude of the tissue response to elevations in dietary molybdenum depends on concomitant inorganic sulfate levels (Dick, 1953a), tungstate (Davies *et al.,* 1960), and copper. The concentrations of molybdenum in liver of animals on normal diets range from 2 to 4 ppm on a dry matter basis and may be as high as 30 ppm if the animals were consuming high levels of molybdenum (Gray and Daniel, 1964). Renal molybdenum concentrations approximate 50 percent of the liver concentrations of molybdenum (Underwood, 1977), and other tissues in a declining order of usual molybdenum concentration are: spleen, lung, brain, bone, and muscle. The total quantity of molybdenum in a skeleton is greater than 50 percent of the total molybdenum in the body (Dick, 1969). The molybdenum levels of whole blood of sheep and cattle on low molybdenum diets can range from 1 to 6 $\mu$g/dl (Beck, cited by Underwood, 1977). The concentration of molybdenum in milk of cattle fed standard diets ranged from 18 to 120 $\mu$g/1 with a mean of 73 $\mu$g (Archibald, 1951) and is primarily associated with the nonlipid fraction of milk, specifically, the xanthine oxidase. Extreme levels of molybdenum in excess of 1 ppm in milk have been associated with high molybdenum pastures.

## MAXIMUM TOLERABLE LEVELS

Estimates of the maximum tolerable levels for molybdenum for several species are presented in Table 25. These range from levels of 5 to 10 ppm, which have been weakly associated with impaired bone development in young horses and cattle, respectively, to very high tolerance levels approximating 1,000 ppm for swine. It must be emphasized that substantially higher levels of molybdenum would be tolerated in the presence of adequate copper and inorganic sulfate.

## SUMMARY

Molybdenum is an essential trace element and a component of xanthine oxidase that is important in purine metabolism. The soils and resulting herbage in some geographic areas have relatively high molybdenum levels that account for a regional incidence of molybdenosis in livestock. This disease is essentially a secondary copper deficiency manifested by diarrhea, anorexia, depigmentation of hair or wool, neurologic disturbances, and premature death. A wide variation in the apparent susceptibility of various livestock species to molybdenum toxicity is due to variations in concurrent dietary levels of copper, zinc, sulfur, silver, cadmium, and sulfur-containing amino acids. The wide tolerance limits range from 6.2 ppm in growing cattle to approximately 1,000 ppm in adult mule deer.

TABLE 25 Effects of Molybdenum Administration in Animals

| Class and Number of Animals[a] | Age or Weight | Administration: Quantity of Element[b] | Source | Duration | Route | Effect(s) | Reference |
|---|---|---|---|---|---|---|---|
| Cattle—50 | 5–12 mo | Up to 6.2 ppm | Natural forage | 5–12 mo | Diet | Abnormal distal metacarpal and tarsal growth plates | Smith *et al.*, 1975 |
| Cattle | Various ages | 25.6 ppm | High Mo pasture | 23 d | Diet | Diarrhea; emaciation; anemia; achromotrichia; death | Cunningham *et al.*, 1953 |
| | Lactating | 100–200 ppm | | 23 d | Diet | Transient increase in Mo level in milk | |
| | | 400 ppm | | 23 d | | Increased milk Mo and toxic effects | |
| Cattle—8 | Lactating | 53–100 ppm | Molybdate | Up to 6 mo | Diet | No effect on liver, blood, or milk Mo levels | Huber *et al.*, 1971 |
| Cattle—4 | | 173–300 ppm | Molybdate | Up to 6 mo | | Diarrhea; inanition; increased milk Cu; decreased liver Cu | |
| Cattle—39 | Primaparous | 5–40 ppm | $(NH_4)_2MoO_4$ | 300 d | Diet | No observed effects | Vanderveen and Keener, 1964 |
| Cattle | | 5–20 ppm | w/o·3% $SO_4$ | | | No observed effects | |
| | | 50 ppm | | | | Alopecia and achromotrichia | |
| | | 200 ppm | | | | Alopecia, achromotrichia, and posterior paresis | |

| | | | | | | | |
|---|---|---|---|---|---|---|---|
| Cattle—25 | Various | 85 ppm | $MoO_3$ | 11 d | Contaminated pasture | Diarrhea and locomotor disturbances within 5 days | Gardner and Hall-Patch, 1962 |
| Cattle—32 | Weanling heifers | 100 ppm | | 1 yr | Diet | Achromotrichia; diarrhea; reduced gains | Lesperance and Bohman, 1961 |
| Cattle—1 | 220 kg female | 2.34 g/d | $Na_2MoO_4$ | 7 mo | Diet | Diarrhea; achromotrichia; 20% weight loss | Britton and Goss, 1946 |
| Cattle—2 | Approx. 200 d | 1–2 g/d | $Na_2MoO_4 \cdot 2H_2O$ | 123 d | Capsule | Reduced gain; achromotrichia; decreased libido; testicular degeneration | Thomas and Moss, 1951 |
| Sheep—2 | Wethers | 10 mg/d with low $SO_4$ diet | $(NH_4)_2MoO_4$ | 34 d | Diet | Increased blood Mo to 2 ppm | Dick, 1953b |
| Sheep—5 | Yearling | 10 mg/d | $(NH_4)_2MoO_4$ | 6 mo | Diet | Decreased liver Cu | Dick and Bull, 1945 |
| Sheep—4 | | 100 mg/d | | | | Decreased liver Cu | |
| Sheep—4 | Lambs | 100 mg/d with 15 ppm Cu | | 13 wk | Diet | Decreased liver Cu from 1,576 to 805 ppm | Ross, 1970 |
| Sheep—16 | 1 yr | 120 mg/d with 7.4 g $SO_4$/d | | 2.5 yr | Diet | Increased plasma Cu; decreased crimp and Cu in wool | Bingley, 1974 |
| Sheep—80 | 27–34 kg | 8 ppm with 10 ppm Cu | $Na_2MoO_4 \cdot 2H_2O$ | 2 mo | Diet | Decreased liver Cu | Goodrich and Tillman, 1966 |
| | | 8 ppm with 40 ppm Cu | | | Diet | Increased liver Cu | |
| | | 8 ppm with 0.1–0.4% $SO_4$ | | | Diet | Decreased liver Cu at 0.4% $SO_4$ | |

## TABLE 25 *Continued*

| Class and Number of Animals[a] | Age or Weight | Administration: Quantity of Element[b] | Source | Duration | Route | Effect(s) | Reference |
|---|---|---|---|---|---|---|---|
| Sheep—4 | 33–36 kg | 50 mg/d | $Na_2MoO_4$ | 18 d | Intraruminal | Rumen $SO_4$ increased from 2.2 to 7.2 g/l | Gawthorne and Nader, 1976 |
| Mule deer | 2–4 yr | 0–1 g/d | $Na_2MoO_4$ | 27 d | Diet | No clinical effect noted | Nagy *et al.*, 1975 |
| | | 2.5–5.0 g/d | | | | Diarrhea | |
| | | 5.0–7.5 g/d | | | | Anorexia | |
| Swine—6 | Growing | 26 ppm | $Na_2MoSO_4$ | 9 wk | Diet | No adverse effect | Gipp *et al.*, 1967 |
| | Growing | 26 ppm with 1,500 ppm Cu | | 9 wk | Diet | Slight decrease in rate of gain; increase in liver copper | |
| | | 26 ppm with 1,000 ppm $SO_4$ | | | Diet | Decreased rate of gain | |
| Swine—208 | Growing | 50 ppm with 500 ppm CU | $Na_2MoSO_4$ | 61 d | Diet | No prevention of Cu toxicosis | Kline *et al.*, 1973 |
| Swine | Growing | 1,000 ppm | | 90 d | Diet | No adverse effects | Davis, 1950 |
| Swine—4 | 11.9-kg barrows | 1,500 ppm with 17.8 ppm Cu and 0.4% S | $Na_2MoO_4 \cdot 2H_2O$ | 69 d | Diet | Depressed growth with time | Standish *et al.*, 1975 |
| | | | | | | Enhanced plasma Cu clearance | |
| Chicken—30 | Young | 200–300 ppm | $Na_2MoO_4$ | 4 wk | Diet | Slight growth reduction | Kratzer, 1952 |
| Chicken—20 | Young | 500 ppm | | 4 wk | Diet | Decreased growth | Davies *et al.*, 1960 |
| | Young | 4,000 ppm | | | | Decreased growth and anemia | |

| | | | | | | | | |
|---|---|---|---|---|---|---|---|---|
| Chicken | Young | 6,000 ppm | $Na_2MoO_4$ and $(NH_4)_2MoO_4$ | | | Diet | 33% mortality | |
| | Young | 8,000 ppm | $Na_2MoO_4$ and $(NH_4)_2MoO_4$ | | | | 61% mortality; $(NH_4)_2MoO_4$ more toxic | |
| Chicken—4 | Adults | 500 ppm | $Na_2MoO_4$ | | 21 d | Diet | Weight loss and reduced hatchability | Lepore and Miller, 1965 |
| | Adults | 1,000 ppm | $Na_2MoO_4$ | | 21 d | | Decreased egg production | |
| Turkey—23 | 1 wk | 300 ppm | $Na_2MoO_4$ | | 4 wk | Diet | Growth rate reduced 25%; no diarrhea or anemia | Kratzer, 1952 |
| Horse | Foals and yearlings | 5–22 ppm | Pasture | Daily | | Diet | Associated with rachitis | Walsh and O'Moore, 1953 |
| Rabbit—31 | Young | 140 ppm with 16.4 ppm Cu | $Na_2MoO_4 \cdot 2H_2O$ | | 4 mo | Diet | No adverse effect | Arrington and Davis, 1953 |
| Rabbit | | Approx. 500 ppm | | | | | No adverse effect | |
| | | Approx. 1,000 ppm | | | 4 wk | | Anorexia; weight loss; dermatosis; reduced bone phosphorus | |
| | | 2,000 ppm | | | | | Splayed forelegs and death | |
| | | 4,000 ppm with 200 ppm Cu | | | 4 mo | Diet | No adverse effect | |
| Rat | Young | 10–100 ppm in Cu-deficient diet | $Na_2MoO_4$ | | 4 wk | Diet | Decreased growth, liver Cu, and hemoglobin levels | Gray and Daniel, 1964 |
| | | 10 ppm with 3 ppm Cu | | | | Diet | No adverse effect | |
| Rat | Young | 75 ppm | $Na_2MoO_4$ | | 5 wk | Diet | Increased liver Cu and Mo | Miller *et al.*, 1972 |
| | | 100 ppm | | | | | Reduced growth (preventable with $SO_4$) | |

TABLE 25 *Continued*

| Class and Number of Animals[a] | Age or Weight | Administration: Quantity of Element[b] | Source | Duration | Route | Effect(s) | Reference |
|---|---|---|---|---|---|---|---|
| Rat | 4 wk | 500 ppm with 6 ppm Cu | $Na_2MoO_4 \cdot 2H_2O$ | 4 wk | Diet | Reduced ceruloplasmin | Whanger and Weswig, 1970 |
| | | 1,000 ppm with 1 ppm Cu and 20 or 80 μg Cu IV | | | Diet | Reduced ceruloplasmin | |
| Rat—6 | Weanling | 880 ppm with 1.2% methionine | $Na_2MO_4$ | 6 wk | Diet | Reduced growth spared by methionine | Gray and Daniel, 1954 |
| | | 800 ppm with 0.6% methionine plus 300 ppm Cu | | | Diet | Reduced toxicosis signs | |
| Rat—10 | Weanling | 800 ppm with 15.6 ppm Cu | $Na_2MoO_4$ | 41 d | Diet | Increased liver Cu | Halverson *et al.*, 1960 |
| | | 800 ppm with 15.6 ppm Cu plus 0.29% $SO_4$ | | | Diet | No adverse effect | |

| | | | | | | | |
|---|---|---|---|---|---|---|---|
| Rat—10 | Approx. 150 g | 14.24 mg/d | $Na_2MoO_4 \cdot 2H_2O$ | 3 wk | Gavage | Increased blood and liver Cu; increased sp. gr. of blood | Comperé *et al.*, 1965 |
| | | 14.24 mg/d plus 4 mg Cu | | | Diet | No adverse effect | |
| Rat | 3 wk | 500–5,000 ppm | $Na_2MoO_4 \cdot 2H_2O$ | 4 wk | Diet | Reduced growth at 500 ppm Mo and above; no diarrhea | Nielands *et al.*, 1948 |
| Rat—48 | Weanling | 400–1,200 ppm | $Na_2MoO_4 \cdot H_2O$ | 7 wk | Diet | Rough hair; reduced growth; increased tissue Mo | Gray and Ellis, 1950 |
| | | 4,000 ppm | | | | Rough hair; reduced growth; increased tissue Mo | |
| Guinea pig | 450–500 g | 100 ppm | | 6 wk | Diet | Increased liver Mo; no other effect | Smith and Wright, 1975 |
| Trout | | 7,340 ppm | $Na_2MoO_4$ | 96 h | Tank water | 50% mortality ($TL_{50}$) | Robitaille and Bilek, 1976 |

[a] Number of animals per treatment.

[b] Quantity expressed in parts per million or as milligrams per kilogram of body weight.

## REFERENCES


Ammerman, C. B., and S. M. Miller. 1975. Molybdenum in ruminant nutrition: A review. (Unpublished).

Archibald, J. G. 1951. Molybdenum in cows' milk. J. Dairy Sci. 34:1026.

Arrington, L. R., and G. K. Davis. 1953. Molybdenum toxicity in the rabbit. J. Nutr. 51:295.

Arthur, D. I., I. Motzok, and H. D. Branion. 1958. Interaction of dietary copper and molybdenum in rations fed to poultry. Poult. Sci. 37:1181.

Bell, M. D., G. B. Diggs, R. S. Lowrey, and P. L. Wright. 1964. Comparison of $Mo^{99}$ metabolism in swine and cattle as affected by stable molybdate. J. Nutr. 84:367.

Bingley, J. R. 1974. Effects of high doses of molybdenum and sulphate on the distribution of copper in plasma and in blood of sheep. Aust. J. Agric. Res. 25:467.

Bortels, H. 1930. Molydan als Katalysator bei der biologischen Stockstoffbundung. Arch. Mikrobiol. 1:333.

Britton, J. W., and H. Goss. 1946. Chronic molybdenum poisoning in cattle. J. Am. Vet. Med. Assoc. 108:176.

Buck, W. B., G. D. Osweiler, and G. A. Van Gelder. 1973. Clinical and Diagnostic Veterinary Toxicology. Kendall/Hunt Publishing Company, Dubuque, Iowa.

Clarke, E. G. D., and M. L. Clarke. 1975. Veterinary Toxicology. Williams & Wilkins, Baltimore, Md.

Comperé, R., A. Burny, A. Riga, E. Francois, and S. Vanuytrecht. 1965. Copper in the treatment of molybdenosis in the rat: Determination of the dose of the antidote. J. Nutr. 87:412.

Cook, G. A., A. L. Lesperance, V. R. Bohman, and E. H. Jensen. 1966. Interrelationship of molybdenum and certain factors to the development of the molybdenum toxicity syndrome. J. Anim. Sci. 25:96.

Cunningham, H. M., J. M. Brown, and A. E. Edie. 1953. Molybdenum poisoning of cattle in the Swan River Valley of Manitoba. Can. J. Agric. Sci. 33:254.

Davies, R. E., B. L. Reid, A. A. Kurnick, and J. R. Couch. 1960. The effect of sulfate on molybdenum toxicity in the chick. J. Nutr. 70:193.

Davis, G. K. 1950. The influence of copper on the metabolism of phosphorus and molybdenum. *In* W. D. McElroy and B. Glass, eds. A Symposium on Copper Metabolism. Johns Hopkins Press, Baltimore, Md.

DeRenzo, E. C., E. Kaleita, P. G. Heytler, J. J. Oleson, B. L. Hutchings, and J. H. Williams. 1953. Identification of the xanthine oxidase factor as molybdenum. Arch. Biochem. Biophys. 45:247.

Dick, A. T. 1953a. The effect of inorganic sulfate on the excretion of molybdenum in sheep. Aust. Vet. J. 29:18.

Dick, A. T. 1953b. The control of copper storage in the liver of sheep by inorganic sulfate and molybdenum. Aust. Vet. J. 29:233.

Dick, A. T. 1956a. Molybdenum in animal nutrition. Soil Sci. 81:229.

Dick, A. T. 1956b. *In* W. D. McElroy and B. Glass (eds.). Inorganic Nitrogen Metabolism. Johns Hopkins Press, Baltimore, Md. 445 pp.

Dick, A. T. 1969. The copper–molybdenum complex in ruminant nutrition. Outlook Agric. 6:14.

Dick, A. T., and L. B. Bull. 1945. Some preliminary observations on the effect of molybdenum on copper metabolism in herbivorous animals. Aust. Vet. J. 21:70.

Ferguson, W. S., A. H. Lewis, and S. J. Watson. 1938. Action of molybdenum in nutrition of milking cattle. Nature (London) 141:553.

Gardner, A. W., and P. K. Hall-Patch. 1962. An outbreak of industrial molybdenosis. Vet. Rec. 74:113.

Gawthorne, J. M., and C. J. Nader. 1976. The effect of molybdenum on the conversion of sulphate to sulphide and microbial–protein–sulphur in the rumen of sheep. Br. J. Nutr. 35:11.

Gipp, W. F., W. G. Pond, and S. E. Smith. 1967. Effects of level of dietary copper, molybdenum, sulfate and zinc on body weight gain, hemoglobin, and liver storage of growing pigs. J. Anim. Sci. 26:727.

Goodrich, R. D., and A. D. Tillman. 1966. Cooper, sulphate and molybdenum interrelationships in sheep. J. Nutr. 90:76.

Goold, G. J., and B. Smith. 1975. The effects of copper supplementation on stock health and production. The effect of parenteral copper on the milk yield characteristics of a dairy herd with hypocuprosis. N.Z. Vet. J. 23:233.

Gray, L. F., and L. J. Daniel. 1954. Some effects of excess molybdenum on the nutrition of the rat. J. Nutr. 53:43.

Gray, L. F., and L. J. Daniel. 1964. Effect of the copper status of the rat on the copper–molybdenum–sulfate interaction. J. Nutr. 84:31.

Gray, L. F., and G. H. Ellis. 1950. Some interrelationships of copper, molybdenum, zinc and lead in the nutrition of the rat. J. Nutr. 40:441.

Halverson, A. W., S. H. Phifer, and K. J. Monty. 1960. A mechanism for the copper–molybdenum interrelationship. J. Nutr. 71:95.

Huber, J. T., N. O. Price, and R. W. Engel. 1971. Response of lactating dairy cows to high levels of dietary molybdenum. J. Anim. Sci. 32:364.

Kline, R. D., M. A. Corzo, V. W. Hays, and G. L. Cromwell. 1973. Related effects of copper, molybdenum and sulfide on performance, hematology, and copper stores of growing pigs. J. Anim. Sci. 37:936.

Kratzer, F. H. 1952. Effect of dietary molybdenum upon chicks and poults. Proc. Soc. Exp. Biol. Med. 80:483.

Kubota, J. 1976. Molybdenum status of United States soils and plants. *In* W. Chappel and K. Peterson (eds.). The Biology of Molybdenum. Marcel Dekker, Inc., New York.

Lepore, P. D., and R. F. Miller. 1965. Embryonic viability as influenced by excess molybdenum in chicken breeder diets. Proc. Soc. Exp. Biol. Med. 118:155.

Lesperance, A. L., and V. R. Bohman. 1961. Criteria for measuring molybdenum toxicity. J. Anim. Sci. 20:940.

Lesperance, A. L., and V. R. Bohman. 1963. Effect of inorganic molybdenum and type of roughage on the bovine. J. Anim. Sci. 22:686.

Marcilese, N. A., R. M. Valsecchi, and H. D. Figueiras. 1976. Studies with $^{67}$Cu and $^{64}$Cu in conditioned copper-deficient ruminants. Preliminary results. *From* International Symposium on Nuclear Techniques in Animal Productions and Health as Related to the Soil-Plant System. Vienna, IAEA/FAO.

Miller, J. R., B. R. Moss, M. C. Bell, and N. N. Snead. 1972. Comparison of $^{99}$Mo metabolism in young cattle and swine. J. Anim. Sci. 34:846.

Mills, C. F., K. L. Monty, A. Ichihara, and P. B. Pearson. 1958. Metabolic effects of molybdenum toxicity in the rat. J. Nutr. 63:129.

Nagy, J. G., W. Chappel, and G. M. Ward. 1975. Effects of high molybdenum intake in mule deer. J. Anim. Sci. 41:412.

Nielands, J. B., F. M. Strong, and C. A. Elvehjem. 1948. Molybdenum in the nutrition of the rat. J. Biol. Chem. 172:431.

Poitevint, A. L., and J. D. Nelson. 1978. Molybdenum in animal nutrition toxicity and requirements, p. 36. *In* Proc. Ga. Nutr. Conf.

Richert, D. A., and W. W. Westerfield. 1953. Isolation and identification of the xanthine oxidase factor as molybdenum. J. Biol. Chem. 203:915.

Robitaille, D. R., and J. G. Bilek. 1976. Molybdate cooling water treatments. Chem. Eng. 83:77

Ross, D. B. 1970. The effect of oral ammonium molybdate and sodium sulfate given to lambs with high liver copper concentrations. Res. Vet. Sci. 11:295.

Schroeder, H. A., J. J. Blasassa, and I. H. Tipton. 1970. Essential trace metals in man: Molybdenum. J. Chron. Dis. 23:481.

Smith, B. P., G. L. Fisher, P. W. Poulos, and M. R. Irwin. 1975. Abnormal bone development and lameness associated with secondary copper deficiency in young cattle. J. Am. Vet. Med. Assoc. 166:682.

Smith, S. W., and H. Wright. 1975. Effect of dietary Mo on Cu metabolism. Evidence of the involvement of Mo in abnormal binding of Cu to plasma protein. Clin. Chem. Acta 62:55.

Standish, J. F., C. B. Ammerman, N. D. Wallace, and G. E. Combs. 1975. Effect of high dietary molybdenum and sulfate on plasma copper clearance and tissue minerals in growing swine. J. Anim. Sci. 40:509.

Stecher, P. G., M. Windholz, and D. S. Leahy (eds.). 1968. The Merck Index. An Encyclopedia of Chemicals and Drugs, 8th ed. Merck and Co., Inc., Rahway, N.J.

Steinberg, R. A. 1936. Relation of accessory growth substances to heavy metals including molybdenum in the nutrition of *Aspergillus niger*. J. Agric. Res. 52: 439.

Suttle, N. F. 1975. The role of organic sulfur in the copper–molybdenum–S interrelationship in ruminant nutrition. Br. J. Nutr. 34:411.

Thomas, J. W., and S. Moss. 1951. The effect of orally administered molybdenum on growth, spermatogenesis and testes histology of young dairy bulls. J. Dairy Sci. 34:939.

Underwood, E. J. 1976. Molybdenum in animal nutrition. *In* W. Chappel and K. Peterson, eds. The Biology of Molybdenum. Marcel Dekker, Inc., New York.

Underwood, E. J. 1977. Trace Elements in Human and Animal Nutrition, 4th ed. Academic Press. New York. 116 pp.

Vanderveen, J. E., and H. A. Keener. 1964. Effects of molybdenum and sulfate sulphur on metabolism of copper in dairy cattle. J. Dairy Sci. 47:1224.

Walsh, T., and L. B. O'Moore. 1953. Excess of molybdenum in herbage as a possible contributory factor in equine osteodystrophia. Nature (London) 171:1166.

Ward, G. M. 1976. Molybdenum toxicity and hypocuprosis in ruminants. A review. Invited paper, 68th Annual Meeting, American Society of Animal Science, College Station, Tex. (Unpublished)

Whanger, P. D., and P. H. Weswig. 1970. Effect of some copper antagonists on induction of ceruloplasmin in the rat. J. Nutr. 100:341.

# Nickel

Nickel (Ni) is a hard silver-white ferromagnetic metal that can be polished to a lustrous finish. It occurs in several ores (chalcopyrite, pyrrhotite, pentlandite, garnierite, niccolite, and millerite) and the element constitutes 0.008 percent of the earth's crust. Nickel is used in alloys, storage batteries, for electroplating, and in Raney nickel, a catalyst used for hydrogenation of organic compounds. In 1971 the United States produced over 15,000 tons of nickel; however, the total consumption in this country during 1972 was estimated to be 159,286 tons (National Research Council, 1975).

With diets very low in nickel, animals maintained in specially clean environments have failed to grow, develop, and reproduce normally; however, nickel deficiencies have not been observed under typical laboratory or practical conditions. Nickel in various forms is relatively nontoxic when consumed orally; however, workers exposed to airborne nickel have an increased incidence of respiratory disease, including cancer. Some individuals develop very marked dermal sensitivity to nickel. The sources, distribution, industrial uses, and biological effects of nickel were reviewed in detail by the National Research Council (1975), and Nielsen (1977) reviewed nickel toxicity.

## ESSENTIALITY

Nielsen and co-workers obtained the first evidence of nickel deficiencies in chicks; the early studies of these and other workers were reviewed (Nielsen, 1974; Nielsen and Ollerich, 1974; Underwood, 1977).

The diets used to produce deficiency have been very low, ranging from 2 to 40 ppb nickel. It was generally necessary to maintain the animals in filtered air environments with nickel sources rigorously excluded and/or to feed the diet throughout a lifetime or through more than one generation. The requirement has been estimated to be 50–80 ppb for the rat and chick (Nielsen and Sandstead, 1974).

Deficiencies of nickel have been produced in chicks, pigs, goats, and rats, as reviewed by Nielsen and Sandstead (1974), and sheep (Spears *et al.,* 1978a,b). Abnormalities observed in deficient animals have varied markedly between species, between laboratories, and within laboratories in successive experiments. The differences appear to be related to degree of deficiency, adequacy of the diet in nutrients other than nickel, other aspects of dietary composition, and inadequacies in the filtered environment. The problems of dietary adequacy and environment have been significantly improved for chicks (Nielsen *et al.,* 1975a) and rats (Nielsen *et al.,* 1975b).

Abnormalities observed in deficient chicks included depressed hematocrits, less yellow lipochrome pigment in the shank skin, and abnormalities in the rough endoplasmic reticulum of the liver (Nielsen *et al.,* 1975a). Similar liver pathology had been described in nickel-deficient chicks by Sunderman *et al.* (1972). Nickel deficiency has caused reproductive problems in goats and swine (Anke *et al.,* 1974). Delayed sexual maturity occurred in the sows; there was high mortality of the young pigs. Nickel-deprived lambs showed depressed growth, total serum proteins, erythrocyte counts, total liver lipids and cholesterol, serum alanine transaminase levels, dietary nitrogen utilization, and liver copper concentration (Spears *et al.,* 1978a,b).

Deficient rats had increased perinatal mortality, a rough hair coat, and ultrastructural changes in the liver (Nielsen *et al.,* 1975b). In deficient rats, Schnegg and Kirchgessner (1978) observed depressed growth, hematocrits, hemoglobin levels, and erythrocyte counts; serum levels of urea, ATP, and glucose; liver levels of triglycerides, glucose, and glycogen; liver, kidney, and spleen levels of zinc, iron, and copper; and activities of several liver and kidney enzymes. They also found marked impairment of iron absorption.

## METABOLISM

Underwood (1977) has reviewed information on nickel absorption and tissue distribution. Absorption of normal dietary nickel levels are 10 percent or less. Daily urinary excretion appears to equal amounts of nickel absorbed when nickel intake is normal. There was little accumulation of nickel in tissues of rats receiving 5 ppm nickel in their drinking water throughout their lifetime (Schroeder *et al.,* 1974).

Nickel was shown to be a component of urease. Lambs fed a low nickel diet had lower levels of ruminal urease than lambs fed 5 ppm nickel (Spears *et al.,* 1977).

The metabolic mechanisms for any essential functions of nickel have not been established.

## SOURCES

Relatively few data are available on the nickel content of animal feeds. Whole oats and rye seeds contained 2–3 ppm nickel, whereas concentrations in wheat ranged from 0 to 0.5 ppm (Schroeder *et al.,* 1962; Zook *et al.,* 1970). Commercial dog and rat diet contained 2.1 and 3.3 ppm nickel, respectively (Schroeder *et al.,* 1962). Nickel in a purified rat diet containing 20 percent casein was 0.21 ppm (Whanger, 1973), and in an unrefined diet for cattle it was 0.9 ppm (O'Dell *et al.,* 1970b). Nickel concentrations in components used in diets for cattle were corn, 0.4 ppm; oats, 1 ppm; soybean meal, 3.6 ppm; alfalfa meal, 1.4 ppm; and cottonseed hulls, 0.6 ppm (O'Dell *et al.,* 1971). Plant foods are generally higher in nickel than foods of animal origin. Nickel in common pasture plants ranged from 0.5 to 3.5 ppm (Underwood, 1977).

The nickel content of water is typically very low. The concentrations in the major river basins and water supplies of the United States were usually less than 10 ppb (National Research Council, 1975). Higher concentrations may occur in water of industrial areas.

There is no need to add nickel to practical animal diets. Nickel from food machinery can contribute significant amounts of nickel to processed foods. Nickel chloride has been used in studies of nickel essentiality. Little is known about the form of nickel in foods and its bioavailability.

Nickel from industrial operations can contribute locally to the nickel level in air, water, and soil. There appear to be no widespread problems for either man or animals from exposure to these sources.

## TOXICOSIS

### LOW LEVELS

Nickel is relatively nontoxic in terms of the quantities required above typical dietary intakes to produce adverse effects in a few weeks. Lactating dairy cows were unaffected by 145, 365, or 1,835 mg nickel per day (Archibald, 1949; O'Dell *et al.*, 1970a). The latter two levels represented 50 and 250 ppm nickel in the diet, respectively. Young calves were not affected by 250 ppm dietary nickel as the carbonate or 50 ppm as the chloride; however, they consumed less food when the diet contained 500 or 1,000 ppm nickel as the carbonate or 100 and 200 ppm as the chloride (O'Dell *et al.*, 1970c). In these experiments, the calves were offered a choice of basal diet or nickel-supplemented diet. A linear depression of palatability, as judged by consumption of nickel-containing diet, was observed as nickel in the diet increased. Total food consumption was unaffected. Levels of 62.5 and 250 ppm nickel as the carbonate caused no adverse health effects or increases in tissue nickel levels of calves after 8 weeks of feeding (O'Dell *et al.*, 1970b, 1971).

The young growing chick responded similarly to the young calf. Dietary levels of 100 and 300 ppm nickel as either the carbonate or the acetate had no adverse effects when fed from hatching to 4 weeks of age (Weber and Reid, 1968). Levels of 500 ppm or more reduced growth.

Dogs fed either 100 ppm or 1,000 ppm nickel as the sulfate showed no adverse effects after 2 years (Ambrose *et al.*, 1976).

Schroeder *et al.* (1974) gave weanling rats 5 ppm nickel in their drinking water for their remaining lifetime. No adverse effects were observed. When weanling rats received 5 ppm nickel in their drinking water and were carried through three generations, more young rats died in each generation (Schroeder and Mitchener, 1971). Significant numbers of runts occurred in the F1 and F3 generations.

Phatak and Patwardhan (1950) fed 250, 500, or 1,000 ppm nickel to rats for 3 to 4 months. They tested nickel carbonate, nickel soap, and nickel catalyst. The soap was prepared from nickel carbonate and mixed fatty acids obtained from refined groundnut oil. The nickel catalyst was finely divided nickel suspended in vegetable oil and supported on kieselguhr. These represent forms that could be present in small amounts in hydrogenated fats. They observed no adverse effects with these levels of nickel. The experimental period encompassed one reproductive cycle. Nickel concentrations in nine tissues were not detectable in the controls; however, all nine tissues of supplemented rats contained nickel concentrations that were generally dose-related.

Nickel in the bodies of newborn rats was measurable only with the highest level of catalyst or the two upper levels of nickel carbonate. Body weights of the pups were not affected by nickel.

Phatak and Patwardhan (1952) fed nickel catalyst at 250 ppm to young rats for 16 months. No adverse effects on growth, gross appearance, or vigor were observed. Soft tissues and bone accumulated nickel, with maximal concentrations attained by 8 months.

No adverse effects occurred in weanling rats fed 100 ppm nickel as the acetate for 6 weeks (Whanger, 1973). Ambrose *et al.* (1976) observed no adverse effects in rats fed 100 ppm nickel as the sulfate for 2 years.

A low level of nickel, 5 ppm, was administered to mice in their drinking water throughout their lifetime (Schroeder *et al.,* 1963, 1964). No adverse effects were observed. Weber and Reid (1969) fed 1,100 ppm nickel to young mice and observed decreased growth of females by 4 weeks. Adult weights were unaffected by this level of nickel and no adverse effects occurred by the end of one reproductive cycle.

The adult monkey, like the cow, was resistant to high dietary levels of nickel. Phatak and Patwardhan (1950) observed no adverse effects of 250, 500, or 1,000 ppm nickel in the diet. They tested the same nickel carbonate, nickel soap, and nickel catalyst that they fed to rats.

## HIGH LEVELS

Due to numerous factors that influence nickel toxicity, as discussed below, there is no sharp demarcation between levels of dietary nickel that produce minimal or no adverse effects and those that produce marked adverse effects. The delineation between this and the previous section is therefore based on severity of response and involves a large overlap of nickel intakes.

O'Dell *et al.* (1970b, 1971) found decreases in feed intake, organ size, and nitrogen retention in calves fed 1,000 ppm nickel as the carbonate for 8 weeks. Even though the calves lost weight, they were not emaciated but simply looked younger than the control group. The concentrations of nickel in 9 of 10 tissues and body fluids was significantly increased above those of control calves. The total nickel intake was not different from that of calves fed 250 ppm nickel, which had no effect on tissue nickel; however, the nickel intake per unit body weight was much higher for calves fed 1,000 ppm nickel. The homeostatic mechanism regulating nickel thus ceased to function at intake levels somewhere between 250 and 1,000 ppm.

Weber and Reid (1968) fed chicks nickel as the sulfate or acetate at seven graded levels from 100 to 1,300 ppm for 4 weeks. Growth rate was decreased in all birds fed 500 ppm or more nickel. Nitrogen retention was decreased by 500 ppm or more nickel as the sulfate and by 900 ppm or more nickel as the acetate. When controls were pair-fed to birds consuming 1,100 ppm nickel, there was no effect of nickel on growth, but there was a significant reduction of nitrogen retention.

When Ambrose *et al.* (1976) fed dogs 2,500 ppm nickel as the sulfate, the dogs vomited and salivated excessively. After return to the control diet followed by a gradual increase to 2,500 ppm nickel, there were no acute problems. The dogs continued for 2 years, exhibiting a moderately reduced growth rate. They developed a mild anemia, granulocytic hyperplastic bone marrow, increased urine volume, and severe lung lesions.

Young rats of both sexes rapidly decreased their food intake and lost weight by 13 days after receiving diets with 1,000 ppm nickel as the chloride (Schnegg and Kirchgessner, 1976). There were increases in many measurements of physiological responses; these included red blood cell counts, hematocrit, hemoglobin, serum protein, nitrogen in tissues, and nickel, copper, zinc, and iron concentrations in some tissues. Somewhat different changes were produced in weanling rats fed 500 and 1,000 ppm nickel as the acetate for 6 weeks (Whanger, 1973). Growth rate was markedly decreased by 500 ppm nickel, but there was a mean 23-g weight loss by rats fed 1,000 ppm nickel. Decreased hemoglobin and heart cytochrome oxidase were found in rats fed 1,000 ppm nickel. They also had consistently increased tissue nickel and iron levels and increased zinc in the liver. In all subcellular fractions of liver and kidney, the concentrations of iron and nickel were increased by 500 and 1,000 ppm dietary nickel. The zinc concentration increased in the nuclei and debris of the kidneys from rats fed excess nickel; however, zinc in the intact kidney was not significantly increased by high nickel.

Ambrose *et al.* (1976) observed mild changes in rats fed 1,000 ppm nickel as the sulfate for 2 years. Females had reduced body weight and liver weight and increased heart weight. Males receiving the same diet exhibited no adverse effect. Increased numbers of stillborn pups occurred in the $F1_a$ generations of rats fed 250, 500, or 1,000 ppm nickel as sulfate through three generations. Decreased numbers of pups were weaned with each of the higher levels of nickel. With 1,000 ppm nickel, weaning weight was decreased; however, the effect was less severe in F2 and F3 generations.

The oral $LD_{50}$ of nickel acetate for rats was 350 mg/kg of body weight (Fairchild *et al.,* 1977).

Young mice fed 1,600 ppm nickel as the acetate had depressed growth by 4 weeks (Weber and Reid, 1969). Activities of cytochrome oxidase, malic dehydrogenase, isocitric dehydrogenase, and succinic dehydrogenase were determined in liver, kidney, and heart and NADH cytochrome c reductase in liver. Values for succinic dehydrogenase were not affected by nickel; however, 1,600 ppm nickel caused decreased levels of the other enzymes in one or all tissues. Almost no enzyme changes were produced by 1,100 ppm nickel. There was no effect of 1,600 ppm nickel on the number of pups born, but there was a marked decrease in the number of pups weaned. The oral $LD_{50}$ of nickel acetate for mice was 136 mg/kg of body weight (Fairchild *et al.*, 1977).

## FACTORS INFLUENCING TOXICITY

The above data show a wide range in response to given levels of dietary nickel. This appears to reflect differences in form of nickel fed, duration of feeding, species, age, reproductive status, and diet composition.

In the studies summarized in Table 26, six forms of nickel were used. All were simple salts except for two. O'Dell *et al.* (1970c) found nickel as the chloride to be approximately 5 times more toxic than nickel as the carbonate. Nickel as the carbonate appeared to have a somewhat greater effect in decreasing nitrogen retention by chicks than did nickel as the acetate (Weber and Reid, 1968). Phatak and Patwardhan (1950) found the following decreasing order of nickel toxicity in rats: carbonate, soap, and catalyst. Overall conclusions regarding the order of toxicity are not possible.

Most studies of nickel toxicity were relatively short, and high levels of nickel were required to produce toxicosis. Nickel at 5 ppm in the drinking water was given to mice and rats throughout their lifetime without ill effects (Schroeder *et al.,* 1963, 1964, 1974). Schroeder and Mitchener (1971) found deaths of the young and/or runts in each of three successive generations of rats given 5 ppm nickel in the drinking water from weaning. Weber and Reid (1969) and Phatak and Patwardhan (1950) gave higher amounts of nickel salts to young rats through one reproductive cycle without difficulties based on gross indices of response (body weight, number of young, etc.).

It is likely that diet composition may have a significant effect on nickel toxicity. When Phatak and Patwardhan (1952) fed diets with protein at 14 or 11 percent, the lower protein was associated with lower concentrations of nickel in some of the rats' tissues. This effect appeared to be greater after 4 months as compared with 8 months of feeding 250 ppm nickel as a catalyst. The effect of dietary protein should be investigated further, particularly since nickel markedly reduces nitrogen retention (Weber and Reid, 1968, 1969; O'Dell *et al.*, 1970b). Changes in tissue concentrations of zinc, iron, manganese, copper, and chromium may mean that alterations in dietary levels of these elements would modify nickel toxicity.

When long-term nickel poisoning is discovered, an immediate switch to diets with low or normal nickel levels should be made. O'Dell *et al.* (1970b) removed excess nickel from the diet of male calves after 8 weeks of feeding. Those previously fed 1,000 ppm nickel as the carbonate had suffered a small weight loss; however, they gained the same amount of weight as the controls during a 6-week recovery period. Phatak and Patwardhan (1952) fed young rats 250 ppm nickel as the catalyst for 8, 12, or 16 months, and then the nickel was removed. Nickel was excreted in urine and feces until nondetectable levels were found for feces by 20 days and urine by 40 days. The kidney still retained significant nickel at this point.

## TISSUE LEVELS

Animals fed basal diets with no added nickel had tissue concentrations of nickel that were generally below 1 ppm fresh weight (Phatak and Patwardhan, 1950, 1952; Schroeder *et al.*, 1963, 1964, 1974; O'Dell *et al.*, 1971; Whanger, 1973). For general purposes of comparison, a 70 percent moisture content was assumed for values reported on a dry weight basis. Tissues or body fluids with the lowest concentrations were bile, serum, vitreous humor, brain, pancreas, red blood cells, skin, and tongue. Tissues in the moderate to high range included heart, kidney, liver, and lung. The spleen and testes varied between studies from low to high and moderate nickel concentrations, respectively. Schroeder *et al.* (1974) observed no significant increases of tissue nickel in rats given 5 ppm nickel in drinking water throughout their lifetime after weaning. There appeared to be higher levels of nickel in some tissues of mice receiving nickel in drinking water for a lifetime (Schroeder *et al.*, 1963, 1964).

With supplemental nickel, the gastrointestinal tract of calves accumulated nickel in relation to dose (O'Dell *et al.*, 1971). Significant increases occurred in the rumen–reticulum, omasum, and abomasum. There was a progressive decline in nickel concentration from the duodenum to the upper half of the remaining small intestine and the lower small intestine. In most studies, high levels of dietary nickel, such as 1,000 ppm, caused significant increases (typically 10-fold) of nickel in the tissues that were assayed. Severe growth depression was also found in these animals, which complicates interpretation of the data. Whanger (1973) found increased nickel in liver and kidneys of rats fed 500 ppm nickel; the limited data of Phatak and Patwardhan (1950) support this observation. When Phatak and Patwardhan (1952) analyzed rats after 4, 8, 12, and 16 months of feeding 250 ppm nickel as the catalyst, maximal concentrations of nickel in the liver, kidney, and spleen were attained by 8 months. Phatak and Patwardhan (1950) found significant nickel in newborn pups of mothers fed 500 or 1,000 ppm nickel as the carbonate.

Nickel was not increased in the milk of dairy cows fed 145 mg nickel per day for 2 months (Archibald, 1949) or 365 or 1,835 mg per day for 6 weeks (O'Dell *et al.*, 1970a).

## MAXIMUM TOLERABLE LEVELS

In only one study, 5 ppm nickel in the drinking water of rats from weaning caused death of young in three generations and runting in the first and third generations. Mice tolerated this level. Although 100 ppm nickel as the chloride decreased food intake of calves, 500 ppm nickel as the carbonate were required for this effect. Five hundred parts per million nickel reduced growth and nitrogen retention of chicks. In most experiments 1,000 ppm had marked adverse effects. These included decreased growth rate or even weight loss, changes in red blood cell numbers and hemoglobin (both increases and decreases were reported), accumulation of nickel, and alterations in tissue concentrations of several essential elements. Emesis was produced in dogs by 2,500 ppm nickel. Adaptive tolerance to high levels of nickel was observed in dogs and rats. The single oral dose $LD_{50}$ of nickel as the acetate was 116 mg per kilogram of body weight for rats and 136 mg per kilogram of body weight for mice. For cattle, the maximum tolerable level was set at 50 ppm, based on the lack of adverse effect with nickel chloride at this level. Additional data are needed for other species.

## SUMMARY

Nickel is an essential element required for growth and iron absorption. There is no evidence that nickel is ever deficient under practical conditions and that nickel supplements would be beneficial. Data on the toxicity of nickel have shown very wide variation in the amounts of nickel to produce harmful effects. The toxicity can be affected by the form of nickel, species, age, reproductive status, duration of administration, and nutrient content of the diet.

## TABLE 26 Effects of Nickel Administration in Animals

| Class and Number of Animals[a] | Age or Weight | Administration: Quantity of Element[b] | Source | Duration | Route | Effect(s) | Reference |
|---|---|---|---|---|---|---|---|
| Cattle—6 | Young | 250 ppm[c] | $NiCO_3$ | 5 d | Diet | No adverse effect | O'Dell *et al.*, 1970c |
| | | 500 ppm | | | | Decreased food intake | |
| | | 1,000 ppm | | | | Decreased food intake | |
| | | 50 ppm[c] | $NiCl_2 \cdot 6H_2O$ | | | No adverse effect | |
| | | 100 ppm | | | | Decreased food intake | |
| | | 200 ppm | | | | Decreased food intake | |
| Cattle—6 | 13 wk | 62.5 ppm | $NiCO_3$ | 8 wk | Diet | No adverse effect | O'Dell *et al.*, 1970b |
| | | 250 ppm | | | | No adverse effect | |
| | | 1,000 ppm | | | | Decreased food intake, growth rate, organ size, and nitrogen retention | |
| Cattle—3 | 13 wk | 62.5 ppm | $NiCO_3$ | 8 wk | Diet | No effect on tissue Ni | O'Dell *et al.*, 1971 |
| | | 250 ppm | | | | No effect on tissue Ni | |
| | | 1,000 ppm | | | | Decreased food intake and growth rate; increased Ni in ten tissues and body fluids | |
| Cattle—5 | Adult | 50 ppm (365 mg/d) | $NiCO_3$ | 6 wk | Diet concentrate | No adverse effect | O'Dell *et al.*, 1970a |
| Cattle—6 | | 250 ppm (1,835 mg/d) | | | | No adverse effect | |

TABLE 26 *Continued*

| Class and Number of Animals[a] | Age or Weight | Administration: Quantity of Element[b] | Source | Duration | Route | Effect(s) | Reference |
|---|---|---|---|---|---|---|---|
| Cattle—6 | Adult | 145 mg/d | $NiCl_2 \cdot 6H_2O$ | 2 mo | Diet | No increase in milk Ni | Archibald, 1949 |
| Chicken—24 | 1 d | 100 ppm | Nickel sulfate | 4 wk | Diet | No adverse effect | Weber and Reid, 1968 |
| | | 300 ppm | | | | No adverse effect | |
| | | 500 ppm | | | | Decreased growth and nitrogen retention | |
| | | 700 ppm | | | | Decreased growth and nitrogen retention | |
| | | 900 ppm | | | | Decreased growth and nitrogen retention | |
| | | 1,100 ppm | | | | Decreased growth and nitrogen retention | |
| | | 1,100 ppm[d] | | | | Decreased growth and nitrogen retention | |
| | | 1,300 ppm | | | | Decreased growth and nitrogen retention | |
| | | 100 ppm | Nickel acetate | | | No adverse effect | |
| | | 300 ppm | | | | No adverse effect | |
| | | 500 ppm | | | | Decreased growth | |
| | | 700 ppm | | | | Decreased growth | |
| | | 900 ppm | | | | Decreased growth and nitrogen retention | |
| | | 1,100 ppm | | | | Decreased growth and nitrogen retention | |
| | | 1,100 ppm[d] | | | | Decreased nitrogen | |

| | | | | | | | |
|---|---|---|---|---|---|---|---|
| | | | | | | retention | |
| | | 1,300 ppm | | | | Decreased growth and nitrogen retention | |
| Dog—6 | 6 mo | 100 ppm | $NiSO_4 \cdot 6H_2O$ | 2 yr | Diet | No adverse effect | Ambrose *et al.*, 1976 |
| | | 1,000 ppm | | | | No adverse effect | |
| | 2 yr | 2,500 ppm | | 3 d | | Emesis; excess salivation; gastrointestinal irritation | |
| | | 2,500 ppm[e] | | 2 yr | | Decreased body weight by 13 wk; decreased hemoglobin; increased urine volume, and liver and kidney weights; granulocytic hyperplasia of bone marrow; lungs: gross multiple subpleural, peripheral cholesterol granulomas, bronchiolectasis, emphysema, and focal cholesterol pneumonia; increased Ni in kidney | |
| Rat—104 | Weanling | 5 ppm | Soluble salt | Lifetime | Water[f] | No adverse effect; no increase in tissue Ni | Schroeder *et al.*, 1974 |
| Rat—10 | Weanling | 5 ppm | Soluble salt | 3 generations | Water[f] | Young: deaths and runts (F1–3 generations) | Schroeder and Mitchener, 1971 |
| Rat—8 | Young | 250 ppm | Nickel carbonate | 3–4 mo, 1 generation | Diet | Increased Ni in 9 tissues by 8 wk | Phatak and Patwardhan, 1950 |
| | | 500 ppm | | | | Increased Ni in nine tissues and pups | |

TABLE 26 *Continued*

| Class and Number of Animals[a] | Age or Weight | Administration: Quantity of Element[b] | Source | Duration | Route | Effect(s) | Reference |
|---|---|---|---|---|---|---|---|
| | | 1,000 ppm | | | | Increased Ni in nine tissues and pups; effects dose-related | |
| | | 250 ppm | Nickel soaps | | | Increased Ni in nine tissues | |
| | | 500 ppm | | | | Increased Ni in nine tissues | |
| | | 1,000 ppm | | | | Increased Ni in nine tissues; effects dose-related and less than with nickel carbonate | |
| | | 250 ppm | Nickel catalyst | | | Increased Ni in nine tissues | |
| | | 500 ppm | | | | Increased Ni in nine tissues | |
| | | 1,000 ppm | | | | Increased Ni in nine tissues and pups; effects dose-related and less than with nickel carbonate | |
| Rat—42 | 4–5 wk | 250 ppm | Nickel catalyst | 16 mo | Diet | Increased Ni in tissues to maximum by 8 mo | Phatak and Patwardhan, 1952 |
| Rat | Young | 1,000 ppm | $NiCl_2$ | 13 d | Diet | Growth cessation and weight loss; increased RBC, hematocrit, hemo- | Schnegg and Kirchgessner, 1976 |

| | | | | | | | |
|---|---|---|---|---|---|---|---|
| | | | | | | globin, serum protein, and tissue Ni, Cu, Zn, and Fe | |
| Rat—6 | Weanling | 100 ppm | Nickel acetate | 6 wk | Diet | No adverse effect | Whanger, 1973 |
| | | 500 ppm | | | | Decreased growth; increased tissue Ni and Fe | |
| | | 1,000 ppm | | | | Weight loss; decreased hemoglobin; increased tissue Ni, Fe, and Zn | |
| Rat—50 | 28 d | 100 ppm | $NiSO_4 \cdot 6H_2O$ | 2 yr | Diet | No adverse effect | Ambrose *et al.*, 1976 |
| | | 1,000 ppm | | | | Females: decreased body weight by 26–100 wk and liver weight; increased heart weight | |
| | | 2,500 ppm | | | | Females: decreased body weight and liver weight; increased heart weight. Males: decreased body weight; increased tissue Ni | |
| Rat—60 | | 250 ppm | | 3 generations[g] | Diet | Increased stillborns in $F1_a$ generation | |
| | | 500 ppm | | | | Increased stillborns in $F1_a$, decreased pups weaned in all generations | |
| | | 1,000 ppm | | | | Decreased body weight by mating; increased stillborns $F1_a$; decreased weaning weights; effects generally less severe F2 and F3 | |

TABLE 26 *Continued*

| Class and Number of Animals[a] | Age or Weight | Administration: Quantity of Element[b] | Source | Duration | Route | Effect(s) | Reference |
|---|---|---|---|---|---|---|---|
| Rat | | 116 mg/kg | Nickel acetate | 1 dose | Oral | $LD_{50}$ | Fairchild *et al.*, 1977 |
| Mouse—104 | Weanling | 5 ppm | Nickelous acetate | Lifetime | Water[f] | Small increase tissue Ni | Schroeder *et al.*, 1963 |
| Mouse—104 | Weanling | 5 ppm | Nickelous acetate | Lifetime | Water[f] | Fewer tumors in females; small increase tissue Ni | Schroeder *et al.*, 1964 |
| Mouse—12 | Young | 1,100 ppm | Nickel acetate | 4 wk | Diet | Decreased growth, females | Weber and Reid, 1969 |
| | | 1,600 ppm | | | | Decreased growth | |
| Mouse—8 | Weanling | 1,100 ppm | | 1 reproductive cycle | Diet | No adverse effect | |
| | | 1,600 ppm | | | | Increased pup mortality during lactation | |
| Mouse | | 136 mg/kg | Nickel acetate | 1 dose | Oral | $LD_{50}$ | Fairchild *et al.*, 1977 |

| | | | | | | | |
|---|---|---|---|---|---|---|---|
| Monkey—2 | Adult | 250 ppm | Nickel carbonate | 6 mo | Diet | No adverse effect | Phatak and Patwardhan, 1950 |
| | | 500 ppm | | | | No adverse effect | |
| | | 1,000 ppm | | | | No adverse effect | |
| | | 250 ppm | Nickel soaps | | | No adverse effect | |
| | | 500 ppm | | | | No adverse effect | |
| | | 1,000 ppm | | | | No adverse effect | |
| | | 250 ppm | Nickel catalyst | | | No adverse effect | |
| | | 500 ppm | | | | No adverse effect | |
| | | 1,000 ppm | | | | No adverse effect | |

[a]Number of animals per treatment.
[b]Quantity expressed as parts per million or as milligrams per kilogram of body weight. All amounts are totals added to the basal diets.
[c]Supplemental nickel fed sequentially to same calves from low to high levels, with 2-day intervals between dose changes.
[d]Controls pair-fed.
[e]After 3-day initial illness, dogs were returned to control diet; nickel was gradually increased during ca. 8 weeks to 2,500 ppm without acute problems.
[f]Drinking water.
[g]Total of $F1_a$, $F1_b$, $F2_a$, $F2_b$, $F3_a$, and $F3_b$ generations; F2 and F3 generations were from $F1_b$ and $F2_b$, respectively.

## REFERENCES

Ambrose, P., P. S. Larson, J. F. Borzelleca, and G. R. Hennigar, Jr. 1976. Longterm toxicologic assessment of nickel in rats and dogs. J. Food Sci. Technol. 13:181.

Anke, M., M. Grün, G. Dittrich, B. Groppel, and A. Hennig. 1974. Low nickel rations for growth and reproduction in pigs, pp. 715–718. *In* W. G. Hoekstra, J. W. Suttie, H. E. Ganther, and W. Mertz (eds.). Trace Element Metabolism in Animals—2. University Park Press, Baltimore, Md.

Archibald, J. G. 1949. Nickel in cow's milk. J. Dairy Sci. 32:877.

Fairchild, E. J., R. J. Lewis, and R. L. Tatken (eds.). 1977. Registry of Toxic Effects of Chemical Substances, vol. II, pp. 590–592. DHEW Publ. No. (NIOSH) 78-104-B.

National Research Council. 1975. Medical and Biological Effects of Environmental Pollutants. Nickel. National Academy of Sciences, Washington, D.C.

Nielsen, F. H. 1974. Essentiality and function of nickel, pp. 381–395. *In* W. G. Hoekstra, J. W. Suttie, H. E. Ganther, and W. Mertz (eds.). Trace Element Metabolism in Animals—2. University Park Press, Baltimore, Md.

Nielsen, F. H. 1977. Nickel toxicity, pp. 129–146. *In* R. A. Goyer and M. A. Mehlman (eds.). Advances in Modern Toxicology. Toxicology of Trace Elements, vol. 2. Hemisphere Publishing Corp., Washington, D.C.

Nielsen, F. H., and D. A. Ollerich. 1974. Nickel: A new trace element. Fed. Proc. 33:1767.

Nielsen, F. H., and H. H. Sandstead. 1974. Are nickel, vanadium, silicon, fluorine, and tin essential for man? Am. J. Clin. Nutr. 27:515.

Nielsen, F. H., D. R. Myron, S. H. Givand, and D. A. Ollerich. 1975a. Nickel deficiency and nickel–rhodium interaction in chicks. J. Nutr. 105:1607.

Nielsen, F. H., D. R. Myron, S. H. Givand, T. J. Zimmerman, and D. A. Ollerich. 1975b. Nickel deficiency in rats. J. Nutr. 105:1620.

O'Dell, G. D., W. J. Miller, W. A. King, J. C. Ellers, and H. Jurecek. 1970a. Effect of nickel supplementation on production and composition of milk. J. Dairy Sci. 53:1545.

O'Dell, G. D., W. J. Miller, W. A. King, S. L. Moore, and D. M. Blackmon. 1970b. Nickel toxicity in the young bovine. J. Nutr. 100:1447.

O'Dell, G. D., W. J. Miller, S. L. Moore, and W. A. King. 1970c. Effect of nickel as the chloride and the carbonate on palatability of cattle feed. J. Dairy Sci. 53:1266.

O'Dell, G. D., W. J. Miller, S. L. Moore, W. A. King, J. C. Ellers, and H. Jurecek. 1971. Effect of dietary nickel level on excretion and nickel content of tissues in male calves. J. Anim. Sci. 32:769.

Phatak, S. S., and V. N. Patwardhan. 1950. Toxicity of nickel. J. Sci. Ind. Res. 9B(3):70.

Phatak, S. S., and V. N. Patwardhan. 1952. Toxicity of nickel—Accumulation of nickel in rats fed on nickel-containing diets and its elimination. J. Sci. Ind. Res. 11B(5):173.

Schnegg, S., and M. Kirchgessner. 1976. [Toxicity of dietary nickel.] Landwirtsch. Forsch. 29(3–4):177; Chem. Abstr. 86:101655y (1977).

Schnegg, S., and M. Kirchgessner. 1978. Ni deficiency and its effects on metabolism, pp. 236–243. *In* M. Kirchgessner (ed.). Trace Element Metabolism in Man and Animals—3. Technische Universität München, Freising-Weihenstephan, West Germany.

Schroeder, H. A., and M. Mitchener. 1971. Toxic effects of trace elements on the reproduction of mice and rats. Arch. Environ. Health 23:102.

Schroeder, H. A., J. J. Balassa, and I. H. Tipton. 1962. Abnormal trace metals in man—Nickel. J. Chron. Dis. 15:51.

Schroeder, H. A., W. H. Vinton, Jr., and J. J. Balassa. 1963. Effects of chromium, cadmium and other trace metals on the growth and survival of mice. J. Nutr. 80:39.

Schroeder, H. A., J. J. Balassa, and W. H. Vinton, Jr. 1964. Chromium, lead, cadmium, nickel, and titanium in mice: Effect on mortality, tumors and tissue levels. J. Nutr. 83:239.

Schroeder, H. A., M. Mitchener, and A. P. Nason. 1974. Life-term effects of nickel in rats: Survival, tumors, interactions with trace elements and tissue levels. J. Nutr. 104:239.

Spears, J. W., C. J. Smith, and E. E. Hatfield. 1977. Rumen bacterial urease requirement for nickel. J. Dairy Sci. 60:1073.

Spears, J. W., E. E. Hatfield, and G. C. Fahey, Jr. 1978a. Nickel depletion in the growing ovine. Nutr. Rep. Int. 18:621.

Spears, J. W., E. E. Hatfield, R. M. Forbes, and S. E. Koenig. 1978b. Studies on the role of nickel in the ruminant. J. Nutr. 108:313.

Sunderman, F. W., Jr., S. Nomoto, R. Morang, M. W. Nechay, C. N. Burke, and S. W. Nielsen. 1972. Nickel deprivation in chicks. J. Nutr. 102:259.

Underwood, E. J. 1977. Trace Elements in Human and Animal Nutrition, 4th ed. Academic Press, New York.

Weber, C. W., and B. L. Reid. 1968. Nickel toxicity in growing chicks. J. Nutr. 95:612.

Weber, C. W., and B. L. Reid. 1969. Nickel toxicity in young growing mice. J. Anim. Sci. 28:620.

Whanger, P. D. 1973. Effects of dietary nickel on enzyme activities and mineral content in rats. Toxicol. Appl. Pharmacol. 25:323.

Zook, E. G., F. E. Green, and E. R. Morris. 1970. Nutrient composition of selected wheat and wheat products. 6. Distribution of manganese, copper, nickel, zinc, magnesium, lead, tin, cadmium, chromium and selenium as determined by atomic absorption spectroscopy and colorimetry. Cereal Chem. 47:720.

# Phosphorus

The element phosphorus (P), although widely distributed in nature, never occurs in the free state. It combines spontaneously and vigorously with oxygen, and even its pentoxide, $P_2O_5$, combines readily with water to form orthophosphoric acid. It is found widely scattered in igneous and sedimentary rocks and in all bodies of water, primarily as salts of orthophosphoric acid (Van Wazer, 1958). Phosphorus was first prepared in the free state by Brandt, an alchemist of Hamburg, Germany, in 1669 and was first recognized as an essential constituent of bones by Gahn, a Swedish chemist, in 1769 (Van Wazer, 1961). It probably plays a more varied role in the chemistry of living organisms than any other single element. This review is limited to naturally occurring phosphorus in feed ingredients and supplemental phosphate compounds. It does not include the organophosphorus compounds, such as certain pesticides, or elemental forms of phosphorus that are highly toxic (Goodman and Gilman, 1975).

## ESSENTIALITY

From the time of its discovery by Brandt, phosphorus was thought to be important in metabolism. In addition to being of major importance as a constituent of bone, phosphorus is an essential component of organic compounds involved in almost every aspect of metabolism. Phosphorus plays an important part in muscle, energy, carbohydrate, amino acid, fat, and nerve tissue metabolism; in normal blood chemistry; and in skeletal growth. Phosphate is an important part of the nucleic acids, DNA and RNA; it is a component of many coenzymes and is found in compounds such as adenosine di- and tri-phosphate (Irving, 1964).

## METABOLISM

Phosphorus has more known functions than any other mineral element in the animal body. In addition to playing a major role, along with calcium, in the formation of bones and teeth, it is located in every cell of the body and is vitally concerned in many metabolic processes, including the buffering of body fluids (White *et al.*, 1968). Practically every energy transfer inside living cells involves the forming or breaking of chemical bonds that link oxides of phosphorus to carbon or to carbon–nitrogen compounds. Since every biological event involves gain or loss of energy, one can readily appreciate the great physiological role of phosphorus in animal metabolism.

Phosphorus in the form of orthophosphate is absorbed chiefly in the upper small intestine, the duodenum. The amount absorbed is dependent on source, calcium to phosphorus ratio, intestinal pH, lactose intake, and dietary levels of calcium, phosphorus, vitamin D, iron, aluminum, manganese, potassium, magnesium, and fat (Irving, 1964). As is the case for most nutrients, the greater the need, the more efficient the absorption. Phosphorus absorbed from the intestine is circulated through the body and is readily withdrawn from the blood for use by the bones and teeth during periods of growth. Some incorporation into the bone occurs at all ages. It may be withdrawn from bones to maintain normal blood plasma levels during periods of dietary deprivation. The plasma phosphorus level, along with calcium, is regulated by the parathyroid hormone and thyrocalcitonin. The plasma phosphorus level is inversely related to the blood calcium level; however, in parturient paresis, both usually decline. Excess phosphorus is excreted primarily by the kidney (Bartter, 1964).

## SOURCES

Phosphorus is present in variable amounts in almost all common feedstuffs (National Research Council, 1971). The phosphorus levels are dependent on the plant species, the level of soil fertility, and the stage of maturity at the time of consumption. The biological availability of the phosphorus in plant material is quite variable and is dependent upon the phosphorus compounds present and the animal species consuming the feedstuffs.

Most animal diets require supplemental sources of phosphorus in addition to that present in the common feedstuffs. There are a number of supplemental sources. The major types include: calcium phosphates (dicalcium phosphate, monocalcium phosphate, defluorinated rock phosphate, bone meal, Guano origin phosphates), ammonium phosphates (monoammonium phosphate, diammonium phosphate, ammonium polyphosphate), sodium phosphates (monosodium phosphate, disodium phosphate, sodium tripolyphosphate), and phosphoric acid.

There is considerable variation in the biological availability of phosphorus from sources within these types, especially within the calcium phosphates. A review on the biological availability of phosphorus to livestock was prepared by Peeler (1972).

## TOXICOSIS

Phosphorus is involved in almost all aspects of metabolism. In addition, it interacts with many of the other essential and nonessential mineral elements making dietary levels of this element critical to optimum animal performance (growth rate, feed efficiency, milk production, egg production).

Optimum animal performance is linked very closely with optimum calcium and phosphorus levels in the diet. Most animals require a fairly narrow calcium to phosphorus ratio—usually no wider than 2:1; however, ruminants can tolerate wider ratios than monogastric animals, providing the phosphorus level is adequate. If the calcium to phosphorus ratio is balanced, the animal can tolerate wider ranges of dietary phosphorus levels. The quantitative aspects of the dietary calcium to phosphorus ratio and subsequent animal performance are discussed in several recent reviews (Hays, 1976; Preston *et al.,* 1977; Waibel *et al.,* 1977).

A relative excess of phosphorus in relation to calcium can result in some very detrimental situations. One of these is a malady of ruminants known as urolithiasis (urinary calculi). This is the formation of stones or calculi in the kidney or bladder with resultant obstruction of urine excretion. The blockage results in a buildup of urine in the bladder with eventual rupture of the bladder or urethra. Bladder rupture results in temporary relief, but this is followed by abdominal distention, depression, and death due to uremia (Blood and Henderson, 1968). The formation of stones may be affected by several factors; however, a series of papers from South Dakota (Emerick and Embry, 1962, 1963, 1964) leaves little doubt regarding the importance of phosphorus level in the ration and the accompanying calcium level upon the development of urinary calculi in wether lambs. Feeding a high level of phosphorus (0.8 percent) resulted in a high incidence of urinary calculi (Table 27). The maximum dietary level of phosphorus that can be tolerated by sheep without the development of calculi lies between 0.37 and 0.69 percent of the diet dry matter. Increasing the calcium level in the ration seemed to provide partial protection against the occurrence of calculi in sheep receiving higher levels of phosphorus. Monosodium phosphate, disodium phosphate, and sodium tripolyphosphate all appeared to be about equal in calculi-producing ability; dicalcium phosphate contributed calcium and significantly reduced the incidence of calculi (Bushman *et al.*, 1965).

Several workers (Singsen *et al.*, 1962; Crowley *et al.*, 1963; Harms *et al.*, 1965; Charles and Jensen, 1975) have observed that high dietary levels of phosphorus (0.8 to 1.2 percent) depressed the performance of laying hens (egg production and egg shell quality). The level necessary to depress performance of hens maintained on the floor was less than for cage layers. It is postulated that this difference is due to the higher dietary phosphorus requirement (approximately 0.2 percent greater) of caged hens. McGillivray and Smidt (1974) investigated the effect of excess phosphorus on broiler performance. They observed decreases in weight gains and efficiency of feed utilization as dietary phosphorus levels were increased above 0.8 percent (calcium level at 1 percent). Significant mortality was noted when the dietary phosphorus level was raised to 2 percent (approximately 4 times requirement).

Another problem associated with a relative excess of phosphorus in relation to calcium is a bone disorder (called osteodystrophia fibrosa, nutritional secondary hyperparathyroidism, osteomalacia, osteoporosis, big head disease) that has been observed in several species (Bartter, 1964). A high phosphorus intake causes an increased concentration of serum phosphorus, which secondarily results in a lowering of the serum calcium. This effect then stimulates the parathyroid gland to increase serum calcium by resorption of the bone and to increase renal phosphate excretion. Thus, pronounced bone loss in adult animals can occur by feeding excess dietary phosphorus or insufficient dietary calcium. In some instances, the demineralized skeleton is replaced by fibrous connective tissue.

Nutritional secondary hyperparathyroidism occurs in horses fed high levels of grain without calcium supplementation (Joyce *et al.*, 1971). It develops in 6 to 12 months when a diet with a calcium to phosphorus ratio of 0.8:1.0 is fed and progresses rapidly when the ratio is 0.6:1.0 (National Research Council, 1973). Studies at Cornell (Schryver *et al.*, 1971; Argenzio *et al.*, 1974) have shown that high phosphate diets (1.2 percent phosphorus, 0.4 percent calcium) depressed the intestinal absorption of calcium, the concentration of calcium in plasma, renal excretion, and calcium retention. The rate of deposition of calcium in bone and the rate of removal of calcium from bone were elevated in response to the high phosphate intake. Phosphate retention and the plasma phosphate concentration increased when the horses were fed the high-phosphate diet.

A purified diet containing 1.20 percent phosphorus and 0.12 percent calcium fed to beagles produced rapid loss of bone and easily detached incisor teeth. This did not occur with dogs fed the same diet containing 0.42 percent phosphorus and 0.54 percent calcium (Henrikson, 1968). Confirmation of the effects of high phosphate intakes on the integrity of adult bone, and direct evidence for a parathyroid hormone-mediated mechanism contributing to increased resorption of bone in dogs was provided by Laflamme and Jowsey (1972) and Krook *et al.* (1971).

In a series of papers, Draper and associates (Anderson and Draper, 1972; Draper *et al.*, 1972; Sie *et al.*, 1974) showed that excess dietary phosphorus has an accelerating effect on bone resorption in aging rats and mice by a mechanism involving the parathyroid hormone. When a diet containing 0.6 percent calcium was fed with 0.3, 0.6, 1.2, or 1.8 percent phosphorus, calcium loss in the 0.6 percent phosphorus group was 16 percent greater than in the 0.3 percent phosphorus group and was 37 percent greater in the 1.2 percent phosphorus group. A similar response was obtained with mice (Krishnarao and Draper, 1972).

High phosphorus levels in laboratory animals have caused hypertrophy of the parathyroid glands. However, a more sensitive criterion of excess phosphate is the appearance of metastatic calcification in soft tissues, especially in the kidney, stomach, and aorta (World Health Organization, 1964; Ellinger, 1972). Kidney calcification may be observed in a few weeks or months, depending on the dose level. The highest levels of phosphorus in diets for rats that did not cause significant kidney damage were 0.90 and 1.30 percent. It is difficult to indicate a borderline between dose levels that do not produce nephrocalcinosis and those that produce early signs of such changes, because the composition of the diet (amount of calcium, acid-base balance, vitamin D) has an important influence on the appearance of renal calcification. From a consideration of the experimental evidence, it is estimated that diets containing 1 percent phosphorus or more may be nephrocalcinogenic in rats. Diets containing 0.9 percent phosphorus and 0.8 percent calcium or higher levels of phosphate produced calcification in the soft tissues of guinea pigs (Hogan *et al.*, 1950; House and Hogan, 1955).

## MAXIMUM TOLERABLE LEVELS

It is difficult, if not impossible, to separate the effects of a dietary calcium–phosphorus imbalance from that of excessive dietary phosphorus levels. Many of the studies conducted with higher levels of phosphorus have simultaneously used inadequate levels of calcium. This accentuates the problem, and one cannot always tell whether the observations are due to the low calcium or the high phosphorus, or both. There are enough studies in the literature, however, to indicate that, even when calcium levels are at requirement or higher levels, high levels of phosphorus can cause increased bone resorption—especially in adult animals. Supplemental phosphates are usually not considered to be highly toxic, since single large oral doses can be tolerated with only minor effects (mild diarrhea, abdominal distress). On the other hand, long-term consumption of dietary levels 2 to 3 times the requirement level will cause severe problems due to induced changes in calcium metabolism. This level is very dependent on the calcium level of the diet, as well as other factors such as vitamin D, aluminum, potassium, and magnesium. Although the effects of excess dietary phosphorus can, to some extent, be counteracted by increasing dietary calcium, the lower efficiency of calcium absorption limits the extent to which the effect of a high phosphorus intake can be offset by increasing the intake of calcium. When tolerance is expressed in terms of a multiple of the requirement level, phosphorus has one of the lowest tolerance factors of any mineral element.

Assuming the presence of adequate levels of dietary calcium, the following phosphorus levels can be tolerated; however, depending on the age and production status of the animal, even these levels may reduce performance: cattle, 1 percent; sheep, 0.6 percent; swine, 1.5 percent; poultry, 1 percent; laying hen, 0.8 percent; horse, 1 percent; rabbit, 1 percent. For the very young of some species, higher levels of phosphorus are tolerated or needed for short periods of time.

## SUMMARY

Phosphorus, although widely distributed in nature, never occurs in the free state. It probably plays a more varied role in the chemistry of living organisms than any other single element. In addition to being of major importance as a constituent of bone, phosphorus is an essential component of organic compounds involved in almost every aspect of metabolism. Animal performance is linked very closely with calcium and phosphorus levels in the diet. Most animals require a fairly narrow calcium to phosphorus ratio—usually no wider than 2:1, however, ruminants can tolerate wider ratios than monogastric animals, providing the phosphorus level is adequate. A relative excess of phosphorus in relation to calcium can result in some very detrimental situations, and, even when calcium is at requirement or higher levels, high levels of phosphorus can cause increased bone resorption in adult animals. Long-term consumption of dietary levels 2 to 3 times the requirement level will cause severe problems due to induced changes in calcium metabolism.

TABLE 27 Effects of Phosphorus Administration in Animals

| Class and Number of Animals[a] | Age or Weight | Administration: Quantity of Element[b] | Source | Duration | Route | Effect(s) | Reference |
|---|---|---|---|---|---|---|---|
| Sheep—16 | Lambs (27 kg) | 0.33, 0.62, and 0.81% (0.44, 0.71, and 0.96% Ca in a 3×3 factorial arrangement) | $Na_2HPO_4$ | 131 d | Diet | No urinary calculi in groups receiving 0.33% P; a 31% incidence of urinary calculi when 0.62% P and 0.44% Ca were fed; when P content of diet was increased to 0.81% and Ca maintained at 0.44%, a 73% incidence of urinary calculi occurred; increasing the level of Ca provided partial protection in sheep receiving the higher levels of P | Emerick and Embry, 1963 |
| Sheep—16 | Lambs (27 kg) | 0.35 and 0.80% (0.51 and 0.90% Ca) | $Na_2HPO_4$ | 90 d | Diet | No urinary calculi in groups receiving 0.35% P; when fed 0.80% P with 0.51 and 0.90% Ca, the incidence of urinary calculi was 68% and 58%, respectively | Emerick and Embry, 1964 |
| Sheep—45 | Lambs (32 kg) | 0.25 and 0.60% (0.31 and 0.58% Ca) | $NaH_2PO_4$<br>$Na_2HPO_4$<br>$Na_5P_3O_{10}$<br>$CaHPO_4$ | 84 d | Diet | No urinary calculi in lambs on 0.25% P diets or in lambs receiving diets containing dicalcium phosphate | Bushman *et al.*, 1965 |

TABLE 27 *Continued*

| Class and Number of Animals[a] | Age or Weight | Administration: Quantity of Element[b] | Source | Duration | Route | Effect(s) | Reference |
|---|---|---|---|---|---|---|---|
| | | | | | | (0.60% P and 0.58% Ca); the percent incidence of urinary calculi was 70, 65, and 60, respectively, for lambs receiving disodium phosphate, monosodium phosphate, and sodium tripolyphosphate when fed with 0.31% Ca; increasing the Ca to 0.58% lowered, but did not prevent, the incidence of urinary calculi | |
| Chicken, hen | 28 wk | 0.68, 0.88, 1.08, 1.28, 1.48, 1.68% (3.3% Ca) | Defluorinated phosphate | 280 d | Diet | A depression of egg production when the P level of the diet was increased above 0.68%; this decrease in egg production was not statistically significant until the diet contained 1.28% P; the amount of feed required to produce a dozen eggs, egg weights, or mortality, was not affected by the P level of the diet | Harms *et al.*, 1965 |
| Chicken | 1 d | 0.53, 0.77, 1.0, | $NaH_2PO_4$ | 21 d | Diet | P levels in excess of 0.8% | McGillivray |

| | | | | | | | |
|---|---|---|---|---|---|---|---|
| | | 1.25, 1.5, and 2.0% (1.0% Ca) | $NH_4H_2PO_4$ | | | depressed weight gain and feed utilization; increased mortality was noted at the highest P levels (2.0%) | and Smidt, 1974 |
| Horse | 2 yr | 1.2% (0.4% Ca) | $NaH_2PO_4$ | 56 d | Diet | Decreased calcium absorption, urinary excretion, and retention but increased fecal calcium excretion; more rapid turnover of bone calcium observed due to high dietary P | Schryver *et al.*, 1971 |
| Dog | Adult (12 kg) | 0.3, 0.9, and 1.0% (0.3% Ca) | | 360 d | Diet | Supplementary P induced greater body retention of calcium; there was deposition of calcium in kidney, tendon, and heart; bone formation rates were not changed, but resorption was significantly increased resulting in greater bone porosity and less bone mass | Laflamme and Jowsey, 1972 |
| Dog | Adult (10 kg) | 1.2% (0.12% Ca) | $K_2HPO_4$ $NaH_2PO_4$ $CaHPO_4$ | 294 d | Diet | Incisor teeth were loose, and there was evidence of severe bone loss | Krook *et al.*, 1971 |
| Rat | Adult | 0.6 and 1.2% (1.2% Ca) | $Ca(H_2PO_4)_2$ | 238 d | Diet | Accelerated rate of bone resorption on the high-P diet | Anderson and Draper, 1972 |
| Rat | Adult | 0.3, 0.6, 1.2, and 1.8% (0.6% Ca) | $Ca(H_2PO_4)_2$ $Mg_3(PO_4)_2$ $NaH_2PO_4$ | 180 d | Diet | Cumulative excretion of $^{45}Ca$ was 16% greater over the 6-month period in the | Draper *et al.*, 1972 |

TABLE 27 *Continued*

| Class and Number of Animals[a] | Age or Weight | Administration: Quantity of Element[b] | Source | Duration | Route | Effect(s) | Reference |
|---|---|---|---|---|---|---|---|
| | | | | | | 0.6% P group than in the 0.3% P group and was 37% greater in the 1.2% P group; urinary Ca was depressed by about 50% in animals that consumed more than 0.3% P in the diet | |
| Mouse | Adult | 0.6 and 1.2% (1.2% Ca) | $Ca(H_2PO_4)_2$ | 480 d | Diet | Femurs of the higher P group had lower breaking strength and a reduced content of ash, Ca, and P, indicating net bone resorption | Krishnarao and Draper, 1972 |

| | | | | | | | |
|---|---|---|---|---|---|---|---|
| Guinea pig | 1 wk (150–175 g) | 0.5 and 0.9% (0.8% Ca) | $CaHPO_4$ $KH_2PO_4$ | 150–600 d | Diet | 90% of the guinea pigs that consumed the diet that contained 0.9% P and 0.8% Ca developed visible deposits of calcium phosphate; if the P was reduced to 0.5%, the incidence of the deposit was less than 10%; the animals on the high-P diet grew more slowly and the survival period was shortened | Hogan *et al.*, 1950 |
| Guinea pig | 200 g | 0.7, 0.9, 1.2, and 1.7% | $CaHPO_4$ $KH_2PO_4$ $NaH_2PO_4$ | 84 d | Diet | Consumption of high-P diets resulted in slow weight gain, stiff joints, calcium phosphate deposits, and high mortality rate; these signs were most severe on diets that contained 1.7% P and 0.9% Ca | House and Hogan, 1955 |

[a]Number of animals per treatment.
[b]Quantity expressed as concentrations in diet. Dietary concentration of calcium also shown.

## REFERENCES

Anderson, G. H., and H. H. Draper. 1972. Effect of dietary phosphorus on calcium metabolism in intact and parathyroidectomized adult rats. J. Nutr. 102:1123.

Argenzio, R. A., J. E. Lowe, H. F. Hintz, and H. F. Schryver. 1974. Calcium and phosphorus homeostasis in horses. J. Nutr. 104:18.

Bartter, F. C. 1964. Disturbances of phosphorus metabolism, pp. 315–339. *In* C. L. Comar and F. Bronner, eds. Mineral Metabolism, vol. 2, part A. Academic Press, New York.

Blood, D. C., and J. A. Henderson. 1968. Veterinary Medicine, 3rd ed. Williams & Wilkins Co., Baltimore, Md.

Bushman, D. H., R. J. Emerick, and L. B. Embry. 1965. Incidence of urinary calculi in sheep as affected by various dietary phosphates. J. Anim. Sci. 24:671.

Charles, O. W., and L. Jensen. 1975. Effect of phosphorus levels on laying hen performance. Poult. Sci. 54:1744.

Crowley, T. A., A. A. Kurnick, and B. L. Reid. 1963. Dietary phosphorus for laying hens. Poult. Sci. 42:758.

Draper, H. H., Ten-Lin Sie, and J. G. Bergan. 1972. Osteoporosis in aging rats induced by high phosphorus diets. J. Nutr. 102:1133.

Ellinger, R. H. 1972. Phosphates as Food Ingredients. CRC Press, Cleveland, Ohio.

Emerick, R. J., and L. B. Embry. 1962. Calcium and phosphorus levels related to urinary calculi in sheep. J. Anim. Sci. 21:995.

Emerick, R. J., and L. B. Embry. 1963. Calcium and phosphorus levels related to the development of phosphate urinary calculi in sheep. J. Anim. Sci. 22:510.

Emerick, R. J., and L. B. Embry. 1964. Effects of calcium and phosphorus levels and diethylstilbestrol on urinary calculi incidence and feedlot performance of lambs. J. Anim. Sci. 23:1079.

Goodman, L. S., and A. Gilman. 1975. The Pharmacological Basis of Therapeutics, 5th ed. The Macmillan Company, New York.

Harms, R. H., B. L. Damron, and P. W. Waldroup. 1965. Influence of high phosphorus levels in caged layer diets. Poult. Sci. 44:1249.

Hays, V. W. 1976. Phosphorus in Swine Nutrition. National Feed Ingredients Association, Des Moines, Iowa.

Henrikson, P. 1968. Periodontal disease and calcium deficiency. Acta Odont. Scandinav. 26(Suppl. 50):1.

Hogan, A. G., W. O. Regan, and W. B. House. 1950. Calcium phosphate deposits in guinea pigs and the phosphorus content of the diet. J. Nutr. 41:203.

House, W. B., and A. G. Hogan. 1955. Injury to guinea pigs that follows a high intake of phosphates. J. Nutr. 55:507.

Irving, J. T. 1964. Dynamics and functions of phosphorus, pp. 249–313. *In* C. L. Comar and F. Bronner, eds. Mineral Metabolism, vol. 2, part A. Academic Press, New York.

Joyce, J. R., K. R. Pierce, W. M. Romane, and J. M. Baker. 1971. Clinical study of nutritional secondary hyperparathyroidism in horses. J. Am. Vet. Med. Assoc. 158:2033.

Krishnarao, G. V. G., and H. H. Draper. 1972. Influence of dietary phosphate on bone resorption in senescent mice. J. Nutr. 102:1143.

Krook, L., L. Lutwak, P. Hendrikson, F. Kallfelz, C. Hirsch, B. Romanus, L. F. Belanger, J. R. Marier, and B. E. Sheffy. 1971. Reversibility of nutritional osteoporosis: Physicochemical data on bones from an experimental study in dogs. J. Nutr. 101:233.

Laflamme, G. H., and J. Jowsey. 1972. Bone and soft tissue changes with oral phosphate supplements. J. Clin. Invest. 51:2834.

McGillivray, J. J., and M. J. Smidt. 1974. Monoammonium phosphate as a phosphorus source for poultry. Poult. Sci. 53:1954. (Abstr.)

National Research Council. 1971. Atlas of Nutritional Data on United States and Canadian Feeds. National Academy of Sciences, Washington, D.C.

National Research Council. 1973. Nutrient Requirements of Domestic Animals. Nutrient Requirements of Horses. National Academy of Sciences, Washington, D.C.

Peeler, H. T. 1972. Biological availability of nutrients in feeds: Availability of major mineral ions. J. Anim. Sci. 35:695.

Preston, R. L., N. L. Jacobson, K.D. Wiggers, M. H. Wiggers, and G. N. Jacobson. 1977. Phosphorus in Ruminant Nutrition. National Feed Ingredients Association, Des Moines, Iowa.

Schryver, H. F., H. F. Hintz, and P. H. Craig. 1971. Calcium metabolism in ponies fed a high phosphorus diet. J. Nutr. 101:259.

Sie, Ten-Lin, H. H. Draper, and R. R. Bell. 1974. Hypocalcemia, hyperparathyroidism and bone resorption in rats induced by dietary phosphate. J. Nutr. 104:1195.

Singsen, E. P., A. H. Spandorf, L. D. Matterson, J. A. Serafin, and J. J. Tlustohowicz. 1962. Phosphorus in the nutrition of the adult hen. 1. Minimum phosphorus requirements. Poult. Sci. 41:1401.

Van Wazer, J. R. 1958. Phosphorus and Its Compounds. vol. I: Chemistry. Interscience Publishers, New York.

Van Wazer, J. R. 1961. Phosphorus and Its Compounds. vol. II: Technology, Biological Functions and Applications. Interscience Publishers, New York.

Waibel, P. E., R. H. Harms, and B. L. Damron. 1977. Phosphorus in Poultry and Game Bird Nutrition. National Feed Ingredients Association, Des Moines, Iowa.

White, A., P. Handler, and E. L. Smith. 1968. Principles of Biochemistry, 4th ed. McGraw-Hill, New York.

World Health Organization Technical Report, Series No. 281. 1964. Specifications for the Identity and Purity of Food Additives and Their Toxicological Evaluation: Emulsifiers, Stabilizers, Bleaching and Maturing Agents. World Health Organization, Geneva.

# Potassium

Potassium (K), a light metal, makes up 2.6 percent of the earth's crust. It is found mainly within the cells of the animal body and is a dietary essential for all animals.

It is widely distributed in feed sources, but the levels are highly variable. Overt deficiencies are seldom encountered in animals. Usually, forages are higher in potassium than concentrates. Certain lush growing forages, such as cereals, may contain high levels of the element. In these cases, the potassium may interfere with magnesium utilization. Forages that have been subjected to weathering may contain low levels of the element.

## ESSENTIALITY

Animals have a dietary requirement for potassium. Deficiencies have been produced in cattle, chicks, swine, and other animals. The quantitative requirement varies among species. Apparently, the requirement is higher for ruminants than nonruminants. For example, the requirements are 0.6 to 0.8 percent for cattle (National Research Council, 1976) and 0.20 to 0.39 percent for swine (National Research Council, 1979). Frequently, the feed sources supply sufficient amounts to meet the dietary requirement. Exceptions are finishing cattle fed high concentrate diets, cattle fed high nonprotein nitrogen diets, dairy cattle on high corn silage feeding programs, and cattle grazing or fed weathered roughages, i.e., late fall and winter range feeding conditions.

## METABOLISM

Potassium is absorbed mainly from the upper small intestine, but some absorption also occurs in the lower small intestine and large intestine (Church and Pond, 1974). Absorption from the intestine appears to be by simple diffusion. The main functions of potassium in the animal are for osmotic equilibrium, maintenance of acid-base balance, enzyme reactions for phosphorylation of creatine, pyruvate kinase activity, cellular uptake of amino acids, carbohydrate metabolism, protein synthesis, and maintenance of normal heart and kidney tissue (Church and Pond, 1974).

Excess potassium is usually excreted through the urine. Aldosterone and sodium intake affect potassium excretion. The hormone increases sodium reabsorption in the kidney, and there is usually an inverse relationship between sodium and potassium excretion. Ward (1966b) reviewed potassium metabolism in ruminants.

## SOURCES

Potassium is present in feedstuffs in different amounts. Generally, it is higher in forages than grains (National Research Council, 1971). For example, corn grain contains 0.35 percent potassium, dry basis, compared to 1.64 percent for corn stover. Alfalfa hay frequently contains over 2 percent potassium. Actively growing wheat pasture forage may contain 5 percent potassium (Miller, 1939). However, weathered range grasses may contain as low as 0.1 percent potassium. Generally, supplemental potassium is supplied as potassium bicarbonate, carbonate, chloride, and sulfate.

## TOXICOSIS

Potassium toxicosis is not likely to occur under practical situations. However, since high dietary levels of potassium interfere with magnesium absorption in ruminants, ingestion of such levels may predispose the animals to hypomagnesemic tetany (Newton *et al.*, 1972). Toxicosis may occur from feeding excess levels of potassium supplements.

### LOW LEVELS

Increasing the level of dietary potassium from 0.7 to 3.0 percent linearly decreased energy and weight gain in lambs (Jackson *et al.*, 1971).

Feeding 1 percent potassium chloride (0.5 percent potassium) reduced the incidence of urinary calculi in lambs (Crookshank, 1966), but had no effect on the incidence in cattle (Hoar *et al.,* 1970). Feeding potassium chloride at 1 or 2 percent (0.5 or 1.0 percent potassium) provided some protection from calculi in lambs when the diet contained 0.5 percent phosphorus, but was without effect when the diet contained 0.7 percent phosphorus (Emerick *et al.,* 1972). Feeding 1 or 2 percent potassium acetate (0.4 or 0.8 percent potassium) increased gain and feed efficiency in swine (Liebholz *et al.,* 1966). The 1 percent supplementation level was more effective than the 2 percent.

## HIGH LEVELS

Intravenous administration of 306 mg potassium chloride per kilogram per dose (160 mg potassium per kilogram) did not cause death in calves, but administration of 629 mg potassium chloride per kilogram (330 mg potassium per kilogram) resulted in death (Bergman and Sellers, 1953). Severe signs resulted with plasma potassium levels of about 31.3 mg/dl, and the only death was in an animal in which potassium reached 50 mg/dl gradually over a period of 168 minutes. When plasma potassium reached 31 mg/dl, the heart rate became slower (Bergman and Sellers, 1954). Respiratory movements increased in rate and amplitude in all potassium experiments. Atrial flutter and complete atrioventricular block with nodal rhythm was observed in two of eight experiments. Administration of 393 g potassium as potassium chloride by stomach tube to cows weighing about 300 kg resulted in one death, two cattle requiring treatment, and two showing no toxic signs (Dennis and Harbaugh, 1948). Two animals administered 182 and 240 g potassium showed no clinical signs, and a third one given 212 g developed milk fever signs that responded to calcium gluconate treatment. Oral administration of 501 mg potassium per kilogram as potassium chloride to a 475-kg dairy cow by stomach tube resulted in death within 10 minutes, apparently from cardiac arrest (Ward, 1966a). The author pointed out that this dose represented about one-half the daily intake of similar cows that were fed 15 kg alfalfa hay per day, and that apparently did not suffer ill effects.

The recognition that incidence of grass tetany seemed to be higher in animals grazing pastures that had been fertilized with potassium (Dryerre, 1932) stimulated interest in the role of potassium in the disturbance. Increasing the potassium level in a liquid diet from 1.2 to 5.8

percent, dry basis, resulted in death of three of eight calves (Blaxter *et al.*, 1960). The clinical signs before death were cardiac insufficiency, edema, severe muscular weakness, and muscular atony. No abnormalities in magnesium metabolism were observed.

Administration of a combination of 157 g potassium chloride and 157 g of trans-aconitic acid or citric acid per 100 kg of body weight resulted in a high incidence of tetany resembling field cases of grass tetany in 237-kg yearling cattle (Bohman *et al.*, 1969). It appeared that the combinations were required for the effects. Plasma magnesium was not affected by the treatments. Acute toxicosis was produced by oral administration of 500 g potassium chloride and 500 g citric acid to 440-kg fistulated steers (Rumsey and Putnam, 1972). Death was preceded by darkening of the blood, frequent urination, frequent attempts to defecate, intense muscular tremors, protruding eyes, and loss of ability to stand. Struggling was noted after the steers were unable to stand. Rapid changes in EKG patterns were noted, and respiration rate was increased.

Plasma levels of potassium and immunoreactive insulin were elevated by intravenous infusion of 51, 64, and 135 mg potassium per kilogram as potassium chloride in calves and intraruminal infusion of 440 mg potassium per kilogram in cows (Lentz *et al.*, 1976). Administering of 64 mg potassium per kilogram resulted in lower plasma potassium and higher plasma glucose and insulin in magnesium deficient calves than in normal calves.

Feeding a diet with approximately 5 percent potassium to ewes did not affect blood serum levels of magnesium, calcium, or potassium (Pearson *et al.*, 1949). Later, including 5 percent potassium as bicarbonate in a ewe diet lowered serum magnesium, but did not produce clinical signs (Kunkel *et al.*, 1953). Feeding a high-potassium hay or supplemental potassium increased serum potassium, but was without effect on serum magnesium in sheep (Eaton and Avampato, 1953). Apparent absorption of magnesium in sheep was severely reduced by feeding a diet containing about 34 percent crude protein and 4.7 percent potassium, dry basis (Fontenot *et al.*, 1960).

Feeding a diet with approximately 4 percent potassium depressed magnesium absorption by about 30 percent in sheep (House and Van Campen, 1971). Magnesium absorption in sheep was reduced 46 percent when the potassium level of the diet was raised from 0.7 to 5.5 percent, dry basis (Newton *et al.*, 1972). Feeding a high potassium level resulted in similar increases in fecal magnesium, expressed as grams, but not as percent of intake over a wide range of dietary magnesium intakes (Frye, 1975). Feeding a diet containing 4 percent potassium resulted in a faster disposal of intravenously administered magnesium, compared to feeding 0.9 percent potassium (House and Bird, 1975).

Single intraperitoneal injections of 0.575 g potassium chloride per kilogram (0.302 g potassium per kilogram) of body weight were fatal to rabbits within 15–30 minutes, with terminal plasma potassium of 58 to 85 mg/dl (Truscoe and Zwemer, 1953).

Including 0.5 to 4.5 g potassium in the diet of adrenalectomized dogs resulted in potassium toxicosis (Allers *et al.*, 1936). The signs of adrenal insufficiency were accompanied by a high level of potassium in the cells (Nilson, 1937). Recovery could be accomplished either by administration of sodium chloride or adrenal cortical extract. Potassium administered by intravenous injection was distributed in a volume greater than that of the extracellular fluid of dogs (Winkler and Smith, 1938). This indicates that it enters some and probably most cells in the body.

Death occurred in dogs in which potassium chloride was injected intravenously when serum potassium reached 47 to 78 mg/dl (Winkler *et al.*, 1939). The toxic effects seemed to be specific for the heart, without similar effects on skeletal muscle. Injecting toxic levels of potassium in dogs resulted in intraventricular block and diastolic arrest (Winkler *et al.*, 1940).

Infusion of a high-potassium solution intravenously prevented induction of atrial fibrillation by acetylcholine or vagal stimulation in dogs (Hashimoto *et al.*, 1970). Dogs exhibited electrocardiographic evidence of prelethal cardiotoxicosis in about 3 hours from infusing 78 mg potassium chloride per kilogram (41 mg potassium per kilogram) of body weight (Hiatt *et al.*, 1975). At that time serum potassium level was 40–41 mg/dl.

Adaptation to high levels of potassium occurred in the rat by gradually increasing the size of the dose (Thatcher and Radike, 1947). Administration of adrenal extract to the rats was helpful in resisting the potassium ion, but the effect was not as beneficial as potassium adaptation.

Feeding a diet with 5 percent potassium as the bicarbonate or 3 percent as the carbonate resulted in similar marked depressions in growth rate of rats (Pearson, 1948). High mortality was observed from feeding the 5 percent potassium diet supplied as carbonate. Increasing dietary magnesium reduced mortality. Feeding a high-potassium (2.9 percent) diet depressed growth in rats (Colby and Frye, 1951). When combined with a high-calcium level (2.5 percent of diet), increased mortality and lowered blood magnesium were noted.

One half of the rats that consumed the 1.04 g of potassium per 100 g of body weight per day died with signs of potassium toxicosis (Drescher *et al.*, 1958). It was concluded that excess dietary potassium produces widening of the zona glomerulosa of the adrenal cortex in rats (Hartroft and Sowa, 1964). High dietary potassium (3.60 g potassium per 100 g feed) decreased plasma renin level (Sealey *et al.*, 1970).

Feeding a diet with 0.68 g potassium per 100 g for several weeks resulted in magnesium depletion in rats (Duarte, 1974). Duarte suggested that the effect may have been due to an enhanced aldosterone secretion or competition for transport between magnesium and potassium. A 200 percent increase in density of cristae mitochondriales in proximal and distal tubules of kidney cortex was observed in rats fed 7.82 mg of potassium per 100 g of diet for 6 weeks (Pfaller *et al.*, 1974). Similar effects were produced by aldosterone injection.

The $LD_{50}$ (5–10 minutes) for immature mice weighing 9–11 g was reported to be 66.5 mg potassium chloride (34.9 mg potassium) per 100 g of body weight and 57.5 mg (30.2 mg), injected intraperitoneally for adrenalectomized mice (Truszkowski and Duszynska, 1940). Resistance of adrenalectomized mice could be raised above that for normal mice by injections of adrenal extract.

## FACTORS INFLUENCING TOXICITY

The deleterious effect of high dietary potassium on magnesium utilization by ruminants is well documented (Fontenot *et al.*, 1973). However, it appears that a high-magnesium level may offer some protection against potassium toxicity. Increasing the magnesium content of a diet containing 5 percent potassium as carbonate reduced mortality in rats (Pearson, 1948). Acute magnesium loading of potassium adapted rats caused an increase in urinary potassium excretion (Duarte, 1974). Adaptation to the potassium ion increases the tolerance of the albino rat to potassium toxicity (Thatcher and Radike, 1947; Drescher *et al.*, 1958). Administration of adrenal extract resulted in increased resistance in rats (Thatcher and Radike, 1947) and mice (Truszkowski and Duszynska, 1940).

Sodium salts, at least under some conditions, reduce effects of high dietary potassium. Recovery of adrenalectomized dogs showing signs of adrenal insufficiency caused by high-potassium intake can be produced by administration of sodium salts (Nilson, 1937). Drops in serum potassium level in dogs of 34 to 53 percent resulted from parenteral administration of magnesium sulfate, indicating the relationship of these elements (Smith, 1949).

## TISSUE LEVELS

Feeding a 5 percent potassium diet resulted in higher potassium levels in skeletal muscle, heart muscle, kidney, and thymus of rats (Meyer *et al.*, 1950), compared to rats fed 0.005 percent potassium. The difference indicated, at least partly, an effect of potassium deficiency. Compared to feeding a 0.50 percent potassium diet, tissue levels were not substantially elevated by feeding 5 percent potassium. No differences occurred in potassium in carcass or heart muscle in rats fed levels ranging from 0.06 to 1.5 mg potassium per 100 g of body weight (Drescher *et al.*, 1958).

## MAXIMUM TOLERABLE LEVELS

No overt signs were produced in ruminants by oral administration of potassium at levels of 3 percent or lower, unless high levels of citric or trans-aconitic acid were administered also. Alfalfa hay, a recognized high-quality roughage, may contain over 2 percent potassium, dry basis (National Research Council, 1971). The maximum tolerable level is set at 3 percent for cattle and sheep. However, a level of 3 percent potassium may lower magnesium absorption. Data for nonruminants are limited but a maximum level of 3 percent appears to be satisfactory.

## SUMMARY

Potassium, found mainly within the cells of the animal body, performs essential metabolic functions. Animals have a dietary requirement for the mineral, which varies with species. It is distributed in feeds, with forages usually containing higher levels than concentrates.

High levels of dietary potassium are toxic to both ruminant and nonruminant animals. Excessive intake of potassium appears to interfere with the absorption and utilization of magnesium. The heart and adrenal glands are adversely affected by excessive intake of potassium. The toxicity of potassium can be mitigated by sodium salts and increased intake of magnesium.

Tissue levels are only slightly affected by dietary potassium levels above the level required in the diet.

TABLE 28 Effects of Potassium Administration in Animals

| Class and Number of Animals[a] | Age or Weight | Administration | | | | Effect(s) | Reference |
|---|---|---|---|---|---|---|---|
| | | Quantity of Element[b] | Source | Duration | Route | | |
| Cattle—1 | Calf | 160 mg/kg | KCl | 99 min | IV | Irritability; excitability; diuresis | Bergman and Sellers, 1953 |
| | | 330 mg/kg | | 168 min | | Irritability; excitability; diuresis; death | |
| Cattle—8 | Calf | 5.8% | KCl | 70 d | Diet | Death of three of eight calves; cardiac insufficiency; edema; muscular weakness; muscular atony | Blaxter *et al.*, 1960 |
| Cattle—1, cow | 300 kg | 61 mg/kg | KCl | Few min | Drench | No adverse effect | Dennis and Harbaugh, 1948 |
| | | 71 mg/kg | | | | Milk fever | |
| | | 80 mg/kg | | | | No adverse effect | |
| Cattle—5, cow | | 131 mg/kg | | | | One death; two requiring treatment; two showing no adverse effect | |
| Cattle—1, cow | 475 kg | 105 mg/kg | KCl | Few min | Drench | Death in 10 min | Ward, 1966b |
| Cattle—12 | Yearling, 237 kg | 820 mg K with 1,570 mg trans-aconitic or | KCl | Few min | Drench | Tetany in 10 of 12 | Bohman *et al.*, 1969 |

TABLE 28 *Continued*

| Class and Number of Animals[a] | Age or Weight | Administration: Quantity of Element[b] | Source | Duration | Route | Effect(s) | Reference |
|---|---|---|---|---|---|---|---|
| | | citric acid/kg | | | | | |
| Cattle—5, steer | 440 kg | 2,598 mg and 1,140 mg citric acid/kg | KCl | Few min | Drench or intraruminal infusion | Death preceded by darkening of blood; frequent urination; frequent attempts to defecate; muscular tremors; not able to stand | Rumsey and Putnam, 1972 |
| Cattle—4, bull calf | 115 kg | 51 or 135 mg/kg | KCl | 90 min | IV | Increase in plasma K and insulin | Lentz *et al.*, 1976 |
| Cattle—4, bull calf | 161 kg | 64 mg/kg | | | | Increase in plasma K and insulin; less effect for K and more for insulin in Mg-deficient calves | |
| Cattle—8, cow | 660 kg | 440 mg/kg | KCl | 40 min | Intraruminal infusion | Increase in plasma K and insulin | |
| Sheep—25 | Lamb | 1.0 and 1.5% | KCl | 119 d | Diet | No effect | Jackson *et al.*, 1971 |
| | | 2.0 and 2.5% | | | | Decreased weight and energy gain | |
| Sheep—6, ewe | | 5% | $KHCO_3$ | 4 mo | Diet | No effect on serum Ca, Mg, K | Pearson *et al.*, 1949 |
| Sheep—22, ewe | 32–45 kg | 5% | $KHCO_3$ | 62 d | Diet | Decrease in serum Mg; no effect on serum | Kunkel *et al.*, 1953 |

| | | | | | | | |
|---|---|---|---|---|---|---|---|
| | | | | | | Ca, Na, K, total protein | |
| Sheep—8 | 1 yr | 4.1% | KCl | 10 d | Diet | Decrease Mg absorption | House and Van Campen, 1971 |
| Sheep—6 | 31.6 kg | 5.5% | $KHCO_3$ | 24 d | Diet | Decrease Mg absorption | Newton *et al.*, 1972 |
| Swine—62 | 2 wk | 0.4% | Potassium acetate | 35 or 45 d | Diet | Increased gain and feed efficiency | Liebholz *et al.*, 1966 |
| Swine—83 | | 0.8% | | | | Increased gain and feed efficiency | |
| Rabbit | 750–2,400 g | 300 mg/kg | | Single | Intraperi-toneal | Death in 15–30 min | Truscoe and Zwemer, 1953 |
| Dog | | 0.5–4.5 g/day | | | Diet | Potassium toxicosis in adrenalectomized dogs | Allers *et al.*, 1936 |
| Dog | 9.6–25.0 kg | 56–183 mg/kg | KCl | 9–53 min | IV | Death when serum K reached 47 to 78 mg/d1 | Winkler *et al.*, 1939 |
| Rat—175 | 160–220 g | Variable | KCl | | Drench | Tolerance to K was in-creased by adaptation to K | Thatcher and Radike, 1947 |
| Rat—26 | 45–60 g | 5% | $KHCO_3$ | 3 wk | Diet | 62% mortality in low Mg and 17% mortality in adequate Mg; decreased gain | Pearson, 1948 |
| Rat—43 | 45–60 g | 5% | $KHCO_3$ | 3 wk | Diet | Decreased gain | Pearson, 1949 |
| Rat | 150–200 g | Increasing | KCl and $KHCO_3$ | | Diet | 50% mortality; high plasma K; electro-cardiograph changes when K intake exceeded 1.0 mg K per 100 g body weight | Drescher *et al.*, 1958 |

TABLE 28 *Continued*

| Class and Number of Animals[a] | Age or Weight | Administration: Quantity of Element[b] | Source | Duration | Route | Effect(s) | Reference |
|---|---|---|---|---|---|---|---|
| Rat—10 | 58–200 g | 7.1% | KCl | 21–33 d | Water | Hypertrophy of adrenal zona glomerulosa | Hartroft and Sowa, 1964 |
| Rat—26 | 43–98 g | 3.1–6.2% | KCl | 2–9 wk | | Hypertrophy of adrenal zona glomerulosa | Hartroft and Sowa, 1964 |
| Rat—30 | 275–285 g | 3.6 mg per 100 g | | 7 wk | Diet | Induced sodium depletion | Sealey *et al.*, 1970 |
| Rat—3 | 350 g | 8% + 0.4% K as KCl in water, *ad lib.* | | 6 wk | Diet | Increased density of cristae mitochondriales in kidney tubules | Pfaller *et al.*, 1974 |

[a] Number of animals per treatment.
[b] Quantity expressed as milligrams per kilogram of body weight.

# REFERENCES

Allers, W. D., H. W. Nilson, and E. C. Kendall. 1936. Studies on adrenalectomized dogs: The toxic action of potassium. Staff Meet. Mayo Clin. 11:283.

Bergman, E. N., and A. F. Sellers. 1953. Studies on intravenous administration of calcium, potassium, and magnesium to dairy calves. I. Some biochemical and general toxic effects. Am. J. Vet. Res. 14:520.

Bergman, E. N., and A. F. Sellers. 1954. Studies on intravenous administration of calcium, potassium and magnesium to dairy calves. II. Some cardiac and respiratory effects. Am. J. Vet. Res. 15:25.

Blaxter, K. L., B. Cowlishaw, and J. A. F. Rook. 1960. Potassium and hypomagnesaemic tetany in calves. Anim. Prod. 2:1.

Bohman, V. R., A. L. Lespernace, G. D. Harding, and D. L. Grunes. 1969. Induction of experimental tetany in cattle. J. Anim. Sci. 29:99.

Church, D. C., and W. G. Pond. 1974. Basic Animal Nutrition and Feeding. D. C. Church, Corvallis, Ore.

Colby, R. W., and C. M. Frye. 1951. Effect of feeding various levels of calcium, potassium and magnesium to rats. Am. J. Physiol. 166:209.

Crookshank, H. R. 1966. Effect of sodium or potassium on ovine urinary calculi. J. Anim. Sci. 25:1005.

Dennis, J., and F. G. Harbaugh. 1948. The experimental alteration of blood potassium and calcium levels in cattle. Am. J. Vet. Res. 9:20.

Drescher, A. N., N. B. Talbot, P. A. Meara, M. Terry, and J. D. Crawford. 1958. A study of the effects of excessive potassium intake upon body potassium stores. J. Clin. Invest. 37:1316.

Dryerre, H. 1932. Lactation tetany. Vet. Res. 12:1163.

Duarte, C. G. 1974. Magnesium loading in potassium-adapted rats. Am. J. Physiol. 227:482.

Eaton, H. D., and J. E. Avampato. 1953. Blood levels and retention of calcium, magnesium and potassium in lambs on normal and high potassium diets. J. Anim. Sci. 11:761.

Emerick, R. J., H. R. King, and L. B. Embry. 1972. Influence of dietary potassium and of transaconitic acid on mineral metabolism related to ovine phosphatic urolithiasis. J. Anim. Sci. 35:901.

Fontenot, J. P., R. W. Miller, C. K. Whitehair, and R. MacVicar. 1960. Effect of a high-protein high-potassium ration on the mineral metabolism of lambs. J. Anim. Sci. 19:127.

Fontenot, J. P., M. B. Wise, and K. E. Webb, Jr. 1973. Interrelationship of potassium, nitrogen, and magnesium. Fed. Proc. 32:1925.

Frye, T. M. 1975. Interrelationship of dietary magnesium and potassium in beef cows and sheep. Ph.D. dissertation. Virginia Polytechnic Institute and State University, Blacksburg.

Hartroft, P. M., and E. Sowa. 1964. Effect of potassium on juxtaglomerular cells and the adrenal zona glomerulosa of rats. J. Nutr. 82:439.

Hashimoto, K., S. Yasuyuki, and S. Chiba. 1970. Effect of potassium excess on pacemaker activity of canine sinoatrial node in vivo. Am. J. Physiol. 218:83.

Hiatt, N., A. Miller, and T. Katayanagi. 1975. Kaluresis and diuresis after administration of antidiuretic hormone to hyperkalemic dogs. Am. J. Physiol. 228:1108.

Hoar, D. W., R. J. Emerick, and L. B. Embry. 1970. Potassium, phosphorous and calcium interrelationships influencing feedlot performance and phosphatic urolithiasis in lambs. J. Anim. Sci. 30:597.

House, W. A., and R. J. Bird. 1975. Magnesium tolerance in goats fed two levels of potassium. J. Anim. Sci. 41:1134.

House, W. A., and D. Van Campen. 1971. Magnesium metabolism of sheep fed different levels of potassium and citric acid. J. Nutr. 101:1483.

Jackson, H. M., R. P. Kromann, and E. E. Ray. 1971. Energy retention in lambs as influenced by various levels of sodium and potassium in the rations. J. Anim. Sci. 33:872.

Kunkel, H. O., K. H. Burns, and B. J. Camp. 1953. A study of sheep fed high levels of potassium bicarbonate with particular reference to induced hypomagnesemia. J. Anim. Sci. 12:451.

Lentz, D. E., F. C. Madsen, J. K. Miller, and S. L. Hansard. 1976. Effect of potassium and hypomagnesemia on insulin in the bovine. J. Anim. Sci. 43:1082.

Liebholz, J. M., J. T. McCall, V. W. Hays, and V. C. Speer. 1966. Potassium, protein and basic amino acid relationships in swine. J. Anim. Sci. 25:37.

Meyer, J. H., R. R. Grunert, M. T. Zepplin, R. H. Grummer, G. Bohstedt, and P. H. Phillips. 1950. Effect of dietary levels of sodium and potassium on growth and on concentrations in blood plasma and tissues of white rat. Am. J. Physiol. 162:182.

Miller, E. C. 1939. A Physiological Study of the Winter Wheat Plant at Different Stages of Its Development. Kans. Agric. Exp. Stn. Tech. Bull. 47.

National Research Council. 1971. Atlas of Nutritional Data on United States and Canadian Feeds. National Academy of Sciences, Washington, D.C.

National Research Council. 1976. Nutrient Requirements of Domestic Animals. No. 4. Nutrient Requirements of Beef Cattle. National Academy of Sciences, Washington, D.C.

National Research Council. 1979. Nutrient Requirements of Domestic Animals. No. 2. Nutrient Requirements of Swine. National Academy of Sciences, Washington, D.C.

Newton, G. L., J. P. Fontenot, R. E. Tucker, and C. E. Polan. 1972. Effects of high dietary potassium intake on the metabolism of magnesium by sheep. J. Anim. Sci. 35:440.

Nilson, H. W. 1937. Corticoadrenal insufficiency: Metabolism studies on potassium, sodium and chloride. Am. J. Physiol. 118:620.

Pearson, P. B. 1948. High levels of dietary potassium and magnesium and growth of rats. Am. J. Physiol. 153:432.

Pearson, P. B., J. A. Gray, and R. Reiser. 1949. The calcium, magnesium, and potassium contents of the serum of ewes fed high levels of potassium. J. Anim. Sci. 8:52.

Pfaller, W., W. M. Fischer, N. Streider, H. Wurnig, and P. Deetjen. 1974. Morphologic changes of cortical nephron cells in potassium-adapted rats. Lab. Invest. 31:678.

Rumsey, T. S., and P. A. Putnam. 1972. EKG, respiratory, saliva flow and serum mineral changes associated with KCl-citric acid induced tetany in cattle. J. Anim. Sci. 35:986.

Sealey, J. E., I. Clark, M. B. Bull, and J. H. Laragh. 1970. Potassium balance and the control of renin secretion. J. Clin. Invest. 49:2119.

Smith, S. G. 1949. Magnesium–potassium antagonism. Arch. Biochem. 20:473.

Thatcher, J. S., and A. W. Radike. 1947. Tolerance to potassium intoxication in the albino rat. Am. J. Physiol. 151:138.

Truscoe, R., and R. L. Zwemer. 1953. Plasma potassium curves in the rabbit following single and repeated injections of potassium chloride. Am. J. Physiol. 175:181.

Truszkowski, R., and J. Duszynska. 1940. Protection of mice against potassium poisoning by corticoadrenal hormones. Endocrinology 27:117.

Ward, G. M. 1966a. Oral potassium chloride fatal to a cow. J. Am. Vet. Med. Assoc. 148:543.

Ward, G. M. 1966b. Potassium metabolism of domestic ruminants—A review. J. Dairy Sci. 49:268.

Winkler, A. W., and P. K. Smith. 1938. The apparent volume of distribution of potassium injected intravenously. J. Biol. Chem. 124:589.

Winkler, A. W., H. E. Hoff, and P. K. Smith. 1939. Factors affecting the toxicity of potassium. Am. J. Physiol. 127:430.

Winkler, A. W., H. E. Hoff, and P. K. Smith. 1940. Cardiovascular effects of potassium, calcium, magnesium and barium. Yale J. Biol. Med. 13:123.

# Selenium

Selenium (Se) is a semimetal (or metalloid), which is very similar to sulfur in its chemical properties. Its allotropic forms include a red powder, red crystals, a dark brown moss, and a silver gray form produced after extended heating at 200–220°C. It is present chiefly in Cretaceous rocks, volcanic material, some seafloor deposits, and glacial drift in central Canada and North Dakota in the form of metallic selenides. These selenides are often associated with sulfides (as in pyrites). Selenium exists in soil as basic ferric selenite [$Fe_2(OH)SeO_3$], calcium selenate ($CaSeO_4$), elemental selenium, and organic compounds derived from plant tissue.

Nearly all primary production of selenium results from treatment of residue slimes generated during electrolytic refining of copper. Annual U.S. use of selenium is approximately 500,000 kg, with substantial use in rectifiers, xerographic copying machines, and photoelectric cells. Selenium is also used in glass, ceramics, rubber, pigments, and plating solutions.

Selenium was once used in insecticides for use on ornamental plants, but this is no longer done. Current primary agricultural use involves the supplementation of animal diets with selenite or selenate to prevent a specific deficiency.

Much of the early interest in selenium among nutritionists concerned its role as a toxic element. Indirect suggestions of its involvement in certain animal disease syndromes have been known for years. Marco Polo referred in his journals to " . . . a poisonous plant . . . which if eaten by (horses) has the effect of causing the hoofs . . . to drop off'' (Polo [Marsden's transl.], 1926). In 1857, Dr. T. C. Madison, a U.S. Army surgeon at Fort Randall, Nebraska Territory, described similar signs in horses which he attributed to ''alkali disease'' (Madison, 1860). Selenium was identified as the toxic principle by scientists from the U.S. Department of Agriculture and the South Dakota and Wyoming State Agricultural Experiment Stations through a series of studies begun in 1929 (Franke, 1934). Not until 1957 (Patterson *et al.*, 1957; Schwarz and Foltz, 1957; Scott *et al.*, 1957; Stokstad *et al.*, 1957) was the role of selenium as an essential nutrient established. Kubota *et al.* (1967) subsequently determined that the selenium-deficient areas of the United States are much larger than those areas that are selenium-toxic.

## ESSENTIALITY

The first evidence that selenium was an essential nutrient involved the discovery that it would prevent liver necrosis in rats (Schwarz and Foltz, 1957) and exudative diathesis in chicks (Patterson *et al.*, 1957). Eggert *et al.* (1957) found selenium would prevent hepatosis dietetica in swine, and DeWitt and Schwarz (1958) found it prevented a number of lesions in mice. Selenium was used successfully by Muth *et al.* (1958) and Hogue (1958) to prevent white muscle disease in young ruminants. Field observations in New Zealand suggested that selenium deficiency may also lead to myopathy in the horse (Dodd *et al.*, 1960; Hartley and Grant, 1961). Scott and Thompson (1968) demonstrated that selenium is essential for the Japanese quail (*Coturnix coturnix*) even in the presence of high dietary levels of vitamin E (100 mg $d$-$\alpha$-tocopheryl acetate per kilogram of diet). Schwarz (1965) reported that administration of selenium to two children with kwashiorkor stimulated growth. Similar results have been reported by Majaj and Hopkins (1966), while Burk *et al.* (1967) noted low blood selenium levels in children with untreated kwashiorkor and an enhanced *in vitro* uptake of radioselenite by the erythrocytes of these affected children.

One of the biochemical functions of selenium in higher animals was defined by Rotruck *et al.* (1973), who discovered that selenium was an integral part of the enzyme glutathione peroxidase (EC1.M.19). This enzyme destroys lipid peroxides and thus functions in protecting cell membranes against peroxidative damage. However, selenium has been shown to be a constituent of other enzyme systems in microorganisms (Stadtman, 1974), and this element may eventually be shown to have additional roles in mammalian metabolism. Diplock and Lucy (1973) have proposed that selenide may occupy an active site in certain non-heme iron proteins. Levander *et al.* (1973, 1974) suggested that selenium plays a role in the electron transport chain, and Whanger *et al.* (1973) found a selenoprotein in lamb muscle containing a heme group identical with that of cytochrome c.

## METABOLISM

Selenium absorption from the gastrointestinal tract and its retention and distribution within the body varies with the chemical form and amount ingested. The amounts, forms, and routes of excretion are also affected by these factors and may be greatly influenced by other elements, notably arsenic.

At toxic or near-toxic levels, selenium is absorbed rapidly and efficiently from naturally seleniferous diets and from soluble selenium salts. Rats consuming a seleniferous wheat diet (18 ppm selenium) retained 63 percent of the ingested selenium within the first week and a similar proportion from the same concentration of selenium as sodium selenite (Moxon, 1937; Anderson and Moxon, 1941). Toxicity studies with rats suggest a higher absorption from seleniferous grains than from selenites and selenates and a very low absorption from selenides and elemental selenium (Franke and Painter, 1938; Smith *et al.*, 1938). Some organic compounds, such as selenodiacetic and selenopropionic acids, are markedly less toxic to rats (per unit of selenium) than is selenite, probably as a consequence of lower absorption (Moxon *et al.*, 1938).

Studies with physiological levels of radioselenium indicate that the duodenum is the main site of absorption, with no absorption from the rumen or abomasum of sheep or the stomach of pigs (Wright and Bell, 1966). Net absorption was about 35 percent in sheep and 85 percent in pigs when the diets contained 0.35 and 0.50 ppm selenium, respectively.

Absorbed selenium is at first carried mostly in plasma (Buescher *et al.*, 1960) in association with plasma proteins (McConnell and Levy, 1962) and is then deposited in all tissues.

Much of the tissue selenium is highly labile, and, following transfer from selenium-adequate or seleniferous diets to low-selenium diets, losses are rapid initially and then slower. The urine is a major pathway of excretion in both ruminants (Lopez *et al.*, 1969) and monogastric species (McConnell, 1941, 1942, 1948). Most of the selenium in the feces is that which has not been absorbed from the diet, plus small amounts excreted in biliary, pancreatic, and intestinal secretions. Levander and Baumann (1966a,b) have shown that biliary selenium excretion is markedly increased when subacute injections of arsenic are given with the selenium. Exhalation of selenium is an important route of excretion at high dietary intakes, but is much less so at low intakes (Olson *et al.,* 1963; Ganther *et al.,* 1966; Handreck and Godwin, 1970).

## SOURCES

Selenium exists in several oxidation states (−2, 0, +4, +6), and its chemical properties are similar to those of sulfur. In its −2 state, it occurs as hydrogen selenide, a highly toxic and reactive gas, that quickly decomposes in the presence of oxygen to elemental selenium and water. Heavy metal selenides are insoluble, and a number of organic selenides have been identified in biological materials, some of which are very volatile. In elemental form (0 oxidation state), selenium is insoluble, not toxic, and not readily oxidized or reduced in nature. When burned, it is oxidized to selenium dioxide, which sublimes, and, when dissolved in water, forms selenious acid. In the +4 state, selenium occurs as inorganic selenites. Those that are soluble are highly toxic. Selenite has an affinity for iron and aluminum sesquioxides and forms stable adsorption complexes with them in soil. In addition, selenite is easily reduced to the elemental form under acid and reducing conditions, and, thus, selenite added to soil may become quite unavailable to plants and is unlikely to pollute water supplies. The formation and stability of selenates (+6 oxidation state) is favored by alkaline and oxidizing conditions. Most selenates are quite soluble and highly toxic. Selenates are not tightly complexed by sesquioxides and, in soils, are available to plants and easily leached.

Biological processes are involved in the reduction of selenium, but reduction also results from burning. Biological reduction can produce volatile organic selenides or hydrogen selenide. Burning can produce particulate elemental selenium or selenium dioxide, and these are the most likely forms in the atmosphere. Oxidation apparently occurs in alkaline soils by chemical weathering.

## TOXICOSIS

Soon after the relationship between high selenium intakes and livestock losses was established, it became apparent that selenium poisoning has more than one form. Rosenfeld and Beath (1946) suggested that three types of selenium toxicosis occur in the field: low level of the blind staggers type, low level of the alkali disease type, and high level.

### LOW LEVELS

Selenium poisoning of the blind staggers type has been ascribed to consumption of limited amounts of accumulator plants over several weeks or months (Rosenfeld and Beath, 1946). Included among the accumulators are many species of *Astragalus,* and some species of *Machaeranthera, Haplopappus,* and *Stanleya.* Affected animals wander aimlessly, stumble, have impaired vision, and exhibit some signs of respiratory failure. Since water extracts of accumulator plants will produce this condition, while pure selenium compounds will not, it is probable that alkaloids rather than selenium may be responsible (Maag and Glenn, 1967).

Low-level selenium toxicosis of the alkali disease type has been described in detail by Moxon (1937) and Rosenfeld and Beath (1946). It is a consequence of consuming feeds ranging from about 5 to 40 ppm selenium over periods of weeks or months. The most prominent signs in cattle and horses include lameness, hoof malformations, loss of hair from mane or tail, and emaciation. Sheep do not usually exhibit hoof or wool lesions, but reproduction is adversely affected, as it may be also in cattle (Minyard, 1961), swine (Wahlstrom and Olson, 1959), and rats (Franke and Potter, 1935). Swine exhibit lameness, hoof malformation, loss of body hair, and emaciation. Poultry show decreased egg hatchability associated with teratogenic effects. Duhamel (1913) described a hemorrhagic exudate in lung alveoli, dilated capillaries, and bronchial exudate. Necrosis, hemorrhage, and fibrosis were seen as hepatic cirrhosis developed. The kidneys exhibited a mild tubular degeneration with acute glomerular injury. Ascites and edema are common. The vascular effects are apparent even in goldfish (Ellis *et al.,* 1937), where marked edema, particularly of gastric submucosa and of perivascular tissues in the kidneys and liver, has been seen.

Herigstad *et al.* (1973) fed selenium as sodium selenite or selenomethionine to pigs at concentrations of 20 to 600 ppm. They noted emesis, anorexia, weight loss, cachexia, central nervous system depression, respiratory distress, coma, subnormal body temperature, and death. Lesions at necropsy were similar for swine given both selenium compounds and included hepatic fatty metamorphosis and centrilobular necrosis; congestion of the renal medulla; necrosis in lymphoid follicles; edema and degenerative changes in cerebrum, cerebellum, and spinal cord; edema and hemorrhagic necrosis of the pancreas;

depletion of hematopoietic cells in bone marrow; hemorrhagic necrosis of the adrenal cortex; serous atrophy of body fat; and degenerative changes in diaphragm and skeletal muscles.

Loew *et al.* (1975) accidentally fed 10 ppm selenium (from sodium selenite) in the diet of cynomolgus monkeys and noted erosions on the tongue, hemorrhagic dermatosis on the tail, loss of nails (onychoptosis), anorexia, lassitude, and leukopenia. These signs developed over 40 days and disappeared when the selenium level was restored to normal.

Suggestions that selenium might induce neoplasia (Nelson *et al.*, 1943), and interest in nutritional requirements for selenium, led to an extensive study at Oregon State University in which 1,437 rats were fed varying levels of selenite or selenate selenium for up to 30 months (Harr *et al.*, 1967; Tinsley *et al.*, 1967). The basal semipurified diet contained 0.1 ppm, and selenium was added at 0.5, 2, 4, 6, 8, or 16 ppm. Acute toxic hepatitis was noted in rats receiving 4 to 16 ppm. These rats were emaciated and pale, and exhibited ascites, edema, and poor-quality hair coats. Most lived less than 100 days. Chronic toxic hepatitis and hyperplastic hepatocytes were reported in rats receiving selenium supplements of 0.5 to 2 ppm. These rats lived 24 to 30 months and many had developed murine pneumonia. Unfortunately, necropsy findings in rats on the basal diet were not adequately described. Although 63 neoplasms were observed in the entire study, none could be attributed to added selenium.

## HIGH LEVELS

In the field, acute poisoning occurs when grazing animals eat sufficient amounts of selenium accumulator plants to cause sudden death or signs of severe distress (labored breathing, ataxia, abnormal posture, prostration, and diarrhea). This type of poisoning is rare, since animals usually avoid these plants. However, when pasture is limited, accumulators may be nearly the only food available, and occasional large losses among sheep and cattle may occur. Acute poisoning has also been produced accidentally or experimentally by the administration of large amounts of selenium compounds to farm animals (Caravaggi and Clark, 1969; Caravaggi *et al.*, 1970; Shortridge *et al.*, 1971; Herigstad *et al.*, 1973).

Rosenfeld and Beath (1946) described vascular manifestations, including petechial hemorrhages in the endocardium and acute congestion and diffuse hemorrhages in the lungs. In ruminants, the omasum was congested and hemorrhagic, and there was desquamation of the mucous epithelium. There was enteritis, intestinal hemorrhage, and occasionally colitis and proctitis. The liver was passively congested, hemorrhagic, and exhibited parenchymatous degeneration with focal necrosis. The kidneys exhibited parenchymatous degeneration, hemorrhages, and nephritis. Steele and Wilhelm (1967) showed that high levels of selenite produced remarkable increases in vascular permeability in the guinea pig.

When Herigstad *et al.* (1973) intravenously injected 3 mg of selenium per kilogram of body weight into two pigs, fatal selenium toxicosis developed in 2½ or 14 hours. The pig dying first received selenium as sodium selenite and exhibited pulmonary edema at necropsy. The pig dying at 14 hours received selenomethionine, and postmortem signs included a yellow-brown mottled liver, pale renal cortex, and congested renal medulla. The clinical course of the toxicosis included vomiting, profound central nervous system depression, weakness, respiratory distress, coma, and death.

## FACTORS INFLUENCING TOXICITY

Halverson *et al.* (1962) have demonstrated that dietary sulfate can decrease the toxicity of selenate but not of selenite or organic selenium. Sellers *et al.* (1950) demonstrated that methionine could protect against selenium toxicity, but only when adequate vitamin E was present in the diet. This was confirmed by Levander and Morris (1970), who also found that several fat-soluble antioxidants could replace vitamin E in potentiating the methionine response.

Starting with the discovery of Moxon (1938) that arsenic could counteract the toxicity of seleniferous grains, a number of interactions between selenium and other elements have been found that render selenium much less toxic than when it is present alone. Levander and Baumann (1966b) found that arsenic functioned by increasing biliary excretion of selenium into the intestine. Moxon and DuBois (1939) reported that tungsten as well as arsenic counteracted the toxicity of selenium. Hill (1975) reported that mercury, cadmium, and copper reduce selenium toxicosis in the chick, and Jensen (1975) found that silver was also effective.

Linseed meal has a unique protective activity against chronic selenium toxicosis (Moxon, 1941), which is not associated with protein and which can be extracted with hot aqueous ethanol (Halverson *et al.*, 1955). Levander *et al.* (1970) showed that this feedstuff resulted in higher and more tightly bound hepatic selenium levels. Recently,

Palmer *et al.* (1980) presented data suggesting that the cyanogenic glycosides, linustatin and neolinustatin, are responsible for the protective action of linseed meal.

Harr *et al.* (1967) noted that rats fed commercial diets showed 2 to 3 times greater resistance to selenium toxicity than when fed a semipurified diet.

Although a number of dietary factors will ameliorate the development of selenium toxicosis, treatment of the poisoned animal is not very satisfactory. Oral administration of 4 to 5 g naphthalene daily for 5 days has been used for chronic selenosis in cattle and horses to increase urinary selenium loss. The dosage is repeated after a 5-day dosage-free interval. Removal of the source of selenium will result in a gradual decline in tissue concentration if kidney function has not been seriously impaired.

## TISSUE LEVELS

Much of the selenium in tissues is highly labile, and transfer of animals from seleniferous to nonseleniferous diets is followed by rapid, and then slow, loss of selenium from the tissues via bile, urine, and/or expired air. Selenium concentrations in tissues tend to reflect dietary selenium concentrations, particularly when provided by natural dietary ingredients as compared to selenate or selenite. Ku *et al.* (1972) found the selenium concentration of swine skeletal muscle (0.034–0.521 ppm, wet basis) was highly correlated (r = 0.95) with that in natural swine diets (0.027–0.493 ppm, air dry) from 13 different U.S. locations. When these workers (Ku *et al.,* 1973) added sufficient selenium (0.4 ppm) from sodium selenite to raise a low-selenium (0.04 ppm) swine diet to the level found in a South Dakota swine diet (0.44 ppm from natural sources), respective skeletal muscle selenium concentrations were 0.12 and 0.48 ppm, wet basis. Corresponding liver selenium concentrations were 0.61 and 0.84 ppm, wet basis. Kidney selenium concentrations were 2.14 and 2.17 ppm, wet basis. When swine diets were selenium-deficient (0.05 ppm), the following tissue selenium concentrations (ppm, wet basis) have been found: longissimus muscle, 0.05; myocardium, 0.11; liver, 0.14; kidney, 1.37 (Groce *et al.,* 1973). A similar pattern of tissue selenium concentrations has been found in cattle and sheep (Ullrey *et al.,* 1977) and in chicks and poults (Scott and Thompson, 1971).

Selenium in serum of swine receiving an unsupplemented natural diet

(0.04 ppm selenium) or this diet supplemented with 0.05, 0.1, or 0.2 ppm selenium from sodium selenite was 0.046, 0.150, 0.164, or 0.168 ppm. Respective erythrocyte selenium concentrations were 0.088, 0.181, 0.193, or 0.207 ppm (Groce *et al.*, 1973).

Cows with hair selenium concentrations between 0.06 and 0.23 ppm produced calves with white muscle disease, while no lesions were seen in calves from cows with hair selenium greater than 0.25 ppm (Hidiroglou *et al.*, 1965). Selenium in hair of yearling cattle on seleniferous range averaged over 10 ppm (Olson *et al.*, 1954).

Allaway *et al.* (1968) reported that cow's milk from a low-selenium area in the United States contained less than 20 ng/ml, compared with 50 ng/ml from a high-selenium area in South Dakota.

Normally, hen's eggs contain a total of 10 to 12 μg selenium, with most in the yolk (Taussky *et al.*, 1963, 1965). Whole-egg selenium concentrations on an adequate diet are about 0.3 ppm (Latshaw, 1975), although very high egg selenium levels can be produced by extremely high dietary intakes (Moxon and Poley, 1938).

## MAXIMUM TOLERABLE LEVELS

Dietary requirements for selenium range from 0.1 to 0.3 ppm in dry matter, and supplements of selenite and selenate are regularly added to animals' diets. Signs of toxicity have been seen in some food animal species when 5 ppm selenium were fed in relatively short-term studies. In rats fed semipurified diets, 4 ppm were toxic. However, 2 ppm selenium has produced no unequivocally toxic signs, and this dietary concentration is suggested as a maximum tolerable level for all species.

## SUMMARY

Selenium is a relatively rare metalloid that is similar to sulfur in its chemical properties. Although early interest among nutritionists was concerned with its potential toxicity, selenium has been established as an essential nutrient and is a constituent of glutathione peroxidase. Selenium toxicity in high selenium areas has been divided into three types. The low-level, blind staggers type results from consumption of limited amounts of selenium accumulator plants over several weeks or months and is probably due to toxic alkaloids. The low-level, alkali disease type results from consuming feeds containing 5 to 40 ppm selenium over weeks or months. High-level, acute poisoning results when grazing animals consume large amounts of accumulator plants sufficient to cause severe distress and sudden death. Since selenite and selenate are now being used to supplement deficient animal diets, a potential for accidental poisoning by this route also exists. Dietary selenium requirements are approximately 0.1 to 0.3 ppm, and toxic dietary levels are about 10 to 50 times greater.

TABLE 29 Effects of Selenium Administration in Animals

| Class and Number of Animals[a] | Age or Weight | Administration: Quantity of Element[b] | Source | Duration | Route | Effect(s) | Reference |
|---|---|---|---|---|---|---|---|
| Cattle—1 | 3–4 yr, 322 kg | 7.0 mg/kg | $Na_2SeO_3$ | Single | Drench | No adverse effect (20 d) | Miller and Williams, 1940a |
| | 3–4 yr, 320 kg | 10.1 mg/kg | | | | Anorexia for 2 days; decreased milk production (observed 5 d) | |
| | 617 kg | 11.2 mg/kg | | | | Death in 30 h | |
| | 476 kg | 14.1 mg/kg | | | | Anorexia; labored breathing; garlic breath; death in 48 h | |
| | 5 d, 30 kg | 20.2 mg/kg | | | | Oral discharge; profuse diarrhea; death in 6 h | |
| Cattle—557 | 6 mo | 100 mg (~0.5 mg/kg) | $Na_2SeO_3$ | Single | S.C. or I.M. inj. | Depression; salivation; respiratory distress; hydrothorax; lung edema; 67% mortality (2 h–5 wk) | Shortridge *et al.*, 1971 |
| Sheep—203 | 2–5 wk, 7–11 kg | 15 mg (~1.7 mg/kg) | $Na_2SeO_3$ | Single | Drench | Lethargy; hyperpnea; frothing from mouth; hydrothorax; lung edema; degenerative changes in liver and kidney; liver Se 40 ppm (dry basis); 72 deaths in 12–48 h | Lambourne and Mason, 1969 |

| | | | | | | | |
|---|---|---|---|---|---|---|---|
| Sheep—20 | 4–14 d, 4.5 kg | 10 mg (~2.2 mg/kg) | $Na_2SeO_3$ | Single | Drench | Seven deaths in 10–16 h; hyperemic and edematous lungs; acute necrotizing nephrosis; diarrhea in eight surviving lambs (observed 2 d) | Morrow, 1968 |
| Sheep—190 | 2–3 mo, 10 kg | 6.4 mg/kg | $Na_2SeO_3$ | Single | Drench | Pulmonary congestion; mild, diffuse fatty changes in liver; liver Se to 64 ppm; 180 deaths in 1–15 d | Gabbedy and Dickson, 1965 |
| Sheep—32 | 8–10 wk, 7–12 kg | 425, 450, 475, and 500 μg/kg | $Na_2SeO_3$ | Single | I.M. inj. | $LD_{50}$ was 455 μg/kg in 1–7 d; marked pulmonary congestion and edema; degenerative changes in liver and kidney; similar to enterotoxemia | Caravaggi *et al.*, 1970 |
| Sheep—20 | 2–4 wk, 5–8 kg | 5 mg (~0.8 mg/kg) | $Na_2SeO_3$ | Single | I.M. inj. | Mortality of 45% in 6 h–7 d | Caravaggi and Clark, 1969 |
| Sheep—1 | Yearling, 45 kg | 1 mg/kg | $Na_2SeO_3$ | Single | I.M. inj. | Depression; ataxia; dyspnea; frequent urination; elevated temperature; increased pulse and respiratory rates; cyanosis; death in 8 h | Morrow, 1968 |
| Sheep—1 | 2 mo, 10 kg | 0.46 mg/kg (initial), 0.46 mg/kg | Se-methionine | Three | I.V. inj. | Retarded weight gain; anorexia; emaciation; hypocythemic, | Neethling *et al.*, 1968 |

TABLE 29 *Continued*

| Class and Number of Animals[a] | Age or Weight | Administration: Quantity of Elements[b] | Source | Duration | Route | Effect(s) | Reference |
|---|---|---|---|---|---|---|---|
| | | (after 1 wk), 0.54 mg/kg (after 3 mo) | | | | macrocytic anemia (observed 6 mo) | |
| | 2 mo, 7 kg | 0.9 mg/kg (initial), 0.9 mg/kg (after 1 wk), 1.0 mg/kg (after 3 mo) | Se-cystine | | | Retarded weight gain; anorexia; emaciation; hypocythemic, macrocytic anemia (observed 6 mo) | |
| Sheep—1 | Adult, 36 kg | 1 mg/kg (initial), 1 mg/kg (after 2.5 mo) | Se-methionine | Two | | Weight loss; cachexia; hypocythemic, macrocytic anemia; death in 5 mo | |
| Sheep—2 | Adult, 35 kg | 3.4 mg/kg | | Single | | Cyanosis; pulmonary edema; death in 10 h | |
| | | 4 mg/kg | $SeO_2$ | | | Hyperpnea; cyanosis; pulmonary edema; tremors; hypersensitivity; opisthotonus; paresis; death due to anoxia in 20 min | |
| Swine—2 | 5 kg | 0.1 ppm | $Na_2SeO_3$ | 35–39 d | Diet | No adverse effect | Herigstad *et al.*, 1973 |
| | 4 kg | | Se-methionine | 35–38 d | | No adverse effect | |

| | | | | | | | |
|---|---|---|---|---|---|---|---|
| | | | Se-methionine | 38 d | | No adverse effect | |
| | | 10 ppm | $Na_2SeO_3$ | 56 d | | No adverse effect | |
| Swine—2 | 5 kg | | Se-methionine | | | | |
| | | 20 ppm | $Na_2SeO_3$ | 84 d | | Anorexia; emesis; weight loss; depression; dyspnea and death of one pig at 32 d; no effect on second | |
| | | | Se-methionine | 63–84 d | | Decreased weight gain in one pig | |
| | | 45 ppm | $Na_2SeO_3$ | 63 d | | Weight loss and death of one pig at 3 d; decreased weight gain and toxic signs in second | |
| | 4 kg | | Se-methionine | 5–9 d | | Weight loss and death | |
| | 5 kg | 60 ppm | $Na_2SeO_3$ | 142–152 h | | Weight loss and death | |
| | 6 kg | | Se-methionine | 100–150 h | | Weight loss and death | |
| | 5 kg | 100 ppm | $Na_2SeO_3$ | 19–28 d | | Weight loss and death | |
| | | | Se-methionine | 11–20 d | | Weight loss and death | |
| | 4 kg | 120 ppm | $Na_2SeO_3$ | 53–56 h | | Weight loss and death | |
| | 5 kg | | Se-methionine | 59–220 h | | Weight loss and death | |
| | 4 kg | 600 ppm | $Na_2SeO_3$ | 24–46 h | | Weight loss and death | |
| | | | Se-methionine | 29–46 h | | Weight loss and death | |
| Swine—12 | 15 kg | 0.1 ppm | $Na_2SeO_3$ | 19 wk | Diet | No adverse effect | Groce *et al.*, 1973 |
| | | 0.2 ppm | | | | No adverse effect | |
| Swine—9 | 9 kg | 0.5 ppm | $Na_2SeO_3$ | 14 wk | Diet | No adverse effect | Groce *et al.*, 1971 |
| Swine—6 | 45 kg | 1.0 ppm | | 8 wk | | No adverse effect | |
| Swine—5 | 35 kg | 5 ppm | Seleniferous corn | | Diet | No adverse effect | Schoening, 1936 |
| | | 10 ppm | | | | Signs of toxicosis in 60% | |

## TABLE 29 *Continued*

| Class and Number of Animals[a] | Age or Weight | Administration: Quantity of Elements[b] | Source | Duration | Route | Effect(s) | Reference |
|---|---|---|---|---|---|---|---|
| Swine—4 | 15 kg | 7 ppm | $Na_2SeO_3$ | 108 d | Diet | Decreased gain; hair loss; cracked hooves; emaciation (by 5 wk); 1 death at 10 wk | Wahlstrom *et al.*, 1956 |
| Swine—8 | 14 kg | 10 ppm | | 120 d | | Decreased gain; prevented by 0.02% arsanilic acid or 0.005% 3-nitro phenylarsonic acid | |
| | 13 kg | 11 ppm | | 128 d | | Decreased gain | |
| Swine—16 | | 13 ppm | | | | Signs of toxicosis in 31% | |
| Swine—10, females | 8 wk, 15 kg | 10 ppm | $Na_2SeO_3$ | Through weaning of 2 litters | Diet | Decreased conception rate; increased services per conception; more small, weak, dead pigs at birth; fewer and lighter pigs at weaning | Wahlstrom and Olson, 1959 |
| Swine—2 | 16–19 kg | 24 ppm | $Na_2SeO_3$ | 79 d | Diet | Anorexia; hair loss; liver degeneration; death | Miller and Schoening, 1938 |
| | 17–22 kg | 49 ppm | | 99 d | | Hair loss; swelling of coronary band; | |

| | | | | | | | | |
|---|---|---|---|---|---|---|---|---|
| | | | | | | | death | |
| | 12–19 kg | 196 ppm | | | 38 d | | Anorexia; weakness; hair loss; swelling of coronary band (feet); liver degeneration; death | |
| | 15–17 kg | 392 ppm | | | 46 d | | Weakness; diarrhea; gastroenteritis; death | |
| Swine—1 | 68 kg | 2.2 mg/kg | $Na_2SeO_3$ | Single | | Drench | No observed effects in 2.5 mo | Miller and Williams, 1940a |
| | 54 kg | 4.4 mg/kg | | | | | Anorexia; depression; recovery in 24 h (observed 2.5 mo) | |
| | | 8.8 mg/kg | | | | | Anorexia; depression; recovery in 5 d (observed 2.5 mo) | |
| Swine—2 | 45–54 kg | 13.2 mg/kg | | | | | Anorexia; depression; recovery in 5–7 d (observed 2.5 mo) | |
| | 59 kg | 17.4 mg/kg | | | | | Emesis; profuse diarrhea; paresis; fatty degeneration of liver (observed 18 d) | |
| Swine—1 | 24 kg | 22.7 mg/kg | | | | | Death in 72 h; vascular congestion; degenerative changes in liver | |
| | 30–70 kg | 0.8 mg/kg | $Na_2SeO_3$ | Single | | S. C. inj. | No adverse effect in 5 d | Orstadius, 1960 |
| | | 0.9–1.1 mg/kg | | | | | Slight to moderate elevation of plasma GOT in half the pigs; | |

TABLE 29 *Continued*

| Class and Number of Animals[a] | Age or Weight | Administration: Quantity of Element[b] | Source | Duration | Route | Effect(s) | Reference |
|---|---|---|---|---|---|---|---|
| | | | | | | returning to normal in 2 d (observed 5 d) | |
| | | 1.2 mg/kg | | | | Trembling; ataxia; paresis; death; muscle dystrophy; increased plasma GOT (observed 5 d) | |
| | | 2 mg/kg | | | | Trembling; ataxia; paresis; death in 4 h; muscle dystrophy; increased GOT | |
| | 7 kg | 3 mg/kg | $Na_2SeO_3$ | Single | I.V. inj. | Pulmonary edema and death in 2.5 h | Herigstad *et al.*, 1973 |
| | | | Se-methionine | | | Pulmonary edema and death in 14 h | |
| Horse—2 | 5–12 yr | 24 ppm (2 mo) 48 ppm (13 mo) 96 ppm (2 mo) | $Na_2SeO_3$ | 17 mo | Diet (drench for 4 mo) | Emaciation; listlessness; loose hair in mane and tail; softening and scaling of hoof wall; hemorrhagic and cirrhotic liver; death | Miller and Williams, 1940b |
| Horse—1 | Aged | 115 ppm | $Na_2SeO_3$ | 5 wk | Diet | Emaciation; listlessness; loose hair in mane and tail; | |

| | | | | | | | |
|---|---|---|---|---|---|---|---|
| | | | | | | softening and scaling of hoof wall; hemorrhagic and cirrhotic liver; death | |
| | 503 kg | 2.7 mg/kg | $Na_2SeO_3$ | Single | Drench | Labored breathing; anorexia for about 1 d; normal thereafter (observed 24 d) | Miller and Williams, 1940a |
| | 658 kg | 4.4 mg/kg | | | | Labored breathing; toxic spasms; death in 26 h | |
| | 651 kg | 8.0 mg/kg | | | | Hemorrhagic gastritis; fatty degeneration of liver; death in 18 h | |
| | 440 kg | 10.1 mg/kg | | | | Cutaneous muscle spasms; dilation of eyes; profuse sweating, labored breathing; death in 22 h | |
| | 538 kg | 12.1 mg/kg | | | | Depressed; weak; trembling; garlic breath; death in 24 h | |
| Chicken—20, hen | 32 wk | 0.1 ppm | $Na_2SeO_3$ | 28 wk | Diet | No adverse effect | Ort and Latshaw, 1978 |
| | | 1 ppm | | | | No adverse effect | |
| | | 3 ppm | | | | No adverse effect | |
| | | 5 ppm | | | | None on egg production, egg weight, or fertility; decreased hatchability | |

TABLE 29 *Continued*

| Class and Number of Animals[a] | Age or Weight | Administration: Quantity of Elements[b] | Source | Duration | Route | Effect(s) | Reference |
|---|---|---|---|---|---|---|---|
| | Laying | 5 ppm | $Na_2SeO_3$ | 16 wk | Diet | No adverse effect | |
| | | 9 ppm | | | | Decreased egg weight, production, and hatchability | |
| Chicken—50, pullet | 1 d | 2 ppm | Selenious acid | 76 wk | Diet | None, except possibly increased weight at 20 wk | Thapar *et al.*, 1969 |
| | | 8 ppm | | | | Reduced body weight, egg weight, production, hatchability, and progeny growth | |
| Chicken | 1 d | 2 ppm | $Na_2SeO_3$ | Several weeks | Diet | No adverse effect | Moxon, 1937 |
| Chicken, hen | Adult | 2.5 ppm | Seleniferous corn, barley, and wheat | | | No adverse effect | |
| Chicken | 1 d | 4 ppm | $Na_2SeO_3$ | | | No adverse effect | |
| Chicken, hen | Adult | 5 ppm | Seleniferous corn, barley, and wheat | | | None on hatchability; wirey down on many hatched chicks and increased mortality | |
| Chicken, pullet | Young, laying | 6.5, then 3.25 ppm | $Na_2SeO_3$ | | | Decreased feed consumption and weight; deformed embryos | |

| | | | | | | | |
|---|---|---|---|---|---|---|---|
| Chicken | 1 d | 8 ppm | $Na_2SeO_3$ | | | Decreased weight gain | |
| Chicken, hen | Adult | 10 ppm | Seleniferous corn, barley, and wheat | | | Embryonic deformities and hatchability declined to zero | |
| Chicken, pullet | Young, laying | 15 ppm | Seleniferous corn, barley, and wheat | 5 wk | | Decreased feed consumption, weight; no decrease in egg production or fertility; deformed embryos and hatchability declined to zero | |
| Chicken—20, hen | Adult | 26 ppm | $Na_2SeO_3$ | 2 wk | | Immediate decrease in feed consumption, weight, and egg production | |
| Chicken—60 | 1 d | 2.5 ppm | $SeO_2$ | 2 wk | Diet | No adverse effect | Hill, 1974 |
| | | 5 ppm | | | | No adverse effect | |
| | | 10 ppm | | | | Gain 72% of controls | |
| | | 20 ppm | | | | Gain 30% of control | |
| | | 40 ppm | | | | Gain 2% of controls | |
| Chicken—10 | 1 d | 5 ppm | $SeO_2$ | 4–5 wk | Diet | Tendency for increased mortality from *S. gallinarum* infection | Hill, 1979 |
| | | 10 ppm | | | | Decreased gain and increased mortality from *S. gallinarum* infection | |
| | | 20 ppm | | | | Decreased gain and increased mortality from *S. gallinarum* infection | |

TABLE 29 *Continued*

| Class and Number of Animals[a] | Age or Weight | Administration | | | | Effect(s) | Reference |
|---|---|---|---|---|---|---|---|
| | | Quantity of Element[b] | Source | Duration | Route | | |
| Chicken—20, hen | Laying | 7 ppm | $Na_2SeO_3$ | 16 wk | Diet | None on egg production; decreased egg weight and hatchability | Ort and Latshaw, 1978 |
| Chicken, hen | Laying | 8 ppm | $Na_2SeO_3$ | 2 wk or longer | Diet | Embryos incubated to 5 d showed no gross pathology but histologically there was pathologic regression of previously well formed parts; the nervous system, limb buds, eyes exhibited necrosis | Gruenwald, 1958 |
| Rat—10 | 28 d | 4.4 ppm | Seleniferous wheat | 100 d | Diet | Slightly decreased gain | Moxon, 1937 |
| | | 8.8 ppm | | | | Moderately decreased gain | |
| | | 13.1 ppm | | | | Moderately decreased gain | |
| | | 17.5 ppm | | | | Markedly decreased gain and weight loss after 70 d | |
| Rat—18 | 70 g | 10 ppm | $K_2SeO_4$ | 18–21 d | Diet | Gain 47% of control; no mortality | Halverson *et al.*, 1962 |

| | | | | | | | |
|---|---|---|---|---|---|---|---|
| | | | $Na_2SeO_3$ | | | Gain 44% of control; 11% mortality | |
| | | | Seleniferous wheat | | | Gain 41% of control; no mortality | |
| Rat—9 | 21 d | 22 ppm | $Na_2SeO_3$ | 359 d | Diet | Decreased gain; liver degeneration; five deaths | Franke and Potter, 1935 |
| | | 34 ppm | | | | Decreased gain; liver degeneration; eight deaths | |
| | | 52 ppm | | 17 d | | Weight loss; death | |
| Rat—48 | 21 d, 40 g | 2 ppm | $Na_2SeO_3$ | 35–63 d | Water | None to slight decrease in gain | Palmer and Olson, 1974 |
| Rat—24 | | | $Na_2SeO_4$ | | | None to slight decrease in gain | |
| | | 3 ppm | $Na_2SeO_3$ | 28–63 d | Water | Slight decrease in gain, feed, and water intake | |
| Rat—22 | | | $Na_2SeO_4$ | | | Slight decrease in gain, feed, and water intake | |
| Rat—12 | | 6 ppm | $Na_2SeO_3$ | | | Decreased gain, feed, and water intake; mortality | |
| | | | $Na_2SeO_4$ | | | Decreased gain, feed, and water intake; mortality | |
| | | 9 ppm | $Na_2SeO_3$ | 28–56 d | | Decreased gain, feed, and water intake; mortality | |
| | | | $Na_2SeO_4$ | 28–63 d | | Decreased gain, feed, and water intake; mortality | |

TABLE 29 *Continued*

| Class and Number of Animals[a] | Age or Weight | Administration | | | | Effect(s) | Reference |
|---|---|---|---|---|---|---|---|
| | | Quantity of Element[b] | Source | Duration | Route | | |
| Rat | 125–175 g | 3.25–3.5 mg/kg | $Na_2SeO_3$ | Single | I.P. inj. | Killed 75% or more in 48 h | Franke and Moxon, 1936 |
| | | 5.25–5.75 mg/kg | $Na_2SeO_4$ | | | Killed 75% or more in 48 h | |
| Hamster—8 | Weanlings, 35 g | 6 ppm | $Na_2SeO_3$ | 4 wk | Water | None on gain; water consumption reduced 30% | Hadjimarkos, 1970 |
| | | 9 ppm | | | | Decreased gain and water consumption 45% | |
| | | 12 ppm | | | | Decreased gain and water consumption 46% | |
| Dog—1 | 60 d | 7.2 ppm | Seleniferous corn | 189 d | Diet | Decreased feed consumption and gain | Rhian and Moxon, 1943 |

| | | | | | | | |
|---|---|---|---|---|---|---|---|
| Dog—6 | 150 d | 10 ppm | $Na_2SeO_3$ | 100–150 d | | Decreased feed consumption and weight | |
| Dog—2 | 72 d | 20 ppm | Seleniferous corn | 150 d | | Decreased feed consumption and gain | |
| Dog—10 | Young | 20 ppm | $Na_2SeO_3$ | Several weeks | Diet | Decreased feed consumption and gain; dull-eyed; sluggish; wandered aimlessly | Moxon, 1937 |
| | | | Seleniferous corn | | | Decreased feed consumption and gain; dull-eyed; sluggish; wandered aimlessly | |
| Monkey—11, cynomolgus (*Macaca fascicularis*) | | 10 ppm | $Na_2SeO_3$ | 40 d | Diet | Tongue erosions; crusty, hemorrhagic tail dermatosis; loss of nails (onychoptosis); anorexia; lassitude; leukopenia | Loew *et al.*, 1975 |

[a] Number of animals per treatment.

[b] Quantity expressed in parts per million or as milligrams per kilogram of body weight.

## REFERENCES

Allaway, W. H., J. Kubota, F. Losee, and M. Roth. 1968. Selenium, molybdenum and vanadium in human blood. Arch. Environ. Health 16:342.

Anderson, H. D., and A.L. Moxon. 1941. The excretion of selenium by rats on a seleniferous wheat ration. J. Nutr. 22:103.

Buescher, R. G., M. C. Bell, and R. K. Berry. 1960. Effect of excessive calcium on selenium-75 in swine. J. Anim. Sci. 19:1251. (Abstr.)

Burk, R., Jr., W. N. Pearson, R. P. Wood II, and F. Viteri. 1967. Blood selenium levels and in vitro red blood cell uptake of $^{75}$Se in kwashiorkor. Am. J. Clin. Nutr. 20:723.

Caravaggi, C., and F. L. Clark. 1969. Mortality in lambs following intramuscular injection of sodium selenite. Aust. Vet. J. 45:383.

Caravaggi, C., F. L. Clark, and A. R. B. Jackson. 1970. Acute selenium toxicity in lambs following intramuscular injection of sodium selenite. Res. Vet. Sci. 11:146.

DeWitt, W. B., and K. Schwarz. 1958. Multiple dietary necrotic degeneration in the mouse. Experientia 14:28.

Diplock, A. T., and J. A. Lucy. 1973. The biochemical modes of action of vitamin E and selenium: A hypothesis. Fed. Eur. Biochem. Soc. Lett. 29:205.

Dodd, D. C., A. A. Blakely, R. S. Thornbury, and H. F. Dewes. 1960. Muscle degeneration and yellow fat disease in foals. N.Z. Vet. J. 8:45.

Duhamel, B. C. 1913. Lésions histologiques dans l'intoxication par le sélénium colloidal et de l'acide sélénieux. C. R. Soc. Biol. 42:742.

Eggert, R. G., E. Patterson, W. T. Akers, and E. L. R. Stokstad. 1957. The role of vitamin E and selenium in the nutrition of the pig. J. Anim. Sci. 16:1037.

Ellis, M. M., H. L. Motley, M. D. Ellis, and R. O. Jones. 1937. Selenium poisoning in fishes. Proc. Soc. Exp. Biol. Med. 36:519.

Franke, K. W. 1934. A new toxicant occurring naturally in certain samples of plant foodstuffs. I. Results obtained in preliminary feeding trials. J. Nutr. 8:597.

Franke, K. W., and A. L. Moxon. 1936. A comparison of the minimum fatal doses of selenium, tellurium, arsenic and vanadium. J. Pharmacol. Exp. Ther. 58:454.

Franke, K. W., and E. P. Painter. 1938. A study of the toxicity and selenium content of seleniferous diets: With statistical consideration. Cereal Chem. 15:1.

Franke, K. W., and V. R. Potter. 1935. A new toxicant occurring naturally in certain samples of plant foodstuffs. IX. Toxic effects of orally ingested selenium. J. Nutr. 10:213.

Gabbedy, B. J., and J. Dickson. 1969. Acute selenium poisoning in lambs. Aust. Vet. J. 45:470.

Ganther, H. E., O. A. Levander, and C. A. Baumann. 1966. Dietary control of selenium volatilization in the rat. J. Nutr. 88:55.

Groce, A. W., D. E. Ullrey, E. R. Miller, D. J. Ellis, and K. K. Keahey. 1971. Selenium and vitamin E in practical swine diets. J. Anim. Sci. 33:230.

Groce, A. W., E. R. Miller, D. E. Ullrey, P. K. Ku, K. K. Keahey, and D. J. Ellis. 1973. Selenium requirements in corn–soy diets for growing–finishing swine. J. Anim. Sci. 37:948.

Gruenwald, P. 1958. Malformations caused by necrosis in the embryo. Illustrated by the effect of selenium compounds on chick embryos. Am. J. Pathol. 34:77.

Hadjimarkos, D. M. 1970. Toxic effects of dietary selenium in hamsters. Nutr. Rep. Int. 1:175.

Halverson, A. W., C. M. Hendrick, and O. E. Olson. 1955. Observations on the protective effect of linseed oil meal and some extracts against chronic selenium poisoning in rats. J. Nutr. 56:51.

Halverson, A. W., P. L. Guss, and O. E. Olson. 1962. Effect of sulfur salts on selenium poisoning in the rat. J. Nutr. 77:459.

Handreck, K. A., and K. D. Godwin. 1970. Distribution in the sheep of selenium derived from $^{75}$Se-labelled ruminal pellets. Aust. J. Agric. Res. 21:71.

Harr, J. R., J. F. Bone, I. J. Tinsley, P. H. Weswig, and R. S. Yamamoto. 1967. Selenium toxicity in rats. II. Histopathology, pp. 153–178. *In* O. H. Muth (ed.). Symposium: Selenium in Biomedicine. AVI Publishing Co., Westport, Conn.

Hartley, W. J., and A. B. Grant. 1961. A review of selenium responsive diseases of New Zealand livestock. Fed. Proc. 20:679.

Herigstad, R. R., C. K. Whitehair, and O. E. Olson. 1973. Inorganic and organic selenium toxicosis in young swine: Comparison of pathologic changes with those in swine with vitamin E-selenium deficiency. Am. J. Vet. Res. 34:1227.

Hidiroglou, M., R. B. Carson, and G. A. Brossard. 1965. Influence of selenium on the selenium contents of hair and on the incidence of nutritional muscular disease in beef cattle. Can. J. Anim. Sci. 45:197.

Hill, C. H. 1974. Reversal of selenium toxicity in chicks by mercury and cadmium. J. Nutr. 104:593.

Hill, C. H. 1975. Interrelationships of selenium with other trace elements. Fed. Proc. 34:2096.

Hill, C. H. 1979. The effects of dietary protein levels on mineral toxicity in chicks. J. Nutr. 109:501.

Hogue, D. E. 1958. Vitamin E, selenium and other factors related to nutritional muscular dystrophy in lambs, pp. 32–39. *In* Proc. Cornell Nutr. Conf. Feed Manuf., Ithaca, New York.

Jensen, L. S. 1975. Modification of a selenium toxicity in chicks by dietary silver and copper. J. Nutr. 105:769.

Ku, P. K., W. T. Ely, A. W. Groce, and D. E. Ullrey. 1972. Natural dietary selenium, $\alpha$-tocopherol and effect on tissue selenium. J. Anim. Sci. 34:208.

Ku, P. K., E. R. Miller, R. C. Wahlstrom, A. W. Groce, J. P. Hitchcock, and D. E. Ullrey. 1973. Selenium supplementation of naturally high selenium diets for swine. J. Anim. Sci. 37:501.

Kubota, J., W. H. Allaway, D. L. Carter, E. E. Cary, and V. A. Lazar. 1967. Selenium in crops in the United States in relation to the selenium-responsive diseases of livestock. J. Agric. Food Chem. 15:448.

Lambourne, D. A., and R. W. Mason. 1969. Mortality in lambs following overdosing with sodium selenite. Aust. Vet. J. 45:208.

Latshaw, J. D. 1975. Natural and selenite selenium in the hen and egg. J. Nutr. 105:32.

Levander, O. A., and C. A. Baumann. 1966a. Selenium metabolism. V. Studies on the distribution of selenium in rats given arsenic. Toxicol. Appl. Pharmacol. 9:98.

Levander, O. A., and C. A. Baumann. 1966b. Selenium metabolism. VI. Effect of arsenic on the excretion of selenium in the bile. Toxicol. Appl. Pharmacol. 9:106.

Levander, O. A., and V. C. Morris. 1970. Interactions of methionine, vitamin E, and antioxidants in selenium toxicity in the rat. J. Nutr. 100:1111.

Levander, O. A., M. L. Young, and S. A. Meeks. 1970. Studies on the binding of selenium by liver homogenates from rats fed diets containing either casein or casein plus linseed oil meal. Toxicol. Appl. Pharmacol. 16:79.

Levander, O. A., V. C. Morris, and D. J. Higgs. 1973. Selenium as a catalyst for the reduction of cytochrome c by glutathione. Biochemistry 12:4591.

Levander, O. A., V. C. Morris, and D. J. Higgs. 1974. Selenium catalysis of swelling of rat liver mitochondria and reduction of cytochrome c by sulfur compounds. Adv. Exp. Biol. Med. 48:405.

Loew, F. M., E. D. Olfert, and B. Schiefer. 1975. Chronic selenium toxicosis in cynomolgus monkeys. Lab. Primate Newsl. 14:7.

Lopez, P. L., R. L. Preston, and W. H. Pfander. 1969. Whole-body retention, tissue distribution and excretion of selenium-75 after oral and intravenous administration in lambs fed varying selenium intakes. J. Nutr. 97:123.

Maag, D. D., and M. W. Glenn. 1967. Toxicity of selenium: Farm animals, pp. 127–140. *In* O. H. Muth (ed.). Symposium: Selenium in Biomedicine. AVI Publishing Co., Westport, Conn.

Madison, T. C. 1860. Sanitary report—Fort Randall, *In* R. H. Coolidge (ed.). Statistical Report on the Sickness and Mortality in the Army in the United States. Senate Exch. Doc. 52:37.

Majaj, A. S., and L. L. Hopkins, Jr. 1966. Selenium and kwashiorkor. Lancet 2:592.

McConnell, K. P. 1941. Distribution and excretion studies in the rat after a single subtoxic subcutaneous injection of sodium selenate containing radioselenium. J. Biol. Chem. 141:427.

McConnell, K. P. 1942. Respiratory excretion of selenium studied with the radioactive isotope. J. Biol. Chem. 145:55.

McConnell, K. P. 1948. Passage of selenium through the mammary glands of the white rat and the distribution of selenium in the milk proteins after subcutaneous injection of sodium selenate. J. Biol. Chem. 173:653.

McConnell, K. P., and R.S. Levy. 1962. Presence of selenium-75 in lipoproteins. Nature 195:774.

Miller, W. T., and H. W. Schoening. 1938. Toxicity of selenium fed to swine in the form of sodium selenite. J. Agric. Res. 56:831.

Miller, W. T., and K. T. Williams. 1940a. Minimum lethal doses of selenium, as sodium selenite, for horses, mules, cattle, and swine. J. Agric. Res. 60:163.

Miller, W. T., and K. T. Williams. 1940b. Effect of feeding repeated small doses of selenium as sodium selenite to equines. J. Agric. Res. 61:353.

Minyard, J. A. 1961. Selenium poisoning in beef cattle. S. Dak. Farm Home Res. 12:1.

Morrow, D. A. 1968. Acute selenite toxicosis in lambs. J. Am. Vet. Med. Assoc. 152:1625.

Moxon, A. L. 1937. Alkali Disease or Selenium Poisoning. S. Dak. Agric. Exp. Stn. Bull. No. 311. South Dakota State College of Agriculture and Mechanic Arts, Agricultural Experiment Station, Brookings. 91 pp.

Moxon, A. L. 1938. The effect of arsenic on the toxicity of seleniferous grains. Science 88:81.

Moxon, A. L. 1941. Some factors influencing the toxicity of selenium. Ph.D. thesis. University of Wisconsin, Madison.

Moxon, A. L., and K. P. DuBois. 1939. The influence of arsenic and certain other elements on the toxicity of seleniferous grains. J. Nutr. 18:477.

Moxon, A. L., and W. E. Poley. 1938. The relation of selenium content of grains in the ration to the selenium content of poultry carcass and eggs. Poult. Sci. 17:77.

Moxon, A. L., H. D. Anderson, and E. P. Painter. 1938. The toxicity of some organic selenium compounds. J. Pharmacol. Exp. Ther. 63:357.

Muth, O. H., J. E. Oldfield, L. F. Remmert, and J. R. Schubert. 1958. Effects of selenium and vitamin E on white muscle disease. Science 128:1090.

Neethling, L. P., J. M. M. Brown, and P. J. DeWet. 1968. The toxicology and metabolic fate of selenium in sheep. J. S. Afr. Vet. Med. Assoc. 39(3):25.

Nelson, A. A., O. G. Fitzhugh, and H. O. Calvery. 1943. Liver tumors following cirrhosis caused by selenium in rats. Cancer Res. 3:230.

Olson, O. E., C. A. Dinkel, and L. D. Kamstra. 1954. A new aid in diagnosing selenium poisoning. S. Dak. Farm Home Res. 6:12.

Olson, O. E., B. M. Schulte, E. I. Whitehead, and A. W. Halverson. 1963. Effect of arsenic on selenium metabolism in rats. J. Agric. Food Chem. 11:531.

Orstadius, K. 1960. Toxicity of a single subcutaneous dose of sodium selenite in pigs. Nature 188:1117.

Ort, J. F., and J. D. Latshaw. 1978. The toxic level of sodium selenite in the diet of laying chickens. J. Nutr. 108:1114.

Palmer, I. S., and O. E. Olson. 1974. Relative toxicities of selenite and selenate in the drinking water of rats. J. Nutr. 104:306.

Palmer, I. S., O. E. Olson, A. W. Halverson, R. Miller, and C. Smith. 1980. Isolation of factors in linseed oil meal protective against chronic selenosis in rats. J. Nutr. 110:145.

Patterson, E. L., R. Milstrey, and E. L. R. Stokstad. 1957. Effect of selenium in preventing exudative diathesis in chicks. Proc. Soc. Exp. Biol. Med. 95:617.

Polo, M. 1926. The Travels of Marco Polo, p. 81. Revised from Marsden's translation and edited with introduction by Manual Komroff. Liveright, New York.

Rhian, M., and A. L. Moxon. 1943. Chronic selenium poisoning in dogs and its prevention by arsenic. J. Pharmacol. Exp. Ther. 78:249.

Rosenfeld, I., and O. A. Beath. 1946. Pathology of Selenium Poisoning. Wyo. Agric. Exp. Stn. Bull. No. 275. University of Wyoming Agricultural Experiment Station, Laramie. 27 pp.

Rotruck, J. T., A. L. Pope, H. E. Ganther, A. B. Swanson, D. G. Hafeman, and W. G. Hoekstra. 1973. Selenium: Biochemical role as a component of glutathione peroxidase. Science 179:588.

Schoening, H. W. 1936. Production of so-called alkali disease in hogs by feeding corn grown in affected area. N. Am. Vet. 17:22.

Schwarz, K. 1965. Selenium and kwashiorkor. Lancet 1:1335.

Schwarz, K., and C. M. Foltz. 1957. Selenium as an integral part of Factor 3 against dietary necrotic liver degeneration. J. Am. Chem. Soc. 79:3292.

Scott, M. L., and J. N. Thompson. 1968. Selenium in nutrition and metabolism, pp. 1–10. *In* Proc. Md. Nutr. Conf. Feed Manuf., College Park, Md.

Scott, M. L., and J. N. Thompson. 1971. Selenium content of feedstuffs and effects of dietary selenium levels upon tissue selenium in chicks and poults. Poult. Sci. 50:1742.

Scott, M. L., J. G. Bieri, G. M. Briggs, and K. Schwarz. 1957. Prevention of exudative diathesis by factor 3 in chicks on vitamin E-deficient torula yeast diets. Poult. Sci. 36:1155. (Abstr.)

Sellers, E. A., R. W. You, and C. C. Lucas. 1950. Lipotropic agents in liver damage produced by selenium or carbon tetrachloride. Proc. Soc. Exp. Biol. Med. 75:118.

Shortridge, E. H., P. J. O'Hara, and P. M. Marshall. 1971. Acute selenium poisoning in cattle. N. Z. Vet. J. 19:47.

Smith, M. I., B. B. Westfall, and E. F. Stohlman. 1938. Studies on the fate of selenium in the organism. U.S. Public Health Rep. 53:1199.

Stadtman, T. C. 1974. Selenium biochemistry. Proteins containing selenium are essential components of certain bacterial and mammalian enzyme systems. Science 183:915.

Steele, R. H., and D. L. Wilhelm. 1967. The inflammatory reaction in chemical injury. II. Vascular permeability changes and necrosis induced by intracutaneous injection of various chemicals. Br. J. Exp. Pathol. 48:592.

Stokstad, E. L. R., E. L. Patterson, and R. Milstrey. 1957. Factors which prevent exudative diathesis in chicks on torula yeast diets. Poult. Sci. 36:1160.

Taussky, H. H., A. Washington, E. Zubillaga, and A. T. Milhorat. 1963. Selenium content of fresh eggs from normal or dystrophic chickens. Nature 200:1211.

Taussky, H. H., A. Washington, E. Zubillaga, and A. T. Milhorat. 1965. Distribution of selenium in tissues of normal or dystrophic chickens. Nature 206:509.

Thapar, N. T., E. Guenthner, C. W. Carlson, and O. E. Olson. 1969. Dietary selenium and arsenic additions to diets for chickens over a half cycle. Poult. Sci. 48:1988.

Tinsley, I. J., J. R. Harr, J. F. Bone, P. H. Weswig, and R. S. Yamamoto. 1967. Selenium toxicity in rats. I. Growth and longevity, pp. 141–152. *In* O. H. Muth, ed. Symposium: Selenium in Biomedicine. AVI Publishing Co., Westport, Conn.

Ullrey, D. E., P. S. Brady, P. A. Whetter, P. K. Ku, and W. T. Magee. 1977. Selenium supplementation of diets for sheep and beef cattle. J. Anim. Sci. 46:559.

Wahlstrom, R. C., and O. E. Olson. 1959. The effect of selenium on reproduction in swine. J. Anim. Sci. 18:141.

Wahlstrom, R. C., L. D. Kamstra, and O. E. Olson. 1956. Preventing Selenium Poisoning in Growing and Fattening Pigs. S. Dak. Agric. Exp. Stn. Bull. No. 456. South Dakota State College, Agricultural Experiment Station, Brookings. 15 pp.

Whanger, P. D., N. D. Pedersen, and P. H. Weswig. 1973. Selenium proteins in ovine tissues. Biophys. Res. Commun. 53:1031.

Wright, P. L., and M. C. Bell. 1966. Comparative metabolism of selenium and tellurium in sheep and swine. Am. J. Physiol. 211:6.

# Silicon

---

Next to oxygen, silicon is the most abundant element on earth, and quartz (crystalline silica) is the most abundant mineral in the earth's crust (Carlisle, 1974). Silicon is found in the ash of most plant and animal tissues in small quantities, but it has been generally considered nonessential for most living organisms. Silica has long been associated with silicosis, a chronic lung disease caused by inhalation of silica-bearing dust, especially in miners, and with certain types of malignant tumors (Allison, 1968). It has been suggested that silicon is an essential nutrient for animals, based on its effect on bone mineralization in the laboratory rat, and on the occurrence of a deficiency in the chick (Carlisle, 1974).

## ESSENTIALITY

It has been shown that silicon is important in formation of young bone in laboratory rats (Carlisle, 1974). The importance of silicon appears to lessen as the bone approaches maturity. The element is also present, along with iron, in blood vessels between metaphyseal trabeculae. It seems to be involved in endochondrial and periosteal bone formation.

A silicon deficiency resulted in a 37 percent depression in growth rate of chicks (Carlisle, 1974). The deficient animals appeared stunted, and all organs were atrophied. Retarded skeletal development was recorded for the deficient chicks. Silicon has been shown to be essential for growth in rats also (Schwarz, 1973). Unicellular microscopic plants, diatoms, require silicon for shell formation and net DNA synthesis (Carlisle, 1974).

## METABOLISM

Silicon is usually found in combination with oxygen as silica. Silicon compounds play an especially significant part in some lower organisms (Carlisle, 1974). At least traces of silicon are found widespread in tissues and fluids of higher animals. The normal human blood level is less than 5 ppm (Carlisle, 1974). Silicic acid is readily absorbed across the intestinal wall and is excreted in the urine. Evidence was obtained that silica was not absorbed in sheep and cattle to a significant extent and appeared to be useful as an inert indicator to determine digestibility (Gallup *et al.,* 1945). The amount of silicon in blood and intestinal tissues in rats is affected by age, sex, castration, adrenalectomy, and thyroidectomy (Carlisle, 1974). Silica can enter in the respiratory tract, go to the lung as silicic acid, and is eventually eliminated.

The quantity of silicon present in the active growth areas of the bone appears to be related to the maturity of the bone (Carlisle, 1974). In the early stages of mineralization, silicon is present in small quantities. As mineralization progresses, silicon increases concomitantly with calcium. Later, the amount of silicon drops, and as the calcium reaches the level in apatite, silicon level is at the detection limit.

## SOURCES

Silicon is taken up by the roots of plants and deposited in cell walls. There is considerable variation in silicon content between plant species (Carlisle, 1974). Cereal grains high in crude fiber are higher in silicon than low-fiber grains. The silicon, present as silica, soluble silicates, and in organic combinations, is bound to the cellulosic cell structure. Silicon does not appear to be essential for plants, but it produces beneficial effects on plant growth (Jones and Handreck, 1967). Kind of soil, plant species, transpiration rate, and nutrient supply affect silica content of plants. Silica is not distributed uniformly among plant parts. It appears to be deposited in largest quantities in parts from which water is lost in greatest quantities. Rice hulls contain high concentrations of silica (Van Soest, 1970). Contamination of feeds with soil, especially in hay and pasture herbage, elevates the level of silica.

Although silicon has been shown to be a dietary essential, it is not likely that deficiency would occur in practice due to the wide distribution of it as a component or contaminant of feed.

## TOXICOSIS

Excess dietary silica appears to be involved in depression in digestibility of forages, depression in growth and reproduction rate, and formation of kidney stones in ruminants. Under practical farm or ranch conditions, silicon toxicosis is not a serious problem.

### LOW LEVELS

Silicon in forages has been shown to depress dry matter digestibility *in vivo* in ruminants (Van Soest and Jones, 1968) and organic matter digestibility *in vitro* (Smith *et al.*, 1971). The depressions amounted to approximately three units of dry matter digestibility *in vivo* (Van Soest and Jones, 1968) and one unit of organic matter digestibility *in vitro* (Smith *et al.*, 1971) per unit of silica in the dry matter of forages. Addition of sodium silicate to rumen cultures depressed *in vitro* organic matter digestibility of siliceous forages and purified cellulose (Smith and Nelson, 1975; Smith and Urquhart, 1975). The effects were ameliorated by addition of glucose in a mixture of minerals to the medium (Smith and Nelson, 1975), which may explain the reason Minson (1971) did not observe a depressing effect of silicon on *in vitro* digestion of neutral detergent solubles.

A siliceous type of urinary calculus was observed in sheep, but added silica did not have any effects (Beeson *et al.*, 1943). Feeding wheat bran appeared to contribute to the disturbance. Beeson *et al.* obtained evidence that urinary magnesium plays an important part in the limitation of urinary silica solubility. Addition of sodium silicate to the diet at the level of 2.2 percent tended to increase the incidence of urinary calculi in lambs fed a calculi-inducing diet (Schneider *et al.*, 1952). Incidence of urinary calculi in grazing steers in Montana was correlated ($r = 0.558$) with the silica content of the forages (Parker, 1957). Negative correlations were reported between the incidence of calculi and the concentration of phosphorus, calcium, magnesium, and potassium in the forage. Silica was the main constituent of uroliths in cattle in the problem area of Canada, but outside the problem area the uroliths contained little or no silica (Connell *et al.*, 1959). Urinary calculi were found in wethers fed rations containing prairie hay, but not in those fed alfalfa hay diets (Emerick *et al.*, 1959). No difference was observed in number of animals with urolithiasis between cattle fed prairie hay from two different sources, but animals fed hay from a ranch with a history of urolithiasis had approximately twice as much urolithic material present in the urinary organs, compared to those fed hay from the urolithiasis-free area (Whiting *et al.*, 1958).

Bailey (1967b) reported a sevenfold greater silica concentration in urine from cows fed prairie hay than from cows fed alfalfa hay. They pointed out that the difference was due more to differences in urine volume than to differences in total silica excreted in the urine. Prairie hay was found to contain 5.7 percent silica, compared to 0.4 percent for alfalfa hay (Bailey, 1976a). Although percent silica absorption was lower in cows fed prairie hay than those fed alfalfa hay, the absolute amount was obviously much higher for the cows fed prairie hay. Bailey (1976c) suggested that low water intake was a necessity, but not a sufficient condition for calculi formations.

Number of offspring born to female rats was decreased by including 600 or 1,200 ppm soluble silica in the water (Smith *et al.*, 1973). Furthermore, the number of offspring surviving until weaning was decreased by the addition of silica to the drinking water. Longevity of rats started on treatments after weaning was not affected by the addition of soluble silica to drinking water.

Administration of dextro- and levorotatory quartz to the lungs of rats by sufflation produced changes in the lungs, consisting of intraalveolar modular foci of macrophages containing quartz crystals (King *et al.*, 1946b).

## HIGH LEVELS

Feeding dogs a semisynthetic diet with 12 percent silicic acid (4.3 percent silicon and 3 percent talc [0.9 percent silicon]) produced fatal urethral blockage, cystitis, uremia, and systemic infection (McCullagh and Ehrhart, 1974). In a subsequent test, 3 of 16 dogs fed the diet developed urethral blockage, but were treated successfully after the diet was changed. Silica has been shown to cause hemolysis of erythrocytes *in vitro* (Stalder and Stober, 1965).

## FACTORS INFLUENCING TOXICITY

Sex of animals appeared to be important in the response to high levels of silica. Adding sodium silicate to the drinking water at a level to supply 374 ppm silicon increased rate and efficiency of gain in growing–finishing wethers, but depressed these in ewes (Smith *et al.*, 1972). The response in wethers was mostly with those fed an alfalfa-hay-based diet, with little effect in those fed different proportions of cottonseed hulls and milo, supplemented with urea. Type of diet did not influence the response in ewes. Adding 600 ppm sodium silicate (280 ppm silicon) to drinking water depressed gain in female albino rats, but increased gain in males (Smith *et al.*, 1973). Phosphorus and nitrogen retention was increased in male rats given access to water with 1,200 ppm soluble silica.

Type of feed appears to affect the formation of calculi. Incidence was higher in animals fed prairie hay than in those fed alfalfa hay as the roughage source (Emerick *et al.*, 1959). This may be due to the higher silica content in prairie hay (Bailey, 1976a) or lower urine volume in cattle fed this roughage (Bailey, 1967b). Also, addition of calcium may have been important, since a lower calcium-to-phosphorus ratio increased incidence of calculi (Schneider *et al.*, 1952). Dietary magnesium level may be involved also. Lower urinary magnesium was observed in sheep fed prairie hay (Emerick *et al.*, 1959). A normal intake of magnesium had an alleviating effect on incidence of calculi (Schneider *et al.*, 1952). A negative correlation was reported between incidence of urinary calculi and magnesium level in the forage (Parker, 1957).

Differences in hemolysis were reported from incubating sheep and human erythrocytes with silica with different crystalline structures (Stalder and Stober, 1965). Alumina was useful in partly alleviating the hemolytic effects of quartz. Aluminum dust did not prevent silicosis in rabbits from administration of silica in the lungs, but the lesions were less advanced in the animals receiving aluminum (King *et al.*, 1946a).

Siliceous calculi were prevented in calves fed a calculi-inducing diet by including 4 percent sodium chloride in the diet, which increased urine volume by 50 percent (Bailey, 1967a). In later work, results were obtained indicating that consumption of 300 g of salt per day in 300-kg calves prevented formation of siliceous calculi (Bailey, 1973). The high-salt intake increased urine volume. Feeding of ammonium chloride did not affect water intake or formation of siliceous calculi (Bailey, 1976b).

## TISSUE LEVELS

Bovine muscle was reported to contain 1.0 to 1.7 ppm silica, dry basis (Van Soest, 1970). Levels in liver were 0.2 to 0.5 ppm, dry basis. Chicken feathers contained 1.6 ppm, dry basis.

## MAXIMUM TOLERABLE LEVELS

It appears that even low levels of silica in forages will depress their digestibility by ruminants. No calculi were observed in sheep fed up to 0.57 percent silicon (Emerick *et al.*, 1959). Thus the maximum tolerable level in sheep appears to be 0.2 percent as soluble salts of high availability. Higher levels of less soluble forms found in natural substances can be tolerated.

## SUMMARY

Silicon is the most abundant element in the earth's crust, next to oxygen, and has been shown to be essential for bone formation in rats and growth of chicks. It is widely distributed in plants in the form of silica. The harmful effects of an excess of silicon in animals include a depression in roughage digestibility by ruminants, abnormal reproduction in rats, and involvement in urinary calculi in sheep. It is toxic when administered in lungs, which appears to be a more serious problem with humans than animals. Toxicity does not appear to be a serious problem in animals.

TABLE 30 Effects of Silicon Administration in Animals

| Class and Number of Animals[a] | Age or Weight | Administration: Quantity of Element[b] | Source | Duration | Route | Effect(s) | Reference |
|---|---|---|---|---|---|---|---|
| Ruminant | | 0.2–2.5% | Natural | | Diet | Decreased digestibility | Van Soest and Jones, 1968 |
| Cattle (*in vitro*) | | 1.6–3.4% | | | | Decreased digestibility | Smith *et al.*, 1971 |
| Cattle—14 | 0 to 10 mo | 3.0% | Natural (prairie hay) | 10 mo | Diet | Calculi in 13 of 14 calves | Bailey, 1967a |
| | | | + 4% NaCl | | Diet | No calculi | |
| Sheep—6 | Lambs | 0.4% | Natural | 115 d | Diet | No calculi | Emerick *et al.*, 1959 |
| | | 0.6% | Natural + 1.0% $Na_2SiO_3$ | | | No calculi | |
| Sheep—6 | Lambs | 1.4% | Natural | | | Calculi in one of six | |
| Sheep—6 | Lambs | 1.6% | Natural + 1.0% $Na_2SiO_3$ | | | Calculi in two of six | |
| Sheep—6, wethers | Lambs | 374 ppm | Sodium silicate | 75 d | Water | Increase gain and feed efficiency | Smith *et al.*, 1972 |
| Sheep—6, ewes | Lambs | 374 ppm | | | | No effect on gain | |
| Rat—6, male | Weanling | 280 ppm | Sodium silicate | 35 d | Water | 6% increase in gain | Smith *et al.*, 1973 |
| Rat—6, female | Weanling | 280 ppm | | 35 d | | 5% decrease in gain | |
| Rat—59, female | Mature | 280 ppm | | 2½ yr | | 33% decrease in young born and 54% decrease in young weaned | |

TABLE 30 *Continued*

| Class and Number of Animals[a] | Age or Weight | Administration: Quantity of Element[b] | Source | Duration | Route | Effect(s) | Reference |
|---|---|---|---|---|---|---|---|
| | | 561 ppm | | 2½ yr | | 20% decrease in young born and 76% decrease in young weaned | |
| Rat—14 | | 50 mg | Quartz (levo- and dextro-rotatory | | Inhalation | Intra-alveolar modular foci of macrophages | King *et al.*, 1946b |
| Dog—16 | | 0.9% | Talc | | Oral (feed) | No adverse effect | McCullagh and Ehrhart, 1974 |
| Dog—24 | | 5.2% | Silicic acid and talc | Up to 14 mo | | Urinary blockage; renal calculi; cystic calculi; bladder ulceration and perforation; death of three dogs | |

[a] Number of animals per treatment.
[b] Quantity expressed in parts per million or as milligrams per kilogram of body weight.

## REFERENCES

Allison, A. C. 1968. Silicon compounds in biological systems. Proc. R. Soc. 171:17.

Bailey, C. B. 1967a. Siliceous urinary calculi in calves: Prevention by addition of sodium chloride to the diet. Science 155:696.

Bailey, C. B. 1967b. Silica excretion in cattle fed a ration predisposing to silica urolithiasis: Total excretion and diurnal variations. Am. J. Vet. Res. 28:1743.

Bailey, C. B. 1973. Formation of siliceous urinary calculi in calves given supplements containing large amounts of sodium chloride. Can. J. Anim. Sci. 53:55.

Bailey, C. B. 1976a. Fate of the silica in prairie hay and alfalfa hay consumed by cattle. Can. J. Anim. Sci. 56:213.

Bailey, C. B. 1976b. Effects of ammonium chloride on formation of siliceous urinary calculi in calves. Can. J. Anim. Sci. 56:359.

Bailey, C. B. 1976c. Relation of water turnover to formation of siliceous calculi in calves given high-salt supplements on range. Can. J. Anim. Sci. 56:745.

Beeson, W. M., J. W. Pence, and G. C. Holm. 1943. Urinary calculi in sheep. Am. J. Vet. Res. 4:120.

Carlisle, E. M. 1974. Silicon as an essential element. Fed. Proc. 33:1758.

Connell, R., F. Whiting, and S. A. Forman. 1959. Silica urolithiasis in beef cattle. I. Observation on its occurrence. Can. J. Comp. Vet. Sci. 23:41.

Emerick, R. J., L. B. Embry, and O. E. Olson. 1959. Effect of sodium silicate on the development of urinary calculi and the excretion of various urinary constituents in sheep. J. Anim. Sci. 18:1025.

Gallup, W. D., C. S. Hobbs, and H. M. Briggs. 1945. The use of silica as a reference substance in digestion trials with ruminants. J. Anim. Sci. 4:68.

Jones, L. H. P., and K. A. Handreck. 1967. Silica in soils, plants, and animals. Adv. Agron. 19:107.

King, E. J., N. Rogers, and M. Gilchrist. 1946a. Attempts to prevent silicosis with aluminum. J. Pathol. Bacteriol. 57:281.

King, E. J., N. Rogers, M. Gilchrist, and G. Nagelschudt. 1946b. A comparison of the effects of laevorotatory and dextro-rotatory quartz on the lungs of rats. J. Pathol. Bacteriol. 57:491.

McCullagh, K. G., and L. A. Ehrhart. 1974. Silica urolithiasis in laboratory dogs fed semisynthetic diets. J. Am. Vet. Med. Assoc. 164:712.

Minson, D. J. 1971. Influence of lignin and silicon on a summative system for assessing the organic matter digestibility of panicum. Aust. J. Agric. Res. 22:589.

Parker, G. 1957. "Water-belly" (urolithiasis) in range steers in relation to some characteristics of rangeland. J. Range Manage. 10:105.

Schneider, B. H., E. D. Tayson, and W. E. Ham. 1952. Urinary Calculi in Male Farm Animals. Wash. Agric. Exp. Stn. Circ. 203.

Schwarz, K. 1973. A bound form of silicon in glycosaminoglycans and polyuronides. Proc. Natl. Acad. Sci. 70:1608.

Smith, G. S., and A. B. Nelson. 1975. Effects of sodium silicate added to rumen cultures on forage digestion with interactions of glucose, urea and minerals. J. Anim. Sci. 41:891.

Smith, G. S., and N. S. Urquhart. 1975. Effect of sodium silicate added to rumen cultures on digestion of siliceous forages. J. Anim. Sci. 41:882.

Smith, G. S., A. B. Nelson, and E. J. A. Boggino. 1971. Digestibility of forages *in vitro* as affected by content of "silica." J. Anim. Sci. 33:466.

Smith, G. S., A. L. Neumann, A. B. Nelson, and E. E. Ray. 1972. Effects of "soluble silica" upon growth of lambs. J. Anim. Sci. 34:839.

Smith, G. S., A. L. Neumann, V. H. Gledhill, and C. Z. Arzola. 1973. Effects of "soluble silica" on growth, nutrient balance and reproductive performance of albino rats. J. Anim. Sci. 34:839.

Stalder, K., and W. Stober. 1965. Haemolytic activity of suspensions of different silica modifications and inert dusts. Nature 207:874.

Van Soest, P. J. 1970. The role of silicon in the nutrition of plants and animals, p. 103. *In* Proc. Cornell Univ. Conf. Feed Manuf.

Van Soest, P. J., and L. H. P. Jones. 1968. Effect of silica in forages upon digestibility. J. Dairy Sci. 51:1644.

Whiting, F., R. Connell, and S. A. Forman. 1958. Silica urolithiasis in beef cattle. Can. J. Comp. Pathol. 22:332.

# Silver

Silver (Ag), a white lustrous metal, is a rare element that is distributed within the earth's crust at a concentration of about 0.1 ppm. It occurs as native silver and in minerals such as cerargyrite, argentite, and several complex sulfides. Most silver is found in association with copper, lead, and zinc, and, therefore, the silver of commercial value is primarily a by-product of the mining of nonferrous base metals. Major industrial uses are in the manufacture of silverware, jewelry, alloys, metallic coatings, coinage (formerly), and in photographic applications (Browning, 1969; Standen, 1969). Domestic silver production in 1978 amounted to about 39 million troy ounces (U.S. Department of the Interior, 1979). Industrial exposure to silver results in a peculiar disease (argyria) in which there is a generalized precipitation of silver in the skin and tissues. Accordingly, the skin assumes a permanent and irreversible pigmentation of a blue-gray color with other health disturbances usually being absent. Research interest in the biological actions of silver stems from its involvement with vitamin E, selenium, and copper metabolism.

## ESSENTIALITY

No known biological function has yet been discovered for silver in animals.

## METABOLISM

Silver metabolism in rats was studied by Scott and Hamilton (1950). Rats were killed at various intervals postdosing with tracer quantities of radiosilver (mixed isotopes) given intramuscularly, intravenously, and by stomach tube. Radiosilver analyses yielded similar results for both the intramuscular and intravenous paths of administration. By 4 days postdosing, most of the administered silver (ca. 93 percent) had been excreted via the feces with relatively little (ca. 0.3 percent) appearing in the urine. The radiosilver was distributed throughout the tissues of the rats with the large intestine having the highest concentration and muscle having the lowest. Kidney and liver were found to contain intermediate quantities. In contrast, stomach tube administration did not result in appreciable quantities of radiosilver in tissues. Most of the dose (ca. 99 percent) was excreted by the fourth day postdosing. These findings were extended by an experiment in which rats were treated with intramuscular doses of radiosilver and with radiosilver plus silver nitrate at 0.4 and 4.0 mg per kg of body weight. At 6 days postdosing, 97 percent of the administered dose appeared in the feces of the rats given the radiotracer alone, whereas similar figures for the 0.4 and 4.0 mg doses were 89 and 37 percent, respectively. Tissue distribution was affected both quantitatively and qualitatively by dosage. As dosage increased, the amount of residual radiosilver in the tissues increased. At the highest dosage, liver and spleen contained the highest concentrations of silver, whereas, with the tracer dose, the gastrointestinal tract had the highest concentrations. It can be concluded from these studies that, in rats, silver elimination occurs via the feces and that silver administration results in a wide and variable tissue distribution.

## SOURCES

No data could be located concerning the silver content of common feedstuffs. Bowen (1966) indicates that soils average 0.1 ppm silver and that land plants in general average 0.06 ppm. Silver can enter feedstuffs, however, due to its natural distribution, its association with mineral feeds, and its loss through environmental dispersion.

## TOXICOSIS

### LOW LEVELS

In a dose–response experiment with chickens grown to 3 weeks of age, Hill *et al.* (1964) demonstrated that 10 to 100 ppm of dietary silver (as $Ag_2SO_4$) was without adverse effects on growth rate, mortality, hemoglobin concentration, and elastin content of the aorta.

### HIGH LEVELS

In growing turkeys, 300 ppm of dietary silver reduced rate of weight gain (Jensen *et al.,* 1974). At 900 ppm, the heart was enlarged, the gizzard musculature was dystrophic, weight gain was severely depressed, and the blood packed cell volume was decreased.

In growing chickens, the signs of silver toxicosis included reduced weight gain, increased mortality, increased heart weight, decreased aortic elastin content, reduced blood hemoglobin, and exudative diathesis (Hill *et al.,* 1964; Bunyan *et al.,* 1968; Peterson and Jensen, 1975a,b).

In growing rats, observed signs of silver toxicosis, in addition to reduced weight gain, liver necrosis, and increased mortality, were a generalized deposition of silver in the tissues (Shaver and Mason, 1951; Diplock *et al.,* 1967). For a further characterization of the effects of long-term administration of silver to rats, the papers of Olcott (1948, 1950) on experimental argyrosis are particularly useful. Essentially, Olcott found that the life term administration of silver at 1,000 ppm in the drinking water of rats produced an intense pigmentation of many of the tissues. More pronounced pigmentation occurred in the basement membrane of the glomeruli, the portal vein and other parts of the liver, the choroid plexus of the brain, the choroid layer of the eye, and in the thyroid gland.

### FACTORS INFLUENCING TOXICITY

Shaver and Mason (1951) were the first to demonstrate that silver (as $AgNO_3$ in the drinking water) was more toxic, as evidenced by mortality, to rats when administered in conjunction with a vitamin E-deficient diet, as compared to a vitamin E-adequate diet. Similarly, Dam *et al.* (1958) noted that 20 ppm silver as silver acetate in the diet promoted exudative diathesis in chickens fed diets deficient in vitamin E. In rats, Diplock *et al.* (1967) noted a vitamin E-deficient diet coupled with 0.15 percent silver acetate in the drinking water produced a high incidence of liver necrosis and mortality that could be totally prevented by tocopherols, partially prevented by selenium at 1 ppm, and only marginally prevented by 0.15 percent DL-methionine. These experiments were extended to chickens by Bunyan *et al.* (1968), who found 0.15 percent silver as silver acetate in the drinking water produced an exudative diathesis that was responsive to treatment with a combination of vitamin E and methionine. The addition of lard to the basal diets of chicks simultaneously treated with silver gave rise to green exudates that were prevented by either vitamin E or selenium. It was concluded that silver and selenium have an antagonistic relationship in rats and chickens.

More recent studies of the silver–selenium antagonism in chicks (Peterson and Jensen, 1975a) compared the effectiveness of vitamin E, selenium, and cystine in preventing the adverse effects of excessive dietary silver. Nine hundred parts per million of silver (as $AgNO_3$) in a diet marginal in vitamin E and selenium resulted in depressed chick growth and caused a high mortality, largely from exudative diathesis. Either 1 ppm selenium or 100 IU vitamin E per kilogram of diet was effective in preventing the growth depression and mortality. Adding 0.15 percent cystine stimulated growth but failed to prevent the mortality. In these studies, all diets were fortified with 50 ppm copper in order to prevent the appearance of a silver-induced copper deficiency.

These studies were extended by Jensen (1975) to reveal the effects of supplementary silver in chickens treated with toxic quantities of selenium. Accordingly, chickens were reared for 2 weeks on diets containing 0, 5, 10, 20, 40, and 80 ppm supplemental selenium with and without 1,000 ppm of silver (as $AgNO_3$). The higher levels of selenium produced a growth depression and an increased mortality that were mitigated by the silver supplement. Radiotracer studies conducted with $^{75}Se$ in these experiments showed that dietary silver interferes with selenium absorption, a finding that could explain the antagonistic relationship between silver and selenium.

Studies were also conducted in turkeys (Jensen *et al.*, 1974). Accordingly, 900 ppm of silver depressed growth rate, reduced packed cell volume, and caused cardiac enlargement. There was a varying incidence of degeneration of the gizzard musculature that could be completely prevented by dietary additions of 1 ppm selenium or partially prevented by 50 IU vitamin E per kilogram of diet. The observed microcytic, hyperchromic anemia was prevented by the addition of 50 ppm of copper but not by the addition of 5 ppm of copper to the diets. The silver-induced cardiac enlargement was also responsive to the copper treatments.

Hill *et al.* (1964) found that copper could reverse the toxic effects of silver in growing chicks as evidenced by growth rate, mortality, hemoglobin, and aorta elastin content. In these studies, 25 ppm of dietary copper were totally effective in preventing the adverse effects of 200 ppm of dietary silver. The silver–copper antagonism was studied further by Peterson and Jensen (1975b) in chickens grown to 4 weeks of age. Adding 900 ppm of silver (as $AgNO_3$) to practical diets significantly depressed growth, increased heart weights, increased mortality, and decreased aorta elastin content. Supplemental copper (50 ppm) was effective in reversing adverse effects except for the growth depression that was only partially corrected.

The effect of dietary silver on the tissue distribution patterns of $^{64}Cu$ in rats has been studied by Van Campen (1966). Liver content of copper was increased as dietary silver increased. No effects on the copper content of heart, kidneys, and spleen were found attributable to the dietary silver.

## TISSUE LEVELS

No data were found concerning the tissue levels of silver which result from dietary administration.

## MAXIMUM TOLERABLE LEVELS

The minimum level of silver observed to have an adverse effect on animals reared under conditions of adequate nutrition is 300 ppm. In this regard, the maximum tolerable level for poultry and swine is set at 100 ppm.

## SUMMARY

The toxicity of silver has been relatively well studied in poultry and rats. In these animals, silver accentuates and/or causes a multiple deficiency of vitamin E, selenium, and copper with the resulting toxic signs approximating the respective deficiency signs. Acute signs of intoxication in rats include increased pigmentation of tissues such as the basement membrane of glomeruli, the portal vein, and the liver.

TABLE 31 Effects of Silver Administration in Animals

| Class and Number of Animals[a] | Age or Weight | Administration: Quantity of Element[b] | Source | Duration | Route | Effect(s) | Reference |
|---|---|---|---|---|---|---|---|
| Chicken—30 | 1 d | 900 ppm | $AgNO_3$ | 28 d | Diet | Increased death rate; decreased rate of weight gain; exudative diathesis | Peterson and Jensen, 1975a |
| Chicken—20 | 1 d | 10 ppm | $Ag_2SO_4$ | 21 d | Diet (low Cu) | No adverse effects | Hill *et al.*, 1964 |
| | | 25 ppm | | | | No adverse effects | |
| | | 50 ppm | | | | Increased death rate | |
| | 1 d | 100 ppm | $Ag_2SO_4$ | 21 d | Diet (low Cu) | Increased death rate; decreased rate of weight gain; decreased hemoglobin concentration | |
| | | 200 ppm | | | | Increased death rate; decreased rate of weight gain; decreased hemoglobin concentration | |
| | | 10 ppm | | | Diet | No adverse effects | |
| | | 25 ppm | | | | No adverse effects | |
| | | 50 ppm | | | | No adverse effects | |
| | | 100 ppm | | | | No adverse effects | |
| Chicken—60 | 1 d | 900 ppm | $AgNO_3$ | 28 d | Diet | Increased death rate; decreased rate of weight gain; increased heart weight; reduced aorta elastin content | Peterson and Jensen, 1975b |

| | | | | | | | |
|---|---|---|---|---|---|---|---|
| | | 900 ppm | | | Diet (+50 ppm Cu) | Decreased rate of weight gain | |
| Chicken—66 | 14 d | 1,500 ppm | Silver acetate | 14 d | Water (diet low in vit. E) | Decreased rate of weight gain; breast and gizzard muscular dystrophy | Bunyan *et al.*, 1968 |
| Turkey—16 | 1 d | 100 ppm | Silver acetate | 28 d | Diet | No adverse effects | Jensen *et al.*, 1974 |
| Turkey—16 | | 300 ppm | | | | No adverse effects | |
| Turkey—56 | | 900 ppm | | | | Decreased rate of weight gain; reduced packed cell volume; dystrophy of gizzard musculature; enlarged hearts | |
| Turkey—20 | | 300 ppm | | 35 d | | Decreased rate of weight gain | |
| Turkey—40 | | 900 ppm | | | | Decreased rate of weight gain; reduced packed cell volume; dystrophy of gizzard musculature; enlarged hearts | |
| Rat—6 | 14 d | 1,000 ppm | Silver acetate | 112 d | Water (diet low in vit. E) | No adverse effects | Bunyan *et al.*, 1980 |
| Rat—7 | | 130 ppm | | 56 d | Water (diet low in protein) | Liver necrosis; increased death rate | |
| Rat—6 | | 1,000 ppm | | | Water (diet low in protein) | Liver necrosis; increased death rate | |

TABLE 31 *Continued*

| Class and Number of Animals[a] | Age or Weight | Administration: Quantity of Element[b] | Source | Duration | Route | Effect(s) | Reference |
|---|---|---|---|---|---|---|---|
| Rat—9 | 37 d | 1,500 ppm | Silver acetate | 72 d | Water (diet low in vit. E) | Liver necrosis; increased death rate | Diplock *et al.*, 1967 |
| Rat | | 1,000 ppm | $AgNO_3$ | Lifetime | Water | Intense pigmentation of the tissues, particularly the basement membrane of glomeruli, the portal vein, the liver, choroid layer of eye, and the thyroid gland; hypertrophy of the left ventricle | Olcott, 1948 |

| | | | | | | | |
|---|---|---|---|---|---|---|---|
| Rat—10 | 55–65 g | 76 ppm | Silver acetate | 46 d | Water (diet low in Se and vit. E) | Decreased rate of weight gain | Wagner *et al.*, 1975 |
| | | 76 ppm | | | Water (diet low in vit. E) | No adverse effect | |
| | | 751 ppm | | | Water (diet low in Se and vit. E) | Decreased rate of weight gain and increased death rate | |
| | | 751 ppm | | | Water (diet low in vit. E) | Slightly decreased rate of weight gain | |

[a] Number of animals per treatment.
[b] Quantity expressed in parts per million in diet or drinking water.

## REFERENCES

Bowen, H. J. M. 1966. Trace Elements in Biochemistry. Academic Press, New York and London.

Browning, E. 1969. Toxicity of Industrial Metals. Butterworths, London.

Bunyan, J., A. T. Diplock, M. A. Cawthorne, and J. Green. 1968. Vitamin E and stress. 8. Nutritional effects of dietary stress with silver in vitamin E-deficient chicks and rats. Br. J. Nutr. 22:165.

Dam, H., C. K. Nielsen, I. Prange, and E. Sondergaard. 1958. Influence of linoleic and linolenic acids on symptoms of vitamin E deficiency in chicks. Nature 182:802.

Diplock, A. T., J. Green, J. Bunyan, D. McHale, and I. R. Muthy. 1967. Vitamin E and stress. 3. The metabolism of D-$\alpha$-tocopherol in the rat under dietary stress with silver. Br. J. Nutr. 21:115.

Hill, C. H., B. Starcher, and G. Matrone. 1964. Mercury and silver interrelationships with copper. J. Nutr. 83:107.

Jensen, L. S. 1975. Modification of a selenium toxicity in chicks by dietary silver and copper. J. Nutr. 105:769.

Jensen, L. S., R. P. Peterson, and L. Falen. 1974. Inducement of enlarged hearts and muscular dystrophy in turkey poults with dietary silver. Poult. Sci. 53:57.

Olcott, C. T. 1948. Experimental argyrosis. Am. J. Pathol. 24:813.

Olcott, C. T. 1950. Experimental argyrosis. Arch. Pathol. 49:138.

Peterson, R. P., and L. S. Jensen. 1975a. Induced exudative diathesis in chicks by dietary silver. Poult. Sci. 54:795.

Peterson, R. P., and L. S. Jensen. 1975b. Interrelationship of dietary silver with copper in the chick. Poult. Sci. 54:771.

Scott, K. G., and J. G. Hamilton. 1950. The metabolism of silver in the rat with radiosilver used as an indicator. Univ. Calif. Publ. Pharmacol. 2:241.

Shaver, S. L., and K. E. Mason. 1951. Impaired tolerance to silver in vitamin E deficient rats. Anat. Rec. 109:382.

Standen, A., ed. 1969. Kirk-Othmer Encyclopedia of Chemical Technology, vol. 18. John Wiley & Sons, New York.

U.S. Department of the Interior. 1979. Bureau of Mines Mineral Industry Surveys, Gold and Silver, March.

Van Campen, D. R. 1966. Effects of zinc, cadmium, silver, and mercury on the absorption and distribution of copper-64 in rats. J. Nutr. 88:125.

Wagner, P. A., W. G. Hoekstra, and H. E. Ganther. 1975. Alleviation of silver toxicity by selenite in the rat in relation to tissue glutathione peroxidase. Proc. Soc. Exp. Biol. Med. 148:1106.

# Sodium Chloride

In ancient times, the distribution of the population centers was predicated essentially by three factors—the availability of salt (NaCl), water, and food (Batterson and Brodie, 1972). Salt, therefore, was among the first of the specific nutrients recognized to be essential for animal nutrition and health. Salt is widely distributed in nature, where it occurs not only in the sea and other saline waters but also in dry deposits as rock salt. The concentration of salt in seawater averages 2.68 percent. Whereas table salt is the most common use of salt, it finds application in literally thousands of commercial processes that yield products containing either sodium or chlorine (Standen, 1970). In 1976 domestic salt production approached 40 million metric tons, of which about 2 million metric tons were used by the feed industry (U.S. Department of the Interior, 1976). Interest in the biological effects of salt is high because of its importance in nutrition, and much literature has been published concerning salt deficiencies and excesses, its relationships with potassium and other minerals, and its potential impacts on human health.

## ESSENTIALITY

Sodium and chlorine have been found to be essential constituents of diets fed to all animals and recommendations for salt supplementation have been prepared and published by the National Academy of Sciences/National Research Council (1974) and by the Salt Institute

(Anonymous, 1974). Sodium and chlorine, along with potassium, in proper concentration and balance are indispensable for a number of important physiologic processes. Sodium, as the chief cation of the extracellular fluids, is the most important ion in maintenance of osmotic pressure, body fluid balance, and hydration of the tissues. Heart action and nerve impulse conduction and transmission are highly dependent upon proper proportions of sodium and potassium. These cations are also essential for the operation of certain enzyme systems and the maintenance of blood pH. Chloride functions mainly to ensure the proper fluid–electrolyte balance.

Signs of a combined sodium and chlorine deficiency in cattle, sheep, swine, and horses include a salt craving evidenced by the animals' licking soil, rocks, wood, and other objects. Eventually, there is a loss of appetite and productive parameters are adversely affected. Cattle and horses take on an unthrifty appearance, and their hair coats roughen. In sheep, wool growth is greatly reduced. In poultry, accompanying the reduction in productive performance, there appear nervous signs and dehydration. The end point of long-term salt deficiency for all animals is death. Generally, the dietary requirements for sodium in animals of economic importance approximate 0.2 percent, with the lowest value (0.10 percent) being required by growing beef calves and the highest value (0.35 percent) being required by horses. Chlorine requirements are less known, but it is apparent that salt supplementation to satisfy the sodium requirements will also satisfy the chlorine requirements.

## METABOLISM

Comprehensive reviews on the metabolism of sodium and chlorine have been published by Forbes (1962), Cotlove and Hogben (1962), and by Tracor-Jitco, Inc. (1974). Meneely and Battarbee (1976) have reviewed the interrelationship between sodium and potassium. These reviews testify to the voluminous literature on sodium and chlorine metabolism, and thus the following information, adapted from Church and Pond (1974), comprises only a brief summary of our knowledge.

Sodium and chloride ions are absorbed by animals principally from the upper small intestine. Approximately 80 percent of the sodium and chloride entering the gastrointestinal tract arises from internal secretions such as saliva, gastric fluids, bile, and pancreatic juice. Thus, large variations in salt intake have relatively small effects on the total amount of sodium and chloride entering the gastrointestinal tract.

The regulation of body concentrations of sodium and chloride ions, as well as potassium ion, is narrowly controlled by, as yet, incompletely defined mechanisms. It is known that increased intakes of each element are accommodated by ready excretion in the kidneys. Plasma levels of sodium are controlled, in part, by aldosterone, which functions to increase sodium reabsorption from the kidney tubule. Other control is exercised by the antidiuretic hormone of the posterior pituitary, which is responsive to changes in osmotic pressure of the extracellular fluid. Both hormones act to maintain a constant ratio of sodium to potassium in the extracellular fluid. Chloride metabolism is controlled in relation to sodium so that excess kidney excretion of sodium is accompanied by chloride. Chloride excretion is also influenced by bicarbonate ion, with a rise in plasma bicarbonate resulting in the excretion of a comparable amount of chloride.

## SOURCES

In general, feedstuffs do not contain sufficient sodium to provide for optimum productive performance in livestock and poultry. Meyer *et al.* (1950) found that most plant and plant products contain relatively small amounts of sodium in comparison to animal products. In their survey, oats contained only 0.008 percent sodium, whereas condensed fish solubles contained 2.52 percent. With regard to chlorine, the broadest range of values occurred between brewer's dried grains (0.03 percent) and condensed fish solubles (4.63 percent). Sodium and chlorine contents, respectively, were for corn (0.004 and 0.06 percent), clover hay (0.14 and 0.15 percent), alfalfa hay (0.07 and 0.19 percent), soybean meal (0.02 and 0.04 percent), and timothy hay (0.008 and 0.14 percent).

## TOXICOSIS

### LOW LEVELS

The toxicity of salt in animals has been reviewed by Tracor-Jitco, Inc. (1974), the National Research Council (1974), and the Salt Institute (Anonymous, 1974). The effects of salt administration are summarized in Table 32.

The effects of relatively low levels (5 percent and less in the feed and 2 percent and less in the drinking water) have been measured in ruminant species. Demott *et al.* (1968) provided lactating dairy cows with 0,

1, 2, and 4 percent salt in the grain (fed at a rate of 1 kg for each 2 kg of fat-corrected milk) for a 2-week period. No adverse effects were noted in the general health, milk production, or average body weight of the treated cows. Jaster *et al.* (1978) studied the effects of saline drinking water in dairy cows. In those studies, high-producing cows were provided with drinking water with and without 2,500 ppm salt for a 28-day period. Although no changes were noted in feed intake and digestibility and milk and blood concentrations of sodium, potassium, calcium, magnesium, chloride, and phosphorus, the treated cows produced less milk and consumed greater quantities of water. In cattle, Weeth *et al.* (1960) and Weeth and Haverland (1961) could produce a toxicosis by administration of salt via the drinking water at levels which ranged from 12 g per liter to 20 g per liter. The signs noted included a severe anorexia, decreased water consumption, anhydremia, weight loss, and collapse. The serum potassium and sodium concentrations were elevated significantly, while both serum magnesium and urea concentrations were depressed. In sheep, the effects of dietary salt administration have been studied by Meyer and Weir (1954), Wilson (1966), Jackson *et al.* (1971), and Kromann (Washington State University, personal communication, 1978). Thus, levels of salt of 5 percent and below were without adverse effects on weight gain, empty body weight, carcass composition and energy gain in growing lambs, and feed consumption in older sheep. Likewise, this level of salt did not affect the productive performance of ewes maintained during growth, fattening, breeding, gestation, and early lactation (253 days), nor did it affect hematocrits, serum albumin and sodium concentrations, or milk protein, sodium, and potassium concentrations.

In a series of studies by Peirce (1957, 1959, 1960, 1962, 1963, 1968a,b), it was found that sheep treated with 2.0 percent salt in the drinking water lost weight continuously throughout the experimental period and were found to be weak and listless. Diarrhea was occasionally observed as was an increase in serum chloride concentration. Certain of Peirce's studies (1959, 1960, 1962, 1963) involved treatments with salt combined with magnesium chloride, sodium sulfate, calcium chloride, and sodium carbonate–bicarbonate, respectively. At an intake of total salts of 1.3 percent or less, little adverse effects were noted in health, feed intake, or body weight. Certain of the combinations (including magnesium chloride) did cause occasional diarrhea and reduced feed consumption, but none affected wool production. Next, Peirce used penned (1968a) or grazing (1968b) ewes and their lambs to assess the effects of synthetic drinking waters composed to resemble underground waters found in Australia. The water compositions tested were 1.3 percent (mostly salt), 1.0 percent (mostly salt), and 0.5 percent (half salt and half sodium bicarbonate). No adverse effects were observed on the health, feed consumption, or wool production of the penned animals, although in one trial a poorer reproductive performance was evidenced by the ewes receiving the 1.3 percent and the 0.5 percent salt treatments. In the grazing animals, 1.3 percent salt decreased body weight gains in lambs and the reproductive rate in ewes, caused diarrhea, and increased mortality in one of the two experiments done. The other treatments either decreased rate of gain and wool production (1.0 percent) or the percentage of ewes that lambed (0.5 percent), but had no other effects. Wilson (1966) also found that salt provided at 2 percent in the drinking water decreased feed consumption. Finally, in sheep, Potter and McIntosh (1974) found that the addition of 1.3 percent salt to the drinking water of pregnant ewes caused neonatal mortality in the resultant lambs. The ewes receiving the saline drinking water had significantly higher levels of plasma potassium and chloride and significantly lower levels of plasma calcium and magnesium.

The single study located for swine was performed by Done *et al.* (1959) and demonstrated that dietary salt at 3 percent combined with drinking water restriction for an 11-day period was without adverse effect.

A group of workers (Quigley and White, 1932; Barlow *et al.*, 1948; Kare and Biely, 1948; Paver *et al.*, 1953; Mohanty and West, 1969) has assessed the effects of relatively low levels of salt administration in chickens. The reported signs of salt toxicosis included decreased rate of weight gain, increased mortality, diarrhea, edema, increased heart size, nervousness, and degenerative changes in kidney, liver, spleen, adrenal, heart, lung, central nervous system, and the gastrointestinal tract. Similarly, the signs noted in turkeys by Matterson *et al.* (1946), Roberts (1957), Robblee and Clandinin (1961), Harper and Arscott (1962), and Morrison *et al.* (1975) were increased mortality, edema, loss of body weight, increased incidence of ascites, pendulous crop, diarrhea, and gross pathologic lesions of the heart, kidney, and lungs.

### HIGH LEVELS

Literature is also available that demonstrates the effects of salt administration to livestock and poultry at relatively high levels (in excess of 5 percent via the diet and 2 percent via the drinking water).

Meyer *et al.* (1955) demonstrated that fattening steers could tolerate dietary levels of salt of 9.33 percent with few adverse effects. Daily gains and dressing percentage were similar for the treated and control steers, although there was a small difference in carcass grade. In sheep, Meyer and Weir (1954) found that a dietary concentration of 13.1 percent salt caused increased weight loss during lactation and a decreased number of lambs raised, as well as an increase in the blood and milk chloride concentration.

In swine, Bohstedt and Grummer (1954) were able to induce salt poisoning by offering diets containing 6–8 percent salt with a restriction in the availability of drinking water. Several days posttreatment, the signs of salt poisoning included nervousness, staggering, weakness, and paralysis. Blindness was observed in one pig with liver changes appearing as the only gross anatomical change. Done *et al.* (1959) and Todd *et al.* (1964) also studied salt toxicosis in swine precipitated by high dietary salt and restricted water intake. Signs noted were muscular tremors, incoordination, convulsions, prostration, coma, and the typical lesions of meningoencephalitis.

In chickens, Blaxland (1946) estimated the lethal dose of salt administered via the crop at about 4 g per kilogram of body weight. Otherwise the toxic signs of salt poisoning in both chickens and turkeys are as they were discussed under Low Levels.

In other poultry, Scott *et al.* (1960) have performed toxicity studies in pheasant and quail. Dietary salt at 7.5 percent in both of these species causes decreased rate of weight gain and increased mortality.

With regard to salt toxicosis in other animal species, the table of acute effects prepared by Tracor-Jitco, Inc. (1974), is particularly pertinent and is herein adapted as Table 33.

### FACTORS INFLUENCING TOXICITY

The major factor that influences salt toxicosis in animals is the availability of drinking water. In the presence of an adequate supply, animals can tolerate relatively large quantities of dietary salt.

## TISSUE LEVELS

Several studies are available concerning the tissue levels of sodium and chloride resulting from excess dietary salt in food-producing animals. In cattle, drinking water concentrations of 1.0 and 1.2 percent salt resulted in serum sodium concentrations of 345 and 356 mg/dl, respectively (Weeth and Haverland, 1961). In swine under conditions of water restriction, Todd *et al.* (1964) reported the following values for various tissues (sodium and chloride, respectively in parts per million fresh weight): cerebrum, 1,886 and 2,769; cerebellum, 1,817 and 2,769; liver, 1,748 and 2,840; kidney, 2,231 and 3,728; heart, 1,265 and 3,195; and spleen, 1,795 and 2,910. Finally, in chickens, Barlow *et al.* (1948) have provided these values for chloride derived from chickens fed 10 percent dietary salt (in parts per million fresh weight): tendon, 3,596; lung, 3,010; liver, 1,924; leg muscle, 983; and kidney, 2,098.

## MAXIMUM TOLERABLE LEVELS

Maximum tolerable levels of dietary salt in animals were established as follows: a level of 4 percent was set for lactating cows, since this level was the maximum level tested by Demott *et al.* (1968); a level of 9.0 percent was set for other cattle and sheep based on the studies of Meyer *et al.* (1955) and Meyer and Weir (1954), respectively; a level of 8 percent was set for swine due to a lack of documentation of toxicosis in swine when given adequate supplies of fresh drinking water; a level of 2 percent was set for poultry based on the studies of Barlow *et al.* (1948) in chickens and Matterson *et al.* (1946) in turkeys; the level of 3 percent for horses and rabbits was obtained by extrapolation.

## SUMMARY

Salt is widely distributed in nature, and its components have been found to be indispensable nutrients for all animals. Sodium and chlorine function in the body to control osmotic pressure, fluid balance, heart action, and nerve impulse conduction and transmission. Salt toxicoses are characterized by increases in water consumption, anorexia, weight loss, edema, nervousness, paralysis, and a variety of signs dependent upon animal species. In many cases, provision of fresh water is effective in reducing the severity of the signs, but there is no doubt that reducing the dietary salt load (in poultry) is also required to mitigate the toxicity.

TABLE 32 Effects of Sodium Chloride Administration in Animals

| Class and Number of Animals[a] | Age or Weight | Administration: Quantity of Element[b] | Administration: Source | Administration: Duration | Administration: Route | Effect(s) | Reference |
|---|---|---|---|---|---|---|---|
| Cattle—2, lactating dairy cows | 619 kg | 1.00% | NaCl | 14 d | Diet | No adverse effects | Demott *et al.*, 1968 |
| | | 2.00% | | | | No adverse effects | |
| | | 4.00% | | | | No adverse effects | |
| Cattle—6, lactating dairy cows | | 0.25% | NaCl | 28 d | Water | Reduced milk production | Jaster *et al.*, 1978 |
| Cattle—14 | 370 kg | 9.33% | NaCl | 84 d | Diet | No adverse effects | Meyer *et al.*, 1955 |
| Cattle—6 | 220 kg | 1.00% | NaCl | 30 d | Water | Increased water consumption | Weeth *et al.*, 1960 |
| | | 2.00% | | | | Anorexia; weight loss; lethargy; anhydremia; collapse | |
| Cattle—6 | 186 kg | 1.25% | NaCl | 30 d | Water | Reduced growth rate | Weeth and Haverland, 1961 |
| | | 1.50% | | | | Reduced growth rate; reduced water intake | |
| | | 1.75% | | | | Weight loss; anorexia; reduced water intake | |
| | 180 kg | 1.00% | NaCl | 30 d | Water | No adverse effects | |
| | | 1.20% | | | | Increased water consumption | |

| | | | | | | | |
|---|---|---|---|---|---|---|---|
| | | | | | | | Weir, 1954 |
| | | 9.10% | | | | No adverse effects | |
| | | 13.10% | | | | Greater weight loss during lactation and decreased lambing rate | |
| Sheep—6 | 55 kg | 1.00% | NaCl | 460 d | Water | No adverse effects | Peirce, 1957 |
| | | 1.50% | | | | Decreased feed consumption and body weight | |
| | | 2.00% | | | | Decreased feed consumption and body weight; weakness | |
| Sheep—2 | 41–51 kg | 5.00% | NaCl | 21 d | Diet | No adverse effects | Wilson, 1966 |
| | | 10.00% | | | | Decreased feed intake | |
| | | 20.00% | | | | Decreased feed intake | |
| | 41–51 kg | 0.50% | NaCl | 21 d | Water | No adverse effects | |
| | | 1.00% | | | | No adverse effects | |
| | | 1.50% | | | | Decreased feed intake | |
| | | 2.00% | | | | Decreased feed intake | |
| Sheep—5 | 27 kg | 1.78% | NaCl | 119 d | Diet | No adverse effects | Jackson *et al.*, 1971 |
| | | 3.31% | | | | No adverse effects | |
| | | 4.83% | | | | No adverse effects | |
| | | 5.85% | | | | No adverse effects | |
| | | 7.63% | | | | Slight reduction in rate of weight gain | |
| Sheep—8 | 7 yr | 1.00% | NaCl | 80+ d | Water | Distress at parturition and neonatal mortality | Potter and McIntosh, 1974 |
| | | 1.30% | | | | Distress at parturition and neonatal mortality | |
| | 3 yr | 1.00% | NaCl | 80+ d | Water | Distress at parturition and neonatal mortality | |
| | | 1.30% | | | | Distress at parturition and neonatal mortality | |

TABLE 32 *Continued*

| Class and Number of Animals[a] | Age or Weight | Administration: Quantity of Element[b] | Source | Duration | Route | Effect(s) | Reference |
|---|---|---|---|---|---|---|---|
| Swine—6 | 30 kg | 6.00% | NaCl | 5 d | Diet | Nervous signs; staggering; weakness; paralysis; liver histopathology | Bohstedt and Grummer, 1954 |
| | | 8.00% | | | | Nervous signs; staggering; weakness; paralysis; liver histopathology | |
| Swine—7 | Mature | 5.30% | NaCl | 10 d | Diet (water restricted) | Tremors; incoordination; convulsions; prostration; death | Done *et al.*, 1959 |
| Swine—4 | Mature | 9.40% | NaCl | 6 d | Diet (water restricted) | Tremors; incoordination; convulsions; prostration; brain lesions; death | |
| Swine—3 | Mature | 3.00% | NaCl | 11 d | Diet (water restricted) | No adverse effects | |
| Swine—4 | 14 kg | 2.5 g/kg body weight | NaCl | 2 d | Drench (water *ad lib*), single | No adverse effects | Todd *et al.*, 1964 |
| Swine—4 | 14 kg | 2.5 g/kg body weight | NaCl | 2 d | Drench (water restricted), single | Tremors; incoordination; convulsions; prostration; coma and death | |
| Chicken—14 | 1 d | 1.30% | NaCl | 72 d | Diet | No adverse effects | Quigley *et al.*, 1932 |
| | | 3.00% | | | | No adverse effects | |
| | | 5.00% | | | | Reduced weight gain | |

| | | | | | | | |
|---|---|---|---|---|---|---|---|
| | | 8.00% | | | | Reduced weight gain and increased mortality | |
| | | 10.00% | | | | Reduced weight gain and increased mortality | |
| | | 15.00% | | | | Reduced weight gain and increased mortality | |
| Chicken—9 | 60–90 d | 5.00% | NaCl | 14 d | Diet | Diarrhea | Blaxland, 1946 |
| | | 10.00% | | | | Diarrhea and reduced rate of weight gain | |
| | | 20.00% | | | | Diarrhea and reduced rate of weight gain | |
| Chicken—10 | 14 d | 5.00% | NaCl | 34 d | Diet | Reduced rate of weight gain; visceral congestion; increased mortality | |
| | | 10.00% | | | | Reduced rate of weight gain; visceral congestion; increased mortality | |
| Chicken—32 | 1 d | 1.00% | NaCl | 72 d | Diet | No adverse effects | Barlow *et al.*, 1948 |
| | | 2.00% | | | | No adverse effects | |
| | | 3.00% | | | | No adverse effects | |
| | | 4.00% | | | | Increased mortality | |
| | | 5.00% | | | | Increased mortality and edema | |
| | | 6.00% | | | | Increased mortality and edema | |
| | | 7.00% | | | | Increased mortality and edema | |
| | | 8.00% | | | | Increased mortality and edema | |
| | | 9.00% | | | | Increased mortality and edema | |
| | | 10.00% | | | | Increased mortality and edema | |

TABLE 32 *Continued*

| Class and Number of Animals[a] | Age or Weight | Administration: Quantity of Element[b] | Source | Duration | Route | Effect(s) | Reference |
|---|---|---|---|---|---|---|---|
| Chicken—12 | 1 d | 1.60% | NaCl | 42 d | Diet | No adverse effects | Kare and Biely, 1948 |
| | | 2.20% | | | | No adverse effects | |
| | | 3.20% | | | | No adverse effects | |
| | | 4.20% | | | | Increased mortality; edema; lung congestion | |
| | | 5.20% | | | | Increased mortality; edema; lung congestion | |
| | | 6.20% | | | | Increased mortality; edema; lung congestion | |
| | | 8.20% | | | | Increased mortality; edema; lung congestion | |
| Chicken—117 | 1 d | 0.40% | NaCl | 28 d | Water | Anorexia | Krista *et al.*, 1961 |
| Chicken—118 | 1 d | 0.70% | | | | Anorexia; decreased rate of weight gain; increased mortality | |
| | | 1.00% | | | | Anorexia; decreased rate of weight gain; nervousness; edema and mortality | |
| Chicken—119 | 1 d | 1.20% | | | | Anorexia; decreased rate of weight gain; nervousness; edema and mortality | |
| Chicken—30 | Laying | 0.40% | NaCl | 112 d | Water | No adverse effects | |

| | | | | | | | |
|---|---|---|---|---|---|---|---|
| | | 0.70% | | | | No adverse effects | |
| | | 1.00% | | | | Decreased egg production | |
| Chicken—20 | 25 d | 2.50% | NaCl | 10 d | Diet | No adverse effects | Eleazer and Bierer, 1964 |
| | | 4.50% | | | | Increased heart size; edema | |
| | | 8.50% | | | | Increased heart size; edema | |
| | | 12.50% | | | | Increased heart size; edema | |
| Chicken—26 | 1 d | 1.00% for 10 d and 1.50% for 32 d | NaCl | 42 d | Water | Anorexia; edema; nervousness; degenerative changes in tissues; increased mortality | Mohanty and West, 1969 |
| Turkey—49 | 1 d | 0.40% | NaCl | 28 d | Water | Anorexia; decreased rate of weight gain; increased mortality | Krista *et al.*, 1961 |
| Turkey—12 | 1 d | 1.00% | NaCl | 23 d | Diet | No adverse effects | Matterson *et al.*, 1946 |
| | | 2.00% | | | | No adverse effects | |
| | | 4.00% | | | | Edema and mortality | |
| Turkey—6 | 217 d | 1.45% | NaCl | 21 d | Diet | No adverse effects | Roberts, 1957 |
| | | 2.45% | | | | No adverse effects | |
| | | 4.45% | | | | No adverse effects | |
| Turkey—6 | 56 d | 1.50% | NaCl | 28 d | Diet | No adverse effects | |
| | | 2.00% | | | | No adverse effects | |
| | | 4.00% | | | | No adverse effects | |
| Turkey—6 | 182 d | 4.00% | NaCl | 28 d | Diet | No adverse effects | |
| | | 6.00% | | | | Weight loss | |
| | | 8.00% | | | | Weight loss | |
| Turkey—30 | 56 d | 4.00% | NaCl | 56 d | Diet | Reduced rate of weight gain and pendulous crop | Harper and Arscott, 1962 |
| Turkey—30 | 1 d | 2.70% | NaCl | 14 d | Diet | Lung congestion; enlarged kidneys and mortality | Morrison *et al.*, 1975 |

TABLE 32 *Continued*

| Class and Number of Animals[a] | Age or Weight | Administration: Quantity of Element[b] | Source | Duration | Route | Effect(s) | Reference |
|---|---|---|---|---|---|---|---|
| Duck—7 | 1 d | 0.40% | NaCl | 21 d | Water | No adverse effects | Krista *et al.*, 1961 |
| | | 0.70% | | | | No adverse effects | |
| | | 1.00% | | | | Anorexia; reduced gain; mortality | |
| | | 1.20% | | | | Anorexia; reduced gain; mortality | |
| Pheasant—100 | 1 d | 2.00% | NaCl | 28 d | Diet | No adverse effects | Scott *et al.*, 1960 |
| | | 3.00% | | | | No adverse effects | |
| | | 4.00% | | | | No adverse effects | |
| | | 5.00% | | | | Reduced rate of gain | |
| | | 7.50% | | | | Reduced rate of gain; increased mortality | |
| Quail—50 | 1 d | 2.50% | NaCl | 28 d | Diet | No adverse effects | |
| | | 5.00% | | | | No adverse effects | |
| | | 7.50% | | | | Reduced rate of gain; increased mortality | |

[a] Number of animals per treatment.
[b] Quantity expressed as concentration in the diet or drinking water or as grams per kilogram of body weight.

TABLE 33 The Acute Toxicity of Sodium Chloride to Animals of Various Species

| Species | No. and Sex | Route | Dosage | Measurement | Reference |
|---|---|---|---|---|---|
| Fish | 10/group | Immersion | 12,946 ppm | $LD_{50}$ | Trama, 1945 |
| Fish | 2–10/group | Immersion | 6,000 ppm | $LD_{50}$ | Ghosh and Pal, 1969 |
| Mouse | | Oral | 6,441 mg/kg body weight | $LD_{50}$ | Behrens, 1924 |
| Rat | 168 (M and F) | Oral | 3,750 mg/kg body weight | $LD_{50}$ | Boyd and Shanas, 1963 |
| Rat | 5 F/group | Subcutaneous | 3,500 mg/kg body weight | MLD | Main, 1939 |
| Rat | | Intravenous | 6,000 mg/kg body weight | MLD | Richter and Mosier, 1954 |
| Guinea pig | | Intravenous | 2,910 mg/kg body weight | $LDL_0$ | Amberg and Helmholx, 1915 |
| Cat | 1–4/group | Intravenous | 4,460–8,500 mg/kg body weight | MLD | Cutting *et al.*, 1939 |

## REFERENCES

Amberg, S., and H. F. Helmholx. 1915. The fatal dose of various substances on intravenous injection in the guinea pig. J. Pharmacol. Exp. Ther. 6:595.

Anonymous. 1974. Salt for Livestock, Poultry and Other Animals. Salt Institute, Alexandria, Va.

Barlow, J. S., S. J. Slinger, and R. P. Zimmer. 1948. The reaction of growing chicks to diets varying in sodium chloride content. Poult. Sci. 27:542.

Batterson, M., and W. W. Brodie. 1972. Salt the Mysterious Necessity. The Dow Chemical Company, Midland, Michigan.

Behrens, B. 1924. Studies on the mechanism of sodium chloride poisoning. I. Significance of osmotic processes. Arch. Exp. Pathol. Pharmacol. 108:39.

Blaxland, J. D. 1946. The toxicity of sodium chloride for fowls. Vet. J. 102:152.

Bohstedt, G., and R. H. Grummer. 1954. Salt poisoning of pigs. J. Anim. Sci. 13:933.

Boyd, E. M., and M. N. Shanas. 1963. The acute oral toxicity of sodium chloride in albino rats. Arch. Int. Pharmacodyn. 144:86.

Church, D. C., and W. G. Pond. 1974. Basic Animal Nutrition and Feeding. D. C. Church. Corvallis, Oregon.

Cotlove, E., and C. A. M. Hogben. 1962. Chloride, pp. 109–157. *In* C. L. Comar and F. Bronner (eds.). Mineral Metabolism. An Advanced Treatise, vol. II. Academic Press, New York.

Cutting, R. A., P. S. Larson, and A. M. Lands. 1939. Cause of death resulting from massive infusions of isotonic solutions. Arch. Surg. 38:599.

Demott, B. J., S. A. Hinton, E. W. Swanson, and J. T. Miles. 1968. Influence of added sodium chloride in grain ration on the freezing point of milk. J. Dairy Sci. 51:1363.

Done, J. T., J. D. J. Harding, and M. K. Lloyd. 1959. Meningo–Encephalitis eosinophilica of swine. II. Studies on the experimental reproduction of the lesions by feeding sodium chloride and urea. Vet. Rec. 71:92.

Eleazar, T. H., and B. W. Bierer. 1964. Effects of added dietary sodium chloride on heart size and weight in chicks. Poult. Sci. 43:1068.

Forbes, G. B. 1962. Sodium. pp. 2–72. *In* C. L. Comar and F. Bronner (eds.). Mineral Metabolism. An Advanced Treatise, vol. II. Academic Press, New York.

Ghosh, A. K., and R. N. Pal. 1969. Toxicity of four therapeutic compounds to fry of Indian major carps. Fish Technol. 6:120.

Harper, J. A., and G. H. Arscott. 1962. Salt as a stress factor in relation to pendulous crop and aortic rupture in turkeys. Poult. Sci. 41:497.

Jackson, H. M., R. P. Kromann, and E. E. Ray. 1971. Energy retention in lambs as influenced by various levels of sodium and potassium in the rations. J. Anim. Sci. 33:872.

Jaster, E. H., J. D. Schuh, and T. N. Wegner. 1978. Physiological effects of saline drinking water on high producing dairy cows. J. Dairy Sci. 61:66.

Kare, M. R., and J. Biely. 1948. The toxicity of sodium chloride and its relation to water intake in baby chicks. Poult. Sci. 27:751.

Krista, L. M., C. W. Carlson, and O. E. Olson. 1961. Some effects of saline waters on chicks, laying hens, poults, and ducklings. Poult. Sci. 40:938.

Kromann, R. P. 1978. Personal communication.

Main, R. J. 1939. Mineral salts as toxic factors in urinary prolan concentrates. Endocrinology 24:523.

Matterson, L. D., H. M. Scott, and E. Jungherr. 1946. Salt tolerance of turkeys. Poult. Sci. 25:539.

Meneely, G. R., and H. D. Battarbee. 1976. Sodium and potassium. Nutr. Rev. 34:225.

Meyer, J. H., and W. C. Weir. 1954. The tolerance of sheep to high intakes of sodium chloride. J. Anim. Sci. 13:443.

Meyer, J. H., R. R. Grunert, R. H. Grummer, P. H. Phillips, and G. Bohstedt. 1950. Sodium, potassium, and chlorine content of feeding stuffs. J. Anim. Sci. 9:153.

Meyer, J. H., W. C. Weir, N. R. Ittner, and J. D. Smith. 1955. The influence of high sodium chloride intakes by fattening sheep and cattle. J. Anim. Sci. 14:412.

Mohanty, G. C., and J. L. West. 1969. Pathologic features of experimental sodium chloride poisoning in chicks. Avian Dis. 13:762.

Morrison, W. D., A. E. Ferguson, J. R. Pettit, and D. C. Cunningham. 1975. The effects of elevated levels of sodium chloride on ascites and related problems in turkeys. Poult. Sci. 54:146.

National Research Council. 1974. Nutrients and Toxic Substances in Water for Livestock and Poultry. National Academy of Sciences, Washington, D.C.

Paver, H., A. Robertson, and J. E. Wilson. 1953. Observations on the toxicity of salt for young chickens. J. Comp. Pathol. 63:31.

Peirce, A. W. 1957. Studies on salt tolerance of sheep for sodium chloride in the drinking water. Aust. J. Agric. Res. 8:711.

Peirce, A. W. 1959. Studies on salt tolerance of sheep. II. The tolerance of sheep for mixtures of sodium chloride and magnesium chloride in the drinking water. Aust. J. Agric. Res. 10:725.

Peirce, A. W. 1960. Studies on salt tolerance of sheep. III. The tolerance of sheep for mixtures of sodium chloride and sodium sulphate in the drinking water. Aust. J. Agric. Res. 11:548.

Peirce, A. W. 1962. Studies on salt tolerance of sheep. IV. The tolerance of sheep for mixtures of sodium chloride and calcium chloride in the drinking water. Aust. J. Agric. Res. 13:479.

Peirce, A. W. 1963. Studies on salt tolerance of sheep. V. The tolerance of sheep for mixtures of sodium chloride, sodium carbonate, and sodium bicarbonate in the drinking water. Aust. J. Agric. Res. 14:815.

Peirce, A. W. 1968a. Studies on salt tolerance of sheep. VII. The tolerance of ewes and their lambs in pens for drinking waters of the types obtained from underground sources in Australia. Aust. J. Agric. Res. 19:577.

Peirce, A. W. 1968b. Studies on salt tolerance of sheep. VIII. The tolerance of grazing ewes and their lambs for drinking waters of the types obtained from underground sources in Australia. Aust. J. Agric. Res. 19:589.

Potter, B. J., and G. H. McIntosh. 1974. Effect of salt water ingestion on pregnancy in the ewes and on lamb survival. Aust. J. Agric. Res. 25:909.

Quigley, G. D., and R. H. White. 1932. Salt tolerance of baby chicks. Md. Agric. Exp. Stn. Bull. 340:343.

Richter, C. P., and H. D. Mosier, Jr. 1954. Maximum sodium chloride intake and thirst in domesticated and wild Norway rats. Am. J. Phys. 176:213.

Robblee, A. R., and D. R. Clandinin. 1961. The effect of levels of sodium salts in the feed and drinking water on the occurrence of ascites and edema in turkey poults. Can. J. Anim. Sci. 41:161.

Roberts, R. E. 1957. Salt tolerance of turkeys. Poult. Sci. 36:672.

Scott, M. L., A. van Tienhoven, E. R. Holm, and R. E. Reynolds. 1960. Studies on the sodium, chloride and iodine requirements of young pheasants and quail. J. Nutr. 71:282.

Standen, A., ed. 1970. Kirk-Othmer Encyclopedia of Chemical Technology, vol. 18. John Wiley & Sons, New York.

Todd, J. R., G. H. K. Lawson, and C. Dow. 1964. An experimental study of salt poisoning in the pig. J. Comp. Pathol. Ther. 74:331.

Tracor-Jitco, Inc. 1974. GRAS monograph series. Sodium chloride, potassium chloride. U.S. Department of Health, Education, and Welfare, Food and Drug Administration.

Trama, F. B. 1945. Acute toxicity of some common salts of sodium, potassium, and calcium to the common bluegill (*Lepomis macrochirus*). Proc. Acad. Nat. Sci. Philadelphia 106:185.

U.S. Department of the Interior. 1976. Bureau of Mines Minerals Yearbook, Salt chapter.

Weeth, H. J., and L. H. Haverland. 1961. Tolerance of growing cattle for drinking water containing sodium chloride. J. Anim. Sci. 20:518.

Weeth, H. J., L. H. Haverland, and D. W. Cassard. 1960. Consumption of sodium chloride water by heifers. J. Anim. Sci. 19:845.

Wilson, A. D. 1966. The tolerance of sheep to sodium chloride in food or drinking water. Aust. J. Agric. Res. 17:503.

# Strontium

Strontium (Sr) is one of the alkaline earth metals, and, because of its high chemical activity, it never occurs free in nature (Pidgeon and Preisman, 1969). It occurs in nature principally in the minerals strontianite ($SrCO_3$) and celestite ($SrSO_4$). The role of strontium in biological systems received little attention until the late 1940's, when it became obvious that $^{90}Sr$ is an abundant and potentially hazardous radioactive by-product of nuclear fission.

## ESSENTIALITY

Strontium has been reported to act as a plant growth stimulant (Underwood, 1977) and to be capable of replacing the calcium required by *Chlorella* (Walker, 1953). It has not, however, been shown to be essential for either plants or animals. A report by Rygh (1949) indicated that the omission of strontium from the mineral supplement fed to rats and guinea pigs consuming a purified diet resulted in growth depression, an impairment of the calcification of the bones and teeth, and a higher incidence of dental caries. This work has not been confirmed.

## METABOLISM

Early studies of the metabolism of strontium showed a close relationship with calcium (Fay *et al.*, 1942). It was shown that strontium behavior in mammals was similar but not identical to that of calcium. The strontium ion ($Sr^{+2}$) is very similar to the calcium ion ($Ca^{+2}$), both chemically and physiologically, and it can substitute for calcium in physiological processes. These include muscular contraction, blood clotting, and bone formation. One difference which has been observed is that the rates of most of these processes are slowed when strontium is substituted for calcium. The major differences between strontium and calcium behavior were found relative to gastrointestinal absorption, renal excretion, lactation, and placental transfer. It appears that whenever there is a metabolically controlled passage of ions across a membrane, calcium is transported more effectively than strontium (Comar and Wasserman, 1964). Once absorbed, the element has a strong affinity for the skeletal system. Strontium is poorly absorbed by adult animals on natural diets, with a large proportion of the ingested strontium appearing in the feces.

## SOURCES

There is limited information on the normal intake of strontium by farm animals. With ruminants, this will be influenced by the strontium status of the soil and by the proportion of legumes to grasses in the herbage consumed. Mitchell (1957) found red clover growing on different soils ranged from 53 to 115 (mean 74) ppm strontium and ryegrass from 5 to 18 (mean 10) ppm strontium (dry basis). Strontium compounds are not added to animal diets and would occur in such diets as contaminants of other ingredients.

## TOXICOSIS

Strontium ($Sr^{+2}$) is less toxic than calcium ($Ca^{+2}$). When strontium is fed along with low levels of calcium to young or growing animals, bone formation is disturbed and a condition known as "strontium rickets" develops (Bartley and Reber, 1961; Colvin and Creger, 1967; Colvin *et al.*, 1972). It has also been reported that strontium in the diet gives rise to insoluble phosphates during digestion and to a phosphorus deficiency (Jones, 1938). The rachitogenic action of strontium is probably related to the levels and ratio of calcium and phosphorus in the diet.

Knight *et al.* (1967) fed growing beef cattle diets containing two levels of calcium (0.13 and 3.1 percent) and three levels of strontium (13, 200, and 2,000 ppm) for a 100-day period. High-calcium diets reduced weight gains and the digestibility of both dry matter and energy. The various strontium levels had no effect on these criteria. The amount of strontium in bone ash increased with each added increment of strontium in the diet. The amount deposited was reduced by high-calcium additions.

Bartley and Reber (1961) reported on the toxic effects of strontium in pigs (3 to 8 weeks old). Dietary treatments included 0.89 or 0.16 percent calcium and "zero" or 6,700 ppm strontium. The pigs fed 6,700 ppm strontium and 0.16 percent calcium were the most severely affected by incoordination and weakness, followed by posterior paralysis.

Weber *et al.* (1968) fed dietary calcium levels of 0.72 and 1.0 percent with 3,000 and 6,000 ppm strontium as $SrCO_3$ to chicks. Feeding 6,000 ppm strontium reduced growth rate more severely at the lower calcium level. Higher levels of strontium appeared to reduce calcium retention, while phosphorus utilization was apparently unaffected. Higher levels of strontium in bone ash were found at the lower calcium level.

In a later study, Doberenz *et al.* (1969) fed high dietary levels of stable strontium to mature hens. Levels of 3,000 to 50,000 ppm of strontium as carbonate were fed for a 4-week period. The diet contained 2.9 percent calcium. The mature hen was able to tolerate considerable amounts of strontium in the diet. Egg weight, egg production, feed consumed, and body weights were unaffected by dietary levels up to 30,000 ppm. At a dietary level of 50,000 ppm strontium, egg weight, egg production, and feed consumed were significantly reduced.

## TISSUE LEVELS

Gerlach and Muller (1934) found the strontium concentration of a wide variety of animal tissues ranged from 0.01 to 0.10 ppm, with no evidence of accumulation in any particular species, organ, or tissue. Studies have demonstrated that as dietary levels of strontium are increased, the strontium absorbed is concentrated in the bones or skeletal system and not the soft tissue. In a study of the strontium content of cow's milk, Jury *et al.* (1960) reported levels ranging from 31 to 65 ppm of the ash, or approximately 0.2 to 0.4 ppm of the whole milk. A report by Murthy *et al.* (1972) on samples of milk collected from cities in the United States gave results in close agreement with those of Jury *et al.* (1960).

## MAXIMUM TOLERABLE LEVELS

The effect of strontium is very dependent on the dietary level of calcium, with more pronounced effects observed at low-calcium levels. When dietary calcium levels are at NRC recommended levels for the respective species, animals have a high tolerance for strontium. Mature animals can tolerate higher levels than the young. Assuming adequate calcium, dietary strontium levels as high as 2,000 ppm (0.2 percent) can be tolerated for extended periods of time, even by the young. Swine and poultry can tolerate 3,000 ppm (0.3 percent) and the laying hen can tolerate levels of 30,000 ppm (3 percent). Limited data indicate that animal diets normally contain about 200 ppm strontium.

## SUMMARY

Strontium is one of the alkaline earth metals, and, because of its high chemical activity, it never occurs free in nature. It has not been shown to be essential for either plants or animals. Strontium is metabolized similarly to calcium by animals, and it can substitute for calcium in physiological processes. When strontium is fed with low levels of calcium to young or growing animals, bone formation is disturbed and a condition known as ''strontium rickets'' develops. Assuming adequate calcium, dietary strontium levels as high as 2,000 ppm (0.2 percent) can be tolerated for extended periods of time, even by the young. Swine and poultry can tolerate 3,000 ppm (0.3 percent), and the laying hen can tolerate levels of 30,000 ppm (3 percent).

TABLE 34 Effects of Strontium Administration in Animals

| Class and Number of Animals[a] | Age or Weight | Administration: Quantity of Element[b] | Source | Duration | Route | Effect(s) | Reference |
|---|---|---|---|---|---|---|---|
| Cattle—6 | 1 yr (240 kg) | 13, 200, and 2,000 ppm (0.16% Ca) | $SrCO_3$ | Once daily | Bolus | No effect on feed efficiency or weight gains; no signs of strontium rickets were observed (100 d) | Knight *et al.*, 1967 |
| | | 13, 200, and 2,000 ppm (3.1% Ca) | | | | No effect on feed efficiency or weight gains; no signs of strontium rickets were observed | |
| Swine—4 | 3 wk | 6,700 ppm (0.89% Ca) | $SrCl_2$ | 5 wk | Diet | Only mild effects (weakness and incoordination) | Bartley and Reber, 1961 |
| | | 6,700 ppm (0.16% Ca) | | | | Incoordination and weakness noted after second week; progressed to complete paralysis by the end of third week | |
| Chicken—18 | 1 d | 3,000 ppm (0.72% Ca) | $SrCO_3$ | 4 wk | Diet | No adverse effect | Weber *et al.*, 1968 |
| | | 6,000 ppm (0.72% Ca) | | | | Reduced growth rate; tibia weight increased; reduced calcium retention | |
| | | 3,000 ppm (1.0% Ca) | | | | No adverse effect | |
| | | 6,000 ppm (1.0% Ca) | | | | Reduced growth rate; reduced calcium retention, but not as severe as at | |

TABLE 34 *Continued*

| Class and Number of Animals[a] | Age or Weight | Administration: Quantity of Element[b] | Source | Duration | Route | Effect(s) | Reference |
|---|---|---|---|---|---|---|---|
| | | | | | | the lower calcium level | |
| | | 12,000 ppm (1.0% Ca) | | | | Reduced growth rate; reduced calcium retention; tibia weight increased | |
| Chicken, hens | 6 mo | 3,000, 6,000, 12,000, 20,000, and 30,000 ppm (2.9% Ca) | $SrCO_3$ | 4 wk | Diet | No adverse effect | Doberenz *et al.*, 1969 |
| | | 50,000 ppm (2.9% Ca) | | | | Egg weight, egg production, and feed consumption significantly reduced | |

[a] Number of animals per treatment.
[b] Quantity expressed in parts per million as concentration in diet. Dietary concentration of calcium also shown.

## REFERENCES


Bartley, J. C., and E. F. Reber. 1961. Toxic effects of stable strontium in young pigs. J. Nutr. 75:21.

Colvin, L. B., and C. R. Creger. 1967. Stable strontium and experimental bone anomalies. Fed. Proc. 26:416.

Colvin, L. B., C. R. Creger, T. M. Ferguson, and H. R. Crookshank. 1972. Experimental epiphyseal cartilage anomalies by dietary strontium. Poult. Sci. 51:576.

Comar, C. L., and R. H. Wasserman. 1964. Strontium, pp. 523–572. *In* C. L. Comar and F. Bronner, eds. Mineral Metabolism, vol. 2. Academic Press, New York.

Doberenz, A. R., C. W. Weber, and B. L. Reid. 1969. Effect of high dietary strontium levels on bone and egg shell calcium and strontium. Calcif. Tissue Res. 4:180.

Fay, M., M. A. Andersch, and V. G. Behrmann. 1942. The biochemistry of strontium. J. Biol. Chem. 144:383.

Gerlach, W., and R. Muller. 1934. The occurrence of strontium and barium in human organs and excreta. Virchows Arch. Pathol. Anat. Physiol. 294:210.

Jones, J. A. 1938. Metabolism of calcium and phosphorus as influenced by addition to the diet of salts of metals which form insoluble phosphates. Am. J. Physiol. 124:230.

Jury, R. V., M. S. Webb, and R. J. Webb. 1960. The spectrochemical determination of total strontium in bone, milk and vegetation. Anal. Chem. Acta 22:145.

Knight, W. M., V. R. Bohman, A. L. Lesperance, and C. Blincoe. 1967. Strontium retention in the bovine. J. Anim. Sci. 26:839.

Mitchell, R. L. 1957. The trace element content of plants. Research (London) 10:357.

Murthy, G. K., U. S. Rhea, and J. T. Peeler. 1972. Copper, iron, manganese, strontium, and zinc content of market milk. J. Dairy Sci. 55:1666.

Pidgeon, L. M., and L. Preisman. 1969. Kirk-Othmer Encyclopedia of Chemical Technology, vol. 19, 2nd ed. John Wiley & Sons, New York.

Rygh, O. 1949. Research on trace elements. 1. Importance of strontium, barium and zinc. Bull. Soc. Chim. Biol. 31:1052.

Underwood, E. J. 1977. Trace Elements in Human and Animal Nutrition, 4th ed. Academic Press, New York.

Walker, J. B. 1953. Inorganic micronutrient requirements of *Chlorella*. I. Requirements for calcium (or strontium), copper, and molybdenum. Arch. Biochem. Biophys. 46:1.

Weber, C. W., A. R. Doberenz, R. W. G. Wyckoff, and B. L. Reid. 1968. Strontium metabolism in chicks. Poult. Sci. 47:1318.

# Sulfur

Sulfur (S) has been used in several forms since antiquity, but not until recently has there been a real understanding of its biological significance. The annual world "harvest" of sulfur is more than 27 billion kg obtained from underground deposits of elemental sulfur and trapped hydrogen sulfide gas, from the manufacture of petroleum products, and from reclamation of combustion products of sulfurous fossil fuels. About 80 percent of the annual production of sulfur is used for the production of sulfuric acid, which in turn is used for the manufacture of phosphate fertilizer, synthetic fibers such as rayon (100 kg cellulose requires 30–50 kg carbon disulfide), pesticides and white pigments (Brieger and Teisinger, 1966), in steel processing, in bleaching agents for paper pulp, sugar and vegetable oils, in preservation of beverages and foods, and in vulcanizing of rubber products (Leclercq, 1972).

While the literature on sulfur chemistry is voluminous, the reviews on this topic by Roy and Trudinger (1970), Senning (1972), Nickless (1968), Young and Maw (1958), and du Vigneaud (1952) are noteworthy. The nutritional aspects of sulfur have been reviewed by Baker (1977), Muth and Oldfield (1970), and Goodrich (1978). While the industrial hazards of organic and inorganic forms of sulfur have been often investigated, limited information is available regarding the potential for or effects of sulfur contamination of feedstuffs.

## ESSENTIALITY

Sulfur is required for the formation of the many sulfur-containing compounds found in essentially all body cells and, therefore, is an essential nutrient. The important body sulfur compounds include the sulfur-containing amino acids (methionine, cysteine, cystine, homocysteine, cystathionine, taurine, and cysteic acid), thiamin, biotin, lipoic acid, coenzyme A, glutathione, chondroitin sulfate, fibrinogen, heparin, ergothionine, and estrogens. The sulfur in the mammalian body represents about 0.15 percent of body weight. Fortunately, all of the above sulfur compounds, except thiamin and biotin, can be synthesized *in vivo* from one essential amino acid, methionine. Approximately 50 percent of the total requirement for sulfur-containing amino acids can be provided by cystine. These amino acids are therefore involved in acid-base balance of intra- and extracellular fluids, protein synthesis, lipid and carbohydrate metabolism, collagen and connective tissue formation through disulfide bonds between and within polypeptide chains, blood-clotting, enzyme synthesis, and endocrine function (Baker, 1977).

## METABOLISM

The metabolism of sulfur differs markedly between monogastrics and ruminants, and an understanding of this difference is basic to an appreciation of the sulfur cycle (Postgate, 1968) and of the nutritional value of sulfur compounds. The major terrestrial source of sulfur is mineral sulfide, which is converted to inorganic sulfate by weathering and to organic sulfur by microbial action in the soil (Young and Maw, 1958). Inorganic sulfate is taken up by higher plants and converted to organic sulfur in the form of the sulfur-containing amino acids, which in turn serve as an organic sulfur source for both monogastric and ruminant animals. Many bacteria, including the microbial flora of the ruminant, are also able to convert inorganic sulfur to organic sulfur in the form of methionine, cysteine, and cystine and hence for the many functions of sulfur in the body. Monogastric animals have few, if any, intestinal assimilatory bacteria to form organic sulfur from inorganic sources and, therefore, must rely upon exogenous sulfur amino acid sources for their requirement of organic sulfur. Both ruminants and nonruminants can utilize inorganic sulfate in the formation of sulfate esters required in the synthesis of mucopolysaccharides. Absorption by active transport of the inorganic sulfate takes place in the small intestine, especially the ileum (Dziewiatkowski, 1970). Organic forms of sulfur are readily absorbed in the small intestine. The absorption mechanism is very efficient. Morrow *et al.* (1952) demonstrated that rats excreted, in the urine, 41 to 64 percent of an oral dose of inorganic sulfate $^{35}S$ within 8 hours of administration. The sulfur status of animals has been measured by balance studies (Walker and Cook, 1967), by serum sulfate levels (Bray, 1965), and by serum amino-acid levels (Schelling and Hatfield, 1968).

## SOURCES

Inorganic sulfur compounds have the potential for contaminating food supplies and/or interfering with biological systems, albeit primarily by industrial exposure of humans. Elemental sulfur (flowers of sulfur) is used topically as a fungicide, karyolytic agent, and parasiticide. Hydrogen sulfide exposure (for humans) is frequent around shale oil plants and from sewer gas developed from the putrefaction of organic matter. Industrial exposure to carbon–sulfur compounds occurs in the manufacture of viscose, rubber, and insecticides (Sorbo, 1972). Sulfur dioxide is a major component of industrial smog as a result of combustion of coal and petroleum. Additional sources of sulfur dioxide are paper bleaches, fumigants, and refrigerants. Sodium sulfides are used as food and pharmaceutical preservatives. The thiosulfates, used in antimycotic and antiparasitic agents, are also used as antidotes for cyanide toxicity and have been used in clinical medicine to measure extracellular fluid space and glomerular filtration rates. Exposure to sulfuric acid, the intermediate in the manufacturing of the many sulfur compounds mentioned, is usually limited to aerosols generated in electroplating and battery charging. Ammonium persulfate is used as an alternative to nitrogen trichloride as a bleaching agent for flour, because the latter causes neurologic problems in dogs consuming products containing American-processed flour (Lewis, 1954). Additional inorganic sulfur sources include sulfate salts used in mineral supplements, dicalcium phosphate, water supplies (Larson, 1959), fish by-products, cement kiln dust, sulfur-coated urea used as a slow-release nitrogen fertilizer for cranberry bogs, and sodium metabisulfide and sulfur dioxide used as silage preservatives.

The organic sources of sulfur include methionine, an essential sulfur-containing amino acid; dimethyl sulfoxide, a powerful solvent to transport medication to poorly vascularized tissues such as the lens and the articulations (Caldwell *et al.*, 1967); and the high-sulfur feed ingredients such as feathers, viscera, and fecal waste used in some contemporary livestock diets.

## TOXICOSIS

The toxicity of sulfur is dependent upon its form and route of administration. Whereas elemental sulfur is considered one of the least toxic elements, hydrogen sulfide rivals cyanide in toxicity. Fitzhugh *et al.* (1946) reviewed the chronic toxicity of sulfites.

### LOW LEVELS

The effects of long-term feeding of sulfur dioxide ($SO_2$) to cattle have been investigated by Weigand *et al.* (1972). Lactating cows fed between 18 and 20 g of sulfur as $SO_2$ daily for 110 days exhibited no deleterious effects on milk production or butterfat level. The mean sulfide levels of the milk or blood were not significantly different between treated and control cattle, and no changes in hematology, behavior, body temperature, heart rate, or rumen motility could be attributed to the $SO_2$ treatments.

The toxic effects of sodium metabisulfite ($Na_2S_2O_5$) and sulfur dioxide, two silage preservatives, have been investigated by Luedke *et al.* (1959) with rumen-fistulated cattle. A cow given 80 g of sodium metabisulfite, equal to 26.5 g sulfur per day for 180 days, seemed unaffected by the treatment and, in fact, completed a gestation and calved normally during the treatment. A second cow given sulfur as sodium metabisulfite on a schedule of 26.5 g per day for 2 days, 40 g per day for the next 2 days, and 53 g per day for the next 2 days stopped eating on the sixth day of the experiment. Supplementation was ceased and the cow appeared normal again within 6 days. Another cow administered sodium metabisulfite on a schedule of 26.5 g per day for 1 week, 40 g per day for the second week, and 53 g per day for the third week was unable to stand on the eighteenth day after receiving 673 g of the sulfur. Other cows and heifers on similar schedules died after 16 to 21 days on experiment and after receiving 568–658 g sulfur as sodium metabisulfite or sulfur dioxide. These cattle had exhibited anorexia, weight loss, constipation, diarrhea, and depression. At necropsy, pulmonary emphysema, cardiac petechiation, congestion of the central nervous system, acute catarrhal enteritis, and hepatic necrosis were present in the treated cows.

Alhassan and Satter (1968) have investigated the effects of intrarumenally administered sodium sulfite in cattle. These investigators administered 15 g of sulfur as $Na_2SO_3$ per day per cow and gradually increased this dosage to 50 g of sulfur per cow daily. These amounts of sulfur caused partial to complete anorexia, depression of milk fat from 3.7 to 2.4 percent, and a depression of the ruminal acetate–butyrate ratio. Weeth and Hunter (1971) provided Hereford cattle with drinking water containing 1,240 ppm sulfur as $Na_2SO_4$ for 30-day trials. This level of sulfur caused reduced feed and water intakes, weight loss, diuresis, and hemoconcentration.

Upton *et al.* (1970) fed 2,200 ppm sulfur as sodium sulfate to sheep without effect and found fecal sulfur correlated well with dry matter intake. Marcilese *et al.* (1969) fed wethers the equivalent of 0.5 g sulfur per day as sodium sulfate for indefinite periods without any adverse effects. Lewis (1954) found that rumen-fistulated wethers given 4, 8, or 15 g of sulfur daily as sodium sulfate via the fistula exhibited no toxic effects from the treatments, even though there was a marked elevation in rumen sulfate concentration.

L'Estrange *et al.* (1969) fed 40 kg wethers either sodium sulfate, sodium bisulfite, ammonium bisulfite, sulfuric acid, or ammonium sulfate to create 1 percent sulfur diets. The sodium sulfate and sodium bisulfite caused a 22 percent decrease in voluntary feed intake, while the other sulfur sources decreased feed intake by 44 percent. The sulfuric acid-supplemented diet caused a decrease in rumen pH. L'Estrange and Murphy (1972) found similar effects of sulfuric acid when added to sheep diets at the rate of 320 mEq/kg of grass. L'Estrange *et al.* (1972) also found graded levels of dietary sulfur (0.5, 1.0, and 1.5 percent) added to pelleted grass meal for wethers decreased feed intake and increased the apparent digestibility of the diets over a 14-day feeding trial.

Sodium disulfite has been used in the processing of sugarcane pulp (Kaemmerer, 1972), which has been incorporated into ruminant rations. Kaemmerer *et al.* (1972) studied the effects of 0.5 percent sulfur as sodium disulfite in sheep diets consisting of hay, sugar pulp, and oatmeal in ratios of 5:5:2. Sheep fed this diet tolerated the supplemental sulfur without signs of ill health, although there was indication of increased incidence of renal cysts in the sulfur-supplemented sheep.

Bird (1972) has studied the effects of ruminal infusions of sulfate solutions in sheep. In one experiment, continuous infusions of sulfate sulfur at rates of 0 to 6 g daily for up to 8 days were given. A significant decrease in the dry matter intake and rumen motility appeared when the infusion rate reached 2.93 g per day. At 6 g per day there was complete anorexia within 3 to 9 days, followed by rumen stasis, impaction, and a foul odor ($H_2S$) on the breath of the sheep. In additional studies with sheep, Peirce (1960) provided 3- to 4-year-old wethers with drinking water containing 0.1 to 0.5 percent sodium sulfate for 15 months. The drinking water for each group was isonatremic. No adverse effects on general health, food intake, weight gain, or wool production were observed in any of the sheep regardless of their level of sulfur intake, which ranged between 1.3 and 6.5 g per day.

Paterson *et al.* (1979) have given sows drinking water containing up to 664 ppm sulfur, as $Na_2SO_4$, from 30 days postbreeding to 28 days of lactation and found no significant effect on reproduction. These investigators have also provided weanling pigs with drinking water containing 600 ppm sulfur as $Na_2SO_4$. This level of sulfur caused loose feces and increased incidence of diarrhea, but no differences in average daily gain or feed conversion ratios.

The effects of inhaled sulfur compounds have been studied in several species. O'Donoghue (1961) reported an experiment in which swine were exposed to 0–470 ppm of $H_2S$ in an inhalation chamber. The pigs appeared barely able to detect the odor of 0.9 ppm $H_2S$, seemed to lose their sense of smell at 28 ppm $H_2S$, were markedly affected by 47–188 ppm $H_2S$, and were killed by 470 ppm $H_2S$. In subsequent studies, O'Donoghue and Graesser (1962) exposed 2-week-old pigs to sulfur dioxide atmospheres equal to 5–20 ppm sulfur for 6.5 hours and sacrificed the pigs after 70 days. Signs exhibited by the pigs in the sulfur dioxide chambers included irritation of the eyes, excessive blinking, serous nasal discharge, and hyperpnea. The pigs, however, appeared to adapt to the conditions. Pigs exposed to the 15–20 ppm levels of sulfur dioxide had pulmonary induration (fibrosis) when examined at necropsy.

The toxicity of ammonium persulfate [$(NH_4)_2S_2O_8$], used as a bleaching agent in processing of flour, has been studied in dogs (Arnold and Goble, 1950). Dietary levels of 0.04 to 0.28 percent sulfur were fed for as long as 16 months without noticeable effect on gross appearance, weight gains, renal function, or hematology and without gross or microscopic lesions observed upon necropsy.

Anderson and Chen (1940) studied the toxicosis of some thiocyanates in dogs. Sodium and potassium thiocyanates appeared to be similarly toxic with the no-effect level as high as 9.1 mg sulfur/kg per day orally for more than 3 months. Sulfur intake of 12.2 mg/kg per day caused death in 45 days, while sulfur at 39–41 mg/kg per day caused deaths of the dogs within 4 days.

The effect of inhaled carbon disulfide ($CS_2$) in rabbits was studied by Scheel (1967). Atmospheres of 926 ppm sulfur as carbon disulfide for 6 hours per day, 5 days per week, for 8 weeks caused reduced weight gains, loss of muscle control, neurologic damage, and marked increases in the copper levels of the thyroid, pancreas, and spinal cord. The altered copper metabolism of the central nervous system tissues may account for the central nervous system disturbances associated with carbon disulfide toxicosis in humans.

Guinea pigs have been exposed to atmospheres of 2.5 to 9.0 ppm sulfur as sulfur dioxide in respiratory chambers for 4–15 days (O'Donoghue and Graesser, 1962). These exposures caused reduced weight gains, but no other pathologic disturbances were detected.

Optimal level of inorganic sulfur, as sodium sulfate, in semipurified diets for male rats has been reported to be 0.02 percent when the total dietary sulfur is maintained at 0.67 percent (Smith, 1973). When the total sulfur-containing amino acid content of test diets for growing rats was 0.14 percent (from peanut protein), Brown and Gamatero (1970) found improved net weight gain, feed efficiency, and protein efficiency ratios as supplemental $SO_4$ levels increased from 0.002 percent to 0.1 percent.

The effects of inhaled sulfur dioxide upon fertility of rats have been evaluated by Mamatashvili (1970), who found exposures of female rats to 0.15 mg $SO_2/mm^3$ equal to 0.075 mg sulfur /$mm^3$ for 72 days caused no changes in estrous cycles or fertility. Concentrations of 4 mg/$mm^3$ (2 mg sulfur/$mm^3$) had some detrimental effect on these parameters.

Relevant studies of toxicity of organic sulfur compounds pertain to excess dietary sulfur-containing amino acids, especially methionine. In fact, it is well accepted that methionine given in excess of its dietary requirement is the most toxic of the amino acids (Snetsinger and Scott, 1961). Daniel and Waisman (1969) fed diets providing 0.21, 0.64, and 1.07 percent sulfur from L-methionine to growing rats. The high L-methionine levels initially caused severe growth depression; however, metabolic adaptation occurred and was believed related to altered hepatic enzyme ratios. Cohen *et al.* (1958) found similar growth depression effects from DL-methionine and homocysteine but concluded the effects were not associated with excess sulfur. A curious hemosiderin deposition in the spleen was associated with the excess dietary methionine (Klavins *et al.*, 1965).

The relative toxicities of organic sulfur compounds for chicks have been evaluated by Katz and Baker (1975). Methionine and homocysteine were very toxic, while cysteine was found relatively nontoxic. When diets contained the minimal required levels of threonine and glycine, 0.52 and 0.51 percent, respectively, methionine at 1.25 percent of the diet (equal to 2,690 ppm sulfur) caused a 40 percent reduction in growth rate of the chicks. Sasse and Baker (1974) studied the effect of increasing levels of $K_2SO_4$ equal to 37–370 ppm sulfur in the diets of chickens and noted a plateauing effect on growth and feed efficiency at the 185 and 370 ppm levels.

### HIGH LEVELS

Acute sulfur toxicosis in cattle has been reported by Coghlin (1944). This incident stemmed from excessive use of liquid sulfur as a top dressing on green chop in an effort to control lice and ringworm. An estimated 1.36 kg of liquid sulfur were added to the feed for 25 head of cattle. Signs of toxicosis were muscular twitching, restlessness, diarrhea, recumbent attitude, and dyspnea. All animals that became recumbent appeared to be blind, became comatose, and died. Luedke *et al.* (1959) found that single oral doses of 530–1,860 g sulfur as sodium bisulfite administered to adult cattle caused death within 16 days.

Acute sulfur toxicosis in sheep was accidentally induced (White, 1964) when excessive quantities of sublimated sulfur (flowers of sulfur) were mixed with a concentrate mix and fed to 480 sheep for treatment of contagious ovine ecthyma. It was estimated that the sheep consumed an average of 62 g sulfur instead of the anticipated 15 g per sheep. Evidence of toxicosis occurred within 24 hours and included colicky pain, depressed attitude, dyspnea, pyrexia, recumbency, $H_2S$ odor of the breath, and about 5 percent mortality within 3 days. Postmortem examination revealed the sulfur-toxic sheep had severe enteritis, peritoneal effusions, darkened kidneys, and generalized petechial hemorrhages.

Several acute sulfur toxicity studies have been conducted with dogs. Dougherty *et al.* (1943) reported on administering hydrogen sulfide ($H_2S$) per rectum to dogs. Hydrogen sulfide administered at the rate of 3.6 cc/min was not lethal, but rates of 4.1 to 10 cc/min were fatal to dogs breathing air. Dogs breathing 90 percent oxygen and 10 percent carbon dioxide tolerated doses up to 22 cc hydrogen sulfide per minute administered per rectum. Gilman *et al.* (1946) intravenously administered single doses of sodium tetrathionate to dogs at rates equal to 59–350 mg sulfur per kilogram. The lowest level was without effect but the highest levels caused death within 3 days.

The acute toxic effects of intraperitoneally administered sodium bisulfite have been investigated in several species (Wilkins *et al.*, 1968), because it is used as an antioxidant in peritoneal dialysis solutions. The dosage range of 174–393 mg/kg of body weight of dog was used, and the $LD_{50}$ for IP-administered sodium bisulfite was calculated to be 244 mg/kg. The $LD_{50}$ for IP-administered sodium bisulfite in other species was found to be 300 mg/kg for rabbits, 650–740 mg/kg for rats, and 675 mg/kg for mice. Hauschild (1960) found the oral administration of 2.5 mg $H_2SO_3$ per kilogram of body weight, intravenous administration of 30 mg $NaHSO_3$ per kilogram of body weight, and the subcutaneous administration of 900–1,000 mg $NaHSO_3$ per kilogram of body weight to be fatal in the dog.

The acute toxic effects of sodium tetrathionate administered intravenously to dogs have been investigated by Gilman *et al.* (1946). Doses of 60 mg sulfur per kilogram as tetrathionate did not cause death, but all higher doses used (250–500 mg/kg) caused death in 2 to 9 days. Acute lethal doses caused vomition, hyperpnea, ataxia, and anorexia, while sublethal doses produced proximal tubular necrosis. The data of Gilman *et al.* (1946) also indicate that rabbits are more sensitive to sodium tetrathionate toxicosis than dogs. All rabbits given 48 mg sulfur per kilogram or more as sodium tetrathionate intravenously died. Saunders and Wills (1954) administered sulfur as sodium tetrathionate to rabbits intravenously at the rates of 4.8 to 48 mg/kg via the femoral vein. Doses between 36 and 48 mg/kg caused proximal tubule necrosis, as was also observed in dogs by Gilman *et al.* (1946).

There appears to be a critical toxic level for sodium bisulfite in rabbits. The $LD_{50}$ for intravenously administered sulfur as sodium bisulfite in rabbits was found to be 20 mg/kg (Hoppe and Goble, 1951), with death due to thrombosis and respiratory failure. Approximately two-thirds of the $LD_{50}$, however, was injected 3 times per day, 5 days per week, for 8 weeks without apparent toxicosis. Rabbits seem more sensitive to sulfur as sodium bisulfite than some other laboratory animals, because the calculated $LD_{50}$ levels for hamsters, rats, and mice are 30, 38, and 42 mg/kg, respectively.

Guinea pigs were found more susceptible to $H_2SO_4$ mists than several other laboratory animal species (Treon *et al.*, 1950). Exposure to $H_2SO_4$ mist concentrations equal to 7.2 ppm sulfur for 2.75 hours caused death in guinea pigs, whereas other species (cats, rabbits, rats, and mice) survived the equivalent of 38 ppm for 7 hours. The morphologic effects of the $H_2SO_4$ mists in the guinea pigs included labored breathing, pulmonary congestion, emphysema, pulmonary edema, and degeneration of the respiratory epithelium.

The $LD_{50}$ for intraperitoneally administered sulfur compounds, $Na_2S \cdot 9H_2O$, $Na_2SO_3 \cdot 7H_2O$, $NaHSO_4 \cdot H_2O$, and $Na_2S_2O_3$ in mice were found by Nofre *et al.* (1963) to be 53, 277, 193, and 226 mg/kg, respectively. These investigators concluded the sulfur ion was 202 times as toxic as the chloride ion.

An accidental case of acute sulfur toxicosis in the horse (Ales, 1907) resulted from the use of a flowers of sulfur gruel given orally as a systemic adjunct to the topical treatment for collar gall in the horse. One and one-half kilograms of flowers of sulfur were given to a total of five horses, and therefore, the individual dosage approximated 300 g per horse. Within 3 hours, the treated horses developed violent colic, collapsed to the ground, and developed foetid diarrhea, $H_2S$ odor of the feces, a feeble, thready pulse, and red-brown urine. Death occurred within a few hours in at least one horse. The surviving horses had unquenchable thirst and diarrhea for 2 days.

## FACTORS AFFECTING TOXICITY

The literature leaves most matters on this topic to speculation. If one determines that factors that increase the requirement for sulfur-containing amino acids in turn decrease the susceptibility of the organisms to sulfur toxicity, then increased dietary protein level, increased dietary energy level, and increased environmental temperature would all be factors which decrease the toxicity to sulfur.

The toxicity of sulfur in large part is determined by the enzyme systems of the exposed organism, and especially by whether the organism has the capacity to form hydrogen sulfide from the inorganic sulfate sources presented. The rate of exposure becomes a critical factor in terms of the organism's ability to safely metabolize the toxic agent or its intermediates (Mudd *et al.*, 1967).

Wolf and Varandani (1960) demonstrated that vitamin A deficiency impairs the ability of the gut mucosa of rats to incorporate sulfate ions and glucose into mucopolysaccharides. This is due to an inhibition of sulfotransferases by the vitamin A deficiency.

The rate of excretion of intraperitoneally injected $^{35}S$ sulfate is greatly increased by the simultaneous subcutaneous administration of 2-naphthylamine or 2-naphthol; therefore, these compounds should decrease the toxicity of sulfur (Laidlaw and Young, 1948). Sodium azide, as well as sodium fluoride, inhibit the transfer of sulfate from the mucosal to the serosal surfaces of the intestine. These compounds, therefore, would also be expected to reduce the toxicity of sulfur (Dziewiatkowski, 1970). Whether high dietary copper levels reduce the toxicity of sulfur in a manner opposite to that in which sulfate and molybdenum induce copper deficiencies appears not to have been demonstrated. It may be that elevated dietary copper and molybdenum levels would also decrease the toxicity of sulfur.

## TISSUE LEVELS

The volatility of sulfur creates analytical errors unless appropriate precautions are taken, especially in analyzing certain vegetables. Masters and McCance (1939) have published some values on the sulfur content of tissues and edible animal by-products. Some selected values include milk, 292 ppm; egg whites, 1,820 to 2,160 ppm; egg yolks, 1,649 to 2,010 ppm; rabbit muscle meat, 1,640 ppm; duck muscle meat, 3,950 ppm.

The nitrogen–sulfur ratio is an important index of the biological value of food substances. For meat and fish samples, the N–S ratios are quite constant; therefore, by nitrogen analysis, one can indirectly calculate quite accurately the sulfur content of feeds. The average N–S ratio is 15.2:1.0 for muscle meats and 13.8:1.0 for fish meats (Masters and McCance, 1939).

## MAXIMUM TOLERABLE LEVELS

The data presented do not establish a very clear-cut, safe upper limit for sulfur for any of the species or for any specific sulfur source. Interpreting the combined data of Lewis (1954), Marcilese *et al.* (1969), and L'Estrange *et al.* (1972), and considering the sheep representative of sulfur metabolism in ruminants, it would appear that 0.4 percent is the maximum tolerable level for dietary sulfur as sodium sulfate.

Data for monogastrics are not definitive either. Arnold and Goble (1950) have fed dogs what would be comparable to 0.28 percent supplemental sulfur in the form of ammonium persulfate for 16 weeks without apparent effect. In the case of the rat, Smith (1973) stated that the optimal total sulfur level, inorganic and organic, for rat diets is about 0.69 percent. The toxic effects of dietary inorganic sulfate are believed due to its conversion to hydrogen sulfide by the gastrointestinal flora in both classes of animals represented, but monogastrics may be less tolerant of this very toxic form of sulfur than ruminants.

## SUMMARY

Sulfur is an abundant and industrially important element essential for plant and animal growth. Its biological importance is primarily due to the fact that it constitutes about 21.5 percent of the essential amino acid, methionine. Plants and the assimilatory gastrointestinal flora of ruminants can form methionine from inorganic sulfur, but monogastrics must rely upon exogenous forms of organic sulfur such as methionine. A variety of sulfur sources from all segments of industry and the environment can subject animals to unacceptable levels of sulfur, but fortunately the toxic dietary levels of inorganic sulfur are at least 3,000 to 5,000 ppm. Manifestations of sulfur toxicosis include anorexia, weight loss, constipation, diarrhea, and depression. In fatal cases of sulfur toxicosis, pulmonary emphysema, cardiac petechiation, congestion of the central nervous system, acute catarrhal enteritis, and hepatic necrosis have been revealed at necropsy. Exposure to forms of sulfur via inhalation causes pulmonary congestion, emphysema, pulmonary edema, and degeneration of the respiratory epithelium. Parenterally administered sulfur ions have been reported to be about 200 times more toxic than chloride ions. The toxicity of ingested sulfur is related to its conversion by gastrointestinal microflora to hydrogen sulfide, which rivals hydrogen cyanide in toxicity. This conversion takes place in both ruminant and monogastric species.

TABLE 35 Effects of Sulfur Administration in Animals

| Class and Number of Animals[a] | Age or Weight | Administration: Quantity of Element[b] | Source | Duration | Route | Effect(s) | Reference |
|---|---|---|---|---|---|---|---|
| Cattle—4 | Adult | 9–10 g/d | Sulfur dioxide | 110 d | Diet | No effect on milk production or butter | Weigand *et al.*, 1972 |
| Cattle—9 | 275 kg | 1,260 ppm | $Na_2SO_4$ | 30 d | Water | | Weeth and Hunter, 1971 |
| Cattle—1 | Adult | 26.5 g/d | Sodium metabisulfite | 180 d | Rumen fistula | No apparent problem with gestation or calving | Luedke *et al.*, 1959 |
| | | 239 g in 6 d | | 6 d | | Complete anorexia; return to normal in 6 d | |
| | | 673 g in 18 d | | 18 d | | Recumbency | |
| Cattle—1 | Adult | 15–51 g/d | $Na_2SO_3$ | | Rumen fistula | Depressed milk fat; impaired acetate formation in rumen | Alhassan and Satter, 1968 |
| Cattle—25 | Adult | 389 g | Liquid sulfur | Single dose | Drench | Muscle twitching; pain and restlessness; diarrhea; recumbency; coma; and death | Coghlin, 1944 |
| Cattle—1 | 275 kg | 530 g | Sodium bisulfite | Single dose | Drench | Death in 16 d | Luedke *et al*, 1959 |
| | 275 kg | 614 g | | | | Death in 16 d | |
| | 275 kg | 599 g | | | | Death in 21 d | |
| | 385 kg | 1,860 g | | | | Death in 37 d; preceded by anorexia, weight | |

| | | | | | | | |
|---|---|---|---|---|---|---|---|
| | | | | | | loss, constipation, diarrhea, and depression | |
| Sheep—5 | | 2,200 ppm | Sodium sulfate | 23 d | Diet | No effect; fecal S related to DM intake | Upton *et al.*, 1970 |
| Sheep—20 | 32 kg | 0.72 g/d | Sodium sulfate | Indefinite | Diet | No effect on serum copper | Marcilese *et al.*, 1969 |
| Sheep—24 | Adult ewes | 5 g/d | Inorganic sulfate | During pregnancy | Drench | No adverse effect | Egan and O'Cuill, 1968 |
| Sheep—5 | 40–60 kg | 4 g/d | $Na_2SO_4 \cdot 10\ H_2O$ | Indefinite | Rumen fistula | No toxic effect | Lewis, 1954 |
| Sheep | | 8 g/d | | | | No toxic effect | |
| | | 15 g/d | | | | Rumen sulfide content reached 14.7 $\mu$mol/ml | |
| Sheep—5 | 40 kg | 1% of DM | Sodium sulfate | 12 d | Diet | Feed intake decreased 22% | L'Estrange *et al.*, 1969 |
| | | 1% of DM | Sodium bisulfate | | | Feed intake decreased 22% | |
| | | 1% of DM | Ammonium bisulfate | | | Feed intake decreased 44% | |
| | | 1% of DM | Sulfuric acid | | | Feed intake decreased 44%; decreased rumen pH | |
| Sheep—2 | 40 kg | 0.56% of DM | Sulfuric acid | 20 d | Diet | Elevation of serum sulfate; decreased urine pH; decreased feed intake 30% | L'Estrange and Murphy, 1972 |
| Sheep—1 | Wethers | 0.5% of DM | Sodium sulfate | 14 d | Diet | Decreased feed intake | L'Estrange *et al.*, 1972 |

## TABLE 35 *Continued*

| Class and Number of Animals[a] | Age or Weight | Administration: Quantity of Element[b] | Source | Duration | Route | Effect(s) | Reference |
|---|---|---|---|---|---|---|---|
| | | 1.0% of DM | | | | Increased rumen sulfate | |
| | | 1.5% of DM | | | | Increased organic matter digestibility | |
| Sheep | | 0.5% of DM | Sodium bisulfite | 90 d | Diet in sugar cane pulp | None; possible increased incidence of renal cysts | Kaemmerer *et al.*, 1972 |
| Sheep—2 | Adult | 2.93 g/d | Sodium sulfate | 8 d | Ruminal infusion | Decreased dry matter intake and rumen motility | Bird, 1972 |
| | | 6.0 g/d | | 3–9 d | | Partial to complete anorexia | |
| Sheep—6 | 40–50 kg | 1.3 g/d | Sodium | 15 mo | Water | No adverse effects | Peirce, 1960 |
| | | 6.5 g/d | | | | No adverse effects | |
| Sheep—480 | Mature | 62 g | Sublimated sulfur | Single dose | Drench | Colicky; depression; recumbent; dyspneic; fever; diarrhea; death | White, 1964 |
| Swine—6 | 12–15 kg | 0.9 ppm | Hydrogen sulfide | | Inhaled in chamber | No adverse effects | O'Donoghue, 1961 |
| | | 28 ppm | | | | Loss of sense of smell | |
| | | 47 ppm | | | | Marked discomfort | |
| | | 188 ppm | | | | Severely toxic | |
| | | 470 ppm | | | | Rapid death | |

| | | | | | | | |
|---|---|---|---|---|---|---|---|
| Swine—4 | 2 wk | 5–10 ppm | Sulfur dioxide | 7½ h | Inhaled | Temporary eye irritation; nasal discharge; shallow breathing | O'Donoghue and Graesser, 1962 |
| Swine—4 | | 15–20 ppm | | | | Fibrous pulmonary induration observed after 70 d | |
| Swine—20 | Adult sow | 664 ppm | $Na_2SO_4$ | 110 d | Water | No effect on reproduction | Paterson *et al.*, 1979 |
| Swine—28 | Weanling | 600 ppm | | 28 d | | Loose feces | |
| Chicken—24 | Growing | 2,690 ppm | L-methionine | 8 d | Diet | 40% reduction in growth rate; decreased feed efficiency | Katz and Baker, 1975 |
| Chicken—21 | Growing | 2,690 ppm | L-methionine | 8 d | Diet | Marked growth depression; hemolytic anemia; splenic hemosiderosis | Harter and Baker, 1978 |
| Chicken—15 | Growing | 14,000 ppm | $Na_2SO_4$ and $K_2SO_4$ | 28 d | Diet | Marked growth depression | Leach *et al.*, 1960 |
| Chicken—7 | 8–21 d | 37 ppm | Potassium sulfate | 7 wk | Diet | No adverse effect | Sasse and Baker, 1974 |
| | | 185 ppm | | | | Plateauing of growth and feed efficiency | |
| | | 370 ppm | | | | Plateauing of growth and feed efficiency | |
| | | 1,840 ppm | | | | Growth depression | |
| Horse—5 | Mature | 300 g | Flowers of S | Single dose | Oral in gruel | Anorexia, colic, and recumbent in 3 h; conjunctival congestion; weak | Ales, 1907 |

TABLE 35 *Continued*

| Class and Number of Animals[a] | Age or Weight | Administration: Quantity of Element[b] | Source | Duration | Route | Effect(s) | Reference |
|---|---|---|---|---|---|---|---|
| | | | | | | pulse; diarrhea; death; myoglobinuria | |
| Rabbit—3 | N/A[c] | 24 mg/kg | Sodium tetrathionate | Single dose | Intravenous | None | Gilman *et al.*, 1946 |
| Rabbit—12 | | 48 mg/kg | | | | Death in 4–14 h | |
| Rabbit—3 | | 480 mg/kg | | | | Death in 1–1.5 h | |
| Rabbit—4 | N/A[c] | 4.8 mg/kg | $Na_2S_4O_6 \cdot 2H_2O$ | Single dose | Intravenous | No adverse effect | Saunders and Wills, 1954 |
| | | 36 mg/kg | | | | Decreased glomerular filtration rate | |
| | | 48 mg/kg | | | | Severe proximal tubular necrosis | |
| Rabbit—4 | 1.1–1.9 kg | 41 mg/kg | Sodium bisulfite | Single period | Intraperitoneal dialysis | Nonfatal | Wilkins *et al.*, 1968 |
| | | 62 mg/kg | | | | Fatal to 50% of rabbits | |
| | | 92 mg/kg | | | | Fatal to 25% of rabbits | |
| | | 139 mg/kg | | | | Fatal to 75% of rabbits | |
| Rabbit—15 | 2–3 kg | 13.5 mg/kg | Sodium bisulfite | 3x/d, 5 d/wk for 8 wk | Intravenous | No apparent toxic effect | Hoppe and Goble, 1951 |
| | | 20 mg/kg | | Single dose | | $LD_{50}$ | |

| Rabbit—2 | N/A[c] | 38.3 ppm | Sulfuric acid | 7 h | Inhaled mist | Labored breathing; hyperemic lungs; emphysema; pulmonary edema; necrosis of respiratory epithelium | Treon *et al.*, 1950 |
|---|---|---|---|---|---|---|---|
| Rabbit—6 | Mature | 926 ppm | Carbon disulfide | 6 h/d, 5 d/wk for 8 wk | Inhalation | Decreased weight gains; loss of muscle control; neurologic damage; increased copper levels of spinal cord, thyroid, and pancreas | Scheel, 1967 |
| Dog—3 | 9–11 kg | 6.6 mg/kg | Potassium thiocyanate | | Diet | No adverse effect | Anderson and Chen, 1940 |
| | | 33 mg/kg | Potassium thiocyanate | | Diet | Recumbency and death | |
| Dog—1 | | 9.1 mg/kg | Sodium thiocyanate | 105 d | Diet | No adverse effect | |
| | | 12.2 mg/kg | | 45 d | Diet | Death | |
| Dog—3 | | 39–41 mg/kg | | 4 d | Diet | Death | |
| Dog—6 | 5–12 kg | 420 ppm | Ammonium persulfate | Up to 16 mo | Diet | No adverse effect | Arnold and Goble, 1950 |
| Dog—4 | | 2,800 ppm | | | | No adverse effect | |
| Dog—4 | 8–15 kg | 54 mg/kg | Sodium bisulfite | Single period | Intraperitoneal dialysis | Nonlethal | Wilkins *et al.*, 1968 |
| | | 71 mg/kg | | | | Fatal to 25% of dogs | |
| | | 92 mg/kg | | | | Fatal to 100% of dogs | |
| | | 121 mg/kg | | | | Fatal to 100% of dogs | |

TABLE 35 *Continued*

| Class and Number of Animals[a] | Age or Weight | Administration | | | | Effect(s) | Reference |
|---|---|---|---|---|---|---|---|
| | | Quantity of Element[b] | Source | Duration | Route | | |
| Dog—7 | | 3.6 cc/min | Hydrogen sulfide | 24 h | Per rectum | Nonlethal | Dougherty *et al.*, 1943 |
| | | 4.1 cc/min | | | | Lethal | |
| Dog—1 | | 59 mg/kg | Sodium tetrathionate | Single dose | Intravenous | No adverse effect | Gilman *et al.*, 1946 |
| Dog—6 | N/A | 236 mg/kg | | | | Vomition; hyperpnea, ataxia; death in 4–9 days; renal necrosis | |
| Dog—1 | | 350 mg/kg | | | | Death in 3 days | |
| Dog—1 | N/A | 0.98 mg/kg | $H_2SO_3$ | Single dose | Diet | Fatal | Hauschild, 1960 |
| | | 9.24 mg/kg | $NaHSO_3$ | Single dose | Intravenous | Fatal | |
| | | 277–308 mg/kg | | Single dose | Subcutaneous | Fatal | |
| Cat—1 | N/A | 38.3 ppm | Sulfuric acid | 7 h | Inhaled mist | Labored breathing; hyperemic lungs; emphysema; pulmonary edema; necrosis of respiratory epithelium | Treon *et al.*, 1950 |
| Rat—5 | Weanling | 4.5 ppm | Sodium sulfate with 0.14% sulfur amino acid | 28 d | Diet | No adverse effect | Brown and Gamatero, 1970 |
| Rat | | 45 ppm | | | | Improved weight gain | |

| | | | | | | | |
|---|---|---|---|---|---|---|---|
| | | 226 ppm | | | | Improved feed efficiency | |
| | | 950 ppm | | | | Improved protein efficiency ratio | |
| Rat—12 | Weanling | 2,150 ppm | L-methionine | 24 d | Diet | No adverse effect | Daniel and Waisman, 1969 |
| | | 6,450 ppm | | | Diet | Growth depression | |
| | | 10,705 ppm | | | Diet | Growth depression—self limiting | |
| Rat—40 | Weanling | 5,160 ppm | DL-methionine | | Diet | Growth depression | Cohen *et al.*, 1958 |
| Rat | Sex mature | 0.075 $mg/m^3$ | Sulfur dioxide | 72 d | Inhaled | No observed changes in estrous cycle | Mamatashvili, 1970 |
| | | 2.0 $mg/m^3$ | | 72 d | | Altered estrous cycle; increased stillbirths | |
| Rat—2 | N/A | 38.3 ppm | Sulfuric acid | 7 h | Inhaled mist | Labored breathing; hyperemic lungs; emphysema; pulmonary edema; necrosis of respiratory epithelium | Treon *et al.*, 1950 |
| Rat—30 | 90–110 g | 35.4 | Sodium bisulfite | Single dose | Intravenous | $LD_{50}$ | Hoppe and Goble, 1951 |
| Rat—4 | 235 g | 92.4 mg/kg | | Single dose | Intra-peritoneal | Nonfatal | Wilkins *et al.*, 1968 |
| | | 139 mg/kg | | | | Fatal to 25% | |
| | | 202 mg/kg | | | | Fatal to 100% | |
| | | 312 mg/kg | | | | Fatal to 100% | |
| Rat | Young | 0.2 μCi/d | $^{35}S$ | 3–6 mo | Gavage | Testicular atrophy; necrosis of germinal epithelium | Organesyan and Gaidova, 1968 |
| Rat | Young | 3.0 μCi/g/d | | Chronic | Gavage | Inhibition of longitudinal bone growth | Levitman, 1968 |
| Mouse—4 | 19–25 g | 92.4 mg/kg | Sodium bisulfite | Single dose | Intra-peritoneal | No adverse effect | Wilkins *et al.*, 1968 |

TABLE 35 *Continued*

| Class and Number of Animals[a] | Age or Weight | Administration: Quantity of Element[b] | Source | Duration | Route | Effect(s) | Reference |
|---|---|---|---|---|---|---|---|
| | | 139 mg/kg | | | | No adverse effect | |
| | | 202 mg/kg | | | | $LD_{50}$ | |
| | | 312 mg/kg | | | | Fatal to 100% | |
| Mouse—60 | 20–24 g | 34.7 mg/kg | Sodium bisulfite | Single dose | Intravenous | $LD_{50}$ respiratory failure | Hoppe and Goble, 1951 |
| Mouse—30 | | 42.7 mg/kg | | | | $LD_{50}$ respiratory failure | |
| | | 326 mg/kg | | | | $LD_{100}$ | |
| Mouse—5 | N/A | 38.3 ppm | Sulfuric acid | 7 h | Inhaled mist | Labored breathing; hyperemic lungs; emphysema; pulmonary edema; necrosis of respiratory epithelium | Treon *et al.*, 1950 |

| | | | | | | | |
|---|---|---|---|---|---|---|---|
| Mouse—30 | 20–25 g | 7.02 mg/kg | Nonahydrate sodium sulfide | 30 days | Intra-peritoneal | $LD_{50}$ | Nofre *et al.*, 1963 |
| | | 25.2 mg/kg | Heptahydrate sodium | Single dose | | | |
| Mouse—30 | 20–25 g | 45.6 mg/kg | Sodium sulfate monohydrate | Single dose | Intra-peritoneal | $LD_{50}$ | Nofre *et al.*, 1963 |
| | | 60.8 mg/kg | Sodium persulfate | | | $LD_{50}$ | |
| Guinea pig—21 | 200–300 g | 2.5 ppm | Sulfur dioxide | 4–15 d | Inhalation | Reduced weight gains | O'Donoghue and Graesser, 1962 |
| Guinea pig—5 | | 5.0 ppm | | 7 d | | Markedly reduced gains | |
| | | 9 ppm | | 4 d | | Weight loss | |
| Guinea pig—3 | N/A[c] | 7.2 ppm | Sulfuric acid | 165 min | Inhaled mist | Death in 165 min | Treon *et al.*, 1950 |
| Hamster—30 | 80–100 g | 29.3 mg/kg | Sodium bisulfite | Single dose | Intravenous | $LD_{50}$ | Hoppe and Goble, 1951 |

[a] Number of animals per treatment.

[b] Quantity expressed in parts per million or as milligrams per kilogram of body weight.

[c] N/A = information not available.

## REFERENCES


Ales. 1907. Case of poisoning by sulphur in the horse. Vet. J. 63:254.

Alhassan, W. S., and L. D. Satter. 1968. Observations on sodium sulfite administration to ruminant. J. Dairy Sci. 51:981.

Anderson, R. C., and K. K. Chen. 1940. Absorption and toxicity of sodium and potassium thiocyanates. J. Am. Pharmacol. Assoc. 29:152.

Arnold, A., and F. C. Goble. 1950. Studies with dogs fed flour treated with ammonium persulfate. Cereal Chem. 27:375.

Baker, D. H. 1977. Sulfur in non-ruminant nutrition. National Feed Ingredients Association, West Des Moines, Iowa.

Bird, P. R. 1972. Sulfur metabolism and excretion studies in ruminants. X. Sulphide toxicity in sheep. Aust. J. Biol. Sci. 25:1087.

Bray, A. C. 1965. Studies on the sulfur metabolism of sheep. Ph.D. thesis. University of Western Australia, Nedlands.

Brieger, H., and J. Teisinger (eds.). 1966. Toxicology of Carbon Disulfide. Excerpta Medica Foundation, Amsterdam.

Brown, R. G., and A. Gamatero. 1970. Effect of added sulfate on the utilization of peanut protein by the rat. Can. J. Anim. Sci. 50:742.

Caldwell, A. D. S., P. G. T. Bye, and M. H. Biggs. 1967. Side effects of dimethyl sulphoxide. Nature 215:1168.

Coghlin, C. L. 1944. Hydrogen sulfide poisoning in cattle. Can. J. Comp. Med. 8:111.

Cohen, H. P., H. C. Choitz, and C. P. Berg. 1958. Response of rats to diets high in methionine and related compounds. J. Nutr. 64:555.

Daniel, R. G., and H. A. Waisman. 1969. Adaptation of the weanling rat to diets containing excess methionine. J. Nutr. 99:299.

Dougherty, R. W., R. Wong, and B. E. Christensen. 1943. Studies of hydrogen–sulfide poisoning. Am. J. Vet. Res. 4:254.

du Vigneaud, V. 1952. A Trail of Research in Sulfur Chemistry and Metabolism. Cornell University Press, Ithaca, N.Y.

Dziewiatkowski, D. D. 1970. Metabolism of sulfate esters, p. 97. *In* O. H. Muth and J. E. Oldfield (eds.). Symposium: Sulfur in Nutrition. AVI Publishing Co., Westport, Conn.

Egan, D. A., and T. O. O'Cuill. 1968. An attempt to produce swayback in lambs born to ewes dosed with high levels of molybdenum, inorganic sulphate and manganese during pregnancy. Irish Vet. J. 22:28.

Fitzhugh, O. G., L. F. Knudsen, and A. A. Nelson. 1946. The chronic toxicity of sulfites. J. Pharmacol. Exp. Ther. 86:37.

Gilman, A., F. S. Philips, E. S. Koelle, R. P. Allen, and E. St. John. 1946. The metabolic reduction and nephrotoxic action of tetrathionate in relation to a possible interaction with sulfhydryl compounds. Am. J. Physiol. 147:115.

Goodrich, R. D. 1978. Sulphur in Ruminant Nutrition. National Feed Ingredients Association, West Des Moines, Iowa.

Harter, J. M., and D. H. Baker. 1978. Factors affecting methionine toxicity and its alleviation in the chick. J. Nutr. 108:1061.

Hauschild, F. 1960. Pharmakologie und Grundlagen der Toxikologie. 2. Aufl. Leipsig. S 368.

Hoppe, J. O., and F. C. Goble. 1951. The intravenous toxicity of sodium bisulfite. J. Pharmacol. Exp. Ther. 101:101.

Kaemmerer, K. Von. 1972. Zur toxikologischen Bedeutung von Sulfit bein Wiederkauer. Zucker 25:123.

Kaemmerer, K. Von, E. Barke, and M. J. Seidler. 1972. Vertraglichkeit von Sulfit in hoher Konzentration auf Trackenschnitzeln bei Schafen. Zucker 25:128.

Katz, R. S., and D. H. Baker. 1975. Toxicity of various organic sulfur compounds for chicks fed crystalline amino acid diets containing threonine and glycine at their minimal dietary requirements for maximal growth. J. Anim. Sci. 41:1355.

Klavins, J. V., T. D. Kinney, and N. Kaufman. 1965. Histopathologic changes in methionine excess. Arch. Pathol. 76:661.

Laidlaw, J. C., and L. Young. 1948. Studies on the synthesis of ethereal sulphates *in vivo*. Biochem. J. 42:1.

Larson, T. E. 1959. Mineral content of public ground water supplies in Illinois. Cir. 31. State of Illinois, Water Survey Division, Urbana.

Leach, R. M., T. R. Zeigler, and L. C. Norris. 1960. The effect of dietary sulphate in the growth rate of chicks fed a purified diet. Poult. Sci. 39:1577.

Leclercq, R. 1972. Commercially important sulfur compounds. *In* A. Senning, ed. Sulfur in Organic and Inorganic Chemistry. Marcel Dekker, New York.

L'Estrange, J. L., and F. Murphy. 1972. Effect of dietary mineral acids on voluntary intake, digestion, mineral metabolism and acid-base balance of sheep. Br. J. Nutr. 28:1.

L'Estrange, J. L., J. J. Clarke, and D. M. McAleese. 1969. Studies on high intake of various sulphate salts and sulphuric acid in sheep. 1. Effects on voluntary feed intake, digestibility and acid base balance. Irish J. Agric. Res. 8:133.

L'Estrange, J. L., P. K. Upton, and D. M. McAleese. 1972. Effects of dietary sulfate on voluntary feed intake and metabolism of sheep. I. A comparison between different levels of sodium sulphate and sodium chloride. Irish J. Agric. Res. 11:127.

Levitman, M. K. L. 1968. Morphological changes in the skeleton of rats under the influence of incorporated Sulfur-35. Mater. Toksikol. Radioak. Veshchestv. 6:65.

Lewis, D. 1954. The reduction of sulphate in the rumen of the sheep. Biochem. J. 56:391.

Luedke, A. J., J. W. Bratzler, and H. W. Dunne. 1959. Sodium metabisulfite and sulfur dioxide gas (silage preservative) poisoning in cattle. Am. J. Vet. Res. 20:690.

Mamatashvili, M. I. 1970. Toxic action of carbon monoxide, sulfur dioxide, and their combinations on the fertility of rats. Gig. Sanit. 35:100.

Marcilese, N. A., C. B. Ammerman, R. M. Valsecchi, B. G. Dunavant, and G. K. Davis. 1969. Effect of dietary molybdenum and sulphate upon copper metabolism in sheep. J. Nutr. 99:177.

Masters, M., and R. A. McCance. 1939. The sulfur content of foods. Biochem. J. 33:1304.

Morrow, P. E., H. C. Hodge, W. F. Neuman, E. A. Maynard, H. J. Blanchet, Jr., D. W. Fassett, R. E. Birk, and S. Manrodt. 1952. The gastrointestinal non-absorption of sodium cellulose sulfate labeled with $S^{35}$. J. Pharmacol. Exp. Therap. 105:273.

Mudd, S. H., F. Irreverre, and L. Laster. 1967. Sulfite oxidase deficiency in man: Demonstration of enzymatic defect. Science 156:1599.

Muth, O. H., and J. E. Oldfield (eds.). 1970. Symposium: Sulfur in Nutrition. AVI Publishing Co., Westport, Conn.

Nickless, G. (ed.). 1968. Inorganic Sulphur Chemistry. Elsevier Publishing Co., New York.

Nofre, C., H. Dufour, and H. Cier. 1963. Toxicite generale comparee des anions mineraux chez la souris. C. R. Acad. Sci. 257:791.

O'Donoghue, J. G. 1961. Hydrogen sulfide poisoning in swine. Can. J. Comp. Med. Vet. Sci. 25:217.

O'Donoghue, J. G., and F. E. Graesser. 1962. Effects of sulphur dioxide on guinea pigs and swine. Can. J. Comp. Med. Vet. Sci. 26:255.

Organesyan, N. M., and E. S. Gaidova. 1968. Morphological changes in rat organs effected by the chronic administration of sulphur[35]. Mater. Toksikol. Radioak. Veshchestv. 6:83.

Paterson, D. W., R. C. Wahlstrom, G. W. Libal, and O. E. Olson. 1979. Effects of sulfate in water on swine reproduction and young pig performance. J. Anim. Sci. 49:664.

Peirce, A. W. 1960. Studies of salt tolerance of sheep. III. The tolerance of sheep for mixtures of sodium chloride and sodium sulphate in the drinking water. Aust. J. Agric. Res. 11:548.

Postgate, J. R. 1968. The sulphur cycle, p. 259. *In* G. Nickless (ed.). Inorganic Sulphur Chemistry. Elsevier Publishing Co., New York.

Roy, A. B., and P. A. Trudinger. 1970. The Biochemistry of Inorganic Compounds of Sulphur. Cambridge University Press, England.

Sasse, C. E., and D. H. Baker. 1974. Factors affecting sulfate sulfur utilization by the young chick. Poult. Sci. 53:652.

Saunders, J. P., and J. H. Wills. 1954. The nephrotoxic action of sodium tetrathionate. J. Pharmacol. Exp. Ther. 112:197.

Scheel, L. D. 1967. Experimental carbon disulfide poisoning in rabbits: Its mechanisms and similarities with human cases. *In* H. Brieger and J. Teisinger (eds.). Toxicology of Carbon Disulfide. Excerpta Medica Foundation, Amsterdam.

Schelling, G. T., and E. E. Hatfield. 1968. Effect of abomasally infused nitrogen sources on nitrogen retention of growing lambs. J. Nutr. 96:319.

Senning, A. (ed.). 1972. Sulfur in Organic and Inorganic Chemistry. Marcel Dekker, New York.

Smith, J. T. 1973. An optimal level of inorganic sulfate for the diet of a rat. J. Nutr. 103:1008.

Snetsinger, D. C., and H. M. Scott. 1961. The relative toxicity of intraperitoneally injected amino acids and the effect of glycine and arginine thereon. Poult. Sci. 40:1681.

Sorbo, B. 1972. The pharmacology and toxicology of inorganic sulfur compounds, p. 143. *In* A. Senning (ed.). Sulfur in Organic and Inorganic Chemistry. Marcel Dekker, New York.

Treon, J. F., F. R. Dutra, J. Cappel, H. Sigmon, and W. Younker. 1950. Toxicity of sulfuric acid mist. Arch. Ind. Hyg. Occup. Med. 2:716.

Upton, P. K., J. L. L'Estrange, and D. M. McAleese. 1970. Studies on high intake of various sulphate salts and sulphuric acid in sheep. 2. Effects on the absorption, excretion and retention of sulphur. Irish J. Agric. Res. 9:151.

Walker, D. M., and L. J. Cook. 1967. Nitrogen balance studies with the milk-fed lamb. 4. Effect of different nitrogen and sulphur intakes on live weight gain and wool growth and on nitrogen and sulphur balances. Br. J. Nutr. 21:237.

Weeth, H. J., and J. E. Hunter. 1971. Drinking of sulfate-water by cattle. J. Anim. Sci. 32:277.

Weigand, E., M. Kirchgessner, W. Granzer, and G. Ranfft. 1972. Zur Futterung hoher Sulfitmengen an Milchkuhe; $SO_2$-Vertraglichkeit und $SO_2$-gehalt der Milch. Zentralbl. Veterinaermed. Reihe A 19:490.

White, J. B. 1964. Sulphur poisoning in ewes. Vet. Rec. 76:278.

Wilkins, J. W., J. A. Greene, and J. M. Weller. 1968. Toxicity of intraperitoneal bisulfite. Clin. Pharmacol. Ther. 9:328.

Wolf, G., and P. T. Varandani. 1960. Studies on the function of vitamin A in mucopolysaccharide biosynthesis. Biochim. Biophys. Acta 43:501.

Young, L., and G. A. Maw. 1958. The Metabolism of Sulphur Compounds. John Wiley & Sons, New York.

# Tin

Tin (Sn) is a soft, white, lustrous, crystalline, malleable metal that has been of great economic importance since the Bronze Age, when early metallurgists found tin–copper alloys very useful in the fabrication of weapons and utensils. Tin is not ubiquitous. The world's largest deposits of tinstone ($SnO_2$), high in the Bolivian Alps, together with the tin from the Federation of Malaya, furnish about 60 percent of the world's annual tin needs of 170,000 tons or 60 g per capita (Mantell, 1949). More than 50 percent of this tin by weight is used in the manufacture of tin plate. The major use of this tin plate, of course, has been in the manufacture of the "tin can." Significant amounts of tin are also used in tinning copper and steel wire and in the manufacture of various alloys including solder, bronze, and babbitt. More recently, tin has been found useful in plasticizers and stabilizers in plastics, as fungicides (especially wood preservatives), disinfectants, and miticides, in the manufacture of cast iron and paints, and in radiopharmaceuticals for nuclear medicine (LaSpada, 1968).

Concern for the biological aspects of tin has historically pertained to its potential of contaminating the contents of the tin cans. Lacquering or resin coating the tin plate has reduced this potential. Substitution materials such as aluminum, tin-free steel, and plastic and cardboard containers for canned, frozen, and dehydrated foods have also lessened the exposure of food to tin. It will be recognized, however, that due to some of the newer uses for tin, concern will continue for tin contamination at various points in the food chain.

## ESSENTIALITY

Schwarz *et al.* (1970) demonstrated tin to be essential for growth in rats. These experiments, conducted in plastic isolators to prevent environmental contamination of the rats with tin, indicated that 1 ppm tin added to meticulously prepared, tin-free, purified control diets enhanced the growth rate of the rats 53 percent. Both inorganic and organic forms of tin were effective, with stannic sulfate giving the best results. Additional studies on the essentiality of tin need to be conducted with other species and in other laboratories.

## METABOLISM

Little is known about the metabolism of tin. The efforts of Perry and Perry (1959), Kehoe *et al.* (1940), and Tipton and Cook (1963) indicate inorganic tin is poorly absorbed, at least in man. Benoy *et al.* (1971) reported tin, obtained from the contamination of canned foods, is only 4 percent absorbed by the oral route and only 2.7 percent absorbed 18 hours after subcutaneous administration. At the cellular level, as a potent inducer of heme oxygenase (Kappas and Maines, 1976), tin enhances heme breakdown in the kidney and impairs heme-dependent cellular functions. Cremer (1962) stated that tetraethyl tin compounds are rapidly converted to triethyl tin compounds in the liver. The triethyl tin compounds are uncouplers of oxidative phosphorylation. Organic forms of tin may also form condensation complexes with several ligands and may therefore contribute to protein structure (Schwarz *et al.,* 1970). Some organic forms of tin may cross the blood–brain barrier.

## SOURCES

There is agreement that the contents of "tin cans," especially acidic foods, may indeed become contaminated with tin from the tin plating, especially when not resin-coated. Levels of 1,370 ppm tin have been recorded in canned fruit juices (Benoy *et al.,* 1971). Canned pet foods are subject to similar contamination potential, but these foods seldom are as strongly acidic as, for instance, tomato juice. deGoeji and Kroon (1973) demonstrated that resin-coating tin plate reduced the amount of tin contamination by a factor of 50, providing the resin remains intact.

The newer uses for tin mentioned above involve both inorganic and organic tin. Inorganic forms, such as stannous chloride, stannic oxide, stannous sulfate, stannous tartrate, and stannous acetate may be present in industrial paints in amounts up to 3 percent. Organic forms, especially the alkyl tin compounds, for example triethyl tin chloride, tributyl tin oxide, dibutyl tin oxide, trioctyl tin dilaurate, di-*n*-octyl tin acetate, tri-*n*-butyl tin, triethyl tin sulfate, and dimethyl tin dichloride, may be present in plastics in concentrations of about 1 percent.

Little information is available on the natural tin content of livestock feedstuffs. Pasture herbage growing in Scotland has been reported to contain 0.3–0.4 ppm tin on a dry basis (Mitchell, 1948). That investigator also reported lichen growing on selicic rocks may contain in excess of 72 ppm tin. Schroeder *et al.* (1964) analyzed a wide variety of substances for their tin contents. Some representative values (parts per million wet tissue) include lean ground beef, 2.76; gelatin, 3.5; beaver meat, 7.28; milk in tinned bulk containers, 0.68; dog chow, 1.0; commercial rat diet, 0.8; and agricultural superphosphate, 3.34.

## TOXICOSIS

### LOW LEVELS

Tin toxicity studies with large animals have been rare. Organotin compounds such as triethyl tin chloride and trimethyl tin chloride, which have shown potential as insecticides for the sheep blowfly, have been found well tolerated by sheep at levels 2 to 3 times the expected volume usage and 2 or more times the necessary concentrations (Hall and Ludwig, 1972).

In chronic studies with orally administered triethyl tin hydroxide (Stoner *et al.,* 1955), domestic fowl were found more tolerant of the drug than other species. Hens tolerated 160 ppm for 15 weeks, whereas, 20 ppm was the approximate tolerance level for rats and rabbits.

Barnes and Stoner (1958) have also studied the toxicity of a wide variety of alkyl tin compounds administered by different routes to rabbits, rats, and mice. The effects of the dialkyl compounds tended to be generalized and involve the biliary tract, while the trialkyl compounds tended to cause edema of the central nervous system. The trialkyl salts were less toxic than the dialkyls, and, of the latter group, butyl was the most toxic. The dioctyl tin salts, which are suitable and effective substitutes for industrial uses of dibutyl salts, were found completely nontoxic per os or per cutaneous.

Animal studies with tin-contaminated canned fruit juices and solid foods have been conducted (Benoy *et al.,* 1971). Consumption of tin from the canned fruit juice source in a single feeding equivalent to approximately 2, 3, 7, and 14 mg tin per kilogram of body weight had no effect on rats, pigeons, cats, and dogs, respectively. An effect of the tin-contaminated food was induced only in cats consuming in excess of 7 mg tin per kilogram of body weight from fruit juice containing 1,370 ppm tin. This caused gastrointestinal disturbances with vomition in 20 to 40 percent of the cats. It was suggested that this was due to gastrointestinal irritation rather than central nervous system toxicosis. Cats acquired a tolerance to dietary tin after continued exposure.

The effects of 0.03, 0.1, 0.3, and 1.0 percent dietary inorganic tin from several sources ranging from 44 to 63 percent tin have been studied in normally fed rats for 4- and 13-week periods (deGroot *et al.*, 1973). Stannic oxide and stannous oxide, sulphide, and oleate had no effect at any level or duration. Stannous chloride, orthophosphate, sulfate, oxalate, and tartrate at 0.3 percent of the diet (equal to more than 1,320 ppm tin) caused growth retardation, decreased feed efficiency, and mild anemia within 4 weeks. Stannous chloride at 1 percent (6,300 ppm tin) for 13 weeks caused pancreatic atrophy, testicular degeneration, renal calcification, and status spongiosis of the brain. In other studies (deGroot, 1973), 50 ppm dietary tin as stannous chloride had no effect for up to 13 weeks in weanling rats; 150 ppm also had no effect provided the dietary copper level was greater than 6 ppm. Levels of tin from 500 to 5,300 ppm caused severe growth depression and anemia. The severity of these effects was diminished by 200 ppm supplemental iron.

The toxicity of oral sodium pentafluorostannite ($NaSn_2F_5$), an active prenatal anticariogenic agent containing 67 percent tin, has been investigated by Conine *et al.* (1976). In rats fed 20, 100, and 175 mg $NaSn_2F_5$ per kilogram of body weight equal to 13.4, 67, and 117 mg tin per kilogram for 30-day periods, there was dose-related growth inhibition and decreased serum glucose. The highest two levels caused proximal renal tubular degeneration and death.

Gaunt *et al.* (1968) have studied the effects of 8–80 ppm di-*n*-butyl tin dichloride, a common stabilizer in polyvinylchloride plastic, in the diets of rats for 90 days. The no-effect level was estimated at 40 ppm or less for a 90-day period. The highest level fed caused a slight reduction in growth rate and a mild anemia.

Another tin compound used as a plasticizer and polyvinylchloride stabilizer, dioctyl tin S,S-*bis* (iso-octylmercapto) acetate, has been fed to rats at the level of 200 ppm for periods up to 3 months. The tin-fed rats experienced significant decreases in body weight and developed increased liver and kidney weights in comparison to controls. This same compound administered by daily gavage to reproducing female rats at rates of 20 to 40 mg per kilogram of body weight caused 17 percent fetal deaths, increased fetal resorption, and diminished birth weights. In 12-month studies, this rate of tin administration caused 20 percent mortality in rats (Nikonorow *et al.,* 1973).

Schroeder *et al.* (1968) provided rats and mice with drinking water containing 5 ppm tin ions for the natural life span of these rodents. This level of tin did not affect the growth rates of either sex of rodent, but it did reduce the life span (longevity) of the female rats. The associated lesions in these rats included severe fatty degeneration of the liver, hepatic necrosis, and vacuolar changes in the renal epithelium. Similar renal changes were also noted in males.

### HIGH LEVELS

In rabbits, stannous chloride, stannous tartrate, or stannous acetate administered orally at the rate of 1 g (equal to 440–630 mg tin) every 6–10 days caused gastritis, posterior paresis, hepatic degeneration, and death in 1 to 2 months (Eckardt, 1909).

The effects of tri-*n*-butyl tin oxide, administered conjunctivally to rabbits, have been evaluated because of its commercial bactericidal, fungicidal, insecticidal, and algicidal uses (Pelikan, 1969). Single doses ranged from 0.46 to 4.6 mg per kilogram of body weight placed onto the left conjunctival sac in a single 0.03 ml dose. The dose-related effects included edema of eye lids, decreased corneal transparency, altered aqueous humor, corneal necrosis in 24 hours, corneal ulcers in 2–5 days, generalized body weakness, and hyperreflexia. The highest doses caused death of the rabbits.

The acute effects of tetra-, tri-, di-, and mono-alkyl tin compounds have been reported for several species (Stoner *et al.,* 1955; Scheinberg *et al.,* 1966; and Robinson, 1969). The triethyl tin compounds were found to be the most toxic. In rats, triethyl tin sulfate was found equally lethal by intravenous, intraperitoneal, or oral routes with an $LD_{50}$ of 5.7 mg per kilogram of body weight. This compound caused secretion of "red tears," and its toxicity was exaggerated by increased ambient temperatures (Stoner *et al.,* 1955). Rabbits were more sensitive to triethyl tin sulfate than rats; however, the reactions were similar including muscular weakness, tremors, and convulsions. These investigators concluded the central nervous system to be the main site of action for the alkyl tin compounds.

Fischer and Zimmerman (1969) have studied the effects of repeated intravenous administrations of insoluble stannic oxide in several species. Rats administered 1–4 injections of 200, 400, 600, or 800 mg tin per kilogram of body weight survived a maximum of 26 months. New Zealand White male and female rabbits administered 1–5 injections of

200 mg tin per kilogram of body weight, via the ear vein, survived 6–26 months. In mongrel dogs intravenously administered, tin stimulated phagocytosis but did not induce fibrosis or neoplasia. Bischoff and Bryson (1976) reported a similar inert character of 4-mg quantities of tin crystals (like asbestos fibers) injected into the thoracic cavity of 3-month-old mice observed for 19 months. The needles initiated a foreign body reaction but no neoplasia or other changes.

Yamaguchi *et al.* (1976) reported that the major effect in rats of 30 mg tin as $SnCl_2$ per kilogram of body weight administered intraperitoneally was a significant inhibition of urinary calcium excretion. In other parenteral experiments, Benoy *et al.* (1971) found that subcutaneously administered tin as tin citrate caused no remarkable changes in rats or mice.

The clinical manifestations and $LD_{50}$ of tin pyrophosphate and polyphosphate compounds in rats have been explored because of the increasing use of $^{99}$technetium metaphosphate compounds as radiopharmaceuticals for bone scanning (Stevenson *et al.,* 1974). Sublethal doses of pyrophosphate (12–20 mg per kilogram of body weight) administered intravenously caused decreased serum ionized and total calcium levels and prolonged QT intervals in electrocardiograms consistent with hypocalcemia. The $LD_{50}$ (5 minutes) for the pyrophosphate and polyphosphate compounds were calculated to be 41 and 29.4 mg/kg of body weight, respectively. In acute studies, di-*n*-butyl tin dichloride at the rate of 50 mg/kg of body weight in single oral doses caused edema and inflammation of the bile duct in rats, and the $LD_{50}$ was 200–400 mg/kg of body weight (Gaunt *et al.,* 1968).

The $LD_{50}$ of $NaSn_2F_5$ administered by several routes to rats and mice has also been determined (Conine *et al.,* 1975). The values were 19, 81, and 573 mg/kg for the IV, IP, and orally administered drug, respectively, in mice and 12.9, 70, and 593 mg/kg, respectively, in weanling rats. Fasting increased the toxicity and the deaths were preceded in both species by ataxia, muscular weakness, and central nervous system depression.

Gaines and Kimbrough (1968) have found the $LD_{50}$ of triphenyl tin (fentin) hydroxide administered by gavage in peanut oil to Sherman strain rats was 36 mg/kg of body weight in females and 240 mg/kg of body weight in males. Signs of toxicosis included sluggishness, unsteady gait, mild diarrhea, anorexia, bloody nose, and death.

The $LD_{50}$ of several organotin compounds of potential value as sheep insecticides have been established for mice immersed for 15 seconds in the solutions of the tin compounds. The $LD_{50}$ for triethyl and trimethyl tin chloride were found to be 35 and 50 mg/kg of body weight, respectively (Hall and Ludwig, 1972).

### FACTORS AFFECTING TOXICITY

The major factors influencing the toxicity of tin relate to its solubility, the acid-base balance of the host, degree of acquired tolerance, and type of diet. The alkyl derivatives are quite soluble, as well as volatile, especially around pH 7. In the case of inorganic forms, as may contaminate canned foods, exposure to air, organic acids, fats, and salt favor the removal of stannous ($Sn^{++}$) ions from tin plate. Based upon the quantities of tin found in urine, Perry and Perry (1959) have suggested that metabolic alkalosis enhances tin absorption and increases urinary tin, while metabolic acidosis tends to reduce the absorption of tin from the gastrointestinal tract. deGroot *et al.* (1973) found that tin in natural-ingredient diets appeared to be less toxic than in semipurified diets. This is probably associated with the sparing action of certain minerals on the toxicity of tin. Copper and iron were found by deGroot (1973) to decrease the toxicity of tin. The species and sex of tin-exposed animals also are important. For example, female rats are several times more susceptible to parenteral triphenyl tin hydroxide than male rats, and guinea pigs are quite refractory to the toxic alkyl for compounds.

## MAXIMUM TOLERABLE LEVELS

The maximum tolerable levels for tin are dependent upon several factors, including source and route of administration. For inorganic $Sn^{++}$ administered daily to rodents over their life span, no effect level appears to be less than 5 ppm (Schroeder *et al.,* 1968). However, provided adequate dietary iron and copper are present, the safe upper levels for oral inorganic tin ($Sn^{++}$) may approximate 150 ppm (deGroot, 1973). For parenterally administered inorganic tin ($Sn^{++}$), the no-effect level (intravenous) is less than 12 mg/kg of body weight in rodents (Stevenson *et al.,* 1974). For one of the most toxic organic tin compounds, triethyl tin dichloride, the no-effect level is less than 20 ppm for periods of up to 4 weeks (Stoner *et al.,* 1955). The safe upper level for intravenously administered triethyl tin sulphate is approximately 5 mg/kg body weight under normal environmental temperatures (Stoner *et al.,* 1955).

## TISSUE LEVELS

Because tin is poorly absorbed, the levels of tin in tin-exposed animals remain remarkably low. Kehoe *et al.* (1940) reported the highest level of tin in long bones (0.8 ppm, wet weight), 14 μg/dl in blood, and a virtual absence of tin in brain tissue of normal humans. Schroeder *et al.* (1968) found the spleens of the rodents on long-term tin toxicity studies to contain the highest levels of tin (1.88 ppm wet weight). In human tissue, Schroeder *et al.* (1964) found the highest levels of tin in the wall of the ileum (range of 53 to 172 ppm in ash), while the range of tin levels in liver, kidney, and lungs approximated 20 to 64 ppm in ash. No tin was found in tissue of the newborn.

## SUMMARY

Despite a variety of commercial uses for tin compounds (tin plate, plasticizers and stabilizers for polyvinylchloride products, fungicides, pesticides, radiomedicine pharmaceuticals) and their frequent direct contact with foods, the potential for tin toxicosis is negligible because the element is poorly absorbed. Of the inorganic tin forms, stannous chloride is among the most toxic, while the triethyl tin compounds appear to be the most toxic organic forms. Inorganic tin induces anorexia with accompanying growth depression, impairs hematopoiesis, and alters calcium metabolism. Pancreatic, hepatic, and renal lesions have also been observed in inorganic tin toxicosis. The organic alkyl tin compounds have a special capacity for inducing inflammation of the biliary tract and edema of the central nervous system, regardless of the route of administration.

TABLE 36 Effects of Tin Administration in Animals

| Class and Number of Animals[a] | Age or Weight | Administration: Quantity of Element[b] | Source | Duration | Route | Effect(s) | Reference |
|---|---|---|---|---|---|---|---|
| Chicken—3 | Adult | 85 ppm | Triethyl tin hydroxide | 15 wk | Diet | Anorexia | Stoner *et al.*, 1955 |
| Chicken | Adult | 1.2 mg/kg | Triethyl tin sulfate | Single dose | Intravenous | Salivation; convulsions; death in min | |
| Pigeon—4 | | 3 mg/kg | Sn-contaminated fruit juice | Single dose | Gavage | No adverse effects | Benoy *et al.*, 1971 |
| Rabbit | | 11 ppm | Triethyl tin hydroxide | 4 wk | Diet | Tolerated | Stoner *et al.*, 1955 |
| | | 21–42 ppm | | | | Progressive weight loss; muscular weakness; paresis and death | |
| Rabbit—6 | 800–900 g | 0.17 mg/kg | Tri-*n*-butyl tin oxide | Single dose | Conjunctival | Weakness; hyperreflexia; edema of eyelids | Pelikan, 1969 |
| | 800–900 g | 0.23 mg/kg | | | | Corneal ulcers; iris edema; necrotic conjunctivitis | |
| | 800–900 g | 1.7 mg/kg | | | | 16% fatal in 11–12 d | |
| | 800–900 g | 2.3 mg/kg | | | | 16% fatal in 11–12 d | |
| Rabbit | | 2–4 mg/kg | Dibutyl tin chloride | Single dose | Intravenous | Pulmonary congestion; death in 24 to 48 h | Barnes and Stoner, 1958 |
| | | 11–22 mg/kg | | Repeated doses | Gavage | Death in 10 d | |
| Rabbit | | 0.4 mg/kg | Triethyl tin sulfate | Single dose | Intravenous | Running movements and vasodilation | Stoner *et al.*, 1955 |

TABLE 36 *Continued*

| Class and Number of Animals[a] | Age or Weight | Administration: Quantity of Element[b] | Source | Duration | Route | Effect(s) | Reference |
|---|---|---|---|---|---|---|---|
| | | 1.2 mg/kg | | | | Convulsions and death within 24 h | |
| | | 3.9 mg/kg | | | Gavage | Convulsions and death 24 h | |
| Rabbit—26 | 2.5–3.5 kg | 0.6 mg/kg/d | Triethyl tin | 6 d | Intraperitoneal | Edema of optic nerve and optic disc; vacuolation of myelin sheaths | Scheinberg *et al.*, 1966 |
| Rabbit | 0.9–2.8 kg | 200 mg | $Sn^{++++}$ oxide | 1 to 5 doses over several wk | Intravenous | Survived 6–26 mo | Fischer and Zimmerman, 1969 |
| Rabbit | | 440 to 630 mg/d | $Sn^{++}$ chloride<br>$Sn^{++}$ tartrate<br>$Sn^{++}$ acetate | Once every 6–10 d | Gavage | Gastritis; posterior paresis; hepatic degeneration; death in 1–2 mo | Eckardt, 1909 |
| Dog—3 | | 200 mg | $Sn^{++++}$ oxide | 1 to 5 doses over several wk | Intravenous | Survived 4–5 yr | Fischer and Zimmerman, 1969 |
| Dog—4 | Adult | 14 mg/kg | Sn-contaminated fruit juice | Single dose | Gavage | No adverse effects | Benoy *et al.*, 1971 |
| Cat—11 | | <7 mg/kg | | | | No adverse effects | |
| | | >7 mg/kg | | | | Gastrointestinal irritation; vomition | |
| Rat—100 | Weanling | 5 ppm | Stannous ions | Natural life span | Water | Reduced life span of female rats | Schroeder *et al.*, 1968 |

| | | | | | | | |
|---|---|---|---|---|---|---|---|
| Rat—18 | | 2.9 mg/kg | Sn-contaminated fruit juice | Single dose | Gavage | No adverse effects | Benoy *et al.*, 1971 |
| Rat—5 | | 11 ppm | Triethyl tin hydroxide | 4 wk | Diet | Anorexia; muscular weakness; death or developed resistance | Stoner *et al.*, 1955 |
| Rat—6 | | 8 ppm | Dibutyl tin dichloride | 6 mo | Diet | No adverse effects | Barnes and Stoner, 1958 |
| | | 19 ppm | | | | Growth depression; anorexia | |
| | | 39 ppm | | | | Death | |
| Rat—12 | | 9 ppm | Tributyl tin acetate | 3 mo | | No effect on growth | |
| | | 18 ppm | | | | Reduced growth | |
| | | 36 ppm | | | | Bile duct injury and brain edema | |
| Rat—32 | Weanling | 16 ppm | Di-*n*-butyl tin dichloride | 90 d | Diet | No adverse effects | Gaunt *et al.*, 1968 |
| | | 31 ppm | | | | Slight decrease in growth; mild anemia | |
| Rat—10 | | 20 mg/kg | | Single dose | Intraperitoneal | Edema and inflammation of bile duct | |
| | | 80–160 mg/kg | | Single dose | Gavage | $LD_{50}$ | |
| Rat | 4–5 mo | 13–25 ppm | Triphenyl tin hydroxide | 64 d | Diet | Reduced feed intake | Gaines and Kimbrough, 1968 |
| | | 25–50 ppm | | 64 d | | Reduced fertility | |
| | | 100 ppm | | 99 d | | Testicular atrophy | |
| Rat | | 4.4 mg/kg/d | Dibutyl tin | 12 d | Topical | Hyperemia of skin; SQ | Barnes and |

TABLE 36 *Continued*

| Class and Number of Animals[a] | Age or Weight | Administration: Quantity of Element[b] | Source | Duration | Route | Effect(s) | Reference |
|---|---|---|---|---|---|---|---|
| | | | chloride | | | edema; bile duct inflammation | Stoner, 1958 |
| | | 3.3 mg/kg | Dibutyl tin chloride | Single dose | Intraperitoneal | Local intense irritation; death in a few hours | |
| | | 1.1–2.2 mg/kg | Dibutyl tin chloride | | Intravenous | Bile duct injury | |
| | | 4.4 mg/kg | Dibutyl tin chloride | | | Death in 18 hours | |
| Rat—4 | | 5.1 mg/kg | Trimethyl tin acetate | | Gavage | $LD_{50}$ | |
| | | 1.9 mg/kg | Triethyl tin acetate | | | $LD_{50}$ | |
| | | 57 mg/kg | Tri-*n*-propyl tin | | | $LD_{50}$ | |
| | | 156 mg/kg | Tri-*n*-butyl tin | | | $LD_{50}$ | |
| | | 320 mg/kg | Tri-*n*-hexyl tin | | | $LD_{50}$ | |
| Rat—22 | | 4.4 mg/kg | Dibutyl tin chloride | | | No adverse effects | |
| Rat | | 8.8 mg/kg | Dibutyl tin chloride | | | Bile duct inflammation | |
| | | 22 mg/kg | Dibutyl tin chloride | | | Illness for 24 to 48 h; dilatation of stomach; inflammation of bile duct | |

| | | | | | | | |
|---|---|---|---|---|---|---|---|
| | | 22 mg/kg | Dibutyl tin chloride | Repeated dose | | Death within 7 d | |
| Rat—5 | | 2 mg/kg | Triethyl tin sulphate | Single dose | Intravenous | Tolerated; aggravated by increased ambient temperature | Stoner *et al.*, 1955 |
| | | 4 mg/kg | | | | Secretion of "red tears"; death in 4 d | |
| | | 2.2 mg/kg | | | Intraperitoneal | $LD_{50}$ within 7 d | |
| Rat—18 | 140–200 g fasted | 2.5 mg/kg | Triethyl tin chloride | Single dose | Intraperitoneal | $LD_{50}$ | Robinson, 1969 |
| | | 2.7 mg/kg | Tributyl tin oxide | | | $LD_{50}$ | |
| | | 19.1 mg/kg | Dibutyl tin oxide | | | $LD_{50}$ | |
| | | 80 mg/kg (cmpd)[c] | Trioctyl tin dilaurate | | | $LD_{50}$ | |
| | | 240 mg/kg | Dioctyl tin acetate | | | $LD_{50}$ | |
| Rat—40, female | 196–312 g | 9 mg/kg | Triphenyl tin hydroxide | Single dose | Gavage | $LD_{50}$ in 1–16 d | Gaines and Kimbrough, 1968 |
| Rat—70, male | 286–522 g | 60 mg/kg | | | | $LD_{50}$ in 1–16 d | |
| Rat | 150 g | 200–800 mg/kg | $Sn^{++++}$ oxide | 1–4 doses over several wk | Intravenous | Survived for 4–26 mo; increased phagocytosis | Fischer and Zimmerman, 1969 |
| Rat | | 30 mg/kg | $Sn^{++}$ chloride | Single dose | Intraperitoneal | Decreased urine calcium | Yamaguchi *et al.*, 1976 |
| Rat—10 | 91–152 g fasted | 146–149 mg/kg | $NaSn_2F_5$ | Single dose | Gavage | $LD_{50}$ in 14 d preceded by muscular weakness, ataxia, and CNS | Conine *et al.*, 1975 |

TABLE 36 *Continued*

| Class and Number of Animals[a] | Age or Weight | Administration: Quantity of Element[b] | Source | Duration | Route | Effect(s) | Reference |
|---|---|---|---|---|---|---|---|
| | | | | | | depression | |
| | 91–152 g fed | 384 mg/kg | | Single dose | | $LD_{50}$ within 14 d | |
| | 91–152 g | 43–50 mg/kg | | Single dose | IP | $LD_{50}$ with renal tubular necrosis | |
| | | 8.7 mg/kg | | Single dose | IV | $LD_{50}$ within 14 d | |
| Rat—12 | 94 g | 13 mg/kg | | Daily for 30 d | Gavage | Growth inhibition | Conine *et al.*, 1976 |
| Rat | | 67 mg/kg | | | | Growth depression; renal tubular degeneration; decreased serum glucose | |
| | | 117 mg/kg | | | | Growth inhibition; renal disease; death | |
| Rat—6 | | 19.3 mg/kg | Sn citrate | Single dose | Subcutaneous | No adverse effects | Benoy *et al.*, 1971 |
| Rat—44 | 170–246 g | 5.2 mg/kg | Sn pyrophosphate | Single dose | Intravenous | Electrocardiographic QT interval changes | Stevenson *et al.*, 1974 |
| | | 8.6 mg/kg | Sn pyrophosphate | | | Decreased serum ionized and total calcium | |
| | | 41 mg/kg | Sn pyrophosphate | | | Acute (5 min) $LD_{50}$ | |
| Rat—57 | 3 mo | 29.4 mg/kg (cmpd)[c] | Sn polyphosphate | | | Acute (5 min) $LD_{50}$ | |

| | | | | | | | |
|---|---|---|---|---|---|---|---|
| Rat—10 | Weanling | 50 ppm | $Sn^{++}$ chloride | 4–13 wk | Diet | No adverse effects | deGroot, 1973 |
| | | 150 ppm | $Sn^{++}$ chloride | | | No effect if dietary Cu 6 ppm | |
| | | 500–6,300 ppm | $Sn^{++}$ chloride | | | Decreased growth; anemia; severity diminished by 200 ppm dietary iron | |
| Rat—10 | 90–120 g | 200 ppm (cmpd)[c] | Di-*n*-octyl tin S–S *bis*-mercapto acetate | 90 d | Diet | Decreased body weight; increased kidney weight and liver weight | Nikonorow *et al.*, 1973 |
| | | 40 mg/kg (cmpd)[c] | | | | Fetal deaths; increased reabsorptions; decreased fetal weights | |
| Rat—20 | Weanling | 8,800 ppm | $Sn^{++}$ chloride | 4 wk | Diet | Decreased growth | deGroot *et al.*, 1973 |
| Rat | | | $Sn^{++}$ ortho$PO_4$ | | | Decreased feed efficiency | |
| | | | $Sn^{++}$ sulphate | | | Mild anemia | |
| | | | $Sn^{++}$ oxalate | | | Anemia | |
| Rat—20 | Weanling | 8,800 ppm | $Sn^{++}$ tartrate | 4 wk | Diet | Anemia | deGroot *et al.*, 1973 |
| | | 8,800 ppm | $Sn^{+4}$ oxide | | | No adverse effects | |
| Rat | | | $Sn^{++}$ sulphide | | | No adverse effects | |
| | | | $Sn^{++}$ oleate | | | No adverse effects | |
| Rat—20 | | | $Sn^{+4}$ oxide | 13 wk | | No adverse effects | |
| | | 6,300 ppm | $Sn^{++}$ chloride | | | Anorexia; abdominal distension; growth retardation; anemia; pancreatic atrophy; testicular degeneration; status spongiosis of brain; renal calcification in males | |

TABLE 36 *Continued*

| Class and Number of Animals[a] | Age or Weight | Administration: Quantity of Element[b] | Source | Duration | Route | Effect(s) | Reference |
|---|---|---|---|---|---|---|---|
| Mouse—100 | Weanling | 5 ppm | $Sn^{++}$ ions | Natural life span | Water | Fatty degeneration and necrosis of the liver | Schroeder *et al.*, 1968 |
| Mouse—20 | | 19.3 mg/kg | Sn citrate | Single dose | Subcutaneous | No effect | Benoy *et al.*, 1971 |
| Mouse—6 | | 35 mg/kg | Triethyl tin chloride | 15 sec | Topical by immersion | $LD_{50}$ | Hall and Ludwig, 1972 |
| | | 50 mg/kg | Trimethyl tin chloride | 15 sec | Immersion | $LD_{50}$ | |
| Mouse | | 22 mg/kg | Dibutyl tin chloride | 3 doses | Gavage | Death in 75% of mice | Barnes and Stoner, 1958 |
| Mouse—10 | 17–24 g | 397 mg/kg | $NaSn_2F_5$ | Single dose | Gavage | $LD_{50}$ in 14 days preceded by muscular weakness, ataxia, and CNS depression | Conine *et al.*, 1975 |

| | | | | | | | |
|---|---|---|---|---|---|---|---|
| Mouse—10 | 17–24 g | 54 mg/kg | $NaSn_2F_5$ | Single dose | Intraperitoneal | $LD_{50}$ within 14 d | Conine *et al.*, 1975 |
| | | 13 mg/kg | $NaSn_2F_5$ | Single dose | Intravenous | $LD_{50}$ within 14 d | |
| Mouse—43 | 3 mo | 4 mg | Sn needles | Single dose | Intrathoracic | Foreign body reaction | Bischoff and Bryson, 1976 |
| Guinea pig | | 2.2 mg/kg | Dibutyl tin chloride | Single dose | Intravenous | Death in 48 h | Barnes and Stoner, 1958 |
| | | 4.4 mg/kg | | | | Death in 24 h | |
| | | 22 mg/kg | | Repeated doses | Gavage | No adverse effects | |
| | | 44 mg/kg | | | | No adverse effects | |
| | | 53 mg/kg | | | Topical | No adverse effects | |

[a] Number of animals per treatment.
[b] Quantity expressed in parts per million or as milligrams per kilogram of body weight.
[c] (cmpd) = concentration of tin in compound not established.

## REFERENCES


Barnes, J. M., and H. B. Stoner. 1958. Toxic properties of some dialkyl and trialkyl tin salts. Br. J. Indust. Med. 15:15.

Benoy, C. J., P. A. Hooper, and R. Schneider. 1971. The toxicity of tin in canned fruit juices and solid foods. Food Cosmet. Toxicol. 9:645.

Bischoff, F., and G. Bryson. 1976. Toxicologic studies of tin needles at the intrathoracic site of mice. Res. Commun. Chem. Pathol. Pharmacol. 15:331.

Conine, D. L., M. Yum, R. C. Martz, G. K. Stookey, J. C. Muhler, and R. B. Forney. 1975. Toxicity of sodium pentafluorostannite, a new anticariogenic agent. I. Comparison of the acute toxicity of sodium pentafluorostannite, sodium fluoride and stannous chloride in mice and/or rats. Toxicol. Appl. Pharmacol. 33:21.

Conine, D. L., M. Yum, R. C. Martz, G. K. Stookey, and R. B. Forney. 1976. Toxicity of sodium pentafluorostannite. A new anticariogenic agent. III. 30-day toxicity study in rats. Toxicol. Appl. Pharmacol. 35:21.

Cremer, J. E. 1962. Tetraethyl lead toxicity in rats. Nature (London) 195:607.

deGoeji, J. J. M., and J. J. Kroon. 1973. IAEA/FAO/WHO Symposium on Nuclear Techniques in Comparative Studies of Food and Environmental Contamination, Otaniemi, Finland. IAEA, Vienna.

deGroot, A. P. 1973. Subacute toxicity of inorganic tin as influenced by dietary levels of iron and cooper. Food Cosmet. Toxicol. 11:955.

deGroot, A. P., V. J. Feron, and H. P. Til. 1973. Short term toxicity studies in some salts and oxides of tin in rats. Food Cosmet. Toxicol. 11:19.

Eckardt, A. 1909. Beitrag zur Frage der Zinnvergiftunger. Z. Unters Nahr.-u Genussmittel 18:193.

Fischer, H. W., and G. R. Zimmerman. 1969. Long retention of stannic oxide. Lack of tissue reaction in laboratory animals. Arch. Pathol. 88:259.

Gaines, T. B., and R. D. Kimbrough. 1968. Toxicity of fentin hydroxide to rats. Toxicol. Appl. Pharmacol. 12:397.

Gaunt, I. F., J. Colley, P. Grasso, M. Creasey, and S. D. Gangolli. 1968. Acute and short-term toxicity studies on di-*n*-butyltin dichloride in rats. Food Cosmet. Toxicol. 6:599.

Hall, C. A., and P. D. Ludwig. 1972. Evaluation of the potential use for several organotin compounds against the sheep blowfly (Lucilia spp.). Vet. Rec. 90:29.

Kappas, A., and M. D. Maines. 1976. Tin: A potent inducer of heme oxygenase in kidney. Science 192:60.

Kehoe, R. A., J. Cholak, and R. V. Storey. 1940. A spectrochemical study of the normal ranges of concentration of certain trace metals in biological materials. J. Nutr. 19:579.

LaSpada, A. 1968. Patterns of World Tin Consumption 1957–1968. The International Tin Council, London.

Mantell, C. L. 1949. Tin: Its Mining, Production, Technology and Applications. Reinhold, New York.

Mitchell, R. L. 1948. The Spectrographic Analysis of Soils, Plants and Related Material. Tech. Commun. Bur. Soil Sci. No. 44.

Nikonorow, M., H. Mazur, and H. Piekacz. 1973. Effect of orally administered plasticizers and polyvinylchloride stabilizers in the rat. Toxicol. Appl. Pharmacol. 26:253.

Pelikan, Z. 1969. Effects of bis (tri-*n*-butyltin) oxide on the eyes of rabbits. Br. J. Ind. Med. 26:165.

Perry, H. M., Jr., and E. F. Perry. 1959. Normal concentrations by some trace metals in human urine: Changes produced by ethylenediaminetetraacetate. J. Clin. Invest. 38:1452.

Robinson, I. M. 1969. Effects of some organotin compounds on tissue amine levels in rats. Food Cosmet. Toxicol. 7:47.

Scheinberg, L. C., J. M. Taylor, I. Herzog, and S. Mandell. 1966. Optic and peripheral nerve response to triethyltin intoxication in the rabbit; Biochemical and ultrastructural studies. J. Neuropathol. Exp. Neurol. 25:202.

Schroeder, H. A., J. J. Balassa, and I. H. Tipton. 1964. Abnormal trace metals in man: Tin. J. Chron. Dis. 17:483.

Schroeder, H. A., M. Kanisawa, D. V. Frost, and M. Mitchener. 1968. Germanium, tin, and arsenic in rats: Effects on growth, survival, pathologic lesions and life span. J. Nutr. 96:37.

Schwarz, K., D. B. Milne, and E. Vinyard. 1970. Growth effects of tin compounds in rats maintained in a trace element-controlled environment. Biochem. Biophys. Res. Commun. 40:22.

Stevenson, J. J., W. C. Eckelman, P. Z. Sobocinski, R. C. Reba, E. L. Barron, and S. G. Levin. 1974. The toxicity of Sn-pyrophosphate: Clinical manifestations prior to acute $LD_{50}$. J. Nucl. Med. 15:252.

Stoner, H. B., J. M. Barnes, and J. I. Duff. 1955. Studies on the toxicity of alkyltin compounds. Br. J. Pharmacol. Chemother. 10:16.

Tipton, I. H., and M. J. Cook. 1963. Trace elements in human tissue. Part II. Adult subjects from the United States. Health Phys. 9:103.

Yamaguchi, M., H. Sato, and T. Yamamoto. 1976. Decrease of calcium concentration in urine of rats treated with stannous chloride. Chem. Pharm. Bull. (Tokyo) 24:3199.

# Titanium

Titanium (Ti) is a dark, gray metal that ranks eighth in abundance in igneous rocks. The concentration in the earth's crust is estimated to be 0.43 percent. The most important titanium ores are ilmenite ($FeTiO_3$) and rutile ($TiO_2$). Titanium is also found as a silicate (sphene), as calcium titanate ($CaTiO_3$), and in association with hematite deposits (Browning, 1969). Titanium forms compounds in which it has an oxidation state of +2, +3, or +4. It is used as a constituent of aluminum, tin, and vanadium alloys and is particularly important as ferrotitanium in the steel industry. Titanium oxide is used as a white pigment in paint and as a constituent of the coating of welding rods. Titanium dioxide is used as an ingredient marker in comminuted meats. Titanium compounds are used as a mordant in the dyeing industry, as a constituent of glass and ceramics, in surgical devices for properties of lightness and tensile strength, and with carbon and tungsten in the manufacture of electrodes and lamp filaments. Cobalt cemented carbides (Carbaloy) used in cutting tools contain titanium, and the military uses titanium chloride as a smoke screen.

## ESSENTIALITY

No essential metabolic function for titanium has been established in either plants or animals.

## METABOLISM

Published evidence on absorption of titanium from the alimentary tract is contradictory. Lehmann and Herget (1927) found no evidence of absorption when titanium oxide was fed. However, when Lloyd *et al.* (1955) attempted to use titanium oxide as an indigestible marker for digestion studies, they were unable to completely recover the element in the feces. They suggested that prolonged retention in some portion of the digestive tract (such as the cecum) may provide the explanation. However, Tipton *et al.* (1966) found significant urinary excretion of titanium in two humans consuming 0.37 and 0.41 mg per day in their diet. Both individuals were in negative titanium balance, and about equal amounts were found in feces and urine. It was not established whether urinary titanium was derived from titanium absorbed during the study or whether it came from previously established tissue stores.

## SOURCES

The titanium concentration of herbage can be used as an index of soil contamination (Barlow *et al.,* 1960), because soil concentrations of titanium are about 10,000 times greater than those in uncontaminated herbage (Swaine, 1955). A variety of plants were assayed by Bertrand and Varonca-Spirt (1929a,b), who found titanium levels ranged from 0.1 to 5 ppm (dry basis) with a majority near 1 ppm. Mitchell (1957) found a mean of 1.8 ppm (range 0.7–3.8) on a dry basis in red clover and a mean of 2.0 ppm (range 0.9–4.6) in ryegrass. Titanium concentrations of individual human foods have not been reported, but Tipton *et al.* (1966) reported the 30-day mean total diet titanium intakes of two individuals were 0.37 and 0.41 mg per day.

The primary titanium exposures in industry are to the metal, the dioxide, and the chloride. There seems to be general agreement that titanium and its compounds are low in toxicity (Browning, 1969). However, since the powders of titanium are pyrophoric and its liquid form burns in air, several explosions have resulted from careless handling. Hydrolysis of titanium chloride will result in release of hydrochloric acid and a consequent hazard from exposure to that chemical. It has been proposed that air titanium concentrations be limited to 15 $mg/m^3$ (Hamilton and Hardy, 1974).

## TOXICOSIS

It is questionable whether a specific toxicity of titanium has been demonstrated. Ereaux (1955) stated that oral administration of large amounts of titanium salts in the diet of experimental animals had no adverse effect. Vernetti-Blina (1928) administered titanium oxide by mouth, by subcutaneous injection, and by inhalation in varying dosages over periods of 1 to 2 months. He concluded this compound was basically inert and innocuous. Even inhalation of the dust for 8 hours a day for 30 days produced no significant clinical illness. At necropsy the peribronchial glands showed some hyperplasia, and the lungs had an increased amount of connective tissue in the stroma, with an exudate in the large and medium bronchi (evidence only of the irritating effect of the dust). Christie *et al.* (1963) exposed rats to titanium dioxide by inhalation for up to 13 months. There was little tissue reaction, but lung ash contained more than 10 percent titanium. Stokinger (1963) reported that inhalation exposure to high levels of titanium chloride produced severe respiratory distress, while lower levels (mean of 8.4 ppm) produced silicosislike lesions. When Bloom and Swensson (1958) injected titanium dioxide intravenously, there was a decrease in circulating thrombocytes. Implantation of small titanium discs into the abdominal muscle of dogs for several months caused no irritation (Beder and Eade, 1956). Titanium plates and screws have also been used in fixation of fractures in dogs and have proved inert (Beder *et al.*, 1957; Gross and Gold, 1957).

Ereaux (1955) has found topical application of titanium salicylate, peroxide, tannate, and oxides beneficial for skin disorders. Káto and Gözsy (1955) suggested that the therapeutic value of organic salts is due to their stimulation of phagocytic activity of capillary endothelial cells, thus increasing defense mechanisms of the skin without causing irritation. Déribéré (1941) found titanium oxides were harmless when used in cosmetics.

Industrial exposure of humans to titanium dust is generally believed not to induce lung fibrosis (Vernetti-Blina, 1928; Lundgren and Ohman, 1954; Moschinski *et al.*, 1959). Inhalation of fumes of titanic acid and titanic oxychloride resulted in marked congestion of the mucous membranes of the pharynx, vocal cords, and trachea, followed by cicatrization and stenosis of the larynx, trachea, and upper bronchi (Heimendinger and Klotz, 1956).

## TISSUE LEVELS

Tipton and Cook (1963) reported that most of the soft tissues of the adult human body contain 0.1 to 0.2 ppm titanium (on a wet basis). However, lungs averaged over 4 ppm, with some samples over 50 ppm. Hamilton *et al.* (1972/1973) found the following mean titanium concentrations (parts per million, wet weight) in human tissues: muscle, 0.2; brain, 0.8; kidney cortex, 1.3; kidney medulla, 1.2; liver, 1.3; and lung, 3.7. Losee *et al.* (1973) also found considerable variability in titanium concentration of 29 samples of human dental enamel. Levels ranged from 0.1 to 4.8 ppm (dry basis), with a mean of 0.46.

## MAXIMUM TOLERABLE LEVEL

No evidence of oral toxicosis has been found.

## SUMMARY

Titanium is a dark, gray metal that is found in the earth's crust at a concentration of about 0.43 percent. It is used in alloys with aluminum, tin, and vanadium and as ferrotitanium in steel. Its compounds are useful as ingredient markers in comminuted meat, as pigments in paint, in the coating of welding rods, in glass and ceramics, in implantable surgical devices, and in electrodes and lamp filaments. Titanium has no known metabolic function in plants or animals, and the metal and its compounds appear to have low toxicity. Implantation of titanium metal during the surgical repair of tissues produces no tissue reaction, and for this reason (plus properties of lightness and tensile strength) titanium has found considerable application in medicine.

## REFERENCES

Barlow, R. M., D. Purves, E. J. Butler, and I. J. McIntyre. 1960. Swayback in south-east Scotland. I. Field aspects. J. Comp. Pathol. Ther. 70:396.

Beder, D. E., and G. Eade. 1956. Tissue tolerance to titanium metal implants in dogs. Surgery 39:470.

Beder, D. E., J. K. Stevenson, and T. W. Jones. 1957. Further investigations of the surgical application of titanium metal in dogs. Surgery 41:1012.

Bertrand, G., and C. Varonca-Spirt. 1929a. Le titane dans les plantes phanerogames. C. R. Hebd. Seances Acad. Sci. 188:1199.

Bertrand, G., and C. Varonca-Spirt. 1929b. Le titane dans les plantes cryptogames. C. R. Hebd. Seances Acad. Sci. 189:73.

Bloom, G., and A. Swensson. 1958. The reaction of thrombocytes to intravenously injected suspensions of submicroscopic particles. Acta Med. Scand. 162:423.

Browning, E. 1969. Toxicity of Industrial Metals, 2nd ed. Butterworth & Co., London. 383 pp.

Christie, H., R. J. MacKay, and A. M. Fisher. 1963. Pulmonary effects of inhalation of titanium dioxide by rats. Am. Ind. Hyg. Assoc. J. 24:42.

Déribéré, M. 1941. Les Composes du Titane et l'Hygiène. Ann. Hyg. Publ. 18:133.

Ereaux, L. P. 1955. Clinical observations on the use of titanium salts in the treatment of dermatitis. Can. Med. Assoc. J. 73:47.

Gross, P. P., and L. Gold. 1957. Compatibility of vitallium and austenium in completely buried implants in dogs. Oral Surg. 10:769.

Hamilton, A., and H. L. Hardy. 1974. Industrial Toxicology, 3rd ed. Publishing Sciences Group, Inc., Acton, Mass. 575 pp.

Hamilton, E. I., M. J. Minski, and J. J. Cleary. 1972/1973. The concentration and distribution of some stable elements in healthy human tissues from the United Kingdom (an environmental study). Sci. Total Environ. 1:341.

Heimendinger, E., and G. Klotz. 1956. Sténose laryngo-trachéo-bronchique consécutive à une brûlure par chlorure de titane. Arch. Otolaryng. 73:645.

Káto, L., and B. Gözsy. 1955. Stimulation of the cell-linked defense forces of the skin. Can. Med. Assoc. J. 73:31.

Lehmann, K. B., and L. Herget. 1927. Studien über die hygienischen Eigenschaften des Titanoxyds und des Titanweiss. Chemiker-ztg. 82:793.

Lloyd, L. E., B. E. Rutherford, and E. W. Crampton. 1955. Ti oxide and Cr oxide as index materials for determining apparent digestibility. J. Nutr. 56:265.

Losee, F., T. W. Cutress, and R. Brown. 1973. Trace elements in human dental enamel. Trace Substances in Environmental Health—VII. University of Missouri, Columbia.

Lundgren, K. D., and H. Ohman. 1954. Pneumokoniose in der Hartmetall-industrie. Virchows Arch. Pathol. Anat. 325:284.

Mitchell, R. L. 1957. The trace element content of plants. Research (London) 10:357.

Moschinski, G., A. Jurisch, and W. Reinl. 1959. Die Lungenversanderungen bei Sinterhartmetall Arbeitern. Arch. Gewerbepath. Gewerbehyg. 16:697.

Stokinger, H. E. 1963. The metals (excluding lead). *In* F. A. Patty, ed. Industrial Hygiene and Toxicology, vol 2, 2nd ed. John Wiley & Sons, New York.

Swaine, D. J. 1955. The Trace Element Content of Soils. Commonw. Bur. Soils Tech. Commun. No. 48.

Tipton, I. H., and M. J. Cook. 1963. Trace elements in human tissue. Part II. Adult subjects from the United States. Health Phys. 9:103.

Tipton, I. H., P. L. Stewart, and P. G. Martin. 1966. Trace elements in diet and excreta. Health Phys. 12:1683.

Vernetti-Blina, L. 1928. Ricerche clinica e sperimentale sull' Assido di Titanio. Riform. med. 47:1516.

# Tungsten

---

Tungsten (W) metal has the highest melting point (3,387°C) of any element, and, because of this property, it is widely used as a filament material in incandescent lamps and as a component of high-temperature structural products. According to Standen (1970), tungsten is one of the rarer elements in the earth's crust, occurring in concentrations that average 5 ppm. Some fairly rich ores are available (e.g., scheelite and wolframite), which contain 2–3 percent of the metal and which permit economical mining and production. In 1977, 3 million kilograms of tungsten were mined domestically to produce alloys, tools, and wear-resistant materials, plating and electrical materials, catalysts, pigments, and corrosion inhibitors (U.S. Department of the Interior, 1977). Interest in the biological effects of tungsten is derived both from its antagonistic action on molybdenum metabolism (De Renzo, 1954) and from potential exposure of industrial workers (Browning, 1969).

## ESSENTIALITY

Although the interrelation between tungsten and molybdenum metabolism has been studied, no known essential role for tungsten has been found in animals.

## METABOLISM

Research on the metabolism of tungsten has dealt with determining its tissue distribution, retention, and excretion after administration to mice, rats, dogs, sheep, and swine.

Wase (1956) reported that mice rapidly eliminate tungsten given as a single intraperitoneal dose (15 mg per kilogram of body weight as $K_2{}^{185}WO_4$) so that, at 24 and 96 hours postdosing, 78 and 98 percent of the administered amount, respectively, were found in the feces. At 8 hours postdosing, tungsten was widely distributed in tissues, with bone and the gastrointestinal tract having the highest concentrations. Kaye (1968) reported similar findings in rats given tungsten (tracer quantity as either $K_2{}^{185}WO_4$ or $K_2{}^{187}WO_4$) by a single gavage. In this case, 40 percent of the tungsten was found (approximately equally divided between urine and feces) in the excreta by 24 hours postdosing, while the like figure at 72 hours approximated 97 percent. Tungsten elimination from the soft tissues was rapid and, again, bone was found to be the principal storage tissue. In dogs, Aamodt (1973) found that tungsten (tracer quantity as $Na_2{}^{181}WO_4$) given by intravenous injection was also rapidly eliminated with 91 percent of the administered dose appearing in the urine at 24 hours postdosing.

Bell and Sneed (1970) evaluated the metabolism of tungsten in sheep and swine. In these tests, growing barrows and mature wethers were dosed with tracer levels of $(NH_4)_2{}^{185}WO_4$. The swine were dosed either by intravenous injection or by gavage, while the sheep were dosed either orally by capsule or abomasally by injection. In swine, urinary excretion appeared to be the principal method of elimination of tungsten regardless of the route of administration. Most of the administered dose was excreted at 24 hours postdosing. Contrariwise, sheep excreted only 15 percent of the administered dose during the same time period. With regard to tissue distribution of tungsten at 48 hours postoral dosing, the concentrations in sheep tissues were in the following order: kidney $>$ liver $>$ bone $>$ muscle. In swine, these relationships were: kidney $>$ bone $>$ liver $>$ muscle.

Further studies on the metabolism of tungsten have concerned the elucidation of its involvement in molybdenum metabolism. De Renzo (1954) first observed that tungsten (as $Na_2WO_4$) fed to rats consuming diets low in molybdenum inhibited the stimulation of intestinal xanthine oxidase (a molybdenum-containing enzyme) caused by molybdenum repletion. In like manner, Higgins *et al.* (1956a,b) found that in chickens tungsten supplementation of molybdenum-low diets resulted in tissue depletion of molybdenum with concomitant decreases in the xanthine oxidase activities of small intestine, liver, kidney, and pancreas. Rats behaved similarly, and the enzyme effects in both chickens and rats could be reversed by molybdenum supplementation. Leach and Norris (1957) confirmed the former observation in growing chickens with regard to the effects on liver xanthine oxidase activity. In breeding chickens, Teekell and Watts (1959) demonstrated that 250 ppm of tungsten (as $Na_2WO_4$) had no effect on the xanthine oxidase activity of liver, kidney, and intestine, whereas 500 ppm caused a steady decline, so that the values resulting after 30 days of feeding were about 10 percent of the original. Dietary tungsten has also been observed to inhibit the activity of xanthine oxidase in the milk from lactating goats and dairy cows and in the liver of growing kids (Owen and Proudfoot, 1968).

Other adverse effects of dietary tungsten on tissue enzyme activity have been observed by Cohen *et al.* (1973), who noted a decrease in liver sulfite oxidase, and Chatterjee *et al.* (1973), who noted an accelerated breakdown of L-ascorbic acid by rat liver enzymes upon supplementation with $Na_2WO_4$.

## SOURCES

No literature is available with regard to the occurrence of tungsten in commonly used feedstuffs. Potential entry may occur through industrial contamination and environmental cycling. As with other elements, the toxicity of tungsten is dependent to a certain extent on the chemical form that is administered. For example, in the studies of Kinard and Van de Erve (1941), ammonium paratungstate was much less toxic (about one-fifth) than either tungstic oxide or sodium tungstate, which were of equivalent toxicity.

## TOXICOSIS

### LOW LEVELS

Several studies are available that involve the effects of tungsten administration to animals at relatively low levels (Table 37). In the experiment performed by Owen and Proudfoot (1968), growing kids were fed 22.5 ppm of dietary tungsten, as $Na_2WO_4 \cdot 2H_2O$, for a 3–5-month period. This treatment caused a marked depression in liver xanthine oxidase activity. In growing chickens, Higgins *et al.* (1956a,b) fed diets low in molybdenum and supplemented with tungsten, as sodium tungstate, at either 45 or 94 ppm. At the end of the 5-week experimental period, these treatments depressed growth rates by 8 and 19 percent, respectively, and increased death rates to 24 and 28 percent, respectively.

Two studies are of pertinence in laboratory animals. Higgins *et al.* (1956a,b) also noted that dietary concentrations of tungsten (45 or 94 ppm) that produced adverse effects in chickens did not adversely affect the livability or rate of gain in growing rats. This species difference in tolerance was hypothesized to be due to the increased need for xanthine oxidation in the chick. In the second pertinent study, Schroeder and Mitchener (1975) did not observe any adverse effects in rats reared for a lifetime on drinking water containing 5 ppm tungsten as sodium tungstate.

### HIGH LEVELS

Owen and Proudfoot (1968) included in their studies observations on the effects of tungsten administered to lactating dairy cows and lactating goats. Thus, a cow was treated with discrete oral doses of tungsten as sodium tungstate in accord with the following regimen: 12.5 mg/kg of body weight on the first day followed 20 days later by two consecutive daily doses of 12.5 mg/kg of body weight. In a second cow, the first treatment was 25.0 mg/kg of body weight, whereas the second treatment was 12.5 mg/kg of body weight. Although the milk production of both cows was unaffected, the milk xanthine oxidase activity was markedly decreased. The experiment in lactating goats yielded similar results. In breeding chickens, Teekell and Watts (1959) did not observe adverse effects on the rate of egg production or hatchability from tungsten, as sodium tungstate, supplementation at levels of 250 and 500 ppm for a 30-day period.

Considerably more data have been collected on the toxicosis of tungsten in laboratory animals. Thus, using rats of both sexes, Selle (1942) observed that tungsten injected subcutaneously in daily doses of 92 mg/kg of body weight caused a weight loss of 11 and 26 percent for females and males, respectively. No corresponding weight loss, however, was observed in rats receiving the same dose of tungsten by gavage.

In a series of studies on the acute toxicity of group VI elements, Pham-Huu-Chanh (1965) evaluated the $LD_{50}$ of tungsten administered as sodium tungstate by intraperitoneal injection in mice and rats. The resulting values were 112 mg/kg of body weight and 79 mg/kg of body weight for adult male rats and adult male mice, respectively. The signs of the acute intoxication include asthenia, adynamia, prostration, coma, and, finally, death. Similar signs were summarized by Browning (1969), who characterized the toxicosis as resulting in nervous prostration, diarrhea, coma, and death with the immediate cause being respiratory paralysis.

In a study designed to assess the toxicity of various tungsten salts, Kinard and Van de Erve (1941) measured the effects of dietary tungsten as tungstic oxide, sodium tungstate, and ammonium paratungstate in growing rats. They found that 1,000 ppm of tungsten as sodium tungstate and tungstic oxide and 5,000 ppm of tungsten as ammonium paratungstate produced a similar and slight growth depression during the 70-day experimental period. The higher doses of these salts tested (see Table 37) all produced extensive mortality.

### FACTORS INFLUENCING TOXICITY

The primary factor influencing chronic tungsten toxicity is the molybdenum content of the animal diet. Accordingly, in the studies of Higgins *et al.* (1956a,b), the growth rate depression, increased mortality, and decreased xanthine oxidase activity caused by tungsten administration to chicks were completely reversed by dietary supplementation with molybdenum (see also Leach and Norris, 1957).

Therapy of acute tungsten intoxication has been the subject of two papers, Lusky *et al.* (1949) and Sivjakov and Braun (1959). Lusky *et al.* (1949) demonstrated that 2,3-dimercaptopropanol could be used to treat rabbits poisoned with sodium tungstate. Similarly, Sivjakov and Braun (1959) demonstrated that calcium disodium ethylenediaminetetraacetate could be used to treat tungsten poisoning in rats.

## TISSUE LEVELS

There is a lack of information on the levels of tungsten in the tissues of food-producing animals. In rodents, however, Kinard and Aull (1944) did evaluate the distribution of tungsten in tissues of rats fed tungsten from various sources. After 100 days of dietary treatment with 1,000 ppm tungsten as tungstic oxide, sodium tungstate, or ammonium paratungstate, the rats were observed to have appreciable quantities of tungsten in bone, skin, and spleen tissues (values ranged from 20–120 ppm fresh weight basis). All other tissues contained trace quantities, i.e., less than 10 ppm. Tungsten, fed in this experiment as the free metal at levels of 2 and 10 percent of the diet, was also found in the tissues at levels comparable to those observed with the various salts.

## MAXIMUM TOLERABLE LEVEL

The available data permit the establishment of 20 ppm as the maximum tolerable level of tungsten in animals. This dose is about twice that observed to have no adverse effects in lifetime studies in rats and well below those causing adverse effects in shorter-term studies. It is less than half of the dose that caused adverse effects in chickens, but it is to be recalled that the chickens were being reared on diets low in molybdenum. Likewise it is below the level fed to growing kids that did not adversely affect production parameters.

## SUMMARY

Acute tungsten intoxication results in death from respiratory paralysis, preceded by nervous prostration, diarrhea, and coma. The most frequently observed sign of chronic intoxication is poor growth, however the most sensitive sign is decreased levels of tissue and/or milk xanthine oxidase activity. In this regard, tungsten is antagonistic to molybdenum in that dietary tungsten will precipitate signs of molybdenum deficiency that can be reversed (within limits) by supplemental molybdenum. Both 2,3-dimercaptopropanol and calcium disodium ethylenediaminetetraacetate have been effective in mitigating acute tungsten toxicosis. Tissue residue data for tungsten in food-producing animals are presently unavailable.

TABLE 37 Effects of Tungsten Administration in Animals

| Class and Number of Animals[a] | Age or Weight | Administration: Quantity of Element[b] | Source | Duration | Route | Effect(s) | Reference |
|---|---|---|---|---|---|---|---|
| Cattle—2, lactating dairy | 450 kg | 37.5 mg/kg in 2 or 3 separate doses | $Na_2WO_4 \cdot 2H_2O$ | 70 d | Capsule | No adverse effects on milk production; decreased milk xanthine oxidase activity | Owen and Proudfoot, 1968 |
| Goat—2, lactating | 50–60 kg | 56 mg/kg in 2 separate doses | $Na_2WO_4 \cdot 2H_2O$ | 21 d | Capsule | Decreased milk xanthine oxidase activity | |
| Goat—3, growing | Unspecified | 22.5 ppm | $Na_2WO_4 \cdot 2H_2O$ | 21–35 d | Diet | Decreased liver xanthine oxidase activity | |
| Chicken—29 | 1 d | 45 ppm | $Na_2WO_4$ | 35 d | Diet (low Mo) | Decreased growth rate; increased death rate | Higgins *et al.*, 1956a,b |
| | | 94 ppm | | | | Decreased growth rate; increased death rate | |
| Chicken—95 | Breeder | 250 ppm | Sodium tungstate | 10 d | Diet | No adverse effects | Teekell and Watts, 1959 |
| | | 500 ppm—10 d prior dosage at 250 ppm | | 20 d | | Decreased xanthine oxidase activity in the tissues | |
| Chicken—52 | 1 d | 500 ppm | Sodium tungstate | 24 d | Diet | Reduced rate of weight gain | |
| Rat | | 92 mg/kg | Sodium tungstate | | Subcutaneous | Weight loss | Selle, 1942 |
| | | | | | Gavage | No adverse effects | |

TABLE 37 *Continued*

| Class and Number of Animals[a] | Age or Weight | Administration | | | | Effect(s) | Reference |
|---|---|---|---|---|---|---|---|
| | | Quantity of Element[b] | Source | Duration | Route | | |
| Rat—10 | Weanling | 1,000 ppm | $WO_3$ | 70 d | Diet | Slight growth depression | Kinard and Van de Erve, 1941 |
| | | 5,000 ppm | | | | Death of all but one rat by the sixty-ninth day of treatment | |
| | | 36,600 ppm | | | | Death of all rats | |
| | | 1,000 ppm | $Na_2WO_4$ | | | Slight growth depression | |
| | | 5,000 ppm | | | | Death of all but two rats by the seventieth day of treatment | |
| | | 20,000 ppm | | | | Death of all rats | |
| | | 5,000 ppm | $(NH_4)_{10}$ | | | Slight growth depression | |

| | | | | | | | |
|---|---|---|---|---|---|---|---|
| | | | $(H_{10}W_{12}O_{46})$ | | | | |
| | | 20,000 ppm | | | | Death of all but two rats by the seventieth day of treatment | |
| | | 50,000 ppm | | | | Death of all rats | |
| Rat—600 | 200 g | 112 mg/kg | $Na_2WO_4 \cdot 2H_2O$ | Single | Intraperitoneal | $LD_{50}$ | Pham-Huu-Chanh, 1965 |
| Mouse—700 | 20 g | 79 mg/kg | $Na_2WO_4 \cdot 2H_2O$ | Single | Intraperitoneal | $LD_{50}$ | |
| Rat—87 | Weanling | 5 ppm | Sodium tungstate | Lifetime | Water | No adverse effects | Schroeder and Mitchener, 1975 |
| Rat | 150 g | 100 ppm | Sodium tungstate | 21 d | Water | Decreased liver sulfite oxidase and xanthine oxidase activities | Cohen *et al.*, 1973 |

[a]Number of animals per treatment.
[b]Quantity expressed in parts per million as concentration in diet or as milligrams per kilogram of body weight.

## REFERENCES

Aamodt, R. L. 1973. Retention and excretion of injected $^{181}W$ labeled sodium tungstate by beagles. Health Phys. 24:519.

Bell, M. C., and N. N. Sneed. 1970. Metabolism of tungsten by sheep and swine. *In* C. F. Mills, ed. Trace Element Metabolism in Animals. E & S Livingstone, Edinburgh and London.

Browning, E. 1969. Toxicity of Industrial Metals. Butterworths, London.

Chatterjee, G. C., R. K. Roy, N. Sasmal, S. K. Banerjce, and P. K. Majumder. 1973. Effect of chromium and tungsten on L-ascorbic acid metabolism in rats. J. Nutr. 103:509.

Cohen, H. J., R. T. Drew, J. L. Johnson, and K. V. Rajagopalan. 1973. Molecular basis of the biological function of molybdenum. The relationship between sulfite oxidase and the acute toxicity of bisulfite and $SO_2$. Proc. Natl. Acad. Sci. 70:3655.

De Renzo, E. C. 1954. Studies on the nature of the xanthine oxidase factor. Ann. N.Y. Acad. Sci. 57:905.

Higgins, E. S., D. A. Richert, and W. W. Westerfeld. 1956a. Competitive role of tungsten in molybdenum nutrition. Fed. Proc. 15:274. (Abstr.)

Higgins, E. S., D. A. Richert, and W. W. Westerfeld. 1956b. Molybdenum deficiency and tungstate inhibition studies. J. Nutr. 59:539.

Kaye, S. V. 1968. Distribution and retention of orally administered radiotungsten in the rat. Health Phys. 15:399.

Kinard, F. W., and J. C. Aull. 1944. Distribution of tungsten in the rat following ingestion of tungsten compounds. J. Pharmacol. Exp. Ther. 83:53.

Kinard, F. W., and J. Van de Erve. 1941. The toxicity of orally-ingested tungsten compounds in the rat. J. Pharmacol. Exp. Ther. 72:196.

Leach, R. M., and L. C. Norris. 1957. Studies on factors affecting the response of chicks to molybdenum. Poult. Sci. 36:1136. (Abstr.)

Lusky, L. B., H. A. Braun, and E. P. Lang. 1949. Effect of BAL on experimental lead, tungsten, vanadium, etc., poisoning. J. Ind. Hyg. 31:301.

Owen, E. C., and R. Proudfoot. 1968. The effect of tungstate ingestion on xanthine oxidase in milk and liver. Br. J. Nutr. 22:331.

Pham-Huu-Chanh. 1965. The comparative toxicity of sodium chromate, molybdate, tungstate and metavanadate. Arch. Int. Pharmacodyn. 154:243.

Schroeder, H. A., and M. Mitchener. 1975. Life-term studies in rats. Effects of aluminum, barium, beryllium, and tungsten. J. Nutr. 105:421.

Selle, R. M. 1942. Effects of subcutaneous injections of sodium tungstate on the rat. Fed. Proc. 1:165. (Abstr.)

Sivjakov, K. I., and H. A. Braun. 1959. The treatment of acute selenium, cadmium, and tungsten intoxication in rats with calcium disodium ethylene-diaminetetraacetate. Toxicol. Appl. Pharmacol. 1:602.

Standen, A., ed. 1970. Kirk-Othmer Encyclopedia of Chemical Technology, vol. 22. John Wiley & Sons, New York.

Teekell, R. A., and A. B. Watts. 1959. Tungsten supplementation of breeder hens. Poult. Sci. 38:791.

U.S. Department of the Interior. 1977. Bureau of Mines Minerals Yearbook, Tungsten chapter.

Wase, A. W. 1956. Absorption and distribution of radio-tungstate in bone and soft tissues. Arch. Biochem. Biophys. 61:272.

# Uranium

Uranium is widely distributed throughout the world, with the average concentration in the earth's crust being about 3–4 ppm (Merritt, 1971). It does not occur in concentrated deposits, and much of the ore from which uranium is recovered contains less than 0.1 percent uranium. More than 100 minerals contain uranium as an important constituent, with the primary minerals being uraninite and pitchblende, both of which consist chemically of uranium oxide ($U_3O_8$). Uranium occurs in both North Carolina and Florida marine sedimentary phosphate minerals and in igneous phosphate minerals from the western states in concentrations up to 250 ppm. It appears to be present in the phosphate minerals as an isomorphous substitution for calcium and is in tetravalent form. Uranium in some phosphate mining districts is presently being extracted as a by-product of the fertilizer industry. Interest in the effect of uranium on biological systems increased significantly during World War II because of the need to process uranium-containing ores for use in various atomic energy projects.

## ESSENTIALITY

There are several reports in the literature that uranium at very low concentrations (0.002 to 0.2 ppm) has a positive effect on the growth of plants and that it is a necessary nutrient in plant life (Dinse, 1953). Uranium has not been demonstrated to be essential in animals.

## METABOLISM

Although uranium is not known to be essential for any metabolic function, a great deal of information is available on how uranium is metabolized. When uranium enters the body, the uranyl ion ($UO_2^{+2}$) is the only stable form present in the oxidation-reduction system (Hodge, 1950). The uranyl ion in the bloodstream or in the extracellular fluid combines reversibly with serum albumin and forms strongly associated complexes with bicarbonate and with a number of organic acids. Uranium is transported to the tissues partly as a nondiffusable protein complex and partly as a diffusable bicarbonate complex. Approximately 40 percent of the uranium is present as the protein complex and 60 percent as the bicarbonate complex.

When uranium enters the bloodstream, it is removed at two principal sites: bone and kidney (Hodge, 1950; Durbin, 1960). This distribution is rapid; within an hour about 30 percent of a parenteral dose of uranium is deposited in the bone, about 15 percent in the kidney, and 20 percent will already have appeared in the urine. After a period of about 1 month, most of the uranium initially found in the bone is still at this site. The kidney may contain 1 or 2 percent of the original dose; the remainder is accounted for in the urine. In the bone, uranium competes with calcium for position on the mineral surface (Neuman *et al.*, 1949). Each uranyl ion reacts with two adjacent surface phosphates with a very stable linkage at sites formerly occupied by two calcium ions.

## SOURCES

Very little information is available on the uranium content of animal diets. The uranium concentration in soils is variable and dependent on the parent geological material; however, most soils contain approximately 1 ppm uranium. Higher uranium concentrations in some soils may result from the heavy usage of phosphate fertilizers (Menzel, 1968; Spalding and Sackett, 1972). Most plants are reported to contain 0.04 ppm or less (Bowen, 1966) and it would appear that plant materials are not a very significant source of uranium in animal diets. Because of the occurrence of uranium in phosphate deposits, the phosphate supplements used in animal feeds would probably be the major source of uranium. Uranium levels in commercial feed grade phosphates containing 18.0 to 18.5 percent phosphorus range from 70 to 180 ppm uranium (Reid *et al.*, 1977). A phosphate supplement of this concentration will be used in complete mixed feeds for farm livestock at a level of about 1 percent. This means that on the average phosphate supplements contribute 0.7 to 1.8 ppm uranium to animal diets.

Uranium compounds were found to have the following comparative toxicities for the mouse when fed in the diet (Tannenbaum, 1951):

$UO_2$, $U_3O_8$: relatively nontoxic, even in large doses (>100 mg uranium per day)
$UO_3$, $UCl_4$: toxic in large doses (80 mg uranium per day)
$UO_2(NO_3)_2$, $UO_4$, $Na_2U_2O_7$: toxic in moderate doses (10 to 20 mg uranium per day)

## TOXICOSIS

Uranium is a highly toxic element when soluble salts are administered by intravenous, subcutaneous, or intraperitoneal injection. The toxicity is dependent upon and modified by many factors and most of the reported studies have been conducted with laboratory animals, primarily mice.

The toxicological effect of uranium appears to be similar in all animals studied (Tannenbaum, 1951). Most of the absorbed uranium is excreted in the urine; however, some of the uranium reacts with the protein of the surface of the columnar cells lining the renal tubule and injures or kills these cells. With small or moderate doses, the distal portion of the proximal convoluted tubule receives the severest injury. If death ensues, it follows a typical uremia caused by kidney dysfunction; if the animal survives, cellular regeneration restores much of the kidney tissue and function.

One-year feeding experiments on dogs have shown that a level of 100 mg of uranyl nitrate hexahydrate per kilogram of body weight per day did not affect body weight (Hodge, 1953). Levels as low as 20 mg/kg of body weight per day of the same compound produced the characteristic histological kidney changes associated with uranium toxicosis.

Tannenbaum and Silverstone (1944) reported on studies in which uranium in the form of uranyl nitrate hexahydrate was fed to mice at levels ranging from 2 to 2,370 ppm in the diet. Over the period (48 weeks) studied, there was no definite indication of toxicosis at any of the levels fed, but there was a decrease in growth in the groups receiving the highest levels.

When mice were fed moderately toxic doses of uranium (1 percent uranyl nitrate hexahydrate—4,740 ppm uranium), the following observations were made (Tannenbaum and Silverstone, 1944): After ingesting the diet for a few days, the animals ate less food, either failed to grow or lost weight, were cold to the touch, huddled together, and had ruffled fur and arched backs. When necropsied during the second or third week, the kidneys were enlarged, pink-gray, and exhibited microscopically an acute necrotizing nephrosis. Some animals died in this period, however, those mice that survived the acute reaction recovered and proceeded to grow at a normal or near-normal rate. This was accompanied by regeneration of the tubular epithelium and a return to normal size and appearance of the kidneys. The striking features of this recovery are: (1) that it occurs despite the continued daily ingestion of the same dose of uranyl nitrate that caused the original acute reaction and (2) that uranium continues to accumulate in the bones and kidneys, reaching levels far in excess of the levels found in these tissues during the acute reaction. These observations and other data tend to support the view that animals acquire a tolerance to uranium. If mice are started on diets containing nontoxic levels of uranyl nitrate hexahydrate (1,422 ppm uranium) and the uranyl nitrate is increased gradually over a period of time, the dose may be increased to a level that would ordinarily cause 100 percent mortality in previously unexposed mice. The preexposed mice do not exhibit the acute reaction clinically or morphologically, but they do proceed to decline in weight and eventually die from the chronic poisoning. The principal morphological changes observed during the chronic poisoning are also renal (Tannenbaum, 1951).

Hodge (1953) reported on a 1-year feeding test utilizing rats in which three dietary levels of uranyl nitrate hexahydrate were fed (474, 2,370, and 9,480 ppm uranium). The rats maintained on the 474 ppm level grew practically as well as the control rats, and there was no difference in mortality in this group, as compared to the controls. At the 2,370 ppm level, a slight depression in body weight was observed; however, the mortality was no different than the control group. At the 9,480 ppm level, there was a marked reduction in growth and a high mortality during the first month of the study. Animals that survived the initial period showed partial recovery.

The solubility of the uranium compound and its rate of absorption from the gastrointestinal tract are probably the most important factors in determining its relative toxicity. Since uranium is a toxic element once it gets into the body tissues and since relatively large amounts of uranium compounds must be ingested to produce toxicosis (in comparison with subcutaneous doses), it is apparent that only a small percentage of an ingested uranium compound is absorbed from the gastrointestinal tract. It has been estimated that even for a soluble compound such as uranyl nitrate hexahydrate less than 0.5 percent of the amount ingested is absorbed (Tannenbaum, 1951).

## TISSUE LEVELS

The most extensive studies on the distribution of uranium in tissues have been performed on animals injected subcutaneously with soluble uranium salts (Ferretti and Schwartz, 1946; Tannenbaum and Silverstone, 1951; Hodge, 1953). Sufficient data have been gathered on mice fed uranium compounds to suggest that the same generalizations as were made for injected animals hold true for those ingesting the material in the diet (Tannenbaum, 1951). Bone and kidney are the principal sites of concentration of uranium following an intake either by subcutaneous injection or the oral route. The liver and spleen contain considerably lower concentrations of uranium than the kidney, yet the concentration in these tissues is higher than in other soft tissues.

Garner (1963) discussed the toxicity of uranium to livestock and its potential transfer to humans via food products. He concluded that it is not accumulated to any appreciable extent in edible tissues or secreted in significant amounts into milk. Chapman and Hammons (1963) indicated that in the dairy cow, milk received only 0.2 percent of the estimated uranium intake per day, whereas greater than 99 percent of the estimated daily intake appeared in the feces.

## MAXIMUM TOLERABLE LEVELS

Although uranium is a toxic element when soluble salts are administered by injection, the amount of dietary uranium that is absorbed from the gastrointestinal tract is very low. The animal can tolerate much higher dietary levels of the element as compared to those administered by injection. Because there are so many factors that influence the toxicity of uranium and all combinations of factors have not been studied, even for a single species, it is difficult to set a safe upper limit for dietary uranium levels. There is very little, if any, information available on the toxicity of uranium in farm livestock. A dietary level of 400 ppm uranium appears to be safe for rats, even when the uranium is present in a highly soluble form such as uranyl nitrate hexahydrate. Except for an inadvertent direct contamination of livestock diets or feed ingredients with a uranium compound, uranium toxicosis does not appear to be a practical problem. Total uranium content of animal diets probably does not exceed 3–4 ppm.

## SUMMARY

Uranium is widely distributed throughout the world, but it does not occur in concentrated deposits. Although uranium is not known to be essential for any metabolic function in animals, a great deal of information is available on how uranium is metabolized. The toxicological effect of uranium appears to be similar in all animals studied and is characterized by kidney dysfunction due to damage of the cells lining the renal tubule. Although uranium is a toxic element when soluble salts are administered by injection, the amount of dietary uranium that is absorbed from the gastrointestinal tract is very low. A dietary level of 400 ppm uranium appears to be safe for rats, even when the uranium is present in a highly soluble form.

TABLE 38 Effects of Uranium Administration in Animals

| Class and Number of Animals[a] | Age or Weight | Administration: Quantity of Element[b] | Source | Duration | Route | Effect(s) | Reference |
|---|---|---|---|---|---|---|---|
| Cattle—2 | Mature | 4 g/d | Uranyl nitrate | | Water | A deterioration in general health for 2 wk with a concomitant decrease in milk yield, followed by a gradual return to an apparently normal state in spite of continued administration | Garner, 1963 |
| Dog | | 9.5 mg/kg | $UO_2(NO_3)_2 \cdot 6H_2O$ | 52 wk | Diet | Lowest level at which histological changes in the kidney were observed | Hodge, 1953 |
| Dog | | 47 mg/kg | | | | No effect on body weight | |
| Mouse | 6 wk | 2 ppm | $UO_2(NO_3)_2 \cdot 6H_2O$ | 48 wk | Diet | No adverse effect | Tannenbaum and Silverstone, 1944 |
| | | 24 ppm | | | | No adverse effect | |
| | | 237 ppm | | | | No adverse effect | |
| | | 2,370 ppm | | | | A slight decrease in growth rate | |
| Mouse | 8–12 wk | 4,740 ppm | $UO_2(NO_3)_2 \cdot 6H_2O$ | 47 wk | Diet | Death of several animals during first 10 d; animals that survived partially recovered from the acute effects | |

TABLE 38 *Continued*

| Class and Number of Animals[a] | Age or Weight | Administration | | | | Effect(s) | Reference |
|---|---|---|---|---|---|---|---|
| | | Quantity of Element[b] | Source | Duration | Route | | |
| Rat | Weanling | 474 ppm | $UO_2(NO_3)_2 \cdot 6H_2O$ | 52 wk | Diet | Grew practically as well as control rats; no difference in mortality from controls | Hodge, 1953 |
| | | 2,370 ppm | | | | Slight depression in body weight and low mortality | |
| | | 9,480 ppm | | | | Marked reduction in growth and high mortality during the first month of the study; animals that survived partially recovered | |

[a] Number of animals per treatment.
[b] Quantity expressed in parts per million as concentration in diet or as milligrams per kilogram of body weight.

## REFERENCES


Bowen, H. J. M. 1966. Trace Elements in Biochemistry. Academic Press, New York.

Chapman, T. S., and S. Hammons, Jr. 1963. Some observations concerning uranium content of ingesta and excreta of cattle. Health Phys. 9:79.

Dinse, A. G. 1953. Effects of uranium on plants, pp. 2257–2269. *In* C. Voegtlin and H. C. Hodge (eds.). Pharmacology and Toxicology of Uranium Compounds. McGraw-Hill Book Co., New York.

Durbin, P. W. 1960. Metabolic characteristics within a chemical family. Health Phys. 2:225.

Ferretti, R. J., and S. Schwartz. 1946. Uranium distribution studies, pp. 247–282. *In* A. Tannenbaum, ed. Toxicology of Uranium. McGraw-Hill Book Co., New York.

Garner, R. J. 1963. Environmental contamination and grazing animals. Health Phys. 9:597.

Hodge, H. C. 1950. Pharmacologic tools in the study of the mechanics of uranium poisoning. Arch. Ind. Hyg. Occup. Med. 2:300.

Hodge, H. C. 1953. *In* C. Voegtlin and H. C. Hodge, eds. Pharmacology and Toxicology of Uranium Compounds. McGraw-Hill Book Co., New York.

Menzel, R. G. 1968. Uranium, radium and thorium content in phosphate rocks and their possible radiation hazard. J. Agric. Food Chem. 16:231.

Merritt, R. C. 1971. The Extractive Metallurgy of Uranium. Colorado School of Mines Research Institute, Atomic Energy Commission.

Neuman, W. F., M. W. Neuman, E. R. Main, and B. J. Muhyan. 1949. The disposition of uranium in bone. VI. Ion composition studies. J. Biol. Chem. 179:341.

Reid, D. F., W. M. Sackett, and R. F. Spalding. 1977. Uranium and radium in livestock feed supplements. Health Phys. 32:535.

Spalding, R. F., and W. M. Sackett. 1972. Uranium in runoff from the Gulf of Mexico distributive province: Anomalous concentrations. Science 175:629.

Tannenbaum, A., ed. 1951. Toxicology of Uranium. McGraw-Hill Book Co., New York.

Tannenbaum, A., and H. Silverstone. 1944. Some aspects of the toxicology of uranium compounds, pp. 59–96. *In* A. Tannenbaum, ed. Toxicology of Uranium. McGraw-Hill Book Co., New York.

Tannenbaum, A., and H. Silverstone. 1951. Distribution in tissues and excretion of uranium, pp. 16–21. *In* A. Tannenbaum, ed. Toxicology of Uranium. McGraw-Hill Book Co., New York.

# Vanadium

Vanadium (V), a bright white metal in the pure state, was named for Vanadis, the Norse goddess of beauty, in 1830 (Busch, 1961). Underwood (1962) noted that vanadium has been considered a rare element because there are few commercially workable deposits; however, it is actually one of the more prevalent trace elements. The element is found in some 50 different naturally occurring minerals, among which carnotite, roscoelite, vanadinite, and patronite are the most important industrial sources (Faulkner-Hudson, 1964).

Vanadium usually occurs in the earth's crust as relatively insoluble salts and is present in some sediments as oxovanadium (IV) anion bound to organic chelates (Yen, 1972). Tool and cutting steel, high-strength structural steel, and wear-resistant cast iron often contain 0.1 to 0.5 percent vanadium. There are several reviews on vanadium (Curran and Burch, 1967; Lillie, 1970; National Research Council, 1974).

## ESSENTIALITY

Hopkins and Mohr (1974) provide evidence that vanadium meets all criteria for essentiality. Schwarz and Milne (1971a,b) reported that vanadium was necessary for growth of rats raised in a trace element-controlled, all-plastic isolator. The addition of sodium orthovanadate ($Na_3VO_4$) to rat diets enhanced growth with 0.1 ppm vanadium optimizing performance. This growth response was confirmed by Strasia

(1971). Hopkins and Mohr (1971b) noted retarded feather growth in chicks consuming a diet containing less than 10 ppb vanadium. In calves, the addition of 0.1 mg sodium metavanadate per kilogram of body weight for 4 months increased growth by 11.5 percent, as well as increasing erythrocyte number 5.5 percent and blood hemoglobin content 7.1 percent (Drebickas, 1966). Nielsen and Ollerich (1973) observed increased hematocrit values, increased epiphyseal plate, and decreased primary spongiosa in the tibia of vanadium-deficient chicks receiving 30–35 ppb of the element.

Hopkins and Mohr (1974) reviewed the vanadium literature and reported that reproductive efficiency is impaired by a deficiency of the mineral. Vanadium's effect in reducing dental caries is still not resolved (Hadjimarkos, 1966).

Vanadium increased oxidation of phospholipids by washed liver suspensions *in vitro* (Bernheim and Bernheim, 1938). Curran (1954) discovered vanadium's ability to inhibit cholesterol biosynthesis for a microbial squalene synthetase system. Several later reports indicate altered blood lipid levels during vanadium deficiency (Hopkins and Mohr, 1971a, 1974; Nielsen and Ollerich, 1973). Recent reports (Nielsen and Myron, 1976; Nielsen and Uthus, 1977) suggest that vanadium has a role in labile methyl metabolism in the chick.

## METABOLISM

Metabolism of vanadium has been discussed in several reviews (Scott *et al.*, 1951; Söremark and Üllberg, 1962; Hopkins and Tilton, 1966). Radioactive tracer studies by Comar and Chevallier (1967) have shown that vanadium is not readily absorbed from the digestive tract of the rat; however, the average body concentration was proportional to dietary intake. Faulkner-Hudson (1964), citing a report by Scott *et al.* (1951), indicated that only 0.5 percent of the radiovanadium administered intragastrically to rats was absorbed. Less that 1 percent vanadium was absorbed following ingestion in man (Curran *et al.*, 1959) and in sheep (Hansard, 1975).

Excretion of the injected or absorbed metal is mainly by the kidneys. Talvitie and Wagner (1954) noted that vanadium, as sodium metavanadate injected intravenously, was excreted rapidly by the kidneys with 60 percent of the dose appearing in the urine within 24 hours in rats and rabbits. Sixty-six percent of vanadium as $VOCl_2$ injected intramuscularly in rats was eliminated in urine within 24 hours (Pepin *et al.*, 1977). Similar values were observed (Hopkins and Tilton, 1966) when approximately 45 percent of the injected isotope ($^{48}VOCl_2^+$) was lost in urine and 9 percent in the feces of rats. Relatively high retention of $^{48}V$ in bones and kidneys of chicks following oral administration of vanadyl dichloride was observed (Hathcock *et al.*, 1964). Radioisotope distribution studies (Hathcock *et al.*, 1964; Comar and Chevallier, 1967; Hansard, 1975) indicate that the element is retained principally by kidney and also by bone, liver, and spleen, although clearance from blood (Hopkins and Tilton, 1966) and soft tissues (Thomassen and Leicester, 1964) is rapid. Growing bone (Hathcock *et al.*, 1964; Parker and Sharma, 1978) and possibly kidney and major organs retain the metal to a small degree. Radioactive $^{45}V_2O_5$ was concentrated in bones and teeth of mice and rats 7 days after injection (Söremark *et al.*, 1962; Söremark and Üllberg, 1962). Other tissues also retained the isotope; the decreasing order of the radioactivity was visceral yolk sac, epithelium, lactating mammary gland, renal cortex, liver, lung, skin, and salivary gland. Vanadium appears to exist in tissues in a protein-bound form (Johnson *et al.*, 1974). Vanadate taken up by the red blood cell is reduced to the +4 oxidation state in the cytoplasm (Cantley and Aisen, 1979). Oxidation state does not appear to affect metabolism of vanadium (Sabbioni *et al.*, 1978).

## SOURCES

Vanadium is distributed widely in nature, occurring in many plants and animals (Curran and Burch, 1967; Underwood, 1977). The level in the earth's crust has been estimated at 110 (Goldschmidt, 1958) to 150 ppm (Vinogradov, 1959). Of special concern is the vanadium content of rock phosphates, which may be used as phosphorus sources for animal diets. Vanadium concentration, although varying with location, may be as high as 6,000 ppm in some rock phosphate deposits (Romoser *et al.*, 1960). Berg (1963) reported that a commercial tricalcium phosphate contained 0.25 percent vanadium pentoxide (1,400 ppm V) and reduced growth in poultry receiving 14–20 ppm vanadium. Grazing animals are exposed to elevated levels of many minerals due to ingestion of soil (Healy, 1973; Thornton, 1974). Most feed sources analyzed by Mitchell (1957) contained less than 0.15 ppm on a dry weight basis. Analysis of animal specimens, vegetables, and fruits (Söremark, 1967) indicated that all samples contained less than 0.5 ppm vanadium (wet weight).

In the natural state, vanadium occurs with positive valences of two, three, four, and five. The pentavalent salts are vanadates and the quadrivalent ion forms vanadites (Faulkner-Hudson, 1964). The most important compound in industry is the pentoxide ($V_2O_5$) (Fairhall, 1949). The bioavailability of various forms of vanadium has been investigated by Schwarz and Milne (1971b). In their studies with mice, sodium orthovanadate was more effective in promoting growth than sodium metavanadate, while the pyrovanadate salt had no activity.

## TOXICOSIS

Numerous reviews on vanadium toxicity are available (Sjöberg, 1950; Stokinger, 1955, 1963; Faulkner-Hudson, 1964; Lillie, 1970; National Research Council, 1974). Vanadium appears to exert its toxic effect through inhibition of enzymes (Underwood, 1977) and cell damage from lysis (Waters *et al.*, 1975). Vanadate has been found to inhibit (Na, K)-ATPase (Cantley *et al.*, 1977, 1978; Nechay and Saunders, 1978; Goodno, 1979; Nieder *et al.*, 1979) and activate cardiac adenylate cyclase (Grupp *et al.*, 1979).

### LOW LEVELS

Industrial exposure to vanadium, through breathing of airborne particulate matter, was found to cause irritation of the nose and throat, anorexia, nausea, and diarrhea in workers (Fairhall, 1949). Toxicosis in humans from ingestion is uncommon except as incidental to high aerial concentrations (National Research Council, 1974). The chick appears to be most susceptible to orally induced vanadium toxicosis (Table 39), consequently the bird is the most studied model for toxicity. Animals that have been shown to adapt to high vanadium levels include the rat (Daniel and Lillie, 1938; Strasia, 1971) and chick (Williams, 1973).

Studies suggest that alterations in rumen function can result from ingestion of vanadium. *In vitro* dry matter digestibility was reduced in rumen fluid inoculum from lambs by 7 ppm vanadium added as sodium ortho- or metavanadate (Williams, 1973). Martinez and Church (1970) reported reduced *in vitro* cellulose digestion by washed suspensions of rumen microorganisms with 5 ppm vanadium as sodium metavanadate. These results were in general agreement with studies by Jha (1966). Lillie (1970), citing a report by Heege (1964), noted cows exposed to vanadium from fuel oil soot on grazing areas showed weakness and ataxia. Vanadium content of liver tissue ranged from 1.5 to 4.7 ppm (wet weight).

Romoser *et al.* (1961) observed that chicks tolerated up to 20 ppm vanadium in a corn–soybean meal diet, while Nelson *et al.* (1962) reported that 35 ppm vanadium was toxic for chick growth. Hathcock *et al.* (1964) reported that 25 ppm vanadium as ammonium metavanadate ($NH_4VO_3$) depressed growth and increased mortality, but 10 ppm vanadium had no effect. Berg (1963) found that two commercial samples of tricalcium phosphate depressed chick growth when compared to other phosphorus sources. The samples contained 0.25 percent vanadium pentoxide ($V_2O_5$) contributing 28 ppm vanadium to the complete diet. Berg *et al.* (1963) also found that 60 ppm vanadium as ammonium metavanadate reduced hatchability of fertile eggs by 10 percent. Egg production was depressed with 30 ppm vanadium, but only 15 to 20 ppm were required to lower egg albumin quality. Other studies with poultry are shown in Table 39.

Since experimental results with domestic animals are limited, laboratory animal studies may supplement available information. Vanadium, as sodium metavanadate ($NaVO_3$) was slightly toxic to rats at 25 ppm and distinctly toxic at 50 ppm (Franke and Moxon, 1937). Daniel and Lillie (1938) observed no effect in rats receiving 11.5 and 22 ppm vanadium as sodium metavanadate over 12 weeks, but 92 ppm was very toxic and 368 ppm was lethal within 10 weeks. Muhler (1957) reported reduced growth and higher mortality in rats receiving 20 ppm vanadium as vanadium pentoxide ($V_2O_5$) in drinking water. At 40 ppm, a 100 percent death rate was observed within 65 days. Schroeder and Balassa (1967) gave mice 5 ppm vanadium as vanadyl sulfate ($VOSO_4$) in drinking water from weaning until natural death and found no toxicosis in terms of growth, survival, or life span, but the metal accumulated in the heart and spleen. A similar long-term study with rats receiving 5 ppm vanadium showed no accumulative toxicity (Schroeder *et al.*, 1970).

## HIGH LEVELS

Few acute vanadium toxicity studies are available. Platonow and Abbey (1968) studied the toxicosis of vanadium in calves. Gelatin capsules of ammonium metavanadate, given orally at daily dosage levels of 1, 3, 5, 7.5, 10, 15, and 20 mg vanadium per kilogram of body weight, produced intoxication at the three higher levels. The dose of 20 mg resulted in adverse effects within 3 days, including diarrhea, dehydration, emaciation, and prostration. Gross pathological changes were congestion of liver and lungs, diffuse hemorrhage covering the kidneys and heart, ruminal ulcers, and hemorrhagic inflammation of the intestinal tract. The greatest concentration of vanadium was found in kidney, liver, and spleen.

Hansard (1975) observed a 65 percent death loss within 80 hours in sheep given 40 mg vanadium per kilogram of body weight as ammonium metavanadate. Vanadium content of kidney, liver, bone, spleen, lung, and muscle was elevated by treatment. Histological examination showed evidence of liver degeneration and nephritis. Hansard *et al.* (1978) also reported reduced growth and feed utilization and diarrhea in sheep fed 400 ppm vanadium as ammonium metavanadate for 84 days (Table 39).

Romoser *et al.* (1960) apparently were the first to show that dietary vanadium retarded the growth rate of chicks. When given as the calcium salt [$Ca_3(VO_4)_2$] the $LD_{50}$ for vanadium was between 300 and 350 ppm, but growth was inhibited at levels above 20 ppm.

Proescher *et al.* (1917) reported the $LD_{50}$ in rats upon subcutaneous injection was 20 to 30 mg vanadium per kilogram of body weight as ammonium metavanadate. Necrosis of renal convoluted tubules, fatty degeneration of the liver, adrenal hemorrhage, constriction of visceral arteries, and inflammatory lesions in the intestinal tract were noted.

## FACTORS INFLUENCING TOXICITY

Various compounds have been credited with alleviating vanadium toxicity. Wright (1968) found that 2,000 ppm supplemental chromium reduced the death rate in chicks from 86.6 to 13.3 percent when fed 20 ppm vanadium. This author suggested that chromium antagonizes vanadium uncoupling of oxidative phosphorylation and retards the intestinal absorption of vanadium. DeMaster (1972) was not able to confirm the role of vanadium in uncoupling of oxidative phosphorylation. Chicks fed 5 ppm vanadium as $NH_4VO_3$ were affected adversely when also supplemented with 500 ppm chromium as $Cr(C_2H_3O_2)$ (Hunt and Nielsen, 1979). Moxon and Rhian (1943) observed that rats given 11 ppm selenium in drinking water died more quickly when selenium was in combination with 5 ppm vanadium.

Mitchell and Floyd (1954) and Berg and Lawrence (1971) tested ascorbic acid and ethylenediaminetetraacetate (EDTA) as antidotes in experimental vanadium poisoning in mice and rats. EDTA may function by preventing vanadium absorption (Hathcock *et al.*, 1964), when fed at twice the molar concentration of vanadium.

Diet composition greatly affected the degree of toxicity of dietary vanadium (Mountain *et al.*, 1959; Hathcock *et al.*, 1964). Berg (1966) and Hill (1979) showed that increasing dietary protein levels gave linear decreases in the mortality rate of chicks, but Hansard (1975) found no difference in performance or tissue levels of rats when protein level was increased from 20 to 30 percent with vanadium levels up to 40 ppm.

Research with poultry (Hafez and Kratzer, 1976a) suggests that vanadium may have a greater adverse effect when fed in purified rather than natural diets.

## TISSUE LEVELS

Vanadium toxicosis has not been studied as extensively as many of the other minerals, although the recent recognition of high concentrations in some phosphates, coals, and petroleum products has increased interest in the movement of the element from the environment to animals and man.

In calves given lethal amounts of vanadium, none was detected in skeletal muscle (detection limit, 0.01 ppm), but up to 5.1 ppm (wet weight) was found in liver (Platonow and Abbey, 1968). It is apparent that animal tissues increase in vanadium content in response to dietary exposure (Comar and Chevallier, 1967; Hopkins and Mohr, 1971b). Söremark (1967) reported that calf liver and muscle contain 10 ppb or less, while milk values were generally lower than 100 ppb (wet weight). Other studies (Hopkins and Tilton, 1966) have shown no particular accumulation in adipose tissue.

In studies with rats (Parker and Sharma, 1978) receiving 50 ppm vanadium as vanadyl sulfate or sodium orthovanadate in drinking water for 90 days, tissue accumulation was greatest for kidney, followed by bone, liver, and muscle. In general, tissue vanadium concentrations were greater in animals receiving the sodium orthovanadate.

Hansard *et al.* (1978) fed 10, 100, and 200 ppm vanadium as ammonium metavanadate to sheep for 84 days and found increased levels of vanadium in bone, liver, kidney, and muscle with 200 ppm vanadium (Table 40). Kidney vanadium levels were also increased when 100 ppm was fed.

## MAXIMUM TOLERABLE LEVELS

The rat and chick appear to have a similar tolerance to vanadium, but ruminants are less susceptible to toxicity. Reduced growth has been obtained in day-old chicks with 8–10 ppm dietary vanadium as ammonium metavanadate. Other studies with the young chick indicate a tolerance of 25 ppm vanadium; a similar level fed to laying hens reduced egg quality. Weanling rats tolerated 20 ppm vanadium as sodium metavanadate but 40 ppm reduced growth. Growing lambs tolerated 200 ppm dietary vanadium as ammonium metavanadate with no effect on growth rate but significant increases in tissue vanadium levels occurred with this intake and also with 100 ppm vanadium in the diet. Lambs did not consume feed containing 400 ppm vanadium. Suggested maximum tolerable dietary levels for vanadium are 50 ppm for cattle and sheep and 10 ppm for poultry. Research with poultry suggests that similar levels of vanadium may have a greater adverse effect when fed in a purified diet than when fed in a diet composed of natural ingredients.

## SUMMARY

Vanadium has been shown to be essential for normal growth and proper physiological function in all species studied. It has also been found to be toxic in all animals studied. Vanadium appears to exert its toxic effect through inhibition of enzymes and cell damage from lysis. Vanadium given daily at 20 mg/kg of body weight produced diarrhea, emaciation, and prostration in calves within 3 days. Sheep showed a 65 percent death rate within 80 hours when given 40 mg vanadium per kilogram body weight as $NH_4VO_3$. Signs of toxicosis in calves and lambs include diarrhea, depressed growth and performance, ataxia, and mortality. EDTA appears to act as an antidote in vanadium toxicosis, possibly by preventing absorption from the intestinal tract. The vanadium content of most feedstuffs is generally low except in some rock phosphate sources. Industrial contamination of air and water appears to be the greatest source of vanadium to the environment.

TABLE 39 Effects of Vanadium Administration in Animals

| Class and Number of Animals[a] | Age or Weight | Administration | | | | Effect(s) | Reference |
|---|---|---|---|---|---|---|---|
| | | Quantity of Element[b] | Source | Duration | Route | | |
| Cattle—1 | 127–185 kg | 1 mg/kg | $NH_4VO_3$ | 28 d | Capsule | No adverse effect | Platonow and Abbey, 1968 |
| | | 3 mg/kg | | | | No adverse effect | |
| | | 5 mg/kg | | | | No adverse effect | |
| | | 7.5 mg/kg | | | | No adverse effect | |
| | | 10 mg/kg | | 14 d | | Diarrhea; prostration | |
| | | 15 mg/kg | | 6 d | | Diarrhea; prostration | |
| | | 20 mg/kg | | 3 d | | Diarrhea; increased vanadium in kidney; prostration | |
| Sheep | Growing lamb | 7 ppm | $NaVO_3$ or $Na_3VO_4$ | | Diet | Reduced dry matter digestibility *in vitro* | Williams, 1973 |
| Sheep—5 | 37 kg | 10 ppm | $NH_4VO_3$ | 84 d | | No effect | Hansard *et al.*, 1978 |
| | | 100 ppm | | | | Increased vanadium in kidney, bone, and liver | |
| | | 200 ppm | | | | Increased vanadium in muscle, kidney, bone, and liver | |
| Sheep—2 | | 400 ppm | | 1 d | | Reduced feed intake; diarrhea | |
| | | 800 ppm | | | | Reduced feed intake; diarrhea | |
| Sheep—3 | 45 kg | 400–450 mg/d | $NH_4VO_3$ $Ca_2V_2O_7$ $Ca_3(VO_4)_2$ | 68–96 h | Capsule | Increased tissue levels; liver and kidney degeneration; death | Hansard, 1975 |

| | | | | | | | |
|---|---|---|---|---|---|---|---|
| Sheep—3 | | 40 mg/kg | $NH_4VO_3$ | Single dose | Capsule | Death | Hansard, 1975 |
| Chicken—12 | Immature | 3 ppm | $NH_4VO_3$ | 27 d | Diet | No adverse effect | Baker and Molitoris, 1975 |
| Chicken—40 | Day old | 2 ppm | $NH_4VO_3$ | 28 d | Diet | No adverse effect | Berg, 1963 |
| | | 4 ppm | | | | No adverse effect | |
| | | 6 ppm | | | | No adverse effect | |
| | | 8 ppm | | | | Reduced growth | |
| | | 10 ppm | | | | Reduced growth | |
| | | 15 ppm | | | | Reduced growth | |
| | | 20 ppm | | | | Reduced growth | |
| Chicken—100 | Laying hen, 15 mo | 10 ppm | $NH_4VO_3$ | 38 d | Diet | No adverse effect | Berg *et al.*, 1963 |
| | | 20 ppm | | | | No adverse effect | |
| | | 30 ppm | | | | Depressed egg production and albumin quality | |
| | | 40 ppm | | | | Depressed egg production and albumin quality | |
| | | 50 ppm | | | | Depressed egg production and albumin quality | |
| | | 60 ppm | | | | Depressed egg production and albumin quality | |
| | | 100 ppm | | | | Depressed egg production and albumin quality | |
| Chicken—200 | Immature | 10 ppm | $NH_4VO_3$, $VOSO_4$, or $VOCl_2$ | 14 d | Diet | Reduced growth | Berg and Lawrence, 1971 |
| | | 20 ppm | | | | Reduced growth; death | |
| Chicken—60 | Immature | 20 ppm | $NH_4VO_3$ | 21 d | Diet | Reduced growth | Berg, 1966 |
| Chicken—10 | Immature | 7 ppm | $NH_4VO_3$ | 28 d | Diet | No adverse effect | Nelson *et al.*, 1962 |

TABLE 39 *Continued*

| Class and Number of Animals[a] | Age or Weight | Administration | | | | Effect(s) | Reference |
|---|---|---|---|---|---|---|---|
| | | Quantity of Element[b] | Source | Duration | Route | | |
| | | 17 ppm | | | | No adverse effect | |
| | | 27 ppm | | | | Reduced growth | |
| | | 47 ppm | | | | Reduced growth | |
| | | 87 ppm | | | | Reduced growth | |
| Chicken—20 | Immature | 7 ppm | $NH_4VO_3$ | 28 d | Diet | No adverse effect | |
| | | 14 ppm | | | | No adverse effect | |
| | | 17 ppm | | | | No adverse effect | |
| | | 21 ppm | | | | No adverse effect | |
| | | 27 ppm | | | | No adverse effect | |
| | | 35 ppm | | | | No adverse effect | |
| | | 47 ppm | | | | Reduced growth | |
| | | 64 ppm | | | | Reduced growth | |
| | | 87 ppm | | | | Reduced growth | |
| | | 120 ppm | | | | Reduced growth | |
| Chicken—52 | Immature | 40 ppm | $Ca_3(VO_4)_2$ | 32 d | Diet | Decreased gain and feed efficiency | Romoser *et al.*, 1961 |
| | | 60 ppm | | | | Decreased gain and feed efficiency | |
| | | 120 ppm | | | | Decreased gain and feed efficiency | |
| | | 200 ppm | | | | 30% death rate | |
| | | 400 ppm | | | | 100% death rate | |
| | | 600 ppm | | | | 100% death rate | |
| Chicken—75 | Immature | 10 ppm | $Ca_3(VO_4)_2$ | 28 d | Diet | No adverse effect | |
| | | 15 ppm | | | | No adverse effect | |
| | | 20 ppm | | | | No adverse effect | |

| | | | | | | | |
|---|---|---|---|---|---|---|---|
| | | 30 ppm | | | | Reduced growth and feed utilization | |
| | | 40 ppm | | | | Reduced growth and feed utilization | |
| Chicken—40 | Immature | 10 ppm | $NH_4VO_3$ | 14 d | Diet | No adverse effect | Hathcock *et al.*, 1964 |
| | | 25 ppm | | | | Reduced growth; 95% death rate | |
| | | 25 ppm | $VOSO_4$ | | | Reduced growth; 85% death rate | |
| | | 25 ppm | $NH_4VO_3$ | | | Reduced growth; 85% death rate | |
| Chicken—320 | Immature | 10 ppm | | 21 d | Diet | Reduced growth | Summers and Moran, 1972 |
| Chicken | Immature | 15 ppm | $NH_4VO_3$ | | Diet | Reduced growth; mortality | Williams, 1973 |
| Chicken | Immature | 20 ppm | $NH_4VO_3$ | | Diet | Reduced growth | Wright, 1968 |
| Chicken | Immature | 25 ppm | $NH_4VO_3$ | 7 d | Diet | Uncoupled oxidative phosphorylation | Hathcock *et al.*, 1966 |
| Chicken—6 | Laying hen, 28 wk | 100 ppm | | 12 d | Diet | No adverse effect | Hafez and Kratzer, 1976b |
| Chicken—50 | Immature | 50 ppm | $VOCl_2$ | 28 d | Diet | Reduced growth and feed utilization | Miller *et al.*, 1961 |
| Duck—10 | Yearling | 1 ppm | $VOSO_4$ | 84 d | Diet | No adverse effect | White and Dieter, 1978 |
| | | 10 ppm | | | | Increased vanadium in bone, liver, and kidney | |
| | | 100 ppm | | | | Increased vanadium in blood, bone, kidney, and brain | |
| Rat | Immature | 25 ppm | $NaVO_3$ | | Diet | Reduced growth; diarrhea | Franke and Moxon, |

TABLE 39 *Continued*

| Class and Number of Animals[a] | Age or Weight | Administration: Quantity of Element[b] | Source | Duration | Route | Effect(s) | Reference |
|---|---|---|---|---|---|---|---|
| | | | | | | | 1937 |
| Rat—12 | 52 g | 5 ppm | $NaVO_3$ | 56 d | Diet | No adverse effect | Hansard, 1975 |
| | | 10 ppm | | | | No adverse effect | |
| | | 20 ppm | | | | No adverse effect | |
| | | 40 ppm | | | | Reduced growth and feed utilization | |
| Rat—6 | 62 g | 20 ppm | $NaVO_3$ | 35 d | Diet | No adverse effect | |
| | | 40 ppm | | | | No adverse effect | |
| | | 80 ppm | | | | Reduced growth and feed utilization | |
| Rat—12 | 60 g | 100 ppm | $NaVO_3$ | 105 d | Diet | Reduced growth | Hansard, 1975 |
| | | 150 ppm | | | Diet | Reduced growth | |
| Rat | Immature | 50–100 ppm | $NaVO_3$ or $V_2O_5$ | | Diet | Reduced growth and feed utilization | Mountain *et al.*, 1959 |
| Rat | Immature | 250 ppm | $NaVO_3$ | | Diet | Reduced growth; death | Strasia, 1971 |

| | | | | | | | |
|---|---|---|---|---|---|---|---|
| Rat | Mature | 400 ppm | $NaVO_3$ | | Diet | Death | Strasia, 1971 |
| Rat | Immature | 368 ppm | $NaVO_3$ | 70 d | Diet | Death | Daniel and Lillie, 1938 |
| Rat—92 | Immature | 500 ppm | $NaVO_3 \cdot 4H_2O$ | | Diet | Reduced liver coenzyme A | Mascitelli-Coriandoi and Citterio, 1959 |
| Rat | Mature | 500 ppm | $V_2O_5$ | | Diet | Reduced growth and feed utilization | Mountain *et al.*, 1953 |
| Mouse—54 | Immature | 5 ppm | $VOSO_4$ | Life term | Diet | No adverse effect | Schroeder and Mitchener, 1975 |
| Guinea pig | Weanling–adult | 5 ppm | $NH_4VO_3$ | 56 d | Diet | Elevated serum cholesterol | Bruffy and Dowdy, 1979 |
| | | 25 ppm | | | | No adverse effects | |
| | | 50 ppm | | | | Decreased weights, hemoglobin, and packed cell volume | |

[a] Number of animals per treatment.
[b] Quantity expressed in parts per million or as milligrams per kilogram of body weight.

TABLE 40 Effect of Dietary Vanadium on Tissue Vanadium of Sheep[a]

| Supplemental Vanadium, ppm | Tissue, ppm dry matter basis | | | |
|---|---|---|---|---|
| | Bone[b] | Liver | Kidney | Muscle |
| 0 | 0.19 ± 0.03[c] | 0.12 ± 0.03[c] | 0.23 ± 0.04[c] | 0.04 ± 0.00[c] |
| 10 | 0.22 ± 0.05[c] | 0.18 ± 0.04[c] | 0.41 ± 0.06[c] | 0.05 ± 0.01[c] |
| 100 | 1.50 ± 0.34[c] | 0.96 ± 0.12[c] | 3.62 ± 0.19[d] | 0.12 ± 0.01[c] |
| 200 | 3.32 ± 0.23[d] | 2.81 ± 0.24[d] | 11.13 ± 0.46[e] | 0.41 ± 0.04[d] |

[a] Means with standard errors. Means represent data from five sheep. Hansard *et al.*, 1978.
[b] Bone, ash weight basis.
[c,d,e] Means in the same column with different superscripts are different ($P < 0.05$).

## REFERENCES


Baker, D. H., and B. A. Molitoris. 1975. Lack of response to supplemental tin, vanadium, chromium and nickel when added to a purified crystalline amino acid diet for chicks. Poult. Sci. 54:925.

Berg, L. R. 1963. Evidence of vanadium toxicity resulting from the use of certain commercial phosphorus supplements in chick rations. Poult. Sci. 42:766.

Berg, L. R. 1966. Effect of diet composition on vanadium toxicity for the chick. Poult. Sci. 45:1346.

Berg, L. R., and W. W. Lawrence. 1971. Cottonseed meal, dehydrated grass and ascorbic acid as dietary factors preventing toxicity of vanadium for the chick. Poult. Sci. 50:1399.

Berg, L. R., G. E. Bearse, and L. H. Merrill. 1963. Vanadium toxicity in laying hens. Poult. Sci. 42:1407.

Bernheim, F., and M. L. C. Bernheim. 1938. Action of vanadium on tissue oxidations. Science 88:481.

Bruffy, G. R., and R. P. Dowdy. 1979. Effect of dietary vanadium on cholesterol metabolism in guinea pigs. Fed. Proc. 38:450. (Abstr.)

Busch, P. M. 1961. Vanadium: A Materials Survey. Bureau of Mines. I. C. 8060. U.S. Department of the Interior, Washington, D.C.

Cantley, L. C., Jr., and P. Aisen. 1979. The fate of cytoplasmic vanadium, implications on (Na, K)-ATPase inhibition. J. Biol. Chem. 254:1781.

Cantley, L. C., Jr., L. Josephson, R. Warner, M. Yanagisawa, C. Lechene, and G. Guidotti. 1977. Vanadate is a potent (Na, K)-ATPase inhibitor found in ATP derived from muscle. J. Biol. Chem. 252:7421.

Cantley, L. C., Jr., L. G. Cantley, and L. Josephson. 1978. A characterization of vanadate interactions with the (Na, K)-ATPase mechanistic and regulatory implications. J. Biol. Chem. 253:7361.

Comar, D., and F. Chevallier. 1967. Concentration du vanadium chez le rat et son influence sur la synthese du cholesterol, etudiees par la technique de radioactivation neutronique et la methode d'equilibre isotopique. Bull. Soc. Chim. Biol. 49:1357.

Curran, G. L. 1954. Effect of certain transition group elements on hepatic synthesis of cholesterol in the rat. J. Biol. Chem. 210:765.

Curran, G. L., and R. E. Burch. 1967. Biological and Health Effects of Vanadium, p. 96. *In* Proceedings of the First Annual Conference on Trace Substances in Environmental Health, University of Missouri, Columbia.

Curran, G. L., D. L. Azarnoff, and R. E. Bolinger. 1959. Effect of cholesterol synthesis inhibition in normocholesteremic young men. J. Clin. Invest. 38:1251.

Daniel, E. P., and R. D. Lillie. 1938. Experimental vanadium poisoning in the white rat. Public Health Rep. 53:765.

DeMaster, E. G. 1972. Inhibition of energy-metabolism by vanadium (V) oxyanions and heteropolytungstates. Ph.D. thesis. Wayne State University, Detroit, Mich.

Drebickas, V. 1966. Effect of additions of vanadium and titanium salts on some physiological indexes of calves. Liet. TSR Aukst. Mokykly Mokslo Darbia, Biol. 6:71.

Fairhall, L.T. 1949. Industrial Toxicology. Williams & Wilkins Co., Baltimore, Md.

Faulkner-Hudson, T. G. 1964. Vanadium. Toxicology and Biological Significance. Elsevier Publishing Co., New York.

Franke, K. W., and A. L. Moxon. 1937. The toxicity of orally ingested arsenic, selenium, tellurium, vanadium and molybdenum. J. Pharmacol. Exp. Ther. 61:89.

Goldschmidt, V. M. 1958. Vanadium. *In* A. Muir (ed.). Geochemistry. Clarendon Press, Oxford.

Goodno, C. C. 1979. Inhibition of myosin ATPase by vanadate ion. Proc. Natl. Acad. Sci. 76:2620.

Grupp, G., I. Grupp, C. L. Johnson, E. T. Wallick, and A. Schwartz. 1979. Effect of vanadate on cardiac contraction and adenylate cyclase. Biochem. Biophys. Res. Commun. 88:440.

Hadjimarkos, D. M. 1966. Vanadium and dental caries. Nature 209:1137.

Hafez, Y. S. M., and F. H. Kratzer. 1976a. The effect of diet on the toxicity of vanadium. Poult. Sci. 55:918.

Hafez, Y. S. M., and F. H. Kratzer. 1976b. The effect of pharmacological levels of dietary vanadium on the egg production, shell thickness and egg yolk cholesterol in laying hens and Coturnix. Poult. Sci. 55:923.

Hansard, S. L. II. 1975. Toxicity and physiological movement of vanadium in the sheep and rat. Ph.D. thesis. University of Florida, Gainesville.

Hansard, S. L. II, C. B. Ammerman, K. R. Fick, and S. M. Miller. 1978. Performance and vanadium content of tissues in sheep as influenced by dietary vanadium. J. Anim. Sci. 46:1091.

Hathcock, J. N., C. H. Hill, and G. Matrone. 1964. Vanadium toxicity and distribution in chicks and rats. J. Nutr. 82:106.

Hathcock, J. N., C. H. Hill, and S. B. Tove. 1966. Uncoupling of oxidative phosphorylation by vanadate. Can. J. Biochem. 44:983.

Healy, W. B. 1973. Nutritional aspects of soil ingestion by grazing animals. *In* G. W. Butler and R. W. Baily (eds.). Chemistry and Biochemistry of Herbage, vol. I. Academic Press, New York.

Heege, J. H. T. 1964. Poisoning of cattle by ingestion of fuel oil soot. Tijdschr. Diergeneesk. 89:1300.

Hill, C. H. 1979. The effect of dietary protein levels on mineral toxicity in chicks. J. Nutr. 109:501.

Hopkins, L. L., Jr., and H. E. Mohr. 1971a. Effect of vanadium deficiency on plasma cholesterol of chicks. Fed. Proc. 30:462. (Abstr.)

Hopkins, L. L., Jr., and H. E. Mohr. 1971b. The biological essentiality of vanadium, pp. 195–213. *In* W. Mertz and W. E. Cornatzer, eds. Newer Trace Elements in Nutrition. Marcel Dekker, New York.

Hopkins, L. L., Jr., and H. E. Mohr. 1974. Vanadium as an essential nutrient. Fed. Proc. 33:1773.

Hopkins, L. L., Jr., and B. E. Tilton. 1966. Metabolism of trace amounts of vanadium[48] in rat organs and liver subcellular particles. Am. J. Physiol. 211:169.

Hunt, C., and F. H. Nielsen. 1979. The interaction between vanadium and chromium in the chick. Fed. Proc. 38:449.

Jha, K. K. 1966. Effect of molybdenum, vanadium, tungsten and cobalt on the growth of *Rhizobium*. Indian J. Microbiol. 6:29.

Johnson, J. L., H. J. Cohen, and K. V. Rajagopalan. 1974. Studies of vanadium toxicity in the rat. Lack of correlation with molybdenum utilization. Biochem. Biophys. Res. Commun. 56:940.

Lillie, R. J. 1970. Vanadium. *In* Air Pollutants Affecting the Performance of Domestic Animals—A Literature Review. Agric. Handb. No. 380. U.S. Department of Agriculture, Washington, D.C.

Martinez, A., and D. C. Church. 1970. Effect of various mineral elements on *in vitro* rumen cellulose digestion. J. Anim. Sci. 31:982.

Mascitelli-Coriandoli, E., and C. Citterio. 1959. Effects of vanadium upon liver coenzyme A in rats. Nature 183:1527.

Miller, E. C., H. Menge, and C. A. Denton. 1961. Effect of type of dietary fat on plasma and liver cholesterol concentration in female chicks. J. Nutr. 75:367.

Mitchell, R. L. 1957. Emission spectrochemical analysis. Determination of trace elements in plants and other biological materials, pp. 398–412. *In* J. H. Yoe and H. J. Koch (eds.). Trace Analysis. John Wiley & Sons, New York.

Mitchell, W. G., and E. P. Floyd. 1954. Ascorbic acid and ethylenediaminetetraacetate (EDTA) as antidotes in experimental vanadium poisoning. Proc. Soc. Exp. Biol. Med. 85:206.

Mountain, J. T., L. L. Delker, and H. E. Stokinger. 1953. Studies in vanadium toxicology. Reduction in the cystine content of rat hair. AMA Arch. Ind. Hyg. Occup. Med. 8:406.

Mountain, J. T., W. D. Wagner, and H. E. Stokinger. 1959. Effects of vanadium on growth, cholesterol metabolism and tissue components in laboratory animals on various diets. Fed. Proc. 18:425. (Abstr.)

Moxon, A. L., and M. Rhian. 1943. Selenium poisoning. Physiol. Rev. 23:305.

Muhler, J. C. 1957. The effect of vanadium pentoxide, fluorides, and tin compounds on the dental experience of rats. J. Dental Res. 36:787.

National Research Council. 1974. Vanadium. National Academy of Sciences, Washington, D.C.

Natusch, D. F. S., J. R. Wallace, and C. N. Evans, Jr. 1973. Toxic trace elements: Preferential concentration in respirable particles. Science 183:202.

Nechay, B. R., and J. P. Saunders. 1978. Inhibition by vanadium of sodium and potassium dependent adenosinetriphosphatase derived from animal and human tissues. J. Environ. Pathol. Toxicol. 2:247.

Nelson, T. S., M. B. Gillis, and H. T. Peeler. 1962. Studies of the effect of vanadium on chick growth. Poult. Sci. 41:519.

Nieder, G. L., C. N. Corder, and P. A. Culp. 1979. The effect of vanadate on human kidney potassium dependent phosphatase. Arch. Pharm. 307:191.

Nielsen, F. H., and D. R. Myron. 1976. Evidence which indicates a role for vanadium in labile methyl metabolism in chicks. Fed. Proc. 35:683 (Abstr.).

Nielsen, F. H., and D. A. Ollerich. 1973. Studies on a vanadium deficiency in chicks. Fed. Proc. 32:929. (Abstr.)

Nielsen, F. H., and E. O. Uthus. 1977. The effect of vanadium deficiency on the activity of some enzymes involved with labile methyl and methionine metabolism in the chick. Fed. Proc. 36:1123. (Abstr.)

Parker, R. D. R., and R. P. Sharma. 1978. Accumulation and depletion of vanadium in selected tissues of rats treated with vanadyl sulfate and sodium orthovanadate. J. Environ. Pathol. Toxicol. 2:235.

Pepin, G., G. Bouley, and C. Boudene. 1977. Toxicological study of vanadium after intramuscular and intratrachial injection in the rat. C. R. Acad. Sci. Paris 285:451.

Platonow, N., and H. K. Abbey. 1968. Toxicity of vanadium in calves. Vet. Rec. 82:292.

Proescher, F., H. A. Seil, and A. W. Stillians. 1917: Contribution to the action of vanadium with particular reference to syphilis. Am. J. Syph. 1:347.

Romoser, G. L., L. Loveless, L. J. Machlin, and R. S. Gordon. 1960. Toxicity of vanadium and chromium for the growing chicken. Poult. Sci. 39:1288.

Romoser, G. L., W. A. Dudley, L. J. Machlin, and L. Loveless. 1961. Toxicity of vanadium and chromium for the growing chick. Poult. Sci. 40:1171.

Sabbioni, E., E. Marafante, L. Amantini, L. Ubertalli, and C. Birattari. 1978. Similarity in metabolic patterns of different chemical species of vanadium in the rat. Bioinorg. Chem. 8:503.

Schroeder, H. A., and J. J. Balassa. 1967. Arsenic, germanium, tin and vanadium in mice: Effects on growth, survival and tissue levels. J. Nutr. 92:245.

Schroeder, H. A., and M. Mitchener. 1975. Life-term effects of mercury, methyl mercury and nine other trace metals on mice. J. Nutr. 105:452.

Schroeder, H. A., M. Mitchener, and A. P. Nason. 1970. Zirconium, niobium, antimony, vanadium and lead in rats: Life term studies. J. Nutr. 100:59.

Schwarz, K., and D. B. Milne. 1971a. Growth effects of vanadium in rats in a trace element controlled environment. Fed. Proc. 30:462. (Abstr.)

Schwarz, K., and D. B. Milne. 1971b. Growth effects of vanadium in the rat. Science 174:426.

Scott, K. G., J. G. Hamilton, and P. C. Wallace. 1951. Deposition of carrier-free vanadium in the rat following intravenous administration. Report UCRL-1318. University of California Radiation Laboratory.

Sjöberg, S. G. 1950. Vanadium pentoxide dust. A clinical and experimental investigation on its effect after inhalation. Acta Med. Scand. 138:238.

Söremark, R. 1967. Vanadium in some biological specimens. J. Nutr. 92:183.

Söremark, R., and S. Üllberg. 1962. Distribution and kinetics of $^{48}V_2O_5$ in mice. *In* N. Fried (ed.). Use of Radioisotopes in Animal Biology and the Medical Sciences, vol. 2. Academic Press, New York.

Söremark, R., S. Üllberg, and L. Appelgren. 1962. Autoradiographic localization of V-48-labelled vanadium pentoxide in developing teeth and bones of rats. Acta Odontol. Scand. 20:225.

Stokinger, H. E. 1955. Organic beryllium and vanadium dusts. A review. AMA Arch. Ind. Health 12:675.

Stokinger, H. E. 1963. Vanadium. *In* F. A. Patty, ed. Industrial Hygiene and Toxicology, vol. II. Toxicology, 2nd ed. Interscience Publishers, New York.

Strasia, C. A. 1971. Vanadium: Essentiality and toxicity in the laboratory rat. Ph.D. thesis. Purdue University, West Lafayette, Ind.

Summers, J. D., and E. T. Moran, Jr. 1972. Interaction of dietary vanadium, calcium and phosphorus for the growing chicken. Poult. Sci. 51:1760.

Talvitie, N. A., and W. D. Wagner. 1954. Studies in vanadium toxicology; distribution and excretion of vanadium in animals. Arch. Ind. Hyg. 9:414.

Thomassen, P. R., and H. M. Leicester. 1964. Uptake of radioactive beryllium, vanadium, selenium, cerium and yttrium in the tissues and teeth of rats. J. Dent. Res. 43:346.

Thornton, I. 1974. Biogeochemical and soil ingestion studies in relation to trace-element nutrition of livestock. *In* W. G. Hoekstra, J. W. Suttie, H. E. Ganther, and W. Mertz, eds. Trace Element Metabolism in Animals—2. University Park Press, Baltimore, Md.

Underwood, E. J. 1962. Trace Elements in Human and Animal Nutrition, 2nd ed. Academic Press, New York.

Underwood, E. J. 1977. Trace Elements in Human and Animal Nutrition, 4th ed., pp. 416–424. Academic Press, New York.

Vinogradov, A. P. 1959. The Geochemistry of Rare and Dispersed Elements in Soil, 2nd ed. Consultants Bureau, Inc., New York.

Waters, M. D., D. E. Gardner, C. Aranyi, and D. L. Coffin. 1975. Metal toxicity for rabbit alveolar macrophages *in vitro*. Environ. Res. 9:32.

White, D. H., and M. P. Dieter. 1978. Effects of dietary vanadium in mallard ducks. J. Toxicol. Environ. Health 4:43.

Williams, D. L. 1973. Biological value of vanadium for rats, chickens, and sheep. Ph.D. thesis. Purdue University, West Lafayette, Ind.

Wright, W. R. 1968. Metabolic interrelationship between vanadium and chromium. Ph.D. thesis. North Carolina State University, Raleigh.

Yen, T. F. 1972. Terrestrial and extraterrestrial stable organic molecules. *In* R. F. Landel and A. Rembaum (eds.). Chemistry in Space Research. Elsevier Publishing Co., New York.

# Zinc

Zinc (Zn) has been used by man for utilitarian or ornamental purposes for almost 2,000 years. Semitic bronzes dating from 1400 to 1000 B.C. have been found to contain as much as 23 percent zinc. The metal is bluish-white, lustrous, relatively soft, and forms distorted hexagonal closely packed structures. Zinc constitutes 0.002 percent of the earth's crust. The principal zinc-bearing ores are sphalerite and wurzite and their weathering products, particularly smithsonite and hemimorphite. Minor ores include zincite and willemite.

The principal commercial uses of zinc are for galvanizing iron, as a component of alloys (bronze, brass, Babbitt metal, German silver, and special alloys for die casting), and in dry cell batteries, castings, and printing plates. Zinc oxide is used in rubber because of its high heat capacity, conductivity, and capacity for scavenging free sulfur. Inorganic salts of zinc are used in ceramic glazes and in dyeing. Two fungicides, zineb and ziram, are organozinc compounds. The zinc on galvanized iron is a sacrificial coating that is stable in dry atmospheres, but forms a film of gray hydrated basic carbonate in moist atmospheres. The American Society of Testing Materials has established maximal levels of lead, iron, and cadmium for five grades of slab zinc. The permissible levels for the highest and lowest grades are 0.003 and 1.60 percent lead, 0.003 and 0.05 percent iron, and 0.003 and 0.50 percent cadmium, respectively. Contact of foods with galvanized metal can lead to contamination, not only with zinc, but also with the toxic elements lead and cadmium. Zinc usage in the United States in 1975 (metric tons) was as follows: for galvanizing, 341,906; in zinc-base alloys, 312,516; in brass, 236,172; as the oxide, 120,048; as rolled zinc, 24,773; and other uses, 82,069.

Zinc is known to be an important essential mineral for numerous species. Many animal diets require supplementation with a concentrated form of the element due either to low total amounts of zinc or to its low bioavailability from the diet. The level of available dietary zinc can influence the net absorption, metabolism, and function of other elements. Signs of zinc toxicity include decreased growth rate, anemia, decreased bone mineralization, bone deformities, and decreased feather pigmentation. The choice of appropriate forms and levels of zinc supplements poses numerous practical problems ranging from adequacy to safety.

The most extensive reviews of zinc are those by the National Research Council (1978), two symposia edited by A. S. Prasad (1966, 1976), and a chapter by Underwood (1977).

## ESSENTIALITY

Zinc is required for normal growth, development, and function in all animal species that have been studied. Severe deficiency can lead to death. Lesser degrees of deficiency are most pronounced during periods of rapid growth and in those cells or tissues that either turn over or grow most rapidly. Characteristics of deficiency include growth retardation, delayed sexual maturation, alopecia, abnormal feathering, skin lesions, hyperkeratinization of the esophagus, reduced numbers of circulating lymphocytes, skeletal abnormalities, impaired reproduction in both males and females, and fetal abnormalities.

The importance of zinc supplementation for a food-producing animal was demonstrated in 1955, when it was shown that parakeratosis in swine was due to inadequate dietary zinc (Tucker and Salmon, 1955). This disease, which caused great economic losses, was precipitated by high levels of dietary calcium in the presence of vegetable proteins containing phytate. The bioavailability of zinc was markedly reduced so that a severe deficiency resulted.

Zinc requirements for young domestic animals and fowl range from approximately 40 to 100 ppm in the diet.

## METABOLISM

Zinc is absorbed from the intestinal tract in relation to need and the primary route of excretion is in the feces. In addition to unabsorbed zinc, small amounts of fecal zinc derive from bile, pancreatic secretions, desquamated epithelial cells, and zinc secreted directly into the gut along its length. Small amounts of zinc are lost in urine, sweat and shed integumental tissues. The highest concentrations of zinc are found in the *tapetum lucidum*—8.5 and 13.8 percent dry weight for the dog and fox, respectively. The iris and choroid also contain high concentrations. The prostate and its secretions are high in zinc. Otherwise, soft tissue concentrations range from approximately 12 to 55 ppm wet weight, varying somewhat between studies and species (Underwood, 1977).

The movement of zinc into and within the body is precisely regulated at levels of intake within the requirement range. Zinc binds to sulfhydryl, amino, imidazole, and phosphate groups; thus, amino acids, proteins, nucleic acids, and other organic molecules bind zinc under physiological conditions. In general, readily available stores of zinc are quite small, as dramatically reflected by drops in the plasma zinc values to the deficiency range within 24 hours after changing to diets very low in zinc.

Zinc activates some enzymes and is a component of a large number of important metalloenzymes (Riordan and Vallee, 1976). The latter include carbonic anhydrase, carboxypeptidases A and B, alcohol dehydrogenase, glutamic dehydrogenase, D-glyceraldehyde-3-phosphate dehydrogenase, lactic dehydrogenase, malic dehydrogenase, alkaline phosphatase, aldolase, superoxide dismutase, ribonuclease, DNA polymerase, and others. The metal is located at the active site of the zinc metalloenzymes and is involved in the catalytic process. Zinc may function in maintaining the secondary, tertiary, or quaternary structure, depending on the enzyme. Zinc also plays a role in the configuration of DNA and RNA. The biochemical functions of zinc have recently been reviewed (Chesters, 1978).

## SOURCES

Underwood (1977) summarized information on the zinc content of animal feeds. Values for pasture herbage ranged between 17 and 60 ppm dry weight, with most values falling between 20 and 30 ppm. Industrial pollution increased the zinc content of grass from 5- to 50-fold (Mills and Dalgarno, 1972). Cereal grains typically contain 20 and 30 ppm zinc, whereas soybean, peanut, and linseed meal contain 50–70 ppm. Fish meal, whale meal, and meat meal may contain 90–100 ppm zinc.

The present drinking water standard is 5 ppm zinc (National Research Council, 1978). This concentration of zinc is almost never found in surface water, municipal drinking water supplies, or in drinking water collected at the home tap. Industrial pollution, such as that derived from dumping plating baths or mining operations, can produce very high concentrations of zinc. Streams tend to become purified by precipitation of zinc with clay sediments or hydrous iron and manganese oxides. A concentration of 25 ppm zinc was recommended as a safe upper limit in drinking water for livestock and poultry (National Research Council, 1974).

Inorganic salts of zinc, such as the oxide, carbonate, acetate, chloride, or sulfate, and even metallic zinc serve as readily available sources for the animal. Even those salts that are insoluble in water are solubilized by gastric juice. Only the zinc in a few ores was found to be unavailable to the chick (Edwards, 1959). Crude sources of zinc should be checked for cadmium and lead. Dietary phytate, which can be supplied by whole seeds and certain seed fractions, decreases the availability of zinc for animals without a functional rumen.

Contamination of food and water with large amounts of zinc can occur upon storage in galvanized containers, particularly under acidic conditions. This type of exposure can be fatal but most frequently produces gastrointestinal distress, emesis, and/or refusal of the animal to eat the food. Other potential sources of excess zinc include pesticides, fungicides, and industrial pollution.

## TOXICOSIS

### LOW LEVELS

In most of the studies summarized in Table 41, no adverse physiological effects were observed at dietary concentrations lower than 600 ppm zinc. In several studies animals appeared grossly normal with far larger dietary zinc concentrations. In studies where graded levels of zinc were given, the most sensitive response was an increase in tissue zinc concentrations. In studies with low zinc levels that did not affect body weight, decreases in rate of gain or losses of body weight were first observed with dietary zinc levels of 900 ppm in cattle (Ott *et al.*, 1966a);

1,500 ppm in sheep (Ott *et al.*, 1966b); 2,000 ppm in swine (Brink *et al.*, 1959); 800, 1,500, and 2,000 ppm in chicks (Berg and Martinson, 1972; Roberson and Schaible, 1960; and Johnson *et al.*, 1962, respectively); 4,000 ppm in turkeys (Vohra and Kratzer, 1968); and 270 ppm in young Japanese quail (Hamilton *et al.*, 1979). Davies *et al.* (1977) found that week-old lambs were very sensitive to high zinc levels.

Miller *et al.* (1965) observed no adverse health effects or changes in milk quantity or composition in cows fed 372 to 1,279 ppm zinc as the oxide for 6 weeks. Small increases of zinc in milk with higher doses were less than corresponding increases in plasma.

Dogs fed excess zinc in meals were unaffected by 400 mg zinc per day, and the only effect of 800 mg zinc was to increase tissue concentrations of zinc (Drinker *et al.*, 1927). Daily intakes of 59–196, 180, or 200 mg zinc did not affect cats (Drinker *et al.*, 1927; Scott and Fisher, 1938; Mannell, 1967).

Mice receiving 500 ppm zinc as the sulfate in their drinking water appeared grossly normal after 1–14 months (Aughey *et al.*, 1977). Histological examination revealed hypertrophy of the adrenal cortex and pancreatic islets. There was also evidence of pituitary hyperactivity. This level of zinc would be similar to about 1,000 ppm in the diet based on total consumption.

## HIGH LEVELS

In most studies, supplemental zinc at 1,000 ppm in the diet or more caused some adverse physiological effect. Reduced weight gains; anemia; reduced bone ash; decreased tissue concentrations of iron, copper, and manganese; and diminished utilization of calcium and phosphorus were observed. Increased consumption of a mineral mixture available *ad libitum* was observed in cattle and sheep fed excess zinc (Ott *et al.*, 1966a,b). The cattle also chewed wood. Sheep fed 750 ppm zinc, beginning with the sixth week of pregnancy, produced almost no viable lambs (Campbell and Mills, 1979). Sheep dosed intraruminally with zinc sulfate developed diarrhea, lost weight, and died (Smith, 1977). These changes were slow to develop with 20 mg zinc per kilogram of body weight, but rapid with 180 mg per kilogram.

Grimmett *et al.* (1937) and Sampson *et al.* (1942) observed arthritis and severe bone and cartilage abnormalities in the joints of the long bones in pigs fed milk containing 268 ppm zinc plus small amounts of grains. Brink *et al.* (1959) also observed an arthritis-like syndrome, internal hemorrhaging, and some mortality in swine receiving 2,000 or

4,000 ppm zinc. Cox and Hale (1962) and Hsu *et al.* (1975) did not observe these changes in swine fed the same high levels of zinc. Willoughby *et al.* (1972) produced severe swelling in the epiphyseal region of long bones in horses receiving high levels of zinc that were increased gradually from 25 to 186 mg per kilogram of body weight. The horses became lame, anemic, had increased tissue levels of zinc, and grew more slowly. Lameness was also observed in mallard ducks fed 3,000 to 12,000 ppm zinc (Gasaway and Buss, 1972). The ducks were anemic, lost weight, and most of them eventually died.

Human beings exposed to excess zinc by the oral route have described an unpleasant taste, gastrointestinal discomfort, and dizziness, responses that animals cannot communicate. Cats were found to vomit or refuse to eat a meal containing 320 or 400 mg zinc as the oxide (Scott and Fisher, 1938). Cows in two dairy herds accidentally received feed in which magnesium oxide was replaced by zinc oxide (Allen, 1968). These high levels of zinc, 72 and 145 g per day, immediately produced scours and declines in food consumption and milk production. With the higher level, pulmonary emphysema, hemolytic anemia, and death were observed.

Table 42 summarizes data on acute toxicosis of single oral doses of zinc salts in small animals.

## FACTORS INFLUENCING TOXICITY

Variability in response to excess levels of zinc is not unexpected when one examines the extensive literature dealing with factors that can affect zinc toxicity. Most of these studies have been carried out in rats rather than food-producing or companion animals, which are the subject of this report.

The first defense against orally administered excess zinc is the homeostatic mechanism(s) that limits absorption. By the use of $^{65}Zn$, it was shown that the initial whole-body retention of zinc by rats after dosing was reduced by a factor of 3 when the dietary intake was 6 to 10 times normal (Furchner and Richmond, 1962). By more extensive studies in Holstein bull calves, Miller *et al.* (1970, 1971) showed that homeostatic mechanisms regulating zinc metabolism became markedly less effective with 600 ppm dietary zinc as compared with 200 ppm. Neither excess level caused any physiological abnormalities. With tracer doses of $^{65}ZnCl_2$ given at various time periods after feeding 600 ppm zinc to calves, Stake *et al.* (1975) showed that deterioration in homeostasis affected specific tissues and organs at different rates.

Numerous studies in rats have shown decreases in tissue levels of iron, copper, and copper-containing enzymes when excess zinc was fed. Supplements of iron and copper were usually beneficial. The antagonism of copper by zinc is very sensitive. This was shown dramatically by Hill and Matrone (1962) in the chick. With a low-copper diet that permitted normal growth but with slightly lowered hemoglobin, the dietary addition of 100 ppm zinc caused growth depression and mortality. With 200 ppm supplemental zinc, there was still further reduction of growth and hemoglobin and an increase in mortality.

Young Japanese quail (*Coturnix coturnix japonica*) fed 1 ppm copper, a marginally deficient level, were more sensitive to the effects of excess zinc than birds fed 1.5 ppm, the requirement, or 3.6 ppm copper (Hamilton *et al.,* 1979). The greater sensitivity with low-copper intake was manifested by decreased body weight, lack of feather pigmentation, and in some cases by perosis. As little as 31.2, 62.5, and 125 ppm zinc (in excess of the 20 ppm in the basal diet) produced significant adverse effects with 1 ppm dietary copper. Concentrations of zinc and manganese in the duodenum and liver were not affected by dietary copper level; however, iron concentration in the liver was consistently lower with the higher level of copper. Supplements of ascorbic acid augmented the adverse effect of excess zinc on growth, feather pigmentation, and bone deformities in young quail fed a diet marginally deficient in copper (Fox *et al.,* 1978). The sensitive bone–joint abnormalities described in swine (Grimmett *et al.,* 1937; Sampson *et al.,* 1942) probably were related to a low-copper intake. Rats fed 1,200 ppm zinc had decreased concentrations of elastin in aorta, skin, and cartilage, changes suggestive of copper antagonism (Philip and Kurup, 1978).

All possible interactions between normal and high levels of zinc, lead, and cadmium, and deficient and normal levels of calcium and vitamin D, were studied in rats (Thawley *et al.,* 1977). High zinc plus high cadmium produced more severe anemia than either alone. A low-serum iron level due to feeding high zinc was further reduced by low calcium or high vitamin D. In rats deficient in calcium, high zinc augmented the porotic process (Ferguson and Leaver, 1972). With a deficiency of calcium and vitamin D, zinc caused both porosis and osteomalacia, whereas in vitamin D deficiency zinc had a porotic effect on bone and a mineralizing effect on dentine.

The level of dietary selenium was shown to be important in zinc toxicity (Jensen, 1975). Chicks fed a natural-ingredient diet containing 0.2 ppm selenium and 8.8 IU added vitamin E showed exudative diathesis, muscular dystrophy, decreased weight gains, and mortality beginning with 2,000 ppm excess zinc. The mortality, exudative diathesis, and muscular dystrophy did not occur when 0.5 ppm selenium was added to the diet with zinc levels up to 4,000 ppm. The supplemental selenium did not affect body weight.

With the exception of the paper by Hamilton *et al.* (1979), diets fed in the studies summarized in Table 41 were composed of nonpurified ingredients supplemented with minerals. The basal diet of Hamilton *et al.* (1979) was a casein gelatin diet that contained required levels only of zinc, iron, manganese, and magnesium. Copper was fed at various levels, as discussed above, and other nutrients were near the required amounts insofar as known. It is thought that sensitivity to excess zinc with this diet reflects the level of essential nutrients, although the rapid growth rate of the young quail may have been a contributory factor. McCall *et al.* (1961) reported that rats fed diets containing 20 and 30 percent protein from soybean oil meal responded more favorably to high dietary zinc than when casein supplied the same levels of protein. Berg and Martinson (1972) reported greater sensitivity of chicks to excess zinc with a sucrose–fish meal diet than with corn–fish meal, sucrose–soybean, or corn–soybean diets. Growth of control birds was somewhat less with the sucrose–fish meal diet. Replacement of sucrose with corn (15 to 67 percent of the diet) effected incremental improvements in growth of birds fed the basal diet alone or with 2,000 ppm zinc. Very young sheep were affected more adversely by zinc in the form of yeast than as zinc sulfate (Davies *et al.,* 1977). The yeast supplied all of the dietary protein, whereas with zinc sulfate the diet was milk.

Numerous investigators have reported improvements in response to toxic levels of zinc with a wide range of dietary supplements in addition to those described above. These include calcium and phosphorus (Stewart and Magee, 1964; Hsu *et al.,* 1975), liver extract (Smith and Larsen, 1946; Magee and Matrone, 1960), distiller's dried solubles (Magee and Spahr, 1964), and ethylenediaminetetraacetic acid (Vohra and Kratzer, 1968). It is possible that dietary phytate and other food components, which can decrease the bioavailability of required levels of zinc, may also decrease the toxicity of zinc.

The effect of feeding low levels of three pesticides (P,P′-DDT, 2.4 ppm; parathion, 0.33 ppm; and carbaryl, 2.3 ppm) with or without 7,000 ppm zinc was studied in female rats for a 4-month period that included a pregnancy (Feaster *et al.,* 1972). The pesticides alone had no adverse effects. High zinc alone did not affect hemoglobin, but when combined with pesticides there was a significant decline of hemoglobin in both maternal and fetal blood. The high zinc level caused increased concentrations of one or more of the pesticides in maternal abdominal fat and liver and in the fetal liver.

The recognition of chronic zinc poisoning in animals should be followed by prompt removal of the source of exposure. Ott *et al.* (1966c) fed calves 2,100 ppm zinc as the oxide for 12 weeks and then with no zinc supplement for 6 weeks. Serum zinc declined rapidly during the first 2 weeks after excess zinc removal; however, by 6 weeks it was still slightly above the normal range. The high concentrations of zinc and iron in the liver at 12 weeks declined when zinc was removed. At the end of the 6-week period liver zinc was approximately 4 times normal and liver iron approximately twice normal levels.

Johnson *et al.* (1962) fed chicks graded levels of zinc from hatching to 10 weeks of age; the excess zinc was deleted between 10 and 16 weeks of age. Birds fed the levels of zinc that depressed growth, 3,000, 4,000, and 5,000 ppm, gained as much weight between 10 and 16 weeks as birds that were unaffected by the initial supplemental zinc. Although some of the initial supplements of zinc increased liver zinc concentrations several-fold, the values were normal by 16 weeks.

Attempts to reduce zinc toxicosis in turkey poults by feeding ethylenediaminetetraacetic acid (EDTA) with very high levels of zinc were unsuccessful (Vohra and Kratzer, 1968). Addition of 15.4 or 30.8 millimoles of EDTA per kilogram of diet did not affect the growth of either controls or zinc-fed birds.

## TISSUE LEVELS

The most sensitive responses to excess dietary zinc were increases in zinc concentrations in serum, liver, kidney, pancreas, and small intestine with bone almost as responsive. Sometimes zinc was higher in the heart, but always remained unchanged in skeletal muscle. Increases of zinc levels in one or more tissues have been observed with dietary zinc levels of 200 and 500 ppm fed to cattle (Miller *et al.*, 1970; Ott *et al.*, 1966c, respectively), 1,000 ppm fed to sheep (Ott *et al.*, 1966d), 2,000 ppm fed to chicks (Johnson *et al.*, 1962), and 125 ppm fed to Japanese quail (Hamilton *et al.*, 1979). In general, tissue concentrations are related to dose level. The wide variations in diet composition and levels of zinc tested preclude precise comparisons to establish differences between species. Murphy *et al.* (1975) summarized data on human foods. Doyle and Spaulding (1978) compiled data on zinc in liver, kidney, heart, and muscle of normal cattle, sheep, swine, and chickens.

## MAXIMUM TOLERABLE LEVELS

The data in Table 41 and the discussion of factors that can either increase or decrease the severity of zinc toxicosis emphasize the difficulty of establishing a maximum safe level. The types of diets of concern here are practical diets composed primarily of natural ingredients. With this type of diet, usually containing most or all nutrients in at least modest excess of requirements, one would not expect any adverse physiological effects of zinc at 500 to 600 ppm. Since a marked decline in the homeostatic control of zinc occurred in cattle fed 600 ppm zinc, it would seem desirable to limit zinc levels to no more than 500 ppm. Sheep fed 750 ppm zinc during pregnancy produced almost no viable young, whereas 150 ppm zinc had no adverse effect. The maximum tolerable level for sheep was set at 300 ppm. Swine, turkeys, and chickens performed normally with 1,000 ppm zinc in nonpurified diets, so this was set as the maximum tolerable level for swine and poultry. With an adequate purified diet that contained most essential elements at only required levels, 125 ppm zinc caused adverse effects in young Japanese quail. Thus, caution is advised for pregnant animals and for animals fed diets with essential nutrients at required levels only.

## SUMMARY

Zinc is an important essential nutrient that is required at every stage of the life cycle. It functions in a large number of zinc metalloenzymes. For most species, overt toxicosis of zinc first appears when levels around 1,000 ppm are incorporated into a natural-ingredient diet with many nutrients above required levels. Lower levels of excess zinc overwhelm the body's mechanism for regulating zinc metabolism and effect changes in tissue concentrations of zinc and several other minerals. With diets containing marginal levels of some minerals, much less zinc produces adverse health effects. Signs of zinc toxicosis may include gastrointestinal distress, emesis, decreased food consumption, pica, decreased growth, anemia, poor bone mineralization, damage to the pancreas, arthritis, white muscle disease, internal hemorrhaging, and nonviable newborn. From considerations of deranged control of zinc metabolism and overt toxicosis, maximum tolerable levels from 300 to 1,000 ppm zinc in the diet appear to be safe, depending on species.

TABLE 41 Effects of Zinc Administration in Animals

| Class and Number of Animals[a] | Age or Weight | Administration | | | | Effect(s) | Reference |
|---|---|---|---|---|---|---|---|
| | | Quantity of Element[b] | Source | Duration | Route | | |
| Cattle—4 | 127 d | 200 ppm | ZnO | 21 d | Diet | Increased liver, rumen, and small intestine Zn | Miller *et al.*, 1970 |
| | | 600 ppm | | | | Increased pancreas, duodenum, rumen, small intestine, liver, hair, rib, and testicle Zn | |
| | | 600 ppm | $ZnSO_4$ | | | Increased pancreas, duodenum, rumen, small intestine, liver, heart, hair, and rib Zn | |
| Cattle—7 | Young | 100 ppm | ZnO | 12 wk | Diet | No adverse effect | Ott *et al.*, 1966a,c |
| | | 500 ppm | | | | Increased tissue Zn | |
| | | 900 ppm | | | | Decreased weight gain; increased tissue Zn and liver Ca; decreased liver Cu | |
| | | 1,300 ppm | | | | Decreased weight gain and feed efficiency; increased tissue Zn and liver Fe and Ca; decreased liver Cu | |
| | | 1,700 ppm | | | | Decreased weight gain and feed efficiency; increased tissue Zn and liver Fe and | |

TABLE 41 *Continued*

| Class and Number of Animals[a] | Age or Weight | Administration: Quantity of Element[b] | Source | Duration | Route | Effect(s) | Reference |
|---|---|---|---|---|---|---|---|
| | | | | | | Ca; decreased liver Cu | |
| | | 2,100 ppm | | | | Decreased weight gain and feed efficiency; increased mineral intake, pica; increased tissue Zn and liver Fe and Ca; decreased liver Cu | |
| Cattle—91 | Adult | 72 g/d | Zinc oxide | 3–4 d | Diet | Scours; decreased milk yield; improved by 5 d; nine cows seriously affected | Allen, 1968 |
| Cattle—102 | Adult | 145 g/d | | | | Scours by 2 d; deaths (seven) by 3–4 d with pulmonary emphysema, flabby myocardium, blood spots on renal cortex, liver pale and friable, and severe hemolytic anemia | |
| Cattle—6 | Young adult | 109 g/d | | 6 d | | Ate diet for 2 d only; no adverse effects | |
| Cattle—6 | Adult | 372 ppm | ZnO | 6 wk | Diet | No adverse effects on cow or milk | Miller *et al.*, 1965 |
| | | 692 ppm | | | | Slightly increased Zn in plasma and milk | |
| | | 1,279 ppm | | | | Small increases in Zn were relatively larger in plasma than in milk | |

| | | | | | | | |
|---|---|---|---|---|---|---|---|
| Sheep—4 | 1 wk | 840 ppm (32.1 mg/ 1,000 kJ) | Yeast | 35 d | Diet | Decreased diet intake and growth; kidneys: enlarged, pale, fibrous tissue in cortex and tubules, and glomerular atrophy | Davies *et al.*, 1977 |
| Sheep—2 | | 32.1 mg/ 1,000 kJ | $ZnSO_4 \cdot 7H_2O$ | 33 d | | Decreased growth; kidneys: enlarged, pale, fibrous tissue in cortex, tubular distortion and glomerular atrophy; elevated Zn and Cu and decreased Fe in cortex and medulla | |
| Sheep—6 | Young | 500 ppm | ZnO | 10 wk | Diet | No adverse effect | Ott *et al.*, 1966b,d |
| | | 1,000 ppm | | | | Decreased feed intake; increased tissue Zn | |
| | | 2,000 ppm | | | | Decreased feed intake, weight gain, hemoglobin, and hematocrit; increased tissue Zn; decreased serum and liver Cu | |
| | | 4,000 ppm | | | | Decreased feed intake, weight gain, hemoglobin, and hematocrit; increased tissue Zn and liver Fe; decreased serum and liver Cu | |
| Sheep—10 | | 500 ppm | | | | No adverse effect | |
| | | 1,000 ppm | | | | Increased tissue Zn | |
| | | 1,500 ppm | | | | Decreased gain and feed intake; increased mineral supplement intake; increased tissue Zn | |
| | | 2,000 ppm | | | | Decraesed gain, feed intake, | |

TABLE 41 *Continued*

| Class and Number of Animals[a] | Age or Weight | Administration: Quantity of Element[b] | Source | Duration | Route | Effect(s) | Reference |
|---|---|---|---|---|---|---|---|
| | | | | | | and hemoglobin; increased mineral intake; increased tissue Zn; decreased serum Cu | |
| | | 2,500 ppm | | | | No gain; decreased feed intake and hemoglobin; increased tissue Zn; decreased serum Cu | |
| | | 3,000 ppm | | | | Decreased gain, feed intake, hemoglobin, and hematocrit; three deaths with acute episode (extension of limbs, convulsions, opisthotonas); increased tissue Zn; decreased serum Cu | |
| | | 3,500 ppm | | | | No gain; decreased feed intake, hemoglobin, and hematocrit; increased mineral intake; 2 deaths (acute episode); increased tissue Zn and liver Fe; decreased serum Cu | |
| Sheep—4 | Young | 2,020 ppm | Zinc sulfate[c] | 14 d | Diet | Decreased net retention, and apparent and true digestibility of Ca; decreased retention of P | Thompson *et al.*, 1959 |
| | | 4,040 ppm | | | | Decreased net retention, and | |

| | | | | | | | |
|---|---|---|---|---|---|---|---|
| | | | | | | apparent and true digestibility of Ca; decreased retention of P | |
| Sheep—2 | 16 mo | 20 mg/kg | Zinc sulfate | 6 wk | Intraruminal | Slow weight loss | Smith, 1977 |
| | | 20 mg/kg | Zinc sulfate | 6 wk | Drench | Continued weight loss, death by 16–35 wk; fatty liver | |
| | | 60 mg/kg | Zinc sulfate | 4 wk | Intraruminal | Diarrhea, weight loss, death; pancreatic necrosis and fibrosis | |
| | | 180 mg/kg | Zinc sulfate | 7–8 d | | Rapid weight loss; death; necrosis lower esophagus, rumen, and abomasal mucosa | |
| | | 480 mg/kg | Zn EDTA | 1 dose | | Rapid rise (by 12 h) serum and urine Zn | |
| | | 480 mg/kg | Zinc sulfate | 1 dose | | High and prolonged (2 d) rise serum Zn, little Zn in urine | |
| | | 480 mg/kg | Zinc oxide | 1 dose | | Gradual (2 d) small rise in serum Zn; no excess Zn in urine | |
| Sheep—6 | Adult, 0–6 wk pregnant | 150 ppm | Zinc sulfate | Gestation | Diet (2.5 ppm Cu) | Almost fourfold increase in lambs' liver Zn | Campbell and Mills, 1979 |
| | | 750 ppm | | | | Reduced weight gain and feed consumption (after 7–10 d); plasma: decreased Cu, ceruloplasmin, and amine oxidase; 85% nonviable lambs; lambs' liver: marked decrease in Cu, eightfold increase in Zn | |

TABLE 41 *Continued*

| Class and Number of Animals[a] | Age or Weight | Administration: Quantity of Element[b] | Source | Duration | Route | Effect(s) | Reference |
|---|---|---|---|---|---|---|---|
| | Adult, 6 wk pregnant | 750 ppm | | ca. 100 d | | No viable lambs; lambs' liver: decreased Cu, 12-fold increase in Zn | |
| | | 750 ppm | | | (10 ppm Cu) | 93% nonviable lambs; lambs' liver: ninefold increase in Zn | |
| Swine—3 | Weanling | 268 ppm | Zinc lactate | 9–12 wk | Milk[d] | Reduced feed intake; rough hair; arthritis; long bones: cartilage erosion, weak and cancellous bone ends; liver necrosis | Grimmett *et al.*, 1937 |
| Swine—5 | 6 wk | 268 ppm | Zinc lactate | 9.5 mo | Milk[d] | Reduced feed intake; rough hair; lame, abnormal bones by 3 mo; autopsy: large joints; thick blood-tinged synovial fluid; thin, dull, irregular cartilage; hemoglobin normal; liver Zn 14 times control | Sampson *et al.*, 1942 |
| Swine—6 | Young | 500 ppm | Zinc carbonate | 6 wk | Diet | No adverse effect | Brink *et al.*, 1959 |
| | | 1,000 ppm | | | | No adverse effect | |
| | | 2,000 ppm | | | | Decreased weight gains; arthritis; internal hemorrhages; two deaths | |
| | | 4,000 ppm | | | | Decreased weight gain and hemoglobin; arthritis and | |

| | | | | | | | |
|---|---|---|---|---|---|---|---|
| | | | | | | internal hemorrhages | |
| Swine—6 | Young | 2,000 ppm | Zinc oxide | 69 d | Diet | Slight scouring; decreased weight gain; increased liver Zn | Cox and Hale, 1962 |
| | | 4,000 ppm | | | | Slight scouring; decreased weight gain and liver Fe; increased liver Zn | |
| Swine—3 | 4 wk | 4,000 ppm | Zinc oxide | 9 wk | Diet | Decreased weight gain; increased tissue Zn | Hsu *et al.*, 1975 |
| Chicken—30 | 1 d | 1,000 ppm | ZnO | 4 wk | Diet | No adverse effect | Roberson and Schaible, 1960 |
| | | 2,000 ppm | | | | No adverse effect | |
| | | 3,000 ppm | | | | Decreased growth | |
| | | 1,000 ppm | $ZnSO_4$ | | | No adverse effect | |
| | | 2,000 ppm | | | | Decreased growth | |
| | | 3,000 ppm | | | | Decreased growth | |
| | | 1,000 ppm | $ZnCO_3$ | | | No adverse effect | |
| | | 2,000 ppm | | | | Decreased growth | |
| | | 3,000 ppm | | | | Decreased growth; mortality | |
| Chicken—40 | | 1,000 ppm | ZnO | | | No adverse effect | |
| | | 1,500 ppm | | | | Decreased growth | |
| | | 1,000 ppm | $ZnSO_4$ | | | No adverse effect | |
| | | 1,500 ppm | | | | Decreased growth | |
| | | 1,000 ppm | $ZnCO_3$ | | | No adverse effect | |
| | | 1,500 ppm | | | | Decreased growth | |
| Chicken—20 | | 252 ppm | Zinc oxide | 9 wk | Diet | No adverse effect | Mehring *et al.*, 1956 |
| | | 371 ppm | | | | No adverse effect | |
| | | 540 ppm | | | | No adverse effect | |
| | | 778 ppm | | | | No adverse effect | |
| Chicken—12 | | 600 ppm | Zinc oxide | 4 wk | Diet | No adverse effect | Kincaid *et al.*, 1976 |

TABLE 41 *Continued*

| Class and Number of Animals[a] | Age or Weight | Administration: Quantity of Element[b] | Source | Duration | Route | Effect(s) | Reference |
|---|---|---|---|---|---|---|---|
| | | 1,200 ppm | | | | No adverse effect | |
| | | 2,400 ppm | | | | Increased liver Zn | |
| Chicken—20 | | 540 ppm | Zinc oxide | 10 wk | Diet | No adverse effect | Johnson *et al.*, 1962 |
| | | 732 ppm | | | | No adverse effect | |
| | | 988 ppm | | | | No adverse effect | |
| | | 1,329 ppm | | | | No adverse effect | |
| | | 1,784 ppm | | | | No adverse effect | |
| | | 2,391 ppm | | | | Decreased growth | |
| | | 3,200 ppm | | | | Decreased growth | |
| | | 1,000 ppm | | | | Increased spleen Cu | |
| | | 2,000 ppm | | | | Decreased growth; increased liver Zn and spleen Cu | |
| | | 3,000 ppm | | | | Decreased growth; increased liver Zn and spleen Cu | |
| | | 4,000 ppm | | | | Decreased growth; increased mortality, liver Zn, and spleen Cu | |
| | | 5,000 ppm | | | | Decreased gain and liver Cu; increased mortality; liver Zn | |
| Chicken—30 | | 200 ppm | ZnO | 2 wk | Diet | No adverse effect | Berg and Martinson, 1972 |
| | | 400 ppm | | | | No adverse effect | |
| | | 800 ppm | | | | Decreased growth[e] | |
| | | 1,200 ppm | | | | Decreased growth and | |

| | | | | | | | |
|---|---|---|---|---|---|---|---|
| | | | | | | bone ash[e] | |
| | | 1,600 ppm | | | | Decreased growth and bone ash[e] | |
| | | 2,000 ppm | | | | Decreased growth and bone ash[e] | |
| Turkey—10 | 5 d | 1,000 ppm | ZnO | 3 wk | Diet | No adverse effect | Vohra and Kratzer, 1968 |
| | | 2,000 ppm | | | | No adverse effect | |
| | | 4,000 ppm | | | | Decreased growth | |
| | | 5,950 ppm | | | | Decreased growth | |
| | | 8,000 ppm | | | | Markedly decreased growth | |
| | | 10,000 ppm | | | | Markedly decreased growth | |
| Duck—6 | 7 wk | 3,000 ppm | Zinc carbonate | 60 d | Diet | Decreased food consumption and loss of body weight; severe paralysis of legs by 20 d; decreased pancreas and gonad weight; low hemoglobin and hematocrit; high Zn in liver and kidney; all effects were dose-related except paralysis (most severe with 3,000 ppm Zn) and tissue Zn (similar for all dietary levels) | Gasaway and Buss, 1972 |
| | | 6,000 ppm | | | | See 3,000 ppm | |
| | | 9,000 ppm | | | | See 3,000 ppm | |
| | | 12,000 ppm | | | | See 3,000 ppm | |
| Japanese quail—10 | 1 d | 62.5 ppm | Zinc carbonate | 2 wk | Diet | Increased duodenum Zn | Hamilton *et al.*, 1979 |
| | | 125 ppm | | | | Decreased hemoglobin and hematocrit; increased duodenum and liver Zn | |
| | | 250 ppm | | | | Decreased growth, hemoglobin, | |

TABLE 41 *Continued*

| Class and Number of Animals[a] | Age or Weight | Administration: Quantity of Element[b] | Source | Duration | Route | Effect(s) | Reference |
|---|---|---|---|---|---|---|---|
| | | | | | | and hematocrit; increased duodenum and liver Zn; decreased duodenum Fe, liver Fe, Mn, and Cu | |
| | | 500 ppm | | | | Decreased growth, hemoglobin, and hematocrit; increased duodenum and liver Zn; decreased duodenum and liver Fe, Mn, and Cu | |
| | | 1,000 ppm | | | | Decreased growth, hemoglobin, and hematocrit; increased duodenum and liver Zn; decreased duodenum and liver Fe, Mn, and Cu | |
| | | 2,000 ppm | | | | Decreased growth, hemoglobin, and hematocrit; mortality; increased duodenal and liver Zn; decreased duodenal and liver Fe, Mn, and Cu | |
| Horse—3 | 3–4 wk | 25 to 186 mg/kg[f] | Zinc oxide | 38 wk | Diet | Swelling at epiphyseal region of long bones; reduced growth; anemia; increased tissue Zn | Willoughby *et al.*, 1972 |
| Dog—2 | 6.5–9.9 kg | 400 mg/d | ZnO | 3–19 wk | Diet | No adverse effect | Drinker *et al.*, 1927 |
| Dog—1 | 9.9 kg | 800 mg/d | | 15 wk | | Increased tissue Zn | |

| | | | | | | | |
|---|---|---|---|---|---|---|---|
| Cat—1 | 3.1 kg | 140–283 mg/d | | 53 wk | | No adverse effects | |
| Cat—3 | 2.9–3.8 kg | 252–800 mg/d | | 16–21 wk | | Loss of appetite; fibrous changes in pancreas; increased tissue Zn | |
| Cat—15 | 3.5 kg | 400 mg/d[g] | Zinc oxide | Few days | Diet | Refusal to eat or vomiting | Scott and Fisher, 1938 |
| | | 200 mg/d | | 4 wk | | No adverse effects | |
| | | 320 mg/d | | Few days | | Refusal to eat or vomiting | |
| | | 240 mg/d | | 12–16 wk | | Weight loss; decreased pancreas weight; fibrotic changes in pancreas; increased liver and pancreas Zn | |
| Cat—5 | 1.7–5.1 kg | 16–69 mg/kg | Oysters | 24 h | 1 meal | No adverse effects | Mannell, 1967 |
| Mouse—150 | 6–8 wk | 500 ppm | Zinc sulfate | 1–14 mo | Water[h] | No gross effects on size or appearance; transient rise in plasma Zn; no effect on Zn in liver, spleen, or skin; histological hypertrophy adrenal cortex and pancreatic islets; evidence of pituitary hyperactivity | Aughey *et al.*, 1977 |

[a] Number of animals per treatment group.

[b] Quantity expressed as parts per million (concentration in diet) or as milligrams per kilogram of body weight. Amounts are the total added to the basal diet.

[c] Hydration of the zinc sulfate was not given. The two dose levels would have been 2,020, 1,810, and 1,130 ppm and 4,040, 3,620, and 2,260 ppm for the anhydrous, monohydrate, and heptahydrate forms, respectively.

[d] Diet was primarily milk plus a small amount of grain.

[e] These adverse effects were obtained with only one diet, which gave slightly reduced growth of control chicks. With three other diets, the only adverse effect was decreased bone ash with 2,000 ppm zinc in one diet.

[f] The zinc intake was gradually increased between 9 and 20 weeks of the experiment.

[g] A single group of 15 cats was fed the four levels of zinc in the sequence shown.

[h] Drinking water.

TABLE 42 Acute Oral Toxicity of Zinc as Various Salts[a,b]

| Species | Salt | $TDL_0$[c] | $LDL_0$[c] | $LD_{50}$[c] |
|---|---|---|---|---|
| Rat | Zinc acetate, dihydrate | — | — | 733 |
| Rat | Zinc chloride | — | — | 168 |
| Mouse | Zinc chloride | — | — | 168 |
| Hamster | Zinc chloride | — | 24 | — |
| Guinea pig | Zinc chloride | — | — | 96 |
| Hamster | Zinc oxide | — | 400 | — |
| Hamster | Zinc stearate | — | 52 | — |
| Rabbit | Zinc sulfate | — | 810 | — |
| Rat | Zinc sulfate | — | 891 | — |
| Hamster | Zinc sulfate | 43 | 20 | — |
| Rabbit | Zinc sulfate, heptahydrate | — | 436 | — |
| Rat | Zinc sulfate, heptahydrate | — | 502 | — |

[a] Fairchild *et al.*, 1977.
[b] $TDL_0$, lowest published toxic dose; $LDL_0$, lowest published lethal dose; $LD_{50}$, lethal dose, 50% mortality.
[c] Milligrams per kilogram of body weight.

## REFERENCES

Allen, G. S. 1968. An outbreak of zinc poisoning in cattle. Vet. Rec. 83:8.

Aughey, E., L. Grant, B. L. Furman, and W. F. Dryden. 1977. The effects of oral zinc supplementation in the mouse. J. Comp. Pathol. 87:1.

Berg, L. R., and R. D. Martinson. 1972. Effect of diet composition on the toxicity of zinc for the chick. Poult. Sci. 51:1690.

Brink, M. F., D. E. Becker, S. W. Terrill, and A. H. Jensen. 1959. Zinc toxicity in the weanling pig. J. Anim. Sci. 18:836.

Campbell, J. K., and C. F. Mills. 1979. The toxicity of zinc to pregnant sheep. Environ. Res. 20:1.

Chesters, J. K. 1978. Biochemical functions of zinc in animals. World Rev. Nutr. Dietet. 32:135.

Cox, D. H., and O. M. Hale. 1962. Liver iron depletion without copper loss in swine fed excess zinc. J. Nutr. 77:225.

Davies, N. T., H. S. Soliman, W. Corrigall, and A. Flett. 1977. The susceptibility of suckling lambs to zinc toxicity. Br. J. Nutr. 38:153.

Doyle, J. J., and J. E. Spaulding. 1978. Toxic and essential trace elements in meat—A review. J. Anim. Sci. 47:398.

Drinker, K. R., P. K. Thompson, and M. Marsh. 1927. An investigation of the effect of the long-continued ingestion of zinc, in the form of zinc oxide, by cats and dogs, together with observations upon the excretion and the storage of zinc. Am. J. Physiol. 80:31.

Edwards, H. W., Jr. 1959. The availability to chicks of zinc in various compounds and ores. J. Nutr. 69:306.

Fairchild, E. J., R. J. Lewis, and R. L. Tatken, eds. 1977. Registry of Toxic Effects of Chemical Substances, vol. 2, pp. 963–965. DHEW Publ. No. (NIOSH) 78-104-B.

Feaster, J. P., C. H. Van Middelem, and G. K. Davis. 1972. Zinc–DDT interrelationships in growth and reproduction in the rat. J. Nutr. 102:523.

Ferguson, H. W., and A. G. Leaver. 1972. The effect of diets high in zinc at different levels of calcium and vitamin D on the rat humerus and incisor. Calcif. Tissue Res. 8:265.

Fox, M. R. S., R. P. Hamilton, A. O. L. Jones, B. E. Fry, Jr., R. M. Jacobs, and J. W. Jones. 1978. Zinc and ascorbic acid antagonism of copper. Fed. Proc. 37:324.

Furchner, J. E., and C. R. Richmond. 1962. Effect of dietary zinc on the absorption of orally administered $Zn^{65}$. Health Phys. 8:35.

Gasaway, W. C., and I. O. Buss. 1972. Zinc toxicity in the mallard duck. J. Wildl. Manage. 36:1107.

Grimmett, R. E. R., I. G. McIntosh, E. M. Wall, and C. S. M. Hopkirk. 1937. Chronic zinc poisoning of pigs; results of experimental feeding of pure zinc lactate. N.Z. J. Agric. 54:216.

Hamilton, R. P., M. R. S. Fox, B. E. Fry, Jr., A. O. L. Jones, and R. M. Jacobs. 1979. Zinc interference with copper, iron and manganese in young Japanese quail. J. Food Sci. 44:738.

Hill, C. H., and G. Matrone. 1962. A study of copper and zinc interrelationships, pp. 219–222. *In* Proc. Twelfth World's Poult. Congr.

Hsu, F. S., L. Krook, W. G. Pond, and J. R. Duncan. 1975. Interactions of dietary calcium with toxic levels of lead and zinc in pigs. J. Nutr. 105:112.

Jensen, L. S. 1975. Precipitation of a selenium deficiency by high dietary levels of copper and zinc. Proc. Soc. Exp. Biol. Med. 149:113.

Johnson, D., Jr., A. L. Mehring, Jr., F. X. Savino, and H. W. Titus. 1962. The tolerance of growing chickens for dietary zinc. Poult. Sci. 41:311.

Kincaid, R. L., W. J. Miller, L. S. Jensen, D. L. Hampton, M. W. Neathery, and R. P. Gentry. 1976. Effect of high amounts of dietary zinc and age upon tissue zinc in young chicks. Poult. Sci. 55:1954.

Magee, A. C., and G. Matrone. 1960. Studies on growth, copper metabolism and iron metabolism of rats fed high levels of zinc. J. Nutr. 72:233.

Magee, A. C., and S. Spahr. 1964. Effects of dietary supplements on young rats fed high levels of zinc. J. Nutr. 82:209.

Mannell, W. A. 1967. Effect of oysters with a high zinc content on cats and man. BIBRA Info. Bull. 6:432.

McCall, J. T., J. V. Mason, and G. K. Davis. 1961. Effect of source and level of dietary protein on the toxicity of zinc to the rat. J. Nutr. 74:51.

Mehring, A. L., Jr., J. H. Brumbaugh, and H. W. Titus. 1956. A comparison of the growth of chicks fed diets containing different quantities of zinc. Poult. Sci. 35:956.

Miller, W. J., C. M. Clifton, P. R. Fowler, and H. F. Perkins. 1965. Influence of high levels of dietary zinc on zinc in milk, performance and biochemistry of lactating cows. J. Dairy Sci. 48:450.

Miller, W. J., D. M. Blackmon, R. P. Gentry, and F. M. Pate. 1970. Effects of high but nontoxic levels of zinc in practical diets on $^{65}Zn$ and zinc metabolism in Holstein calves. J. Nutr. 100:893.

Miller, W. J., E. S. Wells, R. P. Gentry, and M. W. Neathery. 1971. Endogenous zinc excretion and $^{65}$Zn metabolism in Holstein calves fed intermediate to high but nontoxic zinc levels in practical diets. J. Nutr. 101:1673.

Mills, C. F., and A. C. Dalgarno. 1972. Copper and zinc status of ewes and lambs receiving increased dietary concentrations of cadmium. Nature 239:171.

Murphy, E. W., B. W. Willis, and B. K. Watt. 1975. Provisional tables on the zinc content of foods. J. Am. Dietet. Assoc. 66:345.

National Research Council. 1974. Nutrients and Toxic Substances in Water for Livestock and Poultry. National Academy of Sciences, Washington, D.C.

National Research Council. 1978. Metabolic and Biologic Effects of Environmental Pollutants. Zinc. University Park Press, Baltimore, Md.

Ott, E. A., W. H. Smith, R. B. Harrington, and W. M. Beeson. 1966a. Zinc toxicity in ruminants. II. Effect of high levels of dietary zinc on gains, feed consumption and feed efficiency of beef cattle. J. Anim. Sci. 25:419.

Ott, E. A., W. H. Smith, R. B. Harrington, and W. M. Beeson. 1966b. Zinc toxicity in ruminants. I. Effect of high levels of dietary zinc on gains, feed consumption and feed efficiency of lambs. J. Anim. Sci. 25:414.

Ott, E. A., W. H. Smith, R. B. Harrington, H. E. Parker, and W. M. Beeson. 1966c. Zinc toxicity in ruminants. IV. Physiological changes in tissues of beef cattle. J. Anim. Sci. 25:432.

Ott, E. A., W. H. Smith, R. B. Harrington, M. Stob, H. E. Parker, and W. M. Beeson. 1966d. Zinc toxicity in ruminants. III. Physiological changes in tissues and alterations in rumen metabolism in lambs. J. Anim. Sci. 25:424.

Philip, B., and P. A. Kurup. 1978. Dietary zinc and levels of collagen, elastin and carbohydrate components of glycoproteins of aorta, skin and cartilage in rats. Ind. J. Exp. Biol. 16:370.

Prasad, A. S., ed. 1966. Zinc Metabolism. Charles C Thomas, Springfield, Ill.

Prasad, A. S., ed. 1976. Trace Elements in Human Health and Disease, vol I. Zinc and Copper. Academic Press, New York.

Riordan, J. F., and B. L. Vallee. 1976. Structure and function of zinc metalloenzymes, pp. 227–256, *In* A. S. Prasad (ed.). Trace-Elements in Human Health and Disease, vol. I. Academic Press, New York.

Roberson, R. H., and P. J. Schaible. 1960. The tolerance of growing chicks for high levels of different forms of zinc. Poult. Sci. 39:893.

Sampson, J., R. Graham, and H. R. Hester. 1942. Studies on feeding zinc to pigs. Cornell Vet. 32:225.

Scott, D. A., and A. M. Fisher. 1938. Studies on the pancreas and liver of normal and zinc-fed cats. Am. J. Physiol. 121:253.

Smith, B. L. 1977. Toxicity of zinc in ruminants in relation to facial eczema. N.Z. Vet. J. 25:310.

Smith, S. E., and E. J. Larsen. 1946. Zinc toxicity in rats. Antagonistic effects of copper and liver. J. Biol. Chem. 163:29.

Stake, P. E., W. J. Miller, R. P. Gentry, and M. W. Neathery. 1975. Zinc metabolic adaptations in calves fed a high but nontoxic zinc level for varying time periods. J. Anim. Sci. 40:132.

Stewart, A. K., and A. C. Magee. 1964. Effect of zinc toxicity on calcium, phosphorus and magnesium metabolism of young rats. J. Nutr. 82:287.

Thawley, D. G., R. A. Willoughby, B. J. McSherry, G. K. McCleod, K. H. Mackay, and W. R. Mitchell. 1977. Toxic interactions among lead, zinc, and cadmium with varying levels of dietary calcium and vitamin D: Hematological system. Environ. Res. 14:463.

Thompson, A., S. L. Hansard, and M. C. Bell. 1959. The influence of aluminum and zinc upon the absorption and retention of calcium and phosphorus in lambs. J. Anim. Sci. 18:187.

Tucker, H. F., and W. D. Salmon. 1955. Parakeratosis in zinc deficiency disease in pigs. Proc. Soc. Exp. Biol. Med. 88:613.

Underwood, E. J. 1977. Trace Elements in Human and Animal Nutrition. Academic Press, New York.

Vohra, P., and F. H. Kratzer. 1968. Zinc, copper and manganese toxicities in turkey poults and their alleviation by EDTA. Poult. Sci. 47:699.

Willoughby, R. A., E. MacDonald, B. J. McSherry, and G. Brown. 1972. Lead and zinc poisoning and the interaction between Pb and Zn poisoning in the foal. Can. J. Comp. Med. 36:348.